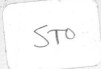

THE UNITED STATES OIL POLICY

A GREAT GUSHER IN THE SANTA FÉ SPRINGS OIL FIELD

THE UNITED STATES
OIL POLICY

BY

JOHN ISE, Ph.D., LL.B.

PROFESSOR OF ECONOMICS IN THE
UNIVERSITY OF KANSAS

NEW HAVEN · YALE UNIVERSITY PRESS

LONDON · HUMPHREY MILFORD · OXFORD UNIVERSITY PRESS

M DCCCC XXVI

PUBLISHED ON THE FOUNDATION ESTABLISHED IN MEMORY OF WILLIAM McKEAN BROWN

THE present volume is the fourth work published by the Yale University Press on the William McKean Brown Memorial Publication Fund. This Foundation was established by gifts from members of his family to Yale University in memory of William McKean Brown, of Newcastle, Pennsylvania, who not only was a leader in the development of his community, but also served the commonwealth as state senator and later as lieutenant-governor of Pennsylvania.

PREFACE

THE United States Oil Policy is the second of what the writer has ventured to hope may be a series of works covering the land policy of the United States. The present volume follows the writer's *United States Forest Policy*, which was published in 1920; and perhaps the writer can do no better, by way of preface, than to repeat a few lines from the earlier work:

"The history of the United States is fundamentally a history of rapid exploitation of immensely valuable natural resources. The possession and exploitation of these resources have given most of the distinctive traits to American character, economic development, and even political and social institutions. Whatever preëminence the United States may have among the nations of the world, in industrial activity, efficiency and enterprise, in standards of comfort in living, in wealth, and even in such social and educational institutions as are dependent upon great wealth, must be attributed to the possession of these great natural resources; and the maintenance of our preëminence in these respects is dependent upon a wise and economical use of remaining resources. Thus the question of conservation is one of the most important questions before the American people, and if the present study throws even a weak and flickering light upon that question, its publication will be abundantly justified."

The writer acknowledges a heavy obligation to some of his friends and colleagues for helpful criticisms and suggestions. Mr. Domenico Gagliardo read most of the chapters and offered many valuable criticisms. Professor R. S. Knappen offered many helpful suggestions concerning questions in geology. My sister, Mrs. M. V. Holmes, read all of the manuscript, and checked the English construction, and my brother, W. J. Ise, solicitor in the Forest Service, checked over the chapters relating to the public lands. Assistant Secretary Edward C. Finney, of the Department of the Interior, generously read and criticized the chapter on Indian lands. Various departments of the government at Washington, including particularly the Geological Survey, the Land Office, the Bureau of Mines, and the Federal Trade Commission, kindly assisted in securing data of various kinds, as did also the American Petroleum Institute. Acknowledgment should be made of the very efficient assistance of Mrs. Frank L. Fleener, and of Mrs. Carroll D. Clark, who did the stenographic work, and of my wife's assistance in reading proofs and in checking English construction.

The writer wishes particularly to acknowledge generous pecuniary assistance from the Graduate Research Committee of the University. He wishes, further, to express his appreciation of the kindness shown by Chancellor E. H. Lindley and Dean J. W. Brandt, in exempting him almost entirely from administrative and committee duties while working on this book.

CONTENTS

ILLUSTRATIONS

THE UNITED STATES OIL POLICY

CHAPTER I

PETROLEUM IN ANCIENT AND MEDIEVAL TIMES

PETROLEUM has been known since the dawn of history, and probably was used by man before the earliest recorded history. According to ancient Chinese and Japanese records, it appears that China and Japan used petroleum for lighting and for medicine, and natural gas for fuel, even before Abraham was called from Ur of Chaldea. The oil wells of Rangoon, in India, have yielded oil since the earliest times, and "Rangoon oil" was an article of commerce in London at the beginning of the nineteenth century. Perhaps the earliest account of the use of bitumen, outside of the Bible, is found in the writings of Herodotus, describing the building of Babylon. The springs called Oyun Hit (the fountains of Hit) were well known among the Arabs and Persians, the latter calling them "Chesmeh Kir" (the fountain of pitch). This liquid bitumen they called "nafta," and the Turks, to distinguish it from pitch, gave it the name Hara sakir (black mastic). Nearly all travellers who went to Persia and the Indies by way of the Euphrates, before the discovery of the Cape of Good Hope, mentioned this fountain of bitumen. Herodotus says: "Eight days' journey from Babylon stands another city called Is, on a small river of the same name, which discharges its stream into the Euphrates. Now this river brings down with its water many lumps of bitumen, from whence the bitumen used in the walls of Babylon was wrought." At the pits of Kir ab ur Susiana, less than a generation ago, bitumen was still collected in the same manner that Herodotus described. Herodotus also describes the method of securing petroleum in Zante, one of the Ionian Islands—then called Zacynthus. This product he calls "pitch," and says it was obtained by thrusting a pole with a bunch of myrtle tied on the end of it to the bottom of the spring, where the pitch adhered to it. These springs of Zante were still producing oil not many years ago.[1]

The oil springs on the shores of the Caspian Sea, at Baku, were known to the Persians two or three thousand years ago, and from the frequency with which the oil and gas there were spontaneously ignited, the place was held sacred by the fire worshippers, who made pilgrimages to it as the Mohammedans do to Mecca.

Pliny and Dioscorides described the oil of Agrigentum in Sicily, which was used in lamps, under the name of "Sicilian Oil." Strabo wrote of the abundant asphaltum found in Babylonia, and quoted Eratosthenes regarding the inflammable "naptha" which was found in Babylonia. He also wrote of the bitumen found in the valley of Judea, and of the trade in this material which an Arab named Nabathenes carried on with the Egyptians, who used it in embalming their mummies. Diodorus, of Sicily, about 50 B.C., described the Lake Asphaltites, and stated that the savages of Sicily had no commerce except with the Egyptians, the commerce consisting of traffic in bitumen, or asphalt. Plutarch saw a petroleum lake on fire near Ecbatana, in Persia. Baku, on the bank of the Caspian, has furnished petroleum for perhaps twenty-five hundred years. It is highly probable that the fires kept perpetually burning in pagan shrines consisted of natural gas jets or were fed with petroleum. Aristotle, Plutarch, Pliny, and Josephus describe in turn the deposits of bitumen in Albania on the eastern shores of the Adriatic.

In Bible history, oil has played a small part, the word translated "salt" having been used indiscriminately for common salt, nitre, and bitumen. In Genesis XI, 3, in the description of the tower of Babel, we are told: "slime had they for mortar"—a statement which is confirmed by Herodotus. Another chapter in Genesis states: "The vale of Siddim was full of slime-pits," the word which is translated "slime" appearing as "bitumen" in the Vulgate. In still another chapter (Genesis VI, 14), the Lord bade Noah to make an ark of gopher wood, and "pitch it within and without with pitch," pitch probably meaning heavy oil or asphalt. In Job XXIX, 6, we find: "And the rock poured me out rivers of oil," and in Deuteronomy XXXII, 13, we find mention of "oil out of the flinty rock." Perhaps both these phrases are merely oriental expressions for olive oil. The ark of bulrushes in which Moses was found was said to be "daubed with slime and with pitch," but whether this refers to bitumen or asphalt is a question. In Maccabees II, 1, 1836, in the Apocrypha, it is stated that the priests hid the fire which they took from the altar, in a deep pit without water. After many years Nehemiah sent some of the posterity of the priests who had hidden it, and "they found no fire, but thick water." This was poured by Nehemiah upon the sacrifices and upon the wood and the altar, and when the sun appeared from behind a cloud it burst into flame, and there was "a great fire kindled." "And Nehemiah called this thing Nephthar, which is as much as to say a cleansing; but many called it Nephai." Perhaps Mark Twain's humorous account of this event in *Roughing It* is substantially correct.

Marco Polo noticed the oil springs of Georgiania in the thirteenth century. As Yule described it: "On the confines toward Georgiania there is a fountain from which oil springs in great abundance, insomuch that a hundred shiploads might be taken from it at one time. This oil is not good to use with food, but 'tis good to burn, and is also used to anoint camels that have the mange. People come from vast distances to fetch it, for in all the countries round about they have no other oil."[2]

The observations of Geoffrey Ducket, agent for the Moscovy Company, on a trip into Persia in 1574, indicate that the oil industry of Baku was a very considerable industry at that time:

There is a very great river which runneth through the plaine of Javat, which falleth into the Caspian Sea by a towne called Bachu, neere unto which towne is a strange thing to behold. For there issueth out of the ground a marveilous quantitie of oile, which oile they fetch from the uttermost bounds of all Persia; it serveth all the countrey to burne in their houses.

This oyle is blacke and is called Nefte: they use to carry it throughout all the countrey upon kine and asses, of which you shall oftentimes meet with foure or five hundred in a company. There is also by the said towne of Bachu another kind of oyle which is white and very precious: and is supposed to be the same that here is called Petroleum. There is also not far from Shamaky, a thing like unto tarre, and issueth out of the ground, whereof we have made the proofe that in our ships it serveth well in the stead of tarre.[3]

The observations of John Cartwright on his voyage from Aleppo to Hispaan, in the early part of the seventeenth century, were much to the same effect:

Neere unto this Towne, (Baku) is a very strange and wonderfull Fountaine under ground, out of which there springeth and issueth a marvellous quantitie of black Oyle which serveth all the parts of Persia to burne in their houses; and they "usually carrie it all over the countrey, upon Kine and Asses, whereof you shall oftentimes meete three or foure hundred in company."[4]

A number of other writers and travellers of the seventeenth and eighteenth centuries speak of petroleum springs around the Caspian Sea; and the manner in which the oil was being exploited at that time indicates that a considerable oil industry had already been developed. James Parkerson, author of *Organic Remains of a Former World*, published in the early part of the nineteenth century, shows by quotations from *Abbe Fortis' Travels* in Dalmatia, and from the works of Captain Cox, that no less than 26,000,000 gallons of petroleum were, even at that early period, shipped annually from Baku, a large proportion of it going to Persia. Walter Raleigh published an account of the Trinidad pitch lake as early as 1595.

Petroleum was in use as a medicine in Europe throughout the last century. In Europe it was produced from springs near Haarlem, in Holland, and was called "Haarlem oil." It was a black, thick fluid with a nauseous odor.[5]

NOTES

1. Information regarding ancient and medieval oil history, and also regarding American oil history, is taken from so many scattered sources that the writer has found it necessary to omit almost all references to sources.

2. Yule, Sir Henry, *Travels of Marco Polo,* London, 1903, I, 46.

3. *Hakluyt's Voyages,* Vol. III, 165, Glasgow, 1903.

4. *Purchas His Pilgrims,* Vol. VIII, 508. "Observations of Master John Cartwright in his voyage from Aleppo to Hispan, and backe againe."

5. On the ancient and medieval history of petroleum, see: Tenth Census of U.S., Vol. X, 3; Bacon and Hamor, *Am. Petroleum Industry,* Ch. V; *Am. Catholic Quarterly,* Vol. 19, 851, Vol. 20, 404; *Harper's Magazine,* Oct., 1890, 723; Redwood, *Treatise on Petroleum,* Vol. I, 1-3; *Natural Gas Jour.,* Sept., 1911, 19; *Oil and Gas Jour.,* Nov. 2, 1916, 33, 37; Indiana Dept. of Geology and Natural Resources, Rept., 1891, 307. Cragun, *Petroleum in History and Finance,* published by U. S. Wolfe & Co., stock brokers, has a very interesting discussion of the ancient and medieval history of petroleum.

CHAPTER II

EARLY HISTORY OF PETROLEUM IN PENNSYLVANIA AND NEW YORK

IN the United States, the Indians doubtless used petroleum for various purposes long before white people came to this country. Numerous peculiar pits have been noted in Pennsylvania and New York, dug in the low grounds adjacent to certain streams, and always in the neighborhood of oil seeps; and, while the Indians denied any knowledge of the origin of these pits, it is possible that they were the work of earlier tribes of Indians or, more probably, of some prehistoric race of men, who used them for collecting petroleum from the springs, and long ago passed away, leaving no other record of their origin or history. Trees growing in some of these pits prove that they were abandoned four or five hundred years ago. The above theory regarding prehistoric pits is repeated in many accounts, although some authorities do not endorse it.

The Indians of the early colonial period used petroleum for various purposes—for toothache, headache, swellings, rheumatism, sprains, and sometimes even internally. They also used it to burn at their religious ceremonies. One of the early writers gave a graphic account of the ceremonies of the Seneca Indians at one of the oil springs near Cuba, New York. Some of the land around the famous Cuba spring was for many years reserved for the Seneca Indians.

A letter of the French missionary, Joseph d'Allion, written in 1627 (some accounts say 1629), describes the territory near the Mohawk, and speaks of good oil as one of the products. This must have been seen in one of the oil springs, probably near Cuba, Allegany County, New York. This letter was published in G. Sagard's *Histoire du Canada et Voyage des Missionnaires Récollets* in 1636. About 1642, according to Charlevoix, several Jesuits penetrated into the same country and found "a thick oily stagnant water which would burn like brandy." Pierre Boucher, who was sent to France in 1662, wrote a little book for prospective emigrants, in which he told of a "spring in the Iroquois country from which exudes a greasy water that is like oil, and that is used in many cases instead of oil."[1]

As early as 1670 an oil spring near Cuba, New York, was shown on a map made by Dollier and Galinee, missionaries of the Order of St. Sulpice, and sent by them to Jean Talon, Intendant of Canada. Nearly a century afterward several maps were drawn showing the location of the oil springs of western Pennsylvania and New York. In 1700 the Governor of New York instructed his chief engineer, Romer, to visit the Five Nations and observe a certain spring which "blazes up in a flame when a lighted coal or firebrand is put into it." Charlevoix visited Canada a second time in 1720, and in 1721 published his journal in which is given perhaps the first record of the oil springs in Pennsylvania—apparently near the head of the Allegheny River.

In 1748 Peter Kalm, a European naturalist, visited the oil springs of Oil Creek, in his travels in America, and on his return to Europe published an account of his travels, with a map showing the location of these springs. A cleverly written letter is often quoted describing certain religious rites which the Seneca Indians performed by the light of burning oil springs. This letter was dated 1750, and was claimed to have been written by the commandant of Fort Duquesne to General Montcalm; but that fort was not built until four years afterward, and McLaurin states that the letter was written in a spirit of fun by two young men in Franklin, and published in the local paper in 1839. In 1775 Lewis Evans, of Philadelphia, published a map showing the oil seepages along the Allegheny and Ohio rivers. Colonel Brodhead, in 1779, wrote about "crossing a creek called Oil Creek." In the oil which the soldiers found floating on the top of a spring they bathed their joints, to "the great relief of the rheumatism with which they were afflicted."[2] In 1783 General Benjamin Lincoln wrote a letter to the president of the University of Cambridge, in which he described the oil springs of Pennsylvania, and the Burning Spring of Virginia, stating that his information regarding the latter had been given him by Washington himself. The number of references to various oil springs about this time indicates considerable public interest in the matter. About 1810 or 1815 Rev. S. J. M. Eaton of Franklin, Pennsylvania, wrote a very interesting account of the early methods of securing and marketing petroleum on Oil Creek, describing the pits and the manner in which blankets were laid on the surface of the water and the oil wrung into tubs or barrels.

In 1833 Benjamin Silliman, Sr., published an article in the *American Journal of Science*, mentioning an oil spring near Lake Seneca, New York, and describing the manner in which people in the vicinity secured the oil. In the same year, another writer, S. P. Hildreth, wrote as follows regarding the early use of petroleum:

From its being found in limited quantities, and its great and extensive demand, a small vial of it would sell for 40 or 50 cents. In neighborhoods where it is abundant it is burned in lamps in place of spermaceti oil, affording a brilliant light, but filling the room with its own peculiar odor. By filtering it through charcoal, much of this empyreumatic smell is destroyed, and the oil greatly improved in quality and appearance. It is also well adapted to prevent friction in machinery, for, being free of gluten, so common to animal and vegetable oils, it preserves the parts to which it is applied for a long time in free motion. . . . This oil rises in greater or less abundance in most of the salt wells of the Kanawha, and, collecting as it rises, in the head of the water, is removed from time to time with a ladle.[3]

In 1837 the Geological Survey of New York described at some length one of the oil springs on the Seneca Indian reservation.[4]

The white people, like the Indians, used petroleum for a variety of purposes, but particularly for lubricating purposes and for medicine. The following appeared on a bottle of petroleum in Pittsburg, in the forties: "Kier's Petroleum, or Rock Oil, Celebrated for its Wonderful Curative Powers. A Natural Remedy! Procured from a Well in Allegheny Co. Pa. Four Hundred Feet below the Earth's Surface. Put up and Sold by Samuel M. Kier, 363 Liberty Street, Pittsburgh, Pennsylvania."

> The healthful balm, from Nature's secret spring,
> The bloom of health and life to man will bring;
> As from her depths the magic liquid flows,
> To calm our sufferings and assuage our woes.

It was used for rheumatism, sprains, and for horse liniment as well. Kier had been a salt manufacturer, but the extended use of his medicine led to the development of a considerable oil industry. The present use of one of the oil products—"paraffine oil"—as a laxative, has its historical antecedent at least as early as 1783 when General Benjamin Lincoln's soldiers, in marching through northern Pennsylvania, were reported to have used this oil as a purge.

The first shipment of petroleum to Pittsburg has been described as follows: "Mr. Cary, one of the first settlers on Oil Creek, possessing perhaps a little more enterprise than his neighbors, would collect or purchase a cargo of oil and proceed to Pittsburg, and exchange it for commodities needed in his home. This cargo consisted of two five-gallon kegs, that were slung on each side of a horse, and thus conveyed by land a distance of seventy or eighty miles. . . . Sometimes the market in Pittsburg became very dull, for a flatboatman would occasionally introduce a barrel or two at once. At other times the demand fell off so that the purchase of a barrel was hazardous."[5] In 1845 Lewis Peterson, of Tarentum, Pennsylvania, entered into a contract with the Hope Cotton Factory at Pittsburg, by which he was to supply two barrels of crude petroleum per week for use as a spindle lubricant in mixture with sperm oil. This mixed lubricant was used at the Hope factory for ten years.

The possibility of using petroleum as an illuminant was early recognized. As early as 1828 a writer in the Pittsburg *Gazette* urged the use of "West Seneca Oil" to light the city. Not long afterward, new conditions developed which called for some new source of illumination. The activity of the New Bedford whalers was reducing the number of whales so much that sperm oil was becoming very scarce and expensive. About 1850 sperm oil sold as high as $2.25 per gallon, a price which was almost prohibitive.

It was the introduction of oil made from coal and shale that paved the way to a recognition of the great value of petroleum deposits. As early as 1739 the attention of the Royal Society of London had been called to the fact that in making gas from coal a black oil was left as a by-product. In 1840 some oil was being made from coal in France, and by 1850 the manufacture of coal oil was a real industry. About this time a similar industry was developing in Scotland. James Young, of Glasgow, was the inventor of a process by which a good illuminant was produced cheaply by the distillation of coal. The material used by him was, for the most part, Torbane Hill cannel, which became so much enhanced in value that litigation arose regarding the leases by which the territory containing it was held. In this litigation, Young testified that in 1849 he had produced 400,000 gallons of this illuminant, and had sold it at one dollar a gallon—largely profit. Young's patents were brought to this country, and it was found upon trial that an excellent illuminant could be produced from our cannel coals and bituminous shales at a cost considerably less than a half dollar per gallon. The trouble with it was that it burned with such an offensive smell as to offend even only fairly sensitive people. However, our coal measures and oil-bearing shales were found to offer good facilities for the manufacture of this oil, and the industry soon spread widely, the factories in 1859 numbering some fifty or sixty.

"Coal oil" was always expensive, however, and there was a great deal of interest in the question of securing a substitute from petroleum; and the first problem emerging here was that of securing an enlarged supply of petroleum. The skimming of small amounts by means of thin boards, blankets, etc., was too slow and expensive. In 1853 a company tried a scheme of trenches, into which the water was pumped and the oil skimmed off; but this proved too expensive, and was abandoned. The company which made these experiments passed through various vicissitudes, and was finally incorporated, in 1854, as the Pennsylvania Rock Oil Company. Some of the oil produced was sent to Professor Crosby, of Dartmouth College, and to Professor Benjamin Silliman, of Yale University, and the report of the latter attracted considerable attention. Professor Silliman reported that eight products could be obtained from oil by distillation, and suggested the use of petroleum for gas production, as a lubricant, and as an illuminant. "It appears to me," he concluded, "that there is much ground for encouragement in the belief that your company have in their possession a raw material from which by a simple and not expensive process they may manufacture very valuable products."

The members of the Pennsylvania Rock Oil Company did not work together well, and little was accomplished for some time. In 1858, however, several of the members organized a new company—the Seneca Oil Company—which leased a plot of land on Oil Creek from the parent company and started operations at Titusville, Pennsylvania. The first plan was to collect oil from the springs on their property, but through a fortunate accident the idea of drilling a well occurred to one of the men, and this plan was put into operation.[6] Colonel E. L. Drake was sent to Titusville as superintendent of the company's operations, and he struck oil at the depth of 69 feet. The well pumped 25 barrels a day for some time.

The Drake well was the first to be drilled in search of oil, and it should not be regarded as a mere accidental circumstance that it was drilled at this particular time. As a matter of fact, oil was discovered at this time because it was needed. In the first place, there was a scarcity of lubricating oils. The rapid growth of factories and development of machine methods required a large amount of lubricating oil, and the animal oils used were neither adequate nor entirely satisfactory. In the second place, as already stated, the problem of a cheap illuminant was becoming serious, in consequence of the exhaustion of the whale fisheries and the inadequacy of the supply of lard which was being used for candles. The development of the coal and shale oil industry met these needs in a measure, but only at a considerable expense. It is significant that many travellers in the decade from 1850 to 1860 described oil springs in various parts of the world, such natural wonders being of particular interest at the time. Petroleum, once discovered, met not only the needs for lubrication and illumination, but also another need which developed during the Civil War. At the beginning of the war, the North was shut off from the turpentine districts of the South, and the small supply of turpentine soon advanced to exorbitant prices; in fact, it was not adequate to the demand for painting and other mechanical purposes. Happily, some of the products discovered in refining crude oil proved a satisfactory substitute for turpentine, for many purposes.

Thus the discovery of petroleum in large quantities was most opportune, and when the famous Drake well was struck in 1859 it created great excitement. Farms were bought or leased on all sides, and preparations were made for drilling wells in many places, though but little was accomplished until 1860. In June, 1861, the first flowing well was struck—the "Fountain" well—with a flow of 300 barrels a day. The discovery that oil would flow from wells upset all calculations; people were willing to believe anything, and great numbers rushed to the oil regions to make their fortunes. In September the Empire well came in, with the amazing flow of 2,500 barrels a day, which persisted almost unabated for six months. In November the Phillips well was brought in with a flow of 3,000 barrels, and other wells followed rapidly. During 1860 and 1861, several hundred wells were drilled along Oil Creek, and while most of them were dry holes or small "pumpers," some were flowing wells, and the production rose rapidly. In June, 1860, the total daily production was estimated at 200 barrels, and within the following year this had risen to thousands of barrels. (The figures as to production of various wells, as well as figures of total production, are to be regarded largely as estimates or guesses, since accurate records were not kept at this time.)

The development of the Mecca field in Ohio, and of the Burning Springs and Oil Springs fields in West Virginia, added to the general over-production; and the price of oil fell from $20 a barrel to 20 cents, then to 15 cents, and even to 10 cents a barrel. Thousands of barrels of oil are reported to have actually been sold as cheap as 5 cents a barrel, and it is said that, even at that price, many purchasers lost money.

The outbreak of the Civil War contributed to the general demoralization. Soon it was impossible to obtain barrels on any terms, for all the coopers in the surrounding country could not make them as fast as the Empire well alone could fill them. Small producers were forced to cease operations, and scores of operators became disheartened and abandoned their wells. The cheapness of the oil, of course, led to very great waste.

The ruinously low price of oil did not greatly discourage drilling, however, and during the latter part of 1861 and 1862 the number of flowing wells was greatly increased, while the daily production mounted steadily.

In 1863 operations were carried on up and down the Allegheny River for eighty miles, as well as along Oil Creek and its tributaries. The Noble and Delamater well on the Farrel farm flowed 3,000 barrels a day, producing in two years, according to one estimate, $3,000,000 worth of oil. The Maple Shade well at Petroleum Center started off at 800 barrels. The McClintock farm of 207 acres, afterward covered by the city of Petroleum Center, was punctured with 150 wells, 80 per cent of them productive. In spite of this, the price of oil increased during the year.

The following year Cherry Run furnished much of the excitement.

A number of good new wells came in: the Lloyd well at 100 barrels, another Phillips well at 800 barrels, the Fertig and Hammond well at 600 barrels, the Log Cabin well at 300 barrels, the Reed and Criswell well at 1,000 barrels a day; and in a short time other heavy producers followed. A number of good wells were brought in on the Widow Mc-Clintock farm. The excitement was intense, and a tremendous rush followed. In a short time all the available property on the Run was taken at high figures. Late comers who failed to secure the fee simple to the land were glad to take leases, and sometimes paid not only a royalty of half the oil, but large bonuses in addition. Soon the valley was planted thickly with derricks, and the pertinacious oil seekers even bored along the hillsides. The two acres on which the Reed and Criswell well was located were sold for $650,000. In spite of the active drilling and large production, the price of oil rose throughout the year, as high as $9.87 per barrel.

In 1865 the famous United States well was struck on Pithole Creek, soon followed by the Grant well, the two producing 2,500 barrels per day. A rush to the vicinity immediately began. The Civil War had just ended, and thousands who had returned home flocked to the oil regions, anxious to invest. Land was sold at fabulous prices, wells were started in all directions, and Pithole City sprang up. In May, 1865, there were two houses on the ground. In August a city of 14,000 inhabitants is said to have been there, with solid business blocks and residences. The daily yield of this section in August was over 5,000 barrels. Many men made fortunes; many more, of course, lost what they had. With the exhaustion of many of the wells in November, the decline of the city began. The population rapidly dwindled; buildings were deserted; $30,-000 houses sold for $1,000 each; and in 1866 much of what was left of the town was swept away by a disastrous fire. Pithole is one of the striking examples of the brief and uncertain life of the oil town. Early in February, 1866, however, the city again went wild with excitement over the discovery that water wells in the borough limits were yielding oil, and town lot operations were begun, with lots selling at fabulous prices. The first discovery was on a lot owned by Mrs. Rickerts, a poor washerwoman. Thousands of speculators and investors flocked into the town again, and the land was riddled with holes. In places oil was gathered in dippers from surface pools and from shallow holes dug in the ground.

A great many speculative companies failed during 1865. In March a flood swept away a part of Oil City, with a great amount of oil property along Oil Creek and other streams. At this time practically all the operations were confined to the low lands and ravines, and floods were always disastrous. But the wrecking of the old companies brought little cessation in the formation of new ones. Each day saw new oil companies organized. These companies were capitalized at from $50,000 to $500,-000, and all classes, from bootblacks to millionaires, became investors. Leases were taken at one-half royalty, seldom less, and the size of the leases ranged from one-quarter of an acre up. A report in March, 1865, stated that within one week twenty oil companies had been formed, with a capital of $12,510,000. At the same time it was estimated that the aggregate capital represented by the oil companies whose certificates were on file in the office of the Secretary of State at Albany, New York, amounted to $350,000,000. The impetus of the oil excitement of 1864 and 1865 carried on into 1866, in spite of many premonitions of a general collapse, but the bursting of the oil bubble was imminent. One company after another failed to meet the expectations of sanguine stockholders. Owners of all kinds of property became anxious to realize, and a rapid reduction of values took place in many sections. In Pithole a destructive flood, and several disastrous fires, helped the general collapse. In the fall and winter of 1866 and 1867, oil companies fell, one after another, like so many rows of bricks; and thousands who had purchased stocks were overwhelmed in the ruin that followed.

The year 1867 was one of general depression and low prices. Thousands of acres of the lands belonging to defunct oil companies, and scores of engines, as well as a vast amount of machinery, were sold by the sheriff of Venango County for debts and taxes. A general thinning out of superintendents, operators, and business men, took place in all the principal localities.

About this time it appeared that the oil industry was settling down to a somewhat more stable basis than had existed previously; yet in 1866 promoters were running free excursions to some of the oil regions, in an effort to sell stock.

For several years after the Drake discovery almost all the oil wells were drilled in the "flats" bordering the principal streams. The impression prevailed that there was some connection, some parallelism, between the streams on the surface and the "oil veins" beneath. But the many failures to strike oil along the streams gradually led drillers to test out the higher ground, and upon lines between good wells. After 1866 it was generally recognized that the chances of striking oil were quite as good on the higher lands.

In 1867 some writers began to show an interest in the question of the life of oil wells, and the probabilities as to their yielding an average profit. The Detroit *Tribune* stated in 1867 that the Canadian oil fields were on their last legs, that the production of the Pennsylvania wells was falling off rapidly, and that many of the wells were being abandoned. A writer in the Titusville *Herald* expressed the same idea: "Nearly all the wells on Dennis Run and above Triumph are rapidly deteriorating, for they are too close to each other, three, four, and even five being on one acre; many of the best last but two and three months and beyond this they have a spasmodic, mocking angry old age."[7]

In 1868, however, the Canadian production was so greatly in excess of the demand, and the price so low that a number of the refineries, and some of the wells and refineries in the Petrolia, Canada, district were closed down. In Pennsylvania there was a decrease in production in the latter months of 1868, due to the smaller average yield of new wells, the exhaustion of the gas in the sand rock, and the exhaustion of wells in the old territory. It was claimed that nearly twice as many wells had to be drilled in 1869 as in 1868, to supply the demand. The scarcity of oil caused an advance in prices. In 1869 the Oil Creek field having been pretty well drilled over, the attention of some capitalists was attracted to the indications of oil near Parkers Landing, eighty miles down the Allegheny from Oil City. Wells had been drilled there before without great success, but in the year 1869, good oil wells were opened, and the crowds of speculators and oil men hurried to the "lower oil field." From Parker City, operations extended to Saint Petersburg and Edenburg on the north, and to Petrolia and Millerstown on the south.

In the early seventies there was another decline in the production of several of the fields. The West Hickory district was reported rapidly declining. The fact that the Canadian production was dwindling rapidly gave additional impetus to prices, and led to some questioning as to the future oil supply. It resulted also in the cleaning out of a great many of the old wells, some of which responded with increased production. In 1872 the Clarion and Butler oil fields were discovered. Numerous gushers were brought in, and the production rose to hitherto unknown figures, with a consequent decline in price which spelled ruin for many producers. Soon a movement developed to organize a shutdown of producers, but very little success attended this movement. Over-production, with agitation for organized curtailment, was a recurrent feature of Pennsylvania history, as of the history of most other oil regions. With this over-production came not only distress to the operators, but waste of the oil, through lack of sufficient tankage.

The following year the Clarion, Butler, and Armstrong fields continued to pour out their oil upon an already over-supplied market, and prices gravitated lower, as low as 50 cents a barrel in October, while the reckless waste continued. Of course operators lost heavily, and many were forced into bankruptcy. We read of one operator who sold at $1.25 over 12,000 barrels of oil for which he had previously been offered $5.70 per barrel. The *National Oil Journal*, commenting on the immense production and low price of oil, stated that the combination responsible for the situation was one of "greedy farm owners and more greedy producers." There were meetings at which organized curtailment of production was discussed; but no important results accrued from such discussions.

In 1874 the great Butler and Armstrong cross belt was opened up, and the flood of oil continued to grow, while prices declined and operators failed by the score. Numerous meetings were held, and shutdown pledges were circulated. In December nearly 700 executions were reported to have been issued in Butler County in two months, and the sheriff found it "necessary to keep 10 or 12 deputies constantly employed cleaning out enthusiastic borers for half-dollar oil."[8]

In spite of bankruptcies and curtailment campaigns, the flood of oil continued. Wells were drilled northward in McKean County, adjoining New York, and the Kinzua pool south of Bradford was opened. The Phillips well, not far from Bradford, spouted 3,000 barrels a day. Another big well was brought in near Stoneham, in Warren County. In January, 1875, stocks of oil were 700,000 barrels greater than they had been a month before. Tanks in the Bradford field were running over. The northern oil field was reported in a "frightful condition," with tankage scarce and oil running down the creeks. The pipe lines refused to run more than one-fourth of any producer's output at the prevailing prices. At Bradford it was reported in January that $50,000 worth of oil men's paper had been protested within a month. Producers held meetings to devise means for storing their oil, and took up seriously the question of building a pipe line to the seaboard.

But the drill went merrily on. In 1876 the Warren field began producing large amounts of oil, and the strike at Bullion Run, in Venango County, aroused great excitement; but the total production of the state increased very little because of the decreased run of other fields. Old wells were declining, and in September there were reports of owners cleaning out old wells to keep up the production. In 1877, however, the Warren field reached its best production, while the Beaver County field, adjacent to the state of Ohio, rose to importance—an importance which it maintained for several years. There were also numerous gusher strikes on Bullion Run and in the Millerstown district of Butler County. From all these sources, and from the fields in other states, production was enormously increased, and the old cry of "stop the drill" was raised.

The next year the Bradford field continued to spread, the Kinzua pool, south of Bradford, was developed further, and there was much drilling in Butler, Clarion, and Venango counties, with the result that the over-production continued. The tanks at the wells were overflowing, the pipe lines were taxed to their utmost capacity, and the surplus constantly increased. At the beginning of 1878, the average production of the Bradford field was above 6,000 barrels a day; in a year's time it had increased to over 26,000 barrels; at the close of 1879 it was over 50,000 barrels; and in 1881 it ran close to 100,000 barrels a day, from nearly 10,000 wells. As a newspaper of the time put it, "Oil running on the ground is a common sight in the Bradford field, but the drill bounces merrily on." A bill was introduced in the state legislature to put a tax of one cent per gallon on all crude oil produced in the state; and there was a movement among the Bradford producers to stop drilling for six months, but nothing came of it. Every man who could secure an oil lease seemed animated by one desire—to get the oil out of the ground as rapidly as he could.

This general situation continued through 1879. As one reporter put it, "More oil! More rigs! More gushers!" The Bradford field continued its growth in production, some of the oil flowing upon the ground for lack of storage facilities, and a small well was opened in Allegany County, New York. The declining production of the Butler and Armstrong and Clarion County fields was not enough to balance the growing output of the northern districts, and the market remained weak throughout the year. In June the United Pipe Lines issued a notice that their tankage was filled to its utmost capacity, and that no oil would be received unless cars or other means were provided for its immediate shipment. Agitation for a tax on the output continued. Several proposals were made at Harrisburg. One was the proposal of a flat tax of one cent per gallon on output. Another was to tax all oil going out of the state to be refined. Another proposal, reported back to the Pennsylvania House of Representatives by the Ways and Means Committee, was for a tax of $500 on each well drilled during 1879, and $250 on each well drilled in 1880, with a tax of five cents a barrel on the crude output. Mass meetings were held at Oil City and elsewhere, denouncing all attempts at taxation of the oil industry; but the producers generally preferred the tax on wells to a tax on output, because the former would have a tendency to curtail production. There was much denouncing, not only of the drillers who were augmenting the production by their efforts, but also of the Standard Oil Company, which was supposed to be partly responsible for the ills of the industry.

The next year was no different. Bradford production continued to grow and swell the surplus stocks in the region, in spite of declines in the "lower country"—Butler, Clarion, and Armstrong counties. The Allegany field in New York attracted increasing attention, and the Clarendon field developed prominence, while a new pool, the Van Scoy or Dew Drop pool, on Kinzua Creek, was opened in the latter part of the year. In some sections of the Bradford field, oil was running to waste along the creeks, while in Clarion County it was reported that over a thousand wells were abandoned during the year, the rigs and machinery, where possible, being moved to the northern field.

In February of the following year it was reported that the waste of oil in the Bradford field was less than a thousand barrels a day, and was decreasing. In the meantime a new well near Richburg, New York, caused a rush of operators to that vicinity, and before the close of the year the new field was being rapidly drained. In 1882 the Bradford field declined rapidly. At the beginning of the year, this field produced 62,000 barrels a day, from 11,630 wells, and in November, 11,128 wells were producing only 37,000 barrels a day. Several firms were engaged in pulling out abandoned wells in the district, the material generally going to the Allegany and Clarendon fields.

In April, 1882, the famous "646," the "Mystery Well," was struck in Cherry Grove township, Warren County. This started a new "white sand" development, which was the forerunner of a new series of white sand pools opened during the next two decades. In July seven wells in the Cherry Grove district were producing 4,350 barrels per day, and by the close of August the daily production had jumped to ten times that amount. Much of this oil was wasted. As early as June the tanks at "646" were overflowing, and the oil was run off to a safe distance and burned. Prices dropped from $1.00 to about 50 cents a barrel. The production of the district, however, fell off about as rapidly as it had risen, and in December was only 3,200 barrels, with prices again over a dollar.

The years 1883, 1884, and 1885 witnessed an important change in the center of Pennsylvania oil production. The Balltown and Cooper pools in Forest County, and the Thorn Creek and Baldridge sections, in Butler County, began to produce heavily, while the Bradford and Allegany fields, and the spectacular Cherry Grove field, dwindled steadily. In the new fields the production overflowed the tanks and ran down the creeks, while in the old fields the pipe line runs were dwindling to nothing. The production of the new fields was so great as to keep the market in a state of depression, and in 1884 a large number of the producers met at Bradford and organized another shutdown movement, which three-fourths of the producers were claimed to have signed. This movement was measurably successful for a while, but early the next year it was abandoned.

In 1886 the gusher wells in Washington County added their output to the increasing flood of petroleum, and the Shannopin field in Beaver County, the Kane field in McKean and Elk counties, the Tarkill pool in Venango County, and the Grand Valley pool in Warren County, supplied new productive areas for enterprising drillers. The annual production increased 5,000,000 barrels. Stocks continued to pile up, prices declined to 59 cents a barrel. This situation continued during the next year, and prices went still lower. The Saxonburg pool in Butler County was being opened up, also the Brush Creek and Bakerstown districts in Allegheny County, and the Taylorstown and Montour districts in the southwestern part of the state. Ohio was becoming an important producer of oil, and the Ohio oil, which sold as low as 15 cents a barrel, helped to increase the over-production. The producers again organized to curtail production, and for a while were more than usually successful.

Although the total production was increasing in 1886, the production per well decreased generally throughout the early eighties, and this fact caused a little concern. The average in 1875 had been 7.8 barrels per day, and in 1885 and 1886 this had declined to 2.80 barrels per day. The increasing output came from the several thousand wells drilled each year.

During almost all of 1888 the producers' curtailment agreement was in effect, and it was successful in greatly reducing stocks and in raising the price of oil, but in November the agreement terminated and was not renewed. The next year production took another bound, and in 1890 made a still greater increase. A number of small pools were discovered, including the Turkeyfoot pool, the Glade Run pool, the Murrinsville pool, and several others, while some of the older pools were extended by new discoveries. The Deer Lick pool, in Warren County, attracted much attention, but soon flashed out. The development of the Mannington and Eureka fields in West Virginia added to the growing stocks, but the price remained fairly steady until the later months of 1890, when it dropped to 67 cents a barrel. Agitation for another curtailment campaign began, but nothing was done until the following year, when another "Producers' Protective Association" was formed. The exact nature of this association is not easy to ascertain, since its sessions were held with closed doors, but its purpose seems to have been to curtail the output, and to fight the Standard Oil Company.

In 1891 the McDonald field in Allegheny County began its sensational career, bringing the production of Allegheny County up to over ten million barrels. The result was a great decline in prices, to an average lower than in any year since 1882, when the Cherry Grove pool had driven the price down to 49 cents.

In 1892 there was a noticeable decline in the production of almost every district in Pennsylvania. With the large stocks that had been accumulated, however, and with the general business depression, prices declined even below those of 1891. The following year the decline in production was quite as marked, but the price did not rise much until the latter part of the year. From 1893 to 1897 the Pennsylvania production remained almost constant, with the older fields declining, and with a few new pools coming in from time to time. The rapidly increasing flow from the Ohio, West Virginia, and Indiana fields greatly increased the total supply of oil, and prices sagged during 1896 and 1897, with resulting distress among the producers, and with much waste of oil.

A great spurt in April, 1895, temporarily revived interest in oil speculations. Again the exchange at Oil City was thronged. Exciting scenes of former years were renewed, as the price climbed ten cents at a jump. From $1.10 on April 4, the price rose to $2.80 on April 17. Certificates were scarce, and credit balances were snapped up eagerly. A few big winnings resulted; then the reaction set in, and the spasm subsided.

Since 1897 the production of Pennsylvania and New York has declined steadily. The production of 19,000,000 barrels in 1897 has dwindled to about 8,000,000 barrels in recent years. The center of oil interest shifted to other regions, and the industry in the Pennsylvania and New York region settled down to a more stable commercial business. The oil exchange, as early as 1900, had ceased to be an important agency, and the speculation in oil certificates had dropped to almost nothing. Hundreds of wells are still drilled each year, but the day of new pools and gushers and quick fortunes in Pennsylvania is about gone. The new wells bored in recent years yield an average of only about two barrels a day, and they are largely "pumpers." Some wells have been pumped profitably in recent years which yielded only a few gallons a day, when they could be grouped and pumped from one central source of power. In New York, some wells are actually being pumped that yield as little as a gallon a day.

The wells of this region are far longer lived than those of many other regions, largely because of the density of the sand rock containing the oil. In the Allegany County field in New York, it was stated in 1909 that more than half of the producing wells had been drilled over twenty-five years before. Some wells forty years old are still producing. Equipment from hundreds of other equally good wells was "pulled out" at the time of the big excitement in other fields, between 1882 and 1892, to be used elsewhere. During recent years some of these abandoned wells have been redrilled and placed on a paying basis. During the early days hundreds of wells were drilled which furnished a few barrels of oil per day, enough to make a profit under present conditions, but they were abandoned because of the low price of oil and the frantic endeavor of the producers to make big strikes elsewhere. No maps were kept, and later, when prices rose, it was expensive, in many instances impossible, to find these wells, and some areas had to be drilled a second time.

A number of devices besides the cleaning out of old wells have been used to increase the flow. In some cases the lower sands have been plugged and an effort made to pump from the upper sands, which were originally neglected. In some cases better pumping facilities have increased the flow, and in recent years compressed air, natural gas, or water have in some districts been forced into strategically located wells, thus forcing oil into adjacent wells. These devices have been used in a number of other regions likewise.

Even in very recent years there has been some speculative excitement in the Appalachian district. In May, 1915, a 125-barrel well was struck on the Lutheran parsonage lot at Evans City, Butler County, and the resulting speculative excitement was strongly suggestive of the earlier period of oil history. A wild scramble for town lots resulted from the strike, and in a very short time 150 wells had been drilled in a small area of the town site, with the result that the pool was practically exhausted before the end of the year. The number of fortunes lost here in the course of a few months suggested that Pennsylvania had not learned her lesson very well.

NOTES

1. Douglas, *Old France in the New World.* On the history of oil development in the United States, the writer has made free use of an unpublished manuscript of G. P. Grimsley, on the "History of Petroleum and Natural Gas Development in the United States."

2. Bryant's *Popular History of the U. S.,* 1881, IV, 6.

3. Bacon and Hamor, *Am. Petroleum Industry,* I, 200.

4. N. Y. Geol. Reports, from the Governor to the Assembly, No. 161 (1837), p. 196; Rept. No. 50 (1840), pp. 55, 449, 464.

5. Pa. Geol. Survey, Ann. Rept., 1886, Pt. II, 590.

6. It is recorded that Bissel, one of the first to become interested in the company, taking refuge from the rain under a druggist's awning one day, happened to see one of Kier's rock-oil circulars, on which there was a picture of a small derrick over a salt well, and from this conceived the idea of boring a well to get the oil.

7. Derrick, *Handbook of Petroleum,* Aug. 13, 1867, p. 92.

8. *Ibid.,* Dec. 9, 1874, 238.

CHAPTER III

SIGNIFICANT FEATURES OF EARLY PENN-SYLVANIA OIL HISTORY: OVER-PRO-DUCTION; AND WASTE OF OIL

IT will be worth while to point out several of the significant features of Pennsylvania oil history as sketched in the preceding pages. In the first place, there was the unfortunate instability of the industry, with the boom and the decline of the various districts as they were exploited, with the recurrent periods of over-production, and with curtailment agreements directed at securing higher and more uniform prices. In the second place, there was the waste of oil, arising from the drilling of wells before storage had been provided, from lack of casings, which permitted seepage into the dry sands, from inefficient transportation facilities, from evaporation, fires, floods, and accidents, from infiltration of water into the oil strata, from incomplete extraction of oil deposits, and from the use of oil for unimportant purposes. In the third place, there was immense waste of capital from lack of scientific knowledge of oil deposits or of their origin, from the operations of quacks and novices in the oil industry, from the secrecy with which operations were carried on, and the consequent lack of generally disseminated information, from speculation and fraud, from the neglect of other industries, and from the drilling of far too many wells. In the fourth place, there was the development of monopoly, in the form of the Standard Oil Company, with its strangle hold on the growing industry; and, finally, there were the unfortunate social conditions which go with every such new, speculative, exploitative, and, for that reason, short-lived industry.

It would be difficult to pick out a more unstable business than the oil business, as it developed in Pennsylvania. Its history is a history of sudden discoveries, feverish exploitation, decline, and then rapid shifting to other regions, where this was repeated over and over again. Thus the first strike of 1861 brought hundreds of people to Titusville, but the decline of oil prices almost immediately brought such a depression that many enthusiastic prospectors were ruined, and by 1869 the output of this district was rapidly dwindling. The history of Pithole in 1865 and 1866 has been recounted. Some of the Clarion and Butler County fields discovered in 1872 were almost entirely abandoned seven years later. The Bradford field, one of the heaviest producers in the state, developed importance in 1875, reached its maximum in 1881, and began a steady decline the next year. The Warren County field, which was developing rapidly in 1876, began to decline within the following year. The Allegany County field of New York, which assumed importance in 1877, was declining a decade later. The Cherry Grove flash of 1882 has been described. The Cogley Run, Red Valley, and Kane fields, discovered in 1885, had declined more than 50 per cent ten years later. The Tarkill pool, opened in 1886, declined 50 per cent the following year. The Grand Valley pool, opened in 1886, was declining a year later. The great McDonald pool, already discussed, was declining the year after it was opened. Most of the fields did not decline so quickly as this, but as one writer expressed it, "The death of every oil field begins the day it is opened."

With such recurring epochs of development and decline, of over-production and dwindling runs, great variations in price were inevitable. In January, 1860, pipe line certificates were selling at nearly $20 per barrel. At the end of the year they had declined to $2.75, and, in December of the next year, had reached 10 cents a barrel—a decline to one two-hundredths of the first price. Within the next six months the price rose 1,000 per cent, to $1.00, and, before the end of the year, to $2.25, the next December to $3.95, and in December, 1864, to $11 a barrel—over one hundred times as high as in December, 1861.[1] The next year the price declined to $6.50, and the following year declined further to $2.12, fluctuated for several years between $1.00 and $5.80, and in November, 1874, went down to 55 cents. Within four months it was again up to $1.75, and a little over a year afterward oil was selling at $3.81. In January, 1877, the price was $3.53, and the following year $1.43, and before the end of 1879 was again below $1.00. For years the price remained below $1.00 most of the time, once going as high, however, as $1.79, and often going near the 50-cent mark. During the Civil War, and even to some extent during the years following, the fluctuations arising from conditions in the oil industry were accentuated by fluctuations in the value of greenbacks.

Prices fluctuated widely, and they were generally far too low. The chance of making great fortunes in the oil fields lured so many into the business that oil was produced far too rapidly, and, although there were months and even years of fairly remunerative prices, the general condition was one of over-production and ruinously low prices. With oil selling for ten cents a barrel, it was not worth the barrels it was put in, and on several occasions coopers refused to barrel the oil for it. When this situation arose, there was nothing for the operators to do but let the oil run to waste until the price rose again.

To overcome this condition, curtailment or shutdown campaigns were organized. In August, 1872, with the price of crude oil at $3.00—a price which not long afterward would have seemed munificent—one of the first of these shutdown campaigns was inaugurated among the producers at Parkers Landing, where 200 producers signed a pledge not to drill any more wells for six months, under penalty of a $2,000 fine. An effort was made to persuade landowners to join the movement, but this was not successful. Shortly afterward, the producers began to agitate for a cessation of all pumping for thirty days, but were unable to secure any agreement. Meanwhile, the producers around Oil City met in September and decided upon the formation of a permanent Producers' Protective Association, and at the same time circulated a shutdown pledge, which was reported to have been generally signed. In October, Derrick estimated that nine-tenths of the production around Oil City was shut in, and at Parkers Landing the report was that very few walking-beams were in motion.

The "producers' agency," through which the producers were to work, was in form a corporation, with a capital of $1,000,000, to be increased if necessary, the stock to be subscribed only by producers or others friendly to the producing interests. The capital subscribed was to be used to purchase oil at a nominal price of $5.00 per barrel, $3.00 down on delivery, and the other $2.00 payable the following month, provided the oil could be sold. Not only operators, but banks subscribed, and some banks announced their willingness to loan money on oil in storage, to prevent its being sold.

For some time it seemed that the success of this movement was assured. In the latter part of October over $200,000 of the stock of the agency had been subscribed, and at a mass meeting in Oil City in November the subscription reached $1,000,000. Even before the agency was formally established, however, there were signs that the shutdown agreement was being violated. Rumors were circulated that the agitation in favor of a shutdown at Parkers Landing had been started by a number of producers who had their tanks full and wanted wells capped until they could sell their stocks at a good price. Some producers at Oil Creek, Pithole, and Shamburg complained that since they had little oil stored, they were closing down just to benefit the other producers, and at the risk of ruining their wells. Late in October it was noted that some pumping was being done in remote sections during the night watches, when the producers thought they would be unobserved. Before the end of the month wells were being started all over the oil region, in many cases with well production greatly reduced by the shutdown. In November there were rumors that some producers were selling directly to the refiners below pool prices, and some oil men began to predict a crash. Before the end of December the failure of the "agency," and of the entire shutdown movement, was admitted in many sections; and this was hardly admitted before agitation arose for another one.

This agitation continued intermittently, and in April, 1874, another curtailment movement was inaugurated at Petrolia. A pledge to refrain from drilling for 90 days was circulated and generally signed. Late in the month a committee reported that 130 drilling wells had been stopped, and that the owners of 27 rigs had agreed not to start drilling for 90 days. For a while it appeared that the movement was a success, but early in May, at a meeting of the producers, it was discovered that some of the signers were violating their agreement, and the movement was called off.

Early in May, 1877, with a serious decline in the price of oil, there arose further agitation for a shutdown, and at a meeting in Oil City the producers worked out a plan to store a certain proportion of the oil produced, and a committee was appointed to secure signatures to an agreement. There was a movement also to stop drilling for six months. Many of the producers of the lower district, recalling the experiences of the early seventies, voiced opposition to the storage plan, and aside from the agitation, nothing seems to have been done, except that the landowners and leaseholders in the Bullion district made an agreement whereby no further leases were to be executed. Agitation for a concerted curtailment of output appeared intermittently during the next several years. In 1884 a shutdown movement was started which was claimed to include 700 operators, but it presently appeared that the signers who lived up to their agreement were simply playing into the hands of those who did not do so, and of those who refused to sign, and the whole movement broke up.

The price of oil remained abnormally low for several years—between 60 cents and 70 cents much of the time in 1887—and on October 31, 1887, another great shutdown movement was inaugurated at Pittsburg. The contract in this case provided for the shutting in of at least 17,500 barrels daily production for the term of one year, and for the entire cessation from drilling during that period. No nitro-glycerine was to be used, and no wells cleaned out. (Nitro-glycerine was generally exploded in the bottom of new wells, to break up the rocks and increase the flow of oil.) The Standard Oil Company joined with the producers, and set aside 5,000,000 barrels of oil at 60 cents a barrel, the profits on this to be divided among the producers who complied with this agreement. In addition, the Standard and the producers together had a pool of 2,000,000 barrels, and the profits accruing from the sale of this was to create a wage fund for laborers thrown out of employment.

The agreement went into effect in all districts except the Saxonburg district, and that district was not considered important enough to prevent the success of the movement. For some months the new movement was a great success. Many producers who at first refused to join were forced by public opinion to come in. Professional drillers were nearly all induced to join, being paid for remaining idle out of funds provided by the association. In December stocks were reported depleted to the extent of about 1,300,000 barrels.

The success of the Producers' Protective Association was so conspicuous that members of the association were summoned to Washington for examination by an investigating committee of the House of Representatives. Members denied that any means except persuasion were used to prevent the drilling of wells, and denied any responsibility for attempts that had been made to blow up drilling rigs.

In the following September, as the time approached for the termination of the agreement, a meeting was held at Bradford, and the producers present agreed to continue the shutdown for six months longer, drilling to be allowed in old territory; but the movement did not spread, and in November the agreement came to an end. It had been the most effective agreement in the history of the oil trade, and its effects outlived the agreement itself, for when the wells were opened again, many of them flowed only about two-thirds as much as before the shutdown, the oil sands having been partly ruined by the infiltration of water, by the escape of the natural gas, and by other causes.

Prices remained below a dollar most of the time, however, and two years later there was a meeting of the independent oil producers at Pittsburg, presumably devoted to the question of curtailment, although no reports of the proceedings were given out. During the next year, 1891, there were a number of such meetings, one of which, at Franklin, was said to have been attended by 100 delegates from 35 local assemblies, representing the entire oil-producing field from New York to West Virginia; but, as the sessions were held behind closed doors, there is no record of the action taken.

One obvious method of discouraging crude oil production did not meet with much favor during these years. During the Civil War a tax had been imposed on crude production, but in 1866 this tax was repealed, largely, it is said, through the efforts of James A. Garfield, then Representative from Ohio. During the succeeding years there were various attempts, by both the federal and state governments, to impose other taxes on oil production, while the producers were working for the abolition of the 10-cent federal excise tax on the manufacture of refined oil, which remained in force under the Act of 1865. As already pointed out, a variety of bills were introduced into the state legislature, but these bills were always killed in committee or voted down. One such bill proposed a tax of from $500 to $1,000 on all oil wells drilled, and an additional tax on crude production, but it met with very determined opposition, and failed to pass. Producers were more favorably inclined toward a tax on wells than toward a tax on production, since a tax on wells would have a stronger tendency to curtail production, but they naturally opposed any sort of taxation. It would be difficult to imagine a more nearly burdenless tax than a tax on drilling when the industry is suffering from great over-production, but few were able to see the logic of it at this time.

Perhaps the most unfortunate feature of Pennsylvania oil history was the waste of oil. Throughout the early history of this region, oil was wasted, recklessly and wantonly, by millions of barrels. The newspapers of the time are full of accounts of oil running on the ground for lack of storage facilities; and frequently, while the oil from some wells was running down the ravines and creeks, drillers were boring other wells to swell the unused production. Some of the following quotations are typical: "So much oil is produced it is impossible to care for it, and thousands of barrels are running into the creek; the surface of the river is covered with oil for miles below Franklin. Some wells are being plugged to save the production. Fears are being entertained that the supply will soon be exhausted if something is not done to prevent the waste" (1861). "Oil flowing on the ground for want of tankage in the Bradford field" (1879). "Much oil is going to waste in the Allegheny field for want of sufficient storage and inability of the lines to handle the output" (1882). "Tanks at '646' are overflowing and the oil is run off to a distance and burned" (1882). "The production of the Reno well running on the ground for want of tankage and pipe line connections" (1883). "In other parts of Pennsylvania a great quantity of oil is running to waste for want of tank facilities" (1891).[2]

In the capping of some of the big gushers considerable waste was inevitable, although the waste was generally greater than would have been considered necessary, had oil been worth saving.

Heavy waste of oil arose from the inefficient transportation methods of the time. At first, the methods of carrying the oil were very primitive. Imperfect barrels were hauled in rough wagons from the wells to the streams, where they were loaded, sometimes on steamboats, but generally on rough flatboats or rafts, by which means they were transported to other places on the river, and reshipped by rail. In some cases a species of boat called a "bulk boat" was used, into which the oil was run in bulk and transported to its destination, there to be transferred to barrels or tanks. These methods could not be used when the rivers were low; and frequently on smaller streams, and notably on Oil Creek, a "pond freshet" would be caused. A number of dams on Oil Creek belonging to abandoned sawmills were utilized to hold back and accumulate a large amount of water, which at length was allowed to escape from all the dams at once, and thus an artificial rise in the stream was secured. This was a perilous business, of course, and a great many oil boats were crushed in these "booms," with the loss of all the oil. This was particularly dangerous, because a wrecked oil boat meant a river covered with oil, and perhaps a fire which burned everything on the river.

Wright described these losses as follows:

The scene is apt to be highly exciting and withal not free from danger, as the newly-emancipated floods rush madly down the valley, bearing scores of huge flat-bottomed boats, all heavily laden with the products of the wells. Sometimes boats laden in bulk collide with those laden with barrels, and both again crash against others which may be moored to the margin of the creek, breaking them into fragments and sending their cargoes adrift.[3]

Another writer presented a similar picture:

The great rush to Oil City will be in the spring, when the rivers open. As soon as the thaw commences the roads are horrible from the continuous travel; some half a dozen abandoned wagons and their oil may be seen lying along the steep banks of the Alleghany, it not paying to have it removed. There is an immense amount of oil wasted every day, and it can be seen flowing on the surface of the streams.[4]

Much of the more volatile elements in the oil was lost by evaporation. The oil was often stored in earthen sumps, or in open tanks, where exposure to the sun caused a very heavy loss. Much was also lost by seepage into the earth, by leakage from the wooden tanks, and by being washed into the creeks by the heavy rains of spring and summer.

Vast quantities of oil were destroyed by fire, for there were many instances of disastrous fires. Not infrequently gushers were set afire in some way, and burned for days, or weeks. When such burning wells were situated near creeks or rivers, terrible conflagrations sometimes resulted. One such disaster in 1865 was vividly described by a contemporary writer:

The boats, loaded with oil, took fire, and burning their lines, went adrift down the stream. As they passed the tank boats, filled with oil in bulk, the flames spread to them with the rapidity of lightning; a single flash and the whole boat was in flames. The burning boats continued in their course of devastation, setting fire to everything they touched on their route. The bursting of tanks covered the stream with oil, which took fire and added to the terrible grandeur of the scene. A great fire of this kind occurred more than a year ago, when blazing boats came down the creek, and, plunging among a large fleet of loaded boats at Oil City, set them on fire; and the whole blazing mass swept down the Allegheny, burning the Franklin bridge as they passed, and spreading terror for miles along the river.[5]

In June, 1892, occurred the appalling disaster of fire and flood at Titusville and Oil City, one of the worst disasters in oil history. The bursting of a dam on the upper waters of Oil Creek let loose a flood that was in itself sufficient to carry death and disaster, but one of the first structures to be carried away was an immense tank of naphtha, which not only covered the water, but filled the air with inflammable gas. A spark from a passing locomotive soon ignited it, and converted the torrent literally into a river of fire, which burned all the buildings in its path. Unfortunate victims caught in their houses were surrounded and engulfed in a lake of fire. The destruction of tanks of petroleum kept up the supply of fuel, so that nothing combustible was left above the surface of the flood. It is estimated that fully three hundred lives were lost.

The waste of oil and property by fire could have been greatly reduced, had any real effort been made, but the temporary and exploitative nature of the whole business did not accord with any great concern over such losses. Since the oil would be exhausted in a few years, and the towns would then be largely abandoned, it was hardly worth while to do anything in a substantial or permanent way. It was cheaper to run the risk. As one writer expressed it:

One-fifth of the immense aggregate (of loss from one great fire and flood), applied to the construction of roads and levees, drains and fences, with reasonable sagacity in forecasting the advent of such a visitation, might have warded off the effects of such a frightful calamity. As it is, companies and individuals must pay the penalty of establishing conditions in which the lowest form of selfishness is the only recognized principle of action.[6]

The amount of oil wasted by seeping into the non-oil strata, especially during shutdown campaigns, and the amount ruined by infiltration of surface water, cannot be even estimated, but it was certainly very great —many millions of barrels. The production in almost all parts of Pennsylvania was largely reduced by the flooding of the oil sands. Some wells were ruined during the shutdown campaigns, and almost all of them flowed a reduced amount after being closed for a while. Such waste as this was, of course, greater where casings were improperly fitted, and casings were not well made or well fitted during the first few years of oil exploitation. At first no casings were used at all, and it took many years to perfect the methods of manufacturing and using them.

The custom of sinking wells in close proximity to each other was responsible for much loss of oil in the underground strata. Whenever an operator struck a good well, other operators immediately sank other wells as close as possible, for the purpose of securing oil or of flooding the sands. If they were unable to secure any oil, they could sometimes compel the owner of the producing well to bribe them not to spoil his production. It was not usually possible for each oil company to secure a large enough block of leases to prevent such robbery, because most oil companies in this early period were small. The landowners could have done much to reduce these evils, but they were seldom inclined to do so. Most landowners leased their lands with a view to obtaining all the available oil in as short a period as possible, often leasing in such small plots as to make over-drilling inevitable.

Much unnecessary waste resulted from the abandonment of wells of small yield. The excited operators were always in search of big gushers, and the small producers were uninteresting, and hard to handle at a profit when oil was selling from 50 cents to $1.00 a barrel. Much of the time, gusher wells governed the market, and it was impossible to pump the small producers profitably, so they were abandoned by the thousand, to be later opened, when methods had improved and prices had gone higher. Abandoned wells served as means whereby surface water and mud and sand seeped into the oil strata, and whereby the oil seeped into other strata, from which much of it was never recovered.

An important cause of incomplete extraction of the oil was the waste of gas from the oil sands. It is the pressure of the gas in the oil sands that drives the oil out, and when gas was wantonly wasted, as it always was in the early history of Pennsylvania, it meant that the extraction of oil was unnecessarily reduced.

Perhaps the most serious waste of oil was its consumption for unimportant uses. When oil is 50 cents a barrel its "marginal use," as economists would say, is a very low use. In other words, much of this oil satisfied very unimportant wants, and in many cases wants that could have been satisfied in other ways. As early as 1866 and 1867, experiments were being carried on in the burning of oil in the boilers of locomotives and steamboats, these experiments, according to some reports, being entirely successful.

It would be vain to try to estimate the amount of oil thus wasted in various ways, but it was certainly very great. One official, as late as 1879, estimated that from 5,000 to 6,000 barrels daily were being wasted outright. In 1880 he estimated that 10,000 to 12,000 barrels daily were being wasted in the Bradford field alone. Outright waste, however, great as it was, constituted but a small share of the total waste through fires, flooding of the oil sands, consumption for low uses, and the like. It should be pointed out, however, that much loss of oil was inevitable in the early period of the oil industry, and that "loss" and "waste" are hardly synonymous terms here. The oil was cheap, and labor and capital were expensive; and, even had they been relatively as cheap as the oil, there was not at this time enough technical knowledge and skill to enable operators to avoid great waste.

An important reason why waste of oil was not regarded more seriously at first was that many operators, and the public in general, had a false theory of the origin of oil. It is now generally believed that oil is the result of the deposit and decomposition of organic material; but at first many believed that oil was being formed in the earth all the time, and that the supply was therefore inexhaustible. This opinion was held by some people many years after the exhaustion and abandonment of some of the Pennsylvania fields should have taught them better. There was little appreciation of the fact that oil is an exhaustible and irreplaceable resource. As Wright expressed it in 1865:

The fact that wells would give out at last was unknown until it had been proved by abundant and most painful evidences. It is still called into question by some, who are either ignorant of the country or interested in creating misconceptions in the public mind.[7]

NOTES

1. The writer does not know whether these are gold prices or greenback prices, but probably they are the latter.

2. The following story, told at an oil men's convention, illustrates the nature of much of the early Pennsylvania development:

Back in the early Pennsylvanian days when oil was selling at 10 cents a barrel, and only selling when the well was near a railroad, an operator went out about 40 miles from a railroad and drilled a 150-barrel wildcat well.

His No. 2 was nearing the sand and the operator decided that he would go to the city for a rest—he couldn't sell his oil anyway, and there was nothing for him to do but rest.

The day after he got to town he received the following message from his driller: "No. 2 is in making 500 barrels. No. 1 has increased to 300 barrels a day. All running on the ground. Rush timbers for No. 3."

3. Wright, *Oil Regions of Pa.* (1865), p. 108; Redwood, *Petroleum and Its Products*, I, 476.

4. *Derrick and Drill,* published by James Miller, N. Y. (1865), p. 69.

5. Wright, *Oil Regions of Pa.,* 109.

6. *Ibid.,* p. 107.

7. *Ibid.,* p. 94.

CHAPTER IV

SIGNIFICANT FEATURES OF EARLY PENN-
SYLVANIA OIL HISTORY (Continued):
WASTE OF CAPITAL
AND ENERGY

OIL was not the only thing wasted. Capital and labor were wasted in a great many ways. In the first place, immense waste arose from the fact that there was little scientific understanding of the exact manner of the occurrence of oil. At first drillers clung tenaciously to the valleys, on the theory that the rivers flowed in chasms rent by upheavals and crackings of the earth's crust, and that the oil streams followed the lines of such fissures; in short, that the oil flowed in streams, which followed the lines of the rivers. Some thought that locations along the creeks and rivers were good because the stratum of earth was thin there, and the lower oil strata could be reached with less difficulty than on the hills. The fact that the first wells were struck along the valleys gave rise to this theory, and it was generally held until about 1865 or 1866. When an operator had experimentally located a well upon a hill and obtained a large flow of oil, a "ridge theory" came into favor, the main feature of which was that the hills had been uplifted, and oil and gas, seeking the higher levels, would be more plentiful on ridges than in valleys. This theory was still held by some operators as late as 1885, and some ridges in Allegheny County were dotted with wells.

A third theory adopted for its surface indications the deer licks and boggy springs which follow the outcrops of the pebbly conglomerate rocks. The "conglomerate" at one time probably incited more wild-catting than any other surface indication, and led to some new developments. It is not now thought that the conglomerate has any definite significance, but its presence led prospectors to drill in places which would otherwise have been neglected, and led to some accidental discoveries.

There was another theory, the "belt-line" theory, which for some years held sway over all others. According to this theory, when oil was found in any region, a belt line drawn to include it, and running northeast to southwest—the general line of geological structure in the Appalachian region—would probably include other oil regions. This theory was later modified so as to become merely an adjunct to the pool theory for oil and the anticlinal theory for gas. The pool theory assumed that the oil-bearing rocks were deposited in detached patches of greater or less extent, and not in long narrow ribbons, as claimed by thorough-going "belt liners." This theory was being discussed as early as 1880, and within a few years had become very popular.

Finally, there appeared the anticlinal theory, now generally held, that gas wells should always be located on anticlines, and not on synclines, because gas is lighter than water or oil, and seeks the highest reservoirs. This theory assumes a permeable sand rock containing water, oil, and gas, or only water and gas, in such proportion and under such conditions that the fluids may stratify themselves as freely and completely as they do in an open tank under air, water and oil at the lower levels, and gas at the higher. Dr. I. C. White was said to be the first geologist to bring this theory into prominence and apply it in a practical way, about 1883, although it had been held by other geologists more than ten years previously.[1]

The organic origin of oil, first suggested by J. S. Newberry in 1873, and substantiated by Edward Orton and others, was generally accepted by 1903. With this theory came greater concern for the future of our oil deposits.[2]

With so many false theories regarding the geological occurrence of oil, and so many operators who were largely ignorant of all that pertained to oil, there was a wide field of opportunity for the "oil smellers," "magnetic finders," and others who sunk wells "under advice from spirit land." Dreams and divining rods and switches had their numerous votaries among the oil producers.

The divining rod has been used for centuries, perhaps for thousands of years, in searching for metal deposits and for hidden treasures of various kinds. Some writers think that it was used by the ancient Scythians, Persians, Medes, and others, but there is no conclusive evidence of this. However, in the sixteenth century and later, it was used, not only in hunting for minerals and water, but also in hunting criminals and in determining the guilt of witches, sorcerers, and criminals. In 1701 the Inquisition found it necessary to issue a decree against the divining rod in criminal prosecution, but for a long time afterward it was much discussed and debated by the churchmen of the time. In modern times it has been used a great deal in seeking water. Many rural communities in America have their "water witches," who "witch" for water whenever a well is to be dug, and there are a great many people who are thoroughly convinced of the efficacy of this method of locating a well.[3]

Thus the "oil smeller" was ready for immediate employment when Drake started the mad search for oil in Pennsylvania, and was employed a great deal in the first years of oil development.

The divining rod was a forked stick commonly of either hazel or peach, held by the extremity of each prong of the fork in a peculiar way—the palms of the hands being upward, and the prongs crossing the palms, and held by the thumbs and tips of the fingers. The wizard walked over the country he was to try, and as he approached the greatest body of oil in the immediate neighborhood, the thick end or handle of the fork was supposed to turn down, in spite of all the efforts of the holder to prevent it. There is no doubt that, owing to the way in which it was held, it had a mechanical tendency to turn, and this increased at the will of the holder, and in such a way as to remain undiscovered by even a close observer. Wells were often discovered by such means, and these "diviners" were sometimes able to realize handsomely on their efforts.

One writer stated in 1873:

There are many operators in whose minds are yet fresh the implicit confidence placed in oil wizards, and their power to successfully locate wells; indeed, they are to be found at the present time, and still a few of our oil men employ hazel twig manipulators to mark the spot to drill upon. That they were skillful, at least so far as manipulating the divining rod—or "dowsing-rod," as it was sometimes called—to the satisfaction of their employers and their own emolument, there can be no doubt, and that they have now almost become extinct is equally true.[4]

Henry gives an interesting account of the location of the Harmonical No. 1 at Pleasantville, which runs as follows:

Mr. James, the spiritualist, in company with a number of gentlemen, was on his way to examine some property a few miles south of Pleasantville. Seated in a buggy with a companion, they had proceeded but a little distance when Mr. James became violently influenced by what is termed his attending spirit-guide. This invisible power increased till, Paul-like, he hardly knew whether he was "in the body or out." The control soon becoming absolute, he was taken over the fence into a lot on the east side of the road, moving rapidly, and his companions following. Nearly unconscious, the locomotion seemed to him like being hurriedly forced over a fence. Proceeding towards the south side, then back and near the north end of the field, he moved more cautiously, as though tracing some lode or vein. Reaching a certain locality, he was thrown heavily upon the ground, and making a mark with his finger, thrust a penny some inches into the earth. He then fell upon his bosom stiff, and apparently lifeless. His eyes were closed, his face pale, the pulse feeble, and the limbs rigid as in death.

In this condition, he was given to understand that they were then upon a superior oil-producing territory, extending many miles in a certain direction—that directly under their feet, were floating streams of oil that if opened would yield rich supplies. This was the spot—the precise location of "Harmonical Well, No. 1" which was struck in February, 1868, and produced upwards of 100 barrels per day. The striking of this well created great activity in oil developments in the district, and thus commenced the famous Pleasantville excitement of 1868. Mr. James has located many wells in the oil region by "spiritual guidance," many of which proved good producing wells. He became prominent as an enterprising and successful operator. Recently, however, he has located a well on the Clarion River, claiming the same manifestation as related in regard to the Pleasantville well. This well is now sunk to the depth of 1600 feet, and no sand and no oil, but drilling still continues.

One of the great obstacles to intelligent oil prospecting and drilling was the fact that operators maintained the strictest secrecy regarding their wells, making it very difficult to secure information of value to the industry generally. The lengths to which this secrecy might go is indicated by the story of one of the great wells in the Cherry Grove district in 1882. Before commencing work, the drillers were put under oath not to divulge anything concerning the well, and the premises were carefully guarded from outside intrusion. When oil was struck, wooden plugs were dropped in and driven down to prevent the oil rising higher, and every vestige of oil was cleaned from the cable and derrick, picket lines were established at a considerable distance from the derrick, and vigilant guardsmen patrolled both day and night to prevent anyone from penetrating within their lines. Thus they held the secret for several weeks, until the well owners had secured adjoining lands and had negotiated a sale of their property. Meanwhile, other operators and speculators had their scouts employed to watch and report upon every wildcat venture approaching completion, the operators with a view to buying leases on territory adjacent to successful wells, and the speculators with a view to using their knowledge of new developments in speculation on the oil exchange. These spies were constantly lying around the well, scanning the derrick and tools with field glasses, examining the stream below for indications of oil, listening to hear if any flows occurred, and hoping for an opportunity to steal in unobserved or to bribe some of the guards so that a "pointer" might be obtained for their principals. But so closely was everything covered and guarded that they could get no satisfactory information, and the well remained a complete mystery for several weeks, until the new owners came to clean the plugs out.

The story of the "Mystery 646" in the Cherry Grove district is somewhat similar. This famous well was boarded up tightly for a long time, and under constant guard, while the owners were buying up territory in the vicinity. Intruders were warned away by the report of shotguns and the whistling of shot around their ears. The well proved to be one of the heaviest producers in the state when it was turned on, and the owners made immense profits on the leases they had secured on adjacent territory. In some cases, drillers were discharged when the horizon of oil rock was approached, and the owners themselves, with guards stationed around the derrick, completed the well.[5]

In speaking of the unfortunate secrecy of operators, State Geologist Peter Lesley wrote in 1883:

It virtually renders impossible the contemporaneous collection of data; for the geologist can learn nothing when new wells, both productive and unproductive, are guarded against investigation, and what is said of them is more probably false than true.[6]

This secrecy was necessary to the individual prospector, however. Every new well was watched by many outsiders eager to secure a hint or a suggestion that might enable them to profit by the results of a venture in which they had not risked a dollar, who would not hesitate to use any information obtained, even to the detriment of the parties furnishing it. Thus by the very necessities of the case, every operator carefully guarded the secrets disclosed by his drill, and when information was given the State Geological Survey it was with the explicit understanding that the facts should be published only in a general way, without any such details as would betray the source of the information.

The well records were generally kept in a very defective manner, anyhow. They were usually only memoranda of a few meager facts which the driller thought worth while to jot down. Even when these records were made available to the Geological Survey, they were often so fragmentary and inaccurate as to be of only limited value.

Many of the men in the oil business did not have even a smattering of the information which was available to experienced and intelligent operators, because so many wholly inexperienced men were constantly streaming into the field, and wasting their substance in unwise operations. As one writer described the situation in 1865:

Under ordinary circumstances we should hardly think of employing a butcher to do our preaching, or a farmer to purchase dry goods at wholesale. Most persons have come to believe that a certain amount of training, of experience, as well as natural capacity, is requisite to employment in any situation requiring knowledge, skill, and judgment. Not so with many of the oil companies. An officer who has done his duty bravely on the field, a glib-tongued lawyer, or politician, a broken-down merchant, anybody, in fact, who is "smart" enough to button-hole a board of ignorant directors, may be safely intrusted with the charge of a heavy interest in Petrolia.[7]

Perhaps even more unfortunate than the waste of oil was the tremendous waste of productive energy in socially profitless speculation. Everyone wanted quick money, everyone wanted to make a fortune in a short time and then "get back to the states." The nominal capital of mushroom oil companies ran into hundreds of millions.

Companies became so numerous that it was difficult to keep the run of their names, their location, or of the names of their officers. So numerous were they, that it was equally difficult to obtain new men to act as officers, hence the same parties figured as presidents and directors of three or four different companies. Offices were called for to an extent that sent up rents in the neighborhood of the Exchange to two-and-three-times the former prices. Even then, the difficulty of procuring accommodations was such as to oblige two or three companies to occupy the same office.[8]   192545

The excitement spread all over the eastern states. The *Philadelphia Ledger*, early in 1865, described the situation as follows:

The oil speculation is now rapidly extending to new classes, and drawing into it an amount of fresh capital every day that is very remarkable, and not a little dangerous. In the first place, those shrewd old capitalists who, six months ago, shook their heads, and predicted a speedy smash, seeing that instead of this, the excitement and prices have doubled, are now the boldest to venture in, and give incredible sums for property they would not have touched then at half the price. In the next place, men of means who never speculated before in any thing, to say nothing of those who, without capital, borrowed from the banks, have been drawn in, so that in every city, gambling in oil-stocks has created a furore that has quite superseded that of the gold-market. . . .

In the meantime, we see increasing multitudes of respectable and industrious men, of small or large means, neglecting some honest and regular business in which they were doing well, and throwing up safe and useful occupations to go up to the oil-regions in quest of a sudden fortune. Or worse still, they are watching the rise and fall of oil-stocks as madly as ever gambler watched the turning up of the dice, and with even less knowledge of the chances for or against them. Visions of wealth dazzle them and unfit them for ordinary duties, and, after being bitten once or twice, and perhaps, ruined themselves, they, who were honest and industrious before, become the greatest and most barefaced swindlers and knaves in the oil-speculations. Those who have begun with nothing, and by hard labor, amid grease and dirt, learned all about putting down wells so as to make them work, or by diligent selections of likely lands have made them known to capitalists, are, as classes, far more reliable than those whose lives are taken up in setting afloat new and visionary joint-stock companies at this moment. In regard to three-fourths of these, it may safely be predicted that some one has got to lose who buys in, and the wholesale and systematic lying put in print, as well as in the mouths of agents, is beyond almost any thing we have ever known.[9]

This universal furor for speculation was referred to by one observer in the following graphic terms:

The millionaire and the mechanic, the artist and the artisan, the scholar and the priest, saint and sinner, pessimists and optimists, all grades, all shades, all opposites and affinities in society, meet on common ground, and turn their thoughts in a common channel towards wealth, through petroleum. The butcher, the baker, and the candle-stick maker, all have "stock" in the —————— run, or —————— farm companies; Croesus, who owns a million shares in the Venango Bengalee Company, is ushered into his palace by a yellow plush, who is represented in the Pactolus Petroleum pumping property by a hundred shares.[10]

"Olei sacra fames!" exclaimed a writer who was visiting the oil regions in 1865. "It leads to every imaginable kind of misrepresentation and cheating. In every transaction involving profit and loss, falsehood is expected, is looked upon as the rule, truth as the exception. This indifference to veracity and honor does not merely extend to matters connected with the oil wells, but to those of everyday life—to engagements entered into by landlord and tenant, by mechanics, laborers, etc., wherever a slight advantage may arise by violating them."[11]

Another writer complained of being "condemned," while in the oil regions, "to see nothing, smell nothing, hear nothing, talk nothing but oil."[12] A scientist who resided in the oil region of Pennsylvania during these years has told the writer of the increase in the amount of insanity which resulted from the oil craze.

Mr. J. P. Lesley, of the Pennsylvania Geological Survey, wrote caustically in 1883 of the speculative spurt of the times:

It is equally manifest that violent fluctuations of the market would be impossible but for a still prevalent ignorance of the Geology of Petroleum, and an equally prevalent epidemic of the gambling spirit. . . . Pennsylvania may be vain of her possession of this most wonderful treasure; but she cannot be proud of the utter demoralization of the crowded population which scrambles for it in so unmanly and thriftless a manner.[13]

Most of the oil companies were of course fraudulent. The sheriff of Venango County was reported as advertising for sale the assets of more than two score of such companies at one time. A Pittsburg man papered his sleeping room with certificates of stock in oil companies, the novel wall covering having originally cost him $53,000. Another man was reported to have come to Oil Creek with 50,000 shares of stock in an Oil Creek company, and to have devoted several days, on foot and horseback, to hunting for his "interests," but without success. There was apparently some sharp practice in connection with the very first well drilled, for when Drake telegraphed the news of his strike to one of the stockholders of the Pennsylvania Rock Oil Company—the company for which he was working,—that enterprising individual "thereupon bought up all the stock of the company that he could get hold of, and immediately visiting Oil Creek, leased large tracts of land that afterward yielded abundantly."[14]

The River Oil Company was one of the ambitious oil schemes of 1865. It was incorporated for the purpose of selling or leasing the bed of the Allegheny River in Warren and Venango counties. This company asked the state legislature to grant them the right to prospect the river bed for oil, salt, and other minerals, as well as the right to construct wharves and cribs and other equipment necessary to market the products. Owing to strenuous opposition in the oil region, the legislature did not grant this request.

During the seventies and early eighties, a number of oil exchanges were established at Petroleum Center, Oil City, Titusville, Franklin, Bradford, Petrolia, Warren, Parker City, St. Petersburg, Pittsburg, Philadelphia, Cleveland, Cincinnati, and New York. In the early eighties, the New York exchange took the lead, the other exchanges gradually falling into insignificance. On these exchanges all sorts of sharp practices were common. Corners were occasionally attempted. The following quotations from Derrick describe one such corner:

December 3, "Corner in oil, engineered by Titusville parties, sends the crude market up 40 cents a barrel." Dec. 5, "The oil corner excites attention; the market advanced $1 a barrel." Dec. 7, "The squeeze of the bears continued; bulls control all the cars. Deliveries to date amount to 22,301 barrels, out of a total of 153,000 bought on options and calls during the first of the month; barely three days remain in which to make deliveries."[15]

Sudden shifts in the market brought queer experiences in the days of wild oil-speculation, enriching some dabblers and impoverishing others. Stories of gains and losses were printed in newspapers, repeated in Europe and exaggerated at home and abroad. A bull-clique at Bradford, acting upon "tips from the inside" dropped four hundred thousand dollars in six months. An Oil City producer cleared three hundred thousand one spring, loaded for a further rise and was bankrupted by the frightful collapse Cherry Grove ushered in. A Warren minister risked three thousand dollars, the savings of his lifetime, which vanished in a style that must have taught him not to lay up treasures on earth. A Pittsburg cashier margined his own and his grandmother's hundred thousand dollars. The money went into the whirlpool and the old lady went to the poor-house. A young Warrenite put up five hundred dollars to margin a block of certificates, kept doubling as the price advanced and quit fifty thousand ahead. He looked about for a chance to invest, but the craze had seized him, and he hazarded his pile in oil. Cherry Grove swept away his fortune in a day. A Bradford hotel-keeper's first plunge netted him a hundred dollars one forenoon. He thought that beat attending bar and haunted the Producers' Exchange persistently. He mortgaged his property in hope of calling the turn, but the sheriff raked in the pot and the poor landlord was glad to drive a beer-wagon. Such instances could be multiplied indefinitely. Hundreds of producers lost in the maelstrom all the earnings of their wells, while the small losers would be like the crowd John beheld in his vision on Patmos, "a great company whom no man can number."[16]

Fortunes were made, not by speculators on the exchanges, but by landowners, lease brokers, speculators, and operators. There was the oft-repeated story of the man who bought a farm for $350 and sold it shortly afterward for $150,000, after receiving $200,000 in royalties; of another man who refused $500,000 for a very modest farm, of another who refused $1,000 per acre; of the Tarr farm, which in 1864 was sold for $800,000 gold (equivalent to about $2,000,000 in greenbacks); of the well-known John Benninghoff, of Benninghoff Run, whose income from land royalties was said to amount to $500,000 per year. One farmer who had bought his farm in the fifties for $3,000, sold it for $75,000 in 1862. The Reed and Criswell well, on Cherry Run, was reported to have been sold for $650,000 in 1864. The plot on which it was struck had been offered for $1,500 a year before. Another well, with four acres of ground, was sold for $220,000. It was stated that Beers and Cornen were offered and refused $4,000,000 for the Smith farm, just above the Reed and Criswell well. This farm had been bought about a year before for $3,500, and had earlier been traded for a yoke of oxen. The income of the owners was reported as $1,500 a day. One half interest in five acres of one farm sold for $200,000. Another five-acre plot sold for $300,000. A twelfth interest in the Maple Shade well was reported as sold for $30,000. The Holmden farm, on which the Frasier well was struck, was sold for $150,000, and later was valued at $4,000,000. One bold operator paid $24,000 bonus for the privilege of drilling a well on a half-acre lease adjacent to the United States well, with a royalty of one-half the oil, and made a handsome profit on a resale. A young teamster who took, as part payment for his services, an interest in a well that was being drilled, was presently able to sell for $150,000 and move to a large farm in Minnesota. The royalties received by landowners were often very high—as high as one-half, or even five-eighths.

Hundreds of such fortunes were made, by those who happened to own land, or who happened to have the energy, or shrewdness, or lack of conscience, which enabled them to profit in such a business. Andrew Carnegie, later the great steel king, got his start in the oil fields of western Pennsylvania. He and some friends purchased the Storey farm, on Oil Creek, for £8,000 sterling, and built a pond to hold 100,000 barrels of oil, which they estimated would be worth 200,000 shillings when the well was exhausted. The pond leaked, and evaporation caused much loss, but they continued to run the oil in. The value of the farm rose to 1,000,000 shillings, and in one year it paid cash dividends of 200,000 shillings—about $40,000.

The story of "Coal Oil Johnny" always serves to illustrate unearned and wildly dissipated wealth. John W. Steele, an inexperienced boy, while working as a teamster, fell heir to the Widow McClintock's farm, on which oil was struck in 1860. His wealth has often been given in terms of millions, but probably that was an exaggeration. However, at one time his income was thought to be near $1,750 a day; and the stories of his reckless spending are part of Pennsylvania history. It is said that when in New York he would sometimes buy a coach and horses to take a ride, and when his ride was completed give the coachman the entire outfit as a gift. It was one of his favorite tricks to light his cigar with a bank note or greenback, and any denomination would serve. He imported a band from New York to play for him while he ate, and was even said to have had his own opera troupe in Oil City, playing in his own opera house. In his innocence, he fell in with rogues and gamblers, and they relieved him of much of his wealth, while the saloon keeper got a fair share of the remainder. In a year or so he dissipated a large fortune—perhaps $500,000—and returned to his home, where he got a job as baggage agent of the Oil Creek Railroad. The notoriety he had got in his years of prosperity followed him, however, and he moved west to escape it, to Iowa, and later to Nebraska.[17]

The chance of making such fortunes in a short time naturally led to a neglect of the ordinary forms of industry. It is said that in many regions farming was almost entirely neglected, partly because owners of leases refused to allow farming operations, and partly because the farmers who should have been tilling the soil were resorting to the more interesting business of watching the oil drill. A great scarcity of local food products, and extremely high prices, resulted. The following description of conditions is taken from a book published in 1870:

Owing to the impetus given by developments, lands around, and even remote from, producing localities, began to appreciate in value. Many of the farmers of Venango County, who had been content in previous years to wrest a bare subsistence from its rugged hills and valleys, were now bewildered at the golden prospects that loomed up before them. Intense as the desire may have been with many, to retain the old homestead, which for generations had sheltered their forefathers, and endeared by all the ties of local attachment, the temptation for acquisition of wealth was too great for the large majority. Their farms were purchased by men eager only to obtain the liquid treasures supposed to lay beneath their surface. The derrick and engine-house frequently occupied the site of the old homestead; orchards were levelled off for fire-wood, or to make room, and the whole face of the country was in a few years so entirely changed as to be scarcely recognizable by its former possessors.[18]

Farmers who happened to own land on which oil was struck were sometimes incapacitated for further useful production, at least until they succeeded in dissipating their unearned gains. Some of them took their money and moved to the richer farm lands farther west, where they worked as before. Others dissipated their royalties in foolish investments, and still others spent their income with the reckless and ostentatious prodigality which sometimes characterizes the newly rich. In general, "oil money" certainly did not stimulate productive industry, and did not bring any general increase in happiness.

The oil industry never developed a healthy social community. Speculation and fraud and lying and gambling were not the only evils. The people were money-mad and largely lacking in culture, and almost wholly lacking in any permanent interest in their communities. There were hangers-on of all kinds: gamblers, "oil smellers," so-called geology "professors," pamphleteers, and liars who tried to glorify all that pertained to Petrolia; prostitutes, thieves, and murderers—perhaps in greater number than in ordinary communities, although some observers commented on the security of life and property, the absence of drunkenness, and even the number of fine people. Occasionally some of these oil towns would become aroused over the number of incendiary fires, or other lawlessness, organize a vigilance committee, and drive a few suspicious characters out of town. Prices were high, streets dirty, houses cheap and badly equipped and badly cared for. The average oil community was emphatically not a pretty place.

Wright describes Oil City in somewhat similar fashion: "As Oil City has the most disgusting *name* in all Petrolia, so everything else is in keeping therewith."

In Oil City 99 out of every 100 persons expect to leave it as soon as they have got money enough. Hence, on the part of nearly every one the understanding is to acquire wealth as rapidly as possible, without being too scrupulous as to the means; then clear out of the country, or "return to the states" as they term it. No impression could be more detrimental to the general well-being of a community than this. It bars all progress, except such as is concerned in momentarily inflating market values. It brings the race-course with its excitement, but not the passable highway with its utility and comfort; the telegraph, with its cooked-up dispatches, not the decent sidewalk; the caravansary, not the well-kept hotel; the mountebank with his "gift enterprises," not the sound public teacher; the "tribe of Moses," not the class of honorable, public-spirited merchants; wild excitements, instead of the purer enjoyments becoming rational and accountable beings. It is an obstruction alike to good morals, pure religion, general education and refinement, as well as to public improvement. Where no person expects to remain except for the briefest possible period, who feels interested in giving *tone* to a community? Who cares for its reputation outside, whether it be good or bad? Who will expend a dime in beautifying house, grounds, etc., adornments which add so much to the sweets of existence in beholders as well as owners? Who will oppose abuses, be they ever so monstrous, or institute reforms, be the necessity ever so urgent, when every step is certain to arouse opposition, and no one feels concerned about the distant future?

If he (the visitor at Petrolia) be a religious man, he will be hourly shocked by profanity; if a humane man, at the brutality with which the lower animals are treated; if a man of generous instincts, at the intense selfishness, the sordid love of gain so widely prevalent; if a man of taste and culture, at the outlandish condition of the houses and the streets, with the indifference of the people toward intellectual pursuits beyond the immediately practical. If he propose to introduce any other topic of conversation beyond the never-ending, still-beginning themes of oil and war, oil and politics, he will presently find his company thinning out. Ten minutes loud conversation on philosophy, literature, science, or religion would give him full command of a parlor, or even a bar-room. For the inhabitants of those large towns removed thither to make money, and do not mean to be turned aside from the one grand object of existence.

Petrolia will become a plague-spot on the score of manners, a great gambling school, a camp of instruction for the whole country in falsehood and rascality. Principles which have lain latent in the human breast elsewhere will there be "developed equally with the oil territory"; and taking their places there as so many cardinal virtues, will come forth to walk the earth with unblushing front, and communicate to old and young the sad distemper.[19]

Yet even this observer was often impressed with the kindness, industry, generosity, intelligence, and even refinement of many of the inhabitants of the oil towns, although he thought the "insane desire of oil" caused a "general indifference to veracity and honor" and even "every imaginable kind of misrepresentation and cheating."

One of the significant sidelights on the character of some of the oil communities is found in the brutal treatment of the horses used in hauling. Wright was constantly harrowed by the sight of "the noblest of God's subordinate creatures being murdered every year by thousands." "I know of no better means for bringing the ferocious drivers to something like feeling than to apply to *them* the blows they are wont to administer to the broken-down quadrupeds, often in the agonies of death. In one instance at least, an earnest threat from a spectator to give blow for blow had the desired effect, and the poor creature was permitted to heave his expiring groans free from the devilish treatment which he was then perhaps too unconscious to feel." In many instances, railroads, or at least passable wagon roads, might have been constructed, but the temporary nature of such an exploitative business did not justify too much permanent investment, in roads or in anything else.

Furthermore, oil operators, in their greedy haste to get the oil out as quickly as possible, were never willing to wait until the work could be done decently or economically or humanely. Deliveries were often by contract, and the burden of prompt delivery, especially when the roads were very bad, fell upon the unfortunate horses. This has been true in other fields than Pennsylvania. Thousands of horses and mules have been killed in the various oil regions of the country—beaten, over-worked, overheated, overloaded, exposed to cold, and to every form of suffering—in order that our oil deposits might be exhausted with the least possible delay. A suggestion as to the quality of the population found in some of the oil centers may be found in the account of a bull and bear fight which was once arranged to furnish amusement at Custer City, Pennsylvania.[20]

McLaurin describes the passing of one of these ephemeral towns:

For two or three years the "Center," called that for convenient brevity, acted as a sort of safety valve to blow off the surplus wickedness of the oil regions. Then the "handwriting on the wall" manifested itself. Clarion and Butler speedily reduced the 4000 population to a mere remnant. The local paper died, houses were removed, and the giddy "Center" became a back number. The sounds of revelry were hushed, flickering lights no longer glared over painted wild women, and the streets were deserted. Bissel's empty bank building, three dwellings, the public schools, two vacant churches, and a dry goods box used as a railroad station—scarcely enough to cast a shadow—are the sole survivors in the plowed fields that were once bustling, booming, surging, foaming Petroleum Center.[21]

Not all oil towns or oil communities were alike, to be sure. Oil City was said not to be a very wicked town, compared with some of the others, as for instance, "Bad Bradford," Petroleum Center, Pithole, or Baby-lon. Yet there were certain general characteristics which were noted in almost all such towns.[22]

In no way was capital wasted more prodigally, perhaps, than by the drilling of too many wells. In some regions of Pennsylvania twenty times as many wells were drilled as would have been necessary to get the oil out, and in other places even thirty to forty times as many wells were drilled as would have been needed. Thus in some fields six and even eight wells per acre were drilled, where one well for every five acres might have been sufficient. In the valley of Oil Creek, above Oil City, leases of one-fourth of an acre were often made, insuring twenty times as many wells as were necessary. In Benninghoff Run, derricks were erected so close together that there was scarcely room for a man to walk between them; and in many other regions the crowding of wells was quite as bad. Over-drilling involved waste of capital in several ways. In the first place, there was the initial cost of the unnecessary wells, perhaps averaging

$3,000 or $4,000 per well. In the second place, there was the very rapid depreciation that had to be charged against all the capital used in a field which lasted only a few years. Had only enough wells been drilled to get all the oil in a term of twenty years, it would have taken only about half as many wells as were drilled in some regions. Also, the pipe lines and other equipment could have been used for the reasonable life of such equipment. With the tremendous zeal for drilling, however, great tank and pipe facilities were needed, and when the oil was gone— perhaps in less than three or four years—much of the tank and pipe line equipment had to be moved at great expense or abandoned. The abandonment of thousands of wells each year was itself a very wasteful business.

The excessive number of wells was of course due to an excessive number of owners and operators. In the Washington, Pennsylvania, oil field, a register of 125 wells showed over forty different companies and individuals as owners, with an average of slightly over three wells for each owner. It was not at all unusual for an individual or company to buy a single well, or even a fraction of a well. Such scattered ownership has in every oil field meant over-drilling and all the attendant evils of over-production and waste.

The question has often been raised as to whether the Pennsylvania oil ever added materially to the wealth of the country, or whether, like gold, it was secured at a loss—whether these deposits, which would have been of immense value under any intelligent system of social control, were really worth anything at all under the wasteful and unintelligent system of exploitation which prevailed. A great many writers have tackled this question, and have used all kinds of figures to prove that the oil was exploited at a profit or at a loss. Joseph Weeks, of the United States Geological Survey, went into some rather elaborate computations in 1889, and concluded that the industry had paid 20 per cent on the capital employed; but Weeks admitted that it was "well nigh impossible" to formulate any accurate conclusions. Other authorities and writers have ventured elaborate statistics and guesses, tending to prove one side or the other of the question.

Even if it were possible to get accurate figures of capital invested in the oil industry, and of the income, it would not be possible to draw very significant conclusions, because there are so many factors in the problem that cannot possibly be calculated in figures. For instance, no figures could possibly be obtained covering the loss in fraudulent and speculative schemes that never resulted in any real development; or the loss due to distracting the minds of the people, not only in the oil regions themselves but all over the country, from the ordinary pursuits of life. On the other hand, no figures could be obtained which would indicate with even approximate accuracy the gains accruing from the stimulus the cheap oil gave to some other industries, or the value of the scientific knowledge gained in exploiting the oil, knowledge which could be used in the development of other fields later. This knowledge had to be obtained somehow, if the oil resources of the country were ever to be utilized, and it could only be obtained at great expense. Thus, any general statement that the oil industry was carried on at a loss or at a profit would not mean much. Most people in the industry had no conception of the total costs of the industry, or of the total gains either, although they were likely to under-estimate the former, and perhaps to over-estimate the latter. Ordinary economic reasoning would suggest that the industry could not, in the long run, be conducted at either a loss or at an unusual profit; but in the oil industry competition operated only in crude and bungling fashion, because those entering the industry did so without any definite or accurate idea as to the chances of gain or loss. There was too much of an element of chance in the business to permit accurate or even approximate calculations of profit and loss.

NOTES

1. Pa. Geol. Survey, Rept., 1885, 42—; Derrick, *Handbook of Petroleum,* 850; *Jour. Franklin Inst.,* Vol. 157, p. 443.

2. On the development of theories of the origin of oil, see an article by Ralph Arnold, *Oil Weekly,* Apr. 5, 1924, 27.

3. U. S. Geol. Survey, Water Supply Paper, 416.

4. Henry, *Early and Later History of Petroleum,* Phil. (1873), 215; Cone and Johns, *Petrolia.*

5. For an interesting account of secrecy surrounding the shooting of the Mystery 646, see Whiteshot, *The Oil Well Driller, A History of the World's Greatest Enterprise,* p. 71.

6. Pa., Second Geol. Survey, 1880 to 1883; Geol. Rept. on Warren Co., p. XIV, IIII.

7. Wright, *Oil Regions of Pa.,* 253.

8. *Derrick and Drill,* 2d ed., 270.

9. *Ibid.,* p. 260.

10. *Ibid.,* p. 269.

11. Wright, *Oil Regions of Pa.,* 60.

12. *Derrick and Drill,* p. 21.

13. Pa., Second Geol. Survey, 1880 to 1883; Geol. Rept. on Warren Co., p. XIV, IIII.

14. Tenth Census, Vol. X, p. 11.

15. Derrick, *Handbook of Petroleum* (1859-1898), 139, 705; *Oil and Gas Jour.,* Dec. 30, 1915, 28; *Manual of Petroleum,* pub. by Financial News Association, N. Y., 1883.

16. McLaurin, *Sketches in Crude Oil,* Harrisburg (1896), p. 284. The

following account of a panic on the Bradford oil exchange indicates that speculative markets have been somewhat the same for a great many decades:

The market opened at $1.12 and very sensitive. The downward tendency was stoutly resisted by the bulls throughout the forenoon, and when the gong struck the game did not stand greatly in favor of the bears. The crowd had hardly gathered around the bull ring in the afternoon before the market began to fall to pieces. Point after point was lost; and, gradually at first, but with mad haste later on, those who had been bulling the market developed into bears. In the corridor of the exchange was a dense throng, in which there were a number of women, whose possessions were growing less at every drop of the market. The crowd extended to the sidewalk and spread out, up and down the street. News of the break had spread throughout the town, and merchants, mechanics, clerks, bookkeepers, day laborers and men of almost every avocation left their duties and hastened to the scene of the excitement. Some had to stay only long enough to hear the first news that came from within the closed doors. It told them that they had lost all they had to lose, and with blanched faces they turned away. Brokers howled themselves hoarse in their efforts to find their customers and call for a renewal of margins. Telegrams of bearish character were flowing into the exchange from the other oil centers, and these contributed to the excitement that was every minute growing wilder about the bull ring. The scene when an excited broker leaned over the ring and shouted his offer to sell fifty thousand barrels at one dollar became indescribable. Another broker had to drop sixty-four thousand barrels because his customers could not or would not make good their margins. Panic reigned in the gallery, where the excited speculators rose to their feet, waved their hands and shrieked in fruitless endeavour to attract the attention of their brokers, who were packed around the ring, struggling for foremost places, gesticulating wildly, and shouting at the top of their voices their offers to sell. (Whiteshot, *The Oil Well Driller,* p. 124.)

17. McLaurin tells an amusing story of another oil magnate:

Neighbors of John Blood, a raw-boned native and his wife, enjoyed an experience not yet forgotten in New York. Selling their farm for big money, the couple concluded to see Manhattanville and set off in high glee, arrayed in homespun clothes of most agonizing country fashion. Wags on the farm advised them to go to the Astor House and insist upon having the finest room in the caravansary. Arriving in New York, they were driven to the hotel, each carrying a bundle done up in a colored handkerchief. Their rustic appearance attracted great attention, which was increased when the man marched to the office-counter and demanded "the best in the shebang, b'gosh." The astounded clerk tried to get the unwelcome guests to go elsewhere, assuring him he must have made a mistake. The rural delegate did not propose to be bluffed by coaxing or threats. At length the representative of petroleum wanted to know "how much it would cost to buy the gol-darned ranche." In despair the clerk summoned the proprietor who soon took in the situation. To humor the stranger, he replied that one-hundred-thousand dollars would buy the place. The chap produced a pile of bills and tendered him the money on the spot! Explanations followed, a parlor and bedroom were assigned the pair and for days they were the lions of the metropolis. Hundreds of citizens and ladies called to see the innocents who had come on their "first tower" as green and unsophisticated as did Josiah Allen's wife twenty years later. (McLaurin, *Sketches in Crude Oil,* p. 140.)

18. Cone and Johns, *Petrolia,* p. 71.

19. Wright, *Oil Regions of Pa.,* 53, 60, 271, 272; McLaurin, in *Oil and Gas Jour.,* July 23, 1920, 64. See also Von Millern, *All about Petroleum and the Great Oil Districts,* N. Y. (1865), p. 51.

20. *Natl. Petroleum News,* March 1, 1922, 67.

21. *Oil and Gas Jour.,* July 23, 1920, 64.

22. *Oil and Gas Jour.,* Nov. 12, 1920, 12.

CHAPTER V

SIGNIFICANT FEATURES OF EARLY PENN-SYLVANIA OIL HISTORY (Continued): MONOPOLY CONDITIONS

MONOPOLY has been one of the natural results of private ownership and exploitation of oil resources, and it appeared early in the history of Appalachian oil development. At the time the Drake well was opened, John D. Rockefeller was engaged in the produce commission business in Cleveland, but, seeing very early the great possibilities of the oil business, he, with his partner, invested in an oil refinery in Cleveland, which had been established by Samuel Andrews; and, when this proved very profitable, he entered into a partnership with Andrews, and associated with this partnership William Rockefeller and several other men of early Standard Oil fame. In 1870, after a number of changes and consolidations, the Standard Oil Company was formed as an Ohio corporation, with a capital of $1,000,000, which was increased to $2,500,000 in 1872, and to $3,500,000 the following year. In this company were consolidated the interests of John and William Rockefeller, Samuel Adams, Henry M. Flagler, and Stephen V. Harkness, and the interests of the three firms of William Rockefeller and Company, Rockefeller and Andrews, and Rockefeller and Company. Even at this time it was the largest oil company in the country, but it produced only about 10 per cent of the total product of refined oil, and there were some 250 competitors in the field. Something of the rapidity of its growth in power may be gleaned from the fact that, in 1879, less than ten years after its formation, it controlled between 90 and 95 per cent of the refining business of the country.

In 1872 another company was organized by the Standard Oil men, the "South Improvement Company," to perform the function of negotiating with the railroads for special rates on oil. In this it was successful, and a series of contracts were secured, providing for special rebates on all oil shipped by the South Improvement Company. Rebates were even to be paid to this company on all oil shipped by the independents.[1]

Such a combination as this created consternation among the independent oil men, who organized an "Oil Men's League" to fight the new monopoly. For some time the fight was carried on with the greatest bitterness, in Congress, in the state legislature, and in the field. An embargo was promptly imposed on sales of oil to the South Improvement Company, and as early as 1872 and 1873 a number of operators were censured by the oil men's association for selling oil to it or to the Standard. The officers of the South Improvement Company were called to Washington by a congressional investigating committee, and the state legislature finally passed a bill repealing its charter. The railroads agreed to abolish their system of rebates.

But the Standard Oil Company continued to receive special rates, and it followed out relentlessly the policy of forcing into the Standard such refineries as were wanted, and of crushing the others. In 1874, having acquired the refineries in the Cleveland district, it proceeded to take over important refineries in Philadelphia, Pittsburg, and New York, and, for the purpose of further acquisitions, established a pool or "alliance," known as the Central Association of Refiners, with John D. Rockefeller as president. Under the association agreement of this company, its members were permitted to operate their own refineries, but the Standard Oil Company was to have the power of determining the quantity of output, to buy all the crude oil and sell all the refined oil, and to negotiate with the railroads and pipe lines as to freight rates. This association was too powerful to be fought successfully, and through it the Standard was able to secure control of almost all the independent refineries within a few years. The Standard Oil Company was apparently stronger than the railroads, and even almost stronger than the federal government, for when in 1876 a congressional committee called the treasurer of the company to testify as to who the stockholders of the company were, and as to the rates the company paid on shipments of oil, answers were refused, and the railroad representatives cited to appear remained away from the hearings.

The Standard Oil men early saw the importance of getting control of the pipe lines. The Standard did not build the first lines, but, only a few years after the construction of the first pipe lines, in 1865, it was reaching out for control, and at the close of 1877, it controlled practically all the pipe lines in the oil regions. This was not accomplished without many bitter contests with the independents, but, with its railway rebates and discriminations, the Standard always won. Even after the Standard had got such control, independent producers tried to free themselves from its dominance by building a pipe line from the oil fields of western Pennsylvania to the seaboard, but the Standard soon got control of this too, by purchase of some of the stock and by pressure exerted through the railroads. This did not end the struggles of the independent producers, for they constantly tried to shake off the stranglehold of the Rockefeller group, but in vain.

The Standard Oil Company—the "trust," as later known—was

formed in 1879, by the adoption of a trust agreement among a number of member companies, but since the history of this "trust" involves more than the Pennsylvania fields, it will be considered later. For many years the company baffled all investigation, and it was almost impossible to learn anything regarding its internal management. It has been said that $25,000 was offered to anyone who would secure a copy of its trust agreement.[2]

How was the Standard Oil Company able to achieve a dominant position in the oil industry so early? As already suggested in several connections, the most important means by which it stamped out competition was through unfair practices—railroad rebates and discriminations, abuse of pipe line control, and unfair methods of competition in the sale of refined products. It is not pertinent here to elaborate on these practices, but it has often been said that the Standard Oil Company has been guilty of about every form of unfair practice known in trust literature. These practices were resorted to very early, and, of course, the advantage arising from their use was to some extent cumulative. The more effective the railroad rebates were, the more wealthy and powerful the Standard Oil group became, and the more powerful they became, the more easily could they secure such discriminatory favors. The railroad situation in the seventies presented rare opportunities for special rates, for this was the period of intense rate wars. Furthermore, the general standard of political morality at this time was so low as to render it very difficult to secure remedial legislation. There were many dishonest men—dishonest according to present standards—in the United States Congress, and in the state legislatures, and when measures for curbing the unfair practices of the Standard Oil Company came before these legislative bodies, these measures did not generally receive fair consideration.[3]

Business morality, as well as political morality, however, was far below that of a half-century later, and many of the practices of the Standard Oil Company were about on a level with those of other business concerns. The history of some of the railroads illustrates the general level of business morality, and political morality as well, in the first decade or two after the Civil War. It should be said, furthermore, that in some ways the Standard Oil Company had higher standards than many other business concerns of the time. At a time when railroad directors were busy fleecing common stockholders, the Standard Oil group were establishing standards of business practice which were in some ways among the highest of the time.

A second factor contributing to the rapid rise of the Standard Oil Company was the undoubted shrewdness and ability of the men composing this group, and the efficiency of the organization they built up.

The "richest man in the world" has always been reputed a man of extraordinary sagacity, and there is no denying that the group he associated with him were men of judgment.[4]

A third reason for the success of the Standard Oil Company, perhaps to be regarded as an illustration of the shrewdness of the Standard Oil men, was the fact that the company always maintained a strong cash position. Rockefeller saw from the first that in so risky a business as the oil business, only those concerns which had immediate cash resources could hope ultimately to dominate the field. Hence, during the period when the business was getting well established, the decade from 1860 to 1870, John D. Rockefeller and his friends year by year added steadily and quietly to their cash, until by 1867 they were in no sense dependent on bankers or financiers, as were the railroads and other large industries of the country. They were their own bankers from the start and were in a position, even in those early days, to snap their fingers at Wall Street and Lombard Street. When the Standard Oil Company of Ohio was formed in 1870, with $1,000,000 cash capital, it was undoubtedly the one great business corporation of America which had no debts and no direct banking alliances or affiliations.[5]

Another factor in the case was the conditions in the oil fields themselves. As already suggested in various connections, conditions in the oil industry were little less than chaotic almost all the time, with hundreds of inexperienced and poorly financed men drilling scores of wells, regardless of market conditions or of the available means of caring for the oil until a market should develop. Periods of over-production and wide fluctuations in price, made production a most hazardous and uncertain business. The refining industry was in somewhat the same general situation. Refineries were being built by the dozen, many of them by irresponsible men who had neither a knowledge of refining methods nor financial backing to insure success. The refineries were small, inefficient, wasteful, weak financially, and far too numerous. There were 30 refineries in Cleveland alone, in 1865; and there were hundreds of them scattered through the oil fields a few years later. Their methods were wasteful of capital and of the oil; and their financial rating was such that a man's connection with the refining business, as with any other branch of the oil business, was detrimental to his credit. Producers had often lost money by turning their product over to some of these concerns.

Out of such chaotic conditions, monopoly developed easily, in the first place, because the many weak and inefficient companies in the producing and refining business were helpless before any strong unit in the industry, and, in the second place, because there was real need of some controlling power. Directors of the South Improvement Company, and later of the Standard Oil Company, claimed that the purpose of their organization and activities was to secure greater stability in the industry, and there can be little doubt that their influence was generally in this direction. One of the first results of the increase in the capacity of the Standard refineries was an advance in refining methods, with an increase in the number and an improvement in the quality of the products, and with a reduction in the waste of the crude resource. There were also decided economies in the marketing of oil products. Many students of the question believe that, in spite of the reprehensible means by which the Standard Oil Company attained its dominating position, its influence upon the industry was generally salutary.[6]

Large scale control gradually appeared, to some extent, not only in the refining and transporting business, but also in the ownership and operation of the wells. A report in 1879 was to the effect that one man had purchased upward of 300,000 acres of the oil lands in West Virginia and Kentucky. The Midland Oil Company claimed to own 2,700 wells in 1893, and another group of interests was said to own nearly 1,000 oil wells in Pennsylvania, besides a one-half royalty interest in 736 other wells. This development of large companies was, of course, fortunate in every way.

Thus the history of oil development in Pennsylvania is a story of rapid and spectacular exploitation, with widely fluctuating prices, but with prices always too low, in spite of repeated curtailment campaigns, which were seldom successful. Oil, almost too cheap to be worth saving, was wasted through delay and carelessness in the capping of wells; through inefficient transportation methods; through evaporation, seepage, and fire; through flooding of the oil sands; through waste of gas, and through abandonment of wells; and, more important than all, through consumption for unimportant purposes. In all these various ways, millions of barrels of oil were wasted, in fact, it is doubtful whether as much as one-tenth of the utility in the oil was ever made available. At the same time, capital was wasted by hundreds of millions of dollars; through lack of understanding of the occurrence and origin of oil; through secrecy, and through exploitation by green and inexperienced men; through fraud and speculation; through neglect of ordinary industry; and through the drilling of far too many wells. Along with the vast waste of oil and capital, an almost absolute monopoly secured a strangle hold on the industry; and, finally, unhealthy social conditions developed almost everywhere in the oil regions.

Perhaps we may repeat that the follies of Pennsylvania oil exploitation, unfortunate as they now seem, should not be viewed too censoriously, for, in the beginning, the industry could hardly have been otherwise than wasteful. The work of the Pennsylvania pioneers may be regarded with a charity which cannot be extended to the criminally wasteful exploitation of later fields to the westward. It is now to be shown that the history of almost every later field has been a dismal repetition of the story of the fields of Pennsylvania.

NOTES

1. See Chapter XVIII. Some authorities have a different opinion as to the nature and functions of the South Improvement Company, and even doubt whether it was originated or promoted by the Standard Oil Company. Standard Oil officials held stock in the company, but so did officials of other companies; and there was really considerable mystery as to its exact purpose and functions. In this last respect it certainly resembled the Standard.

2. *Eng. and Min. Jour.*, March 3, 1888, 161. A copy of the original agreement is given here.

3. Some letters which William Randolph Hearst secured from the files of Senator Joseph Foraker in 1908, indicate that the Standard Oil Company spent a great deal of money to influence legislation, although Senator Foraker denied that the payments of money were for that purpose. (Foraker, *Notes of a Busy Life,* II, 330. For the story of an attempt by the Standard Oil Company to buy a Select Committee of the House of Commons, see the *Contemporary Review,* Aug., 1898, 232.)

4. *R. of R.'s,* June, 1901, 711.

5. Moody, *The Masters of Capital,* 54.

6. For bibliography on the Standard Oil Company, see p. 258.

CHAPTER VI

OIL EXPLOITATION IN OTHER EASTERN
AND CENTRAL-EASTERN STATES

OHIO

THE petroleum industry of Ohio, like that of several other states, dates from just about the time of the Drake well in Pennsylvania. Oil was known in Ohio, however, long before this. About 1806 drilling for salt brine was started, and in less than a decade oil and gas were found in several places. A brine well drilled in Noble County, in 1814, encountered such a strong flow of gas that it threw oil and water to a height of thirty or forty feet, according to contemporary accounts. About the same time, both oil and gas were discovered in Washington County, farther south. This oil was used in a small way for medicinal, illuminating, and lubricating purposes. Similar discoveries were recorded at many points in the southeastern part of the state, but the oil and gas were merely regarded as a nuisance. The oil ruined the brine for the manufacture of salt, and the gas was regarded as dangerous. Sometimes ordinary water wells would liberate small quantities of oil or gas, and occasionally these products were found in still shallower excavations. At a few points oil was found as a very thin film on the surface of streams. In 1853 an oil spring was discovered at Mecca, and as early as 1859 Professor J. S. Newberry made a report on the "Rock Oils of Ohio."

About 1860 oil was discovered at several points in southeastern Ohio. A well drilled near Macksburg, on the banks of Duck Creek, near the Ohio River, brought a flow of between 100 and 200 barrels per day, and created great excitement. Operators and speculators flocked in, especially from Pennsylvania, and a feverish speculation in leases resulted. The production of this field even influenced the Pennsylvania oil exchanges.

About the same time that the Macksburg district was being tried, a company was formed at Lancaster, to drill near Joy, Morgan County, a short distance westward, and this company secured a 20-barrel well early in 1861. Other wells followed, and the resulting excitement brought many speculators to the district. In the summer of 1861 probably 25 wells were completed in the field. The valley was soon dotted with wells, most of which were presently abandoned, but a number of these, drilled during the Civil War, were still producing in 1903.

The Buck Run oil field was also discovered in the early sixties, and furnished its share of speculative excitement, with a minimum of oil. In the fall of 1860 a well was drilled on the east bank of Wolf Creek, and this well yielded a small amount of oil, which was used in McConnelsville for lubricating purposes. Another well struck what was called the "First Cow Run Sand" at 120 feet, and yielded 20 barrels per day. Another well, opened in 1861, flowed so much oil that dams were constructed to hold the production, and from eight to ten teams were kept busy hauling it to McConnelsville. A number of wells were drilled here during the Civil War, and considerable excitement prevailed at times. In 1865, however, the first boom was largely over, and, a decade later, little was left but the crumbling derricks and broken bull wheels that mark the decadent oil field.

The Cow Run field, in Washington County, was also discovered in the early sixties, the first well coming in early in 1861. Oil was at one time worth $15.25 a barrel here, and large fortunes were paid for small tracts of oil lands. At the crest of the excitement there were thousands of people at the Run, and the hungry prospector had a hard time to find rations. In 1864, the oil fever broke out at Macksburg again, but when no important gushers appeared, the excitement soon subsided. In 1866 an oil man who was sent into the Mecca district to investigate the oil possibilities there reported his surprise at finding "so many old wells and abandoned openings."

A great many wells were drilled in other parts of Ohio during the sixties, in Meigs and Trumbull counties, and elsewhere; in fact, a careful study of Ohio history reveals a surprising amount of prospecting and drilling in various parts of the state, even during the Civil War. It was not, however, until 1878, that the first well was struck in the geological horizon known as the "Berea Grit," and not until 1885 that there was any important production.

The Findlay field, in northwestern Ohio, was discovered in 1884, and the following year the famous Lima field was discovered, and rapidly developed a large production. In 1889 there were 1,200 producing wells in this field, and the next year there were 165 new wells completed, 188 drilling and 239 derricks ready for drilling. In 1891 this field produced over 17,000,000 barrels of oil. This "Lima oil" was, however, an inferior oil, heavy, charged with sulphur, for a long time considered unfit for anything but fuel use; and for this reason it had little effect on the price of Pennsylvania oil, even when it sold for as little as 15 cents a barrel. Better refining methods, developed by the Standard Oil Company, finally made possible a better utilization of this heavy sulphurous oil, and led to a great increase in its value.[1]

With the opening of the Lima field, Ohio soon became an important oil-producing state, surpassing within a decade Pennsylvania and

New York. The Lima field alone for several years produced more than one-third of the total production of the United States. In 1889 several single wells in Ohio blew in with a production of 10,000 barrels per day. In 1891 and 1892 the Sistersville pool was opened. The first well in this field was drilled on the Ohio side of the Ohio River in July, 1891, and at the end of the year 1892 the production reached 20,000 barrels a day. Within the next ten years a number of important pools were opened in the eastern part of the state.

In August, 1898, the Scio Oil and Gas Company drilled a small producer on a town lot at Scio. It attracted but little attention, but when the company struck another a month later, it started one of the most spectacular booms in oil history. Not long afterward a strike was made on the John Bricker farm, in Harrison County, and many operators with experience and good bank accounts were soon on the ground, gathering in leases, for which they paid large bonuses. In all directions operations were extended, but Nature had stacked the cards against them, and when the "dusters" began to come in on every side, oil men withdrew from the field. In 1902 the limits of the various pools had been pretty definitely determined, and during the next decade no large pools were found, except in the "Clinton sand" which disclosed a large pool near Bremen, in 1907, in the south-central part of the state. About 1907 the production of the Lima field began to decline, in spite of extensive drilling, and in 1911 an operator reported that in several parts of the field, "the principal occupation has been pulling old wells." In spite of the discovery of the Bremen field in 1907, and the Junction City field two years later, and in spite of the boring of nearly 2,000 wells each year, the total oil production of Ohio declined 60 per cent between 1903 and 1917.

INDIANA

The history of oil production in Indiana is to some extent connected with that of the Lima field in Ohio, which extends westward into Indiana. There are, however, a number of sections of the state in which oil has been found, or at least drilled for. Tar springs were known in Crawford County in the southern part of the state, from the earliest settlement of the country, and during the oil excitement from 1862 to 1866, oil seekers flocked there in considerable numbers. A number of wells were bored near Terre Haute, and a small amount of oil was found, but nothing of commercial importance. In 1865 a well was drilled near Francesville, Pulaski County, and a considerable flow of gas was found, with some water, but at that time the use of gas was not known, and the well was abandoned, and continued to flow water and gas until 1888, when it was plugged. Much of the oil found in Indiana in the early period was not used commercially, because the oil was of inferior grade, and because many of the wells were too scattered to justify pipe lines to gather it; and much oil was wasted in various ways. Such waste was, however, a conspicuous feature of all the early development.

Oil production on a large scale in Indiana dates from the early nineties, the production coming largely from what is called the Lima-Indiana field, which extends from Ohio into northeastern Indiana. In the year 1896 the production reached nearly five million barrels. In 1897 the Alexandria field, in Madison County, was brought in, with a 250-barrel well. This created great excitement, and hundreds of operators and speculators flocked to Alexandria to buy leases. Further drilling brought little oil, although the drillers managed to tap large gas deposits, and waste enormous quantities of it. In the same year that the Alexandria field was discovered, a good well was brought in near Peru, by 100 citizens of Peru, organized into a "People's Oil and Gas Company."

From 1897 to 1904 the oil production of Indiana increased very greatly, but not long afterward the Lima-Indiana field began to decline, and with it the production of the state as a whole. In 1906 and 1907 a great many drillers began to leave for the more productive fields of Illinois. In 1908 five wells were abandoned for every well drilled.

About this time, however, there was active drilling in the southeastern part of the state, by operators who hoped to extend the Illinois fields into Indiana. The result was the opening of several small pools of fairly high-grade oil. Discoveries in this district, and the discovery, a few years ago, of what is known as the "Stone River Horizon," below the old Trenton horizon, were important enough to keep the production of the state around one million barrels during recent years, in spite of the decline of the northern fields.

The oil industry has been more closely connected with the gas industry in Indiana than in any other state, and since gas was relatively very important, oil development was delayed, in order that the gas might be utilized. In the first few years of oil development, vast amounts of gas were wasted, but the persistent work of the state geologist, Blatchley, and others, resulted in the passage of a law prohibiting the waste of gas and oil. Many of the operators refused to observe this law and fought it to the Supreme Court of the United States, where it was upheld.

WEST VIRGINIA

The oil history of West Virginia goes back to Revolutionary days, when there was only one Virginia, and when a grateful country gave

George Washington a tract of land near the mouth of the Little Kanawha, at what was called Burning Springs, because of the gas that bubbled up through the waters and was sometimes ignited. As early as 1771 Jefferson wrote an interesting description of this burning spring. Several years later David and Joseph Ruffner, while drilling for salt in the Kanawha Valley, not far from Charleston, encountered a flow of petroleum, much to their disgust. In the forties "Seneca Oil" from Virginia was widely sold as medicine. In 1860 Pittsburg capitalists drilled the first commercial well near Burning Springs, and soon this place became a lively oil center, but the Confederate army destroyed many of the wells.

For twenty years after the close of the Civil War, West Virginia produced only a small amount of oil. Some drilling was done at various points: at Burning Springs, Oil Rock, California House, Volcano, Sand Hill, and Horse Neck, and some light oil was found; but the production was unimportant. The Pennsylvania fields were producing more oil than could be handled profitably most of the time, and there was little incentive for vigorous exploration. In the late eighties, however, Pennsylvania began to show signs of a decline, and when new discoveries were made in the northern panhandle of West Virginia, a wildcatting campaign was inaugurated which, in 1889, developed gushers in the Eureka and Belmont pools, and then brought in the great Mannington pool. These pools marked the real beginning of modern oil development in the state. This development was particularly important because of the high quality of West Virginia oil, perhaps even higher than Pennsylvania oil.

About 1891 the production of West Virginia began to increase very rapidly, and within three years had increased 1,600 per cent. Early in 1892 the Sistersville pool loomed into prominence. The first well was drilled in July, 1891, on the Ohio side, and in less than two years wells were extended for ten or twenty miles in each direction. About this time the Mannington field began to assume considerable importance. In 1894 the first well was opened in the Flat Run pool, and soon 100 wells were drilled here. Other pools followed: Campbell's Run, St. Leo Cross Roads, Klondike, Bula, and other minor pools. The production of the state increased from 3,800,000 barrels in 1892 to 8,400,000 barrels the following year, and to 16,000,000 barrels in 1900.

Pool after pool was opened. In 1897 the Elk Fork pool, in Tyler County, was opened, and in a few months was producing 45,000 barrels a day. Late in the same year the Whisky Run pool, in Ritchie County, was opened with a 250-barrel well. Other pools followed: Milton, Green, Point Pleasant, Rowell's Run, and others, mainly in the northwestern part of the state, adjacent to the Ohio River. The Paw Paw district was opened in 1903, and soon punctured with a hundred wells. By the drilling of hundreds of wells each year, the production of the state was maintained at about 10,000,000 barrels a year from 1896 until very recent years. In 1911 the Blue Creek pool in Kanawha County was opened—one of the largest pools in the history of West Virginia—in territory that had been condemned by most oil men. This pool produced at one time 25,000 barrels a day, but began to decline within a year. A number of pools were opened in 1911 and in the years following, one of the most important of which was the Cabin Creek pool. In 1912, the Falling Rock Creek pool, adjoining the Blue Creek pool, was opened with a 100-barrel well. A little later the Big Sandy pool came in with a gusher estimated at 2,000 barrels. In December, 1914, a new pool was opened near the mouth of Laurel Run, in the Cabin Creek district, which proved moderately productive. The following year a 1,500-barrel well opened the Dent's Run pool, near Mannington.

In spite of these discoveries, the production of the state declined from 9,680,000 barrels in 1914 to 5,899,000 barrels in 1924. The state has developed a tremendous production of natural gas, but the output of oil is declining steadily.

KENTUCKY AND TENNESSEE

Kentucky produced oil in considerable amounts, perhaps as early as Pennsylvania did. The oil springs in Allen County were known from the earliest times, and, as early as 1806 or 1808, oil was found by drillers who were seeking salt brine. In 1818 or 1819 Martin Beatty, while drilling for salt, struck a good flow of oil on the Big Fork of the Cumberland River.

About ten years later another prospector for salt water opened at Burkesville what was known as the "Great American" well, which flowed out upon the creek and upon the Cumberland River, where, according to some accounts, fifty miles from the mouth of the well it was ignited and burned the entire distance back to the well itself. This "burning river" is graphically described by some historical references of the time. The product of this well was sold as medicine, under the name "American Rock Oil," for fifty cents a bottle. In 1840 oil was discovered in quantity on Little Richland Creek in Knox County, in southeastern Kentucky, near Barbourville. This well was said to have flowed 100 barrels per day. As early as 1848 some 200 barrels of oil from Kentucky springs were said to have been sold for lubrication uses in Cincinnati. In 1850 a few barrels of oil from near Bowling Green were shipped to Pittsburg where they were sold as medicine. In 1859 and 1860 a small amount was found near Paint Creek, on a branch of Paint Creek, in Johnson County.

During the Civil War, while the Confederate armies had possession of much of Kentucky, including Bowling Green, notice was taken of these native sources of oil, and all that could be got was wagoned across the country and shipped to Memphis. It appears that some shallow wells were drilled near Bowling Green, each of them yielding five or six barrels of heavy oil, but the accounts of these are not clear.

Later in the war some of Kentucky was overrun by the Federal armies, and to some of these northern men, who noticed the oil springs, was due the later opening up of oil wells. At the close of the war the district in which the oil springs were situated, particularly along the Cumberland River and adjacent creeks, was leased by prospectors on a large scale, and in 1865, 1866, and 1867 operators began to drill for oil, instead of for salt. Nothing important resulted, except in one locality—the Porter farm, three miles west of Scottsville—where in 1866 or 1867 a well was struck which produced between 100 and 500 barrels a day, according to tradition. The oil was largely wasted at first, but some was later sent to Louisville and St. Louis, where it did not sell readily because of the sulphur content. Transportation was too expensive to make this traffic profitable, in any event.

Some of these early oil discoveries caused great excitement. Wells were drilled in all sorts of places, but the operators were not shrewd in their judgment. Slight and unimportant resemblance to certain aspects of the oil fields of Pennsylvania was often seized upon and magnified into sure promise of fortune. An oil seep or a feeble gas spring was regarded as an infallible guide to a new Oil Creek. This desultory yet enthusiastic exploitation brought no important results. Not a single valuable oil field was brought to light by all the outlay of money and energy. Even the records of the work done were nearly all lost.

In 1887 another wave of excitement swept the state when a well reached the so-called "Trenton Limestone," the horizon which had proved so productive in northern Ohio and Indiana. In 1892 a Pennsylvania oil operator opened a successful well in the now famous Beaver Creek pool. The news of the strike spread rapidly, and caused a great influx of capital and enthusiasm. In 1895 oil was discovered in the Cooper field, at the forks of Beaver Creek, and in the next few years a large number of wells were drilled into this pool, some of them more than 100-barrel wells. This field had more than ordinary lasting qualities, but by 1912 many of the wells had been abandoned and the production had dwindled to 100 barrels a day. In October, 1900, the Ragland oil field was discovered, and within a couple of years was punctured with some 200 holes. This field produced a black, heavy oil of relatively low value.

The beginning of the twentieth century saw an oil boom in Knox County, Kentucky, but after a short boom period, the production declined rapidly, and within a few months the field was almost abandoned.

The Sunnybrook field was discovered in 1901. In 1902 the Dry Fork field was opened, and attained a production of 275 barrels a day. In March, 1903, the Campton field was opened in eastern Kentucky, with a 100-barrel well. Over 300 wells were drilled in this field, and in a short time it was exhausted. In 1904 the Steubenville field was discovered. Within the next few years the rather unimportant Mt. Pisgah and Griffin pools, in Wayne County, were opened and developed, and in 1912 the Cannel City field, in Morgan County, was ushered in with a 700-barrel gusher.

During these years the oil production of Kentucky was never very important. In only two years, 1905 and 1906, did it rise above 1,000,-000 barrels annually; and toward the end of the year 1914 and during the early months of 1915 production declined rapidly, to less than 500,-000 barrels in 1915. In 1915, however, drillers opened an important extension to the old Irvine pool, in Estill County, which soon developed into an important producer. The Scottsville district, in Allen County, had been exploited somewhat in 1914 and 1915, and was further extended in 1916. In 1917, while the Irvine extension was still going strong, the Bowling Green pool, farther to the west, in Warren County, the Ashley pool, and a promising pool at Pilot, in the Irvine district, were opened. During the following years several pools were discovered: the Big Sinking pool, Gainesville, Angie McReynolds, and Jake Moulder pools. The Big Sinking pool has been one of the important fields in the development of the state, the oil being of very high quality. Recently there has been active exploitation in the shallow pool areas along the Cumberland River, in Clinton and Cumberland counties, and across the border in Clay County, Tennessee.

The result of all these discoveries has been a rapid rise in the production of the state, from a little more than 1,000,000 barrels in 1916 to nearly 10,000,000 barrels in 1922. Since 1922 it has declined somewhat.

Oil exploration and drilling in Tennessee date back to the close of the Civil War. Previous to that time indications of oil and gas had been noted in a number of brine wells drilled by settlers in southern Kentucky and northern Tennessee. In one or two instances it was reported that wells, exhausted of brine by constant pumping, began to furnish oil instead of salt water, and were abandoned in disgust by their owners. As early as 1820 a well was said to have flowed enough oil out on Wolf River to produce a "terrible conflagration" when set on fire. Another well dug in 1837 on the Obey River was said to have produced oil. A great many test wells were drilled in the years following the discovery of oil in Pennsylvania, largely by northern men who had been through this region during the Civil War.

Some of these wells were reported to have flowed small amounts of oil. In the Spring Creek field, for instance, 11 wells were drilled between

1866 and 1870, and several thousand barrels of oil were estimated to have been produced. The cost of hauling the oil to the market was so great, however, that this field was presently abandoned, as were all other fields in which wells had been drilled.

Within recent years a number of wells have been drilled in Tennessee, but no important results have yet accrued. Tennessee, like the other southern states east of the Mississippi River, has been a disappointing field for oil prospectors.

ILLINOIS

Illinois should logically have been developed before the states to the west. During the early years of Pennsylvania excitement, some wells were drilled in Clark County, and oil was said to have been found near Chicago. There was even a small production of heavy oil from a few wells at Litchfield previous to 1904, but it was not until 1904 that Pittsburg interests brought in commercially successful wells. During the next few years, exploitation in the southeastern part of the state was extremely rapid, the total production rising to over 33,000,000 barrels in 1908. During the next few years a number of fields were opened: the important Carlyle field, east of St. Louis, the Carlinville pool, and the Staunton pool, farther north in Macoupin County, and the Flat Rock, Allendale, and Dennison township pools, in southeastern Illinois. In spite of these discoveries, most of which were of small importance, the total production of the state has declined steadily since 1910, and at the present time the production is only slightly over 8,000,000 barrels.

The history of oil development in Illinois presents more encouragement for the advocate of private ownership of oil lands than the history of almost any other oil region, for development here was accomplished with less wild speculation, conscienceless promotion, fraud and waste, than developments in most other regions. Leases sold at very good prices, and bonuses of $150 to $200 an acre, with a royalty of one-eighth, were not uncommonly demanded in the productive district; but there was little speculation by those not familiar with the oil business and its risks. Few of the usual stock-peddling companies were organized, and the sentiment was strongly against them. Experienced men found the field an unusually profitable one, despite the high bonuses asked. Probably the main reason why exploitation has proceeded more sanely than in most fields is that the fields have been more nearly monopolized than the fields of almost any other state. The Illinois Geological Survey reported in 1913 that the Ohio Oil Company, one of the Standard companies, controlled 70 per cent of the total lease holdings, and bought and stored 90 per cent of the oil produced in the state.[2]

MICHIGAN

Michigan has been the scene of a great deal of wildcatting, but no discoveries of importance have resulted. Early oil discoveries in Canada naturally led to exploration in Michigan. In the early sixties oil was discovered in the Petrolia and Oil Springs fields, in southern Ontario, and in 1886 a third Canadian field was discovered in Euphemia township, southeast of the Petrolia and Oil Springs pools. The axes of these fields seemed to point to Sarnia (Ontario) and Port Huron as prospects, and explorations were begun in 1886, small discoveries of oil and gas resulting. Little was done thereafter until 1898, when an operator from the Ontario fields drilled some wells in and about Port Huron, and got small but commercially important amounts of oil. In 1914 21 wells had been drilled here, and 15 wells were being operated. Other wells were drilled later, here and in other parts of Michigan, but, except for some very modest discoveries in the Allegan and Saginaw districts, no important results accrued.[3]

NOTES

1. The development of refining processes for Lima oil—some of the oils produced in Indiana and Canada were of similar quality—is often attributed to the faith and persistence of John D. Rockefeller. It has been said that he consistently increased his holdings in the heavy oil fields, in spite of the apparent worthlessness of this oil, and in spite of the opposition of other Standard Oil men to such a policy. Miss Ida Tarbell tells a somewhat different story. According to her story, the Standard Oil Company, which was practically the only refiner and transporter of oil from the Lima field, built a pipe line from Lima to Chicago, while a chemist was working on a method of utilizing this sulphurous oil. Oil producers were sufficiently discouraged about their oil anyhow, and when the Standard found a means of refining it, the company kept that fact secret, closed its pipe line, as if hopeless of utilizing the oil, and sent out an agent to buy up all the production and leases in the field. When the field had been bought, at very low prices, of course, the new process was put into operation. The Standard realized handsomely on its strategy. (*McClure's Magazine,* Vol. 25, 480.)

2. *Mineral Industry,* 1907, 751; 1912, 635; 1913, 539; 1914, 559; *Mineral Resources,* 1905, 848; 1907, 373; Ill. State Geol. Survey, *Bull. 22,* 146.

3. *Publication 14,* Geol. Series 11, Mich. Geol. and Biol. Survey.

CHAPTER VII

OIL EXPLOITATION IN THE SOUTHERN STATES

TEXAS

TEXAS oil history dates from about 1895—about the same time that important production began in Indiana, although the existence of oil in Texas was known much earlier—at least as early as 1860. There is a story that in 1543 the companions of De Soto, after his death, sailed down the Mississippi and along the coast to the westward, where they found numerous evidences of the presence of bitumen. Taking refuge from a storm in one of the creeks emptying into the Gulf, they found a pitchy scum cast up by the sea, and it is recorded that they painted the bottoms of their boats with it.[1]

In 1868 oil was found in an oil spring near Nacogdoches, in eastern Texas, and 100 shallow wells were sunk. A pipe line was even constructed, and a storage tank erected, but the oil proved to be a poor illuminant, and the field was abandoned. At least as early as 1879 Pennsylvania operators became interested in some of the indications of petroleum at Sour Lake and Beaumont. In 1893 drilling began at Sour Lake, in the Beaumont district, not far from Port Arthur, and within a year or two, oil had been found in several wells. In 1894 Mayor Alexander Beaton discovered oil at Corsicana, in east-central Texas, about forty miles south of Dallas; and in the same year an organization of the citizens of Corsicana also found oil. People flocked in from the oil regions of Pennsylvania, Ohio, Indiana, and elsewhere; but no great oil production resulted.[2]

The Beaumont district, which includes the Sour Lake and Spindletop districts, did not become important until 1901, when the famous Lucas well was struck in Spindletop. For many years surface indications—the escape of gas in pools after a rain, and in some shallow wells, the presence of sulphur, etc.—had pointed to the elevated mound known as Spindletop as a possible oil field. Patillo Higgins, one of the early Texas oil men, always insisted there was petroleum under Spindletop, and even laid out Gladys City in anticipation of the coming importance of the locality. In 1894 a well was drilled to a depth of 400 feet, but it did not reach the oil. In 1901, however, Captain A. F. Lucas, who had become convinced from his work in Louisiana that salt, sulphur, gypsum, dolomite, sulphureted hydrogen, and oil were in some way associated with such mounds or "islands," brought in a

great well, which spouted oil 160 feet high—one of the greatest wells
ever struck, with a daily flow which has been variously estimated at from
5,000 to 100,000 barrels.[3] This well blew off for ten days before any
of the oil was saved. The discovery marked the beginning of a new
era in the history of our petroleum development, and startled operators
in all parts of the world. Several other great wells came in soon afterward in the Beaumont field—the Beatty, the Haywood, and the Higgins,
with production of from 20,000 to 70,000 barrels each. Four wells
here, including the Lucas, produced more than all the wells in Pennsylvania were producing at the time.

An unprecedented rush of speculators and operators followed. Oil
experts and operators came from the oil fields of the United States
and Canada, but many of the operators, unable to secure territory on
any favorable terms, withdrew. Then followed a swarm of promoters,
who promised the landowners almost anything they asked, or took
options on the land at high prices. Promoters ran railroad excursions
from New York, St. Louis, New Orleans, Galveston, and other cities,
to see the new wonder oil field. The city of Spindletop soon became
famous, and of course greatly over-populated. The hotels were inadequate, and many who were unable to find lodging slept on billiard tables,
in barber chairs, in fact anywhere that a little rest could be secured,
and paid handsomely for the privilege. The promoters had little trouble
convincing people that the whole plain surrounding the Spindletop
mound was saturated with petroleum, that the available territory
reached far into the interior; and much of the land within thirty miles
of Spindletop was leased.

Within a few months after the Lucas well spouted, 400 companies
were organized to bore for oil, to sell land, and to build refineries, with
a total nominal capital of $175,000,000. Vast quantities of stock in all
kinds of foolish and fantastic ventures were sold. One oil company,
organized soon after the opening of the Lucas gusher, when the supply
of oil was thought to be practically inexhaustible, acquired over 1,000,-
000 acres of territory, issuing $15,000,000 in stock to pay for it. When
Spindletop proved to be only a 170-acre affair, this company had a
very large amount of "unproductive acreage." In another instance,
four companies, each capitalized at $1,000,000, owned jointly a tract
45 feet square. These companies contributed equally to a fund for
drilling a well in the center of the lot, each company to own one-fourth
of the production. The well was completed, and each advertised the
ownership of a well.

Acreage sold for fabulous prices. Land worth $5 per acre before was
now sold for as high as $100,000 per acre, and there is a record of one
acre sold for $120,000. It was said that a New York broker offered
$100,000 for any acre of proved ground and was laughed at. Tracts

as small as 25 feet square were sold for ridiculous prices, and were even made the basis of million-dollar companies. At the height of the boom, local newspapers offered premiums for the most "catchy" names for new oil companies. One man was said to have offered Captain Lucas $100,000 for permission to name his company the Lucas Oil Company. Men came even from England. One English oil man in Bombay heard of the discovery "the day before it was made," and hurried back to Liverpool and then to Texas, where he later made a handsome fortune.

At the close of 1901 there were 138 producing wells on Spindle-top mound, and a year later over 380 wells had been drilled, wells so close together that the derricks almost touched. When the "dusters" began to come in along the edge of this mound, however, the boom collapsed, and millions of dollars of stock in the numerous "paper" companies were thrown into wastebaskets with other rubbish. Very soon, too, with the excessively close drilling on the mound, the production began to dwindle, and at the close of 1904, 95 wells, of the 1,200 drilled in Spindletop, were altogether producing less than 6,000 barrels daily—one-tenth as much as the first well alone produced. One writer estimated that $4.23 was expended for every barrel of oil produced in 1901, not including the very large sums paid for the land; and some of this oil sold for 17 cents a barrel, or even less.

Speculation was not the only evil attendant on the exploitation of Spindletop, for the waste of oil was very great. Such a well as the Lucas gusher had never been struck in this country, and, of course, it caught the drillers wholly unprepared, with the result that it was only with the greatest difficulty, and after considerable delay, that the well was brought under control. Thousands of barrels of oil were lost before it was controlled, and afterward a disastrous fire broke out, which burned 300,000 barrels of oil that had been dammed up near the well, and destroyed millions of dollars' worth of equipment. This pool, to the close of 1904, yielded 33,000,000 barrels of oil, of which probably 1,500,000 barrels were wasted or destroyed by fire.

In 1901 deep drilling was begun in the Sour Lake district, and notable success presently attended this, with a 20,000-barrel well, in May, 1903. In August, 1903, 220 wells had been completed or were in process of drilling, and the output of the field exceeded 100,000 barrels per day. But there were too many wells. From that time the production steadily declined until July, 1904, when it was only 11,000 barrels a day. This was presently raised somewhat by drilling in deeper sands, but early in 1905 the entire group of deep sand wells went to salt water, and the production declined rapidly. It has been claimed that more money was lost than made in the Sour Lake district, in spite of the fact that the ownership was mainly in one large company. The price of oil was ridiculously low in periods of flush production.

Just at the time when the waning production of Sour Lake had set free a large number of men and drills, oil was discovered in the Batson field—only 13 miles from Sour Lake—and another sensational chapter in oil history was opened. The very first well pierced almost the center of the field, and as a result the wells were from the start phenomenally successful. By the close of the year 1903 there were 28 derricks in the new field, and by March, 1904, the daily production had reached 151,000 barrels. At this time salt water began to make rapid advances, and in April the production of the field had fallen to 35,000 barrels a day. In December nearly 500 wells had been drilled, and the production had declined to 10,000 barrels a day. It has since declined still further.

The Saratoga field, only 12 miles from Sour Lake, had been exploited to some extent as early as 1896, and after the opening of the Lucas well in 1901, it was again probed with energy, and developed a production of 150,000 barrels for 1903, which was increased nearly 500 per cent in 1904, and increased still further early in 1905, daily records of 20,000 barrels being common; but in February, 1905, the advent of salt water began to make serious inroads on the production. Since 1908 the decline in production has been steady.

In the Saratoga field were illustrated some of the most spectacular "blowouts" seen in Texas. In the Rio Bravo No. 211, for instance, a pocket of gas was struck at approximately 500 feet, with a blowout so violent that the surrounding ground was affected by cracks to a distance of 250 feet. From these cracks there were eruptions of mud at distances of at least 200 feet from the mouth of the well. A crater was formed a few rods in diameter and more than 20 feet deep. The machinery and the wreck of the derrick were swallowed. To save the boiler, which stood some 200 feet away, the workmen took the precaution to anchor it to a tree. Such occurrences as this make drilling expensive, but have not prevented rapid—too rapid—exploitation of the Texas fields. Of course, they involved a great waste of natural gas.

Not only in the Beaumont region of the coastal plain, but in the Matagorda field one hundred and fifty miles southwest, along the coast, considerable drilling was done in 1903 and 1904; but only a small production was ever attained. The Humble field, 18 miles north of Houston, had been suspected of oil possibilities for years, but owing to frequent blowouts in drilling, the first successful well was not opened until early in 1905—an 8,500-barrel well. Other large wells followed, some of them 10,000-barrel producers, and by March 1 the daily production was 90,000 barrels. On March 3 water began to appear in some of the great gushers, and in ten days the production fell from nearly 90,000 barrels to less than 25,000 barrels. Partly through the bringing in of new wells, and partly through the renewal of flow from the wells which had been "drowned out," the field recovered its former production, so that the yield in April was nearly 2,000,000 barrels— an average of 66,000 barrels a day. The production soon began to decline again, and in November, 1905, was only 20,000 barrels a day.

The Humble field was remarkable for its large gas flow, and terrible blowouts. In the Higgins No. 2 occurred, on December 13, 1904, one of the most violent eruptions recorded in the Texas fields. After the well had emitted gas, water, sand, and chunks of clay for some hours, the casing became choked, but apparently the gas rose outside the casing to the higher sands, and by the morning of the fourteenth this abnormal pressure sought relief through wells on the neighboring properties. Every well within a considerable radius became involved in the general eruption of gas, water, and sand.

In 1904 the oil fields of northern Texas, in the Mid-Continent area, began to assume importance, with the discovery of the Petrolia or Henrietta field in Clay County, very near the Oklahoma line. There were no very significant developments for several years after this, but in 1911 and 1912 the great Wichita County fields were discovered. The first of these fields, the Electra field, was discovered in 1911, and produced several million barrels of oil the following year. Hundreds of speculators and operators flocked to the new field, and between sixty and seventy drilling rigs were soon in operation. In 1912 the Iowa Park and Old Burkburnett fields were opened, and the following year the production of Wichita County was over eight million barrels.

During the next few years a number of new fields were discovered in the northern section; the Moran field, in Shackelford County, in 1913; the Thrall field, in Williamson County, in 1914; the Strawn field, in Palo Pinto County, in 1915; the Breckenridge and Caddo fields in Stephens County, in 1916; and several important fields in 1917: the Parks field, in Stephens County; the Brownwood field, in Brown County, and the far more famous Ranger field, in Eastland County. The Brownwood discovery was followed by great excitement, and over 200 wells were drilled before the close of the year, and over 600 wells within two years. The Ranger field was opened with a 2,000-barrel well, which stirred up a great fever of speculative enthusiasm. There had been some small wells in this region before, but they had settled down to small pumpers, when the McClesky well was brought in. This initiated a great boom, and perhaps the greatest period of wildcatting in Texas oil history ensued. Within a short time a thousand wells were drilled, in an area between Bosque County in eastern Texas and the extreme Trans-Pecos in the West. In the Ranger field itself, so great was the excitement, and so large were the crowds of speculators and operators, that it was practically impossible to secure lodging anywhere within a hundred miles of the field. It was said that some operators established headquarters a night's ride from the oil field, so they could sleep on their way back and forth. The exploitation of the Ranger field was the outstanding feature of the oil history of 1918.

The year 1918 was marked by other development, however. The Desdemona field, a short distance south of Ranger, was discovered, and a number of large wells were opened—some of them yielding as high as 7,000 to 10,000 barrels a day. At this time the Burkburnett field entered a period of hectic town-lot drilling which will be treated in a later connection.[4]

But it was not alone in the Mid-Continent fields of northern Texas that important things were being done. In 1914 the Sour Lake production rose 286 per cent, as a result of deeper drilling in extensions of the pool. The Humble output likewise nearly doubled, with deeper drilling, and the flow of several less important fields increased somewhat. In 1916 there were sensational developments at Goose Creek, in Harris County. In 1917 the Goose Creek district alone produced over 7,000,-000 barrels—almost equal to the total production of Pennsylvania at the present time.

The Hull field, which first registered production in 1918, produced over 4,000,000 barrels two years later, and 8,000,000 barrels in 1921. The West Columbia field, opened in the same year, was exploited even more rapidly, producing over 8,000,000 barrels in 1919, and 12,000,000 barrels two years later.

In 1921 another great pool was opened. The Mexia-Groesbeck field had been known as a gas field since 1912, and many wells had been drilled but no oil secured. In 1921, however, Colonel A. E. Humphreys struck oil, and inaugurated one of the most spectacular oil booms ever staged. From a quiet town of 3,000, Mexia grew within a year to a considerable little city—some enterprising journalists credited it with a population of 30,000—with hundreds of oil companies, two stock exchanges, congested hotels and restaurants, millionaires and retired farmers, tool dressers and mule drivers, speculators and promoters, swindlers and sharpers and gamblers and profiteers, thugs and bootleggers and prostitutes—a typical western oil town. As one writer described it: "An oil gusher every day or two, murders, robberies, holdups, gambling, bootlegging, military control, a one-half-million dollar fire—Mexia has had them all inside of a year."[5] The production of oil increased prodigiously. The output from this one field alone—over 35,000,000 barrels in 1922—was so great as to break the oil market and depress prices for some time. In February, 1922, nearly 100 wells had been completed, and in April, the number of completions had risen to 160. "Development" was too aggressive, however. In April the field was already declining, and in June the production had declined 50 per cent. A total of 899 wells were drilled in this one field in the year 1922.

In 1921 and 1922 other fields were discovered: Pierce Junction, Orange, South Electra, Bunger, Pioneer, Mirando, Luling, and Kosse. The Kosse field, about 28 miles south of Mexia, was opened in August, 1922, with a 12,000-barrel well, and was heralded as a second Mexia. It created considerable excitement, but soon fizzled out.[6]

The Powell field, a few miles east of Corsicana, had produced small quantities of oil since 1900, but the production began to dwindle about 1906, and for over a decade the field produced only insignificant amounts of oil. In May, 1923, however, a deeper sand was opened, and the production of the field rose with startling rapidity. In September, 1923, the output, from a total of 190 wells, was nearly 200,000 barrels a day, and new wells were coming in at the rate of four a day. But the extraordinary drilling campaign brought its usual result, and within ten days after the peak of more than 200,000 barrels a day had been reached, the production dropped 50,000 barrels, in spite of the addition during the ten days of 42 new wells.

The very rapid increase in production at Powell meant great waste from inadequate storage, for large amounts of oil were run into open sumps. The fire hazard was also very great. The first gusher completed in the field, the McKie No. 1, was ignited and burned for days, with the loss of fourteen lives; and in July, 1923, another fire burned a great quantity of oil before it could be put out. The Texas Railroad Commission finally issued an order that no more oil in the Powell field should be run to earthen storage.[7]

For a brief period it appeared that the Powell field was rapidly approaching extinction, but presently, within one week, 33 new wells brought over 100,000 barrels additional flush production. The flush production of the wells in this field was proving so short-lived that a tremendous drilling campaign was staged. Early in November it was reported that 435 wells had been drilled in the Powell field. The total for the Powell-Corsicana region reached 646 before the end of the year. Production climbed with astonishing rapidity during October, reaching a daily flow of over 300,000 barrels. For a very short time this field produced 357,000 barrels daily—more than any of the great fields of southern California.

The question of storage became very critical. Pipe lines running from the field were unable to care for the flood of oil. Dozens of storage tanks were erected. The Humble Oil and Refining Company, the largest producer, shipped a solid trainload of tanks into the field, for the care of its 100,000 barrels a day. Meantime prices declined considerably, reaching 75 cents a barrel early in November. In October, stocks of crude oil in storage in this field and in the Mexia field were over 23,000,-000 barrels.[8]

This tremendous drilling campaign was carried on at a heavy cost, not only in money but in lives as well. As already stated, the field was ushered in with a fire which cost fourteen lives. The rush in drilling called many inexperienced men into the work, and these men were subject to a very high accident and death rate. During the first few months, ten men were either killed or seriously injured by falling from derricks, and many were injured by timbers, boards, and tools falling from the upper part of the derricks. In August, the list of accidents of all kinds ran from 15 to 30 daily, many of them due to the haste with which the field was being exploited.[9]

Just about the time the Powell production seemed destined to break the crude market and throw the entire oil industry into hopeless confusion, it suddenly began one of the most extraordinary declines ever seen in any oil field. From over 350,000 barrels daily on November 13, the flow declined to 81,000 barrels on December 20. In a little over six months an intensive drilling campaign had brought the output of this field up to 357,000 barrels—almost a record for any single oil field in the United States—and exhaustion of the sands had brought it down again to the modest figure of 81,000 barrels. Dow, Jones, and Company, a financial agency of New York, estimated that over $30,000,000 had been spent in the exploitation of the field before the end of December. So rapid was the decline of production that a large amount of storage which had been ordered was never needed, and many orders for tanks were cancelled.

Such rapid exploitation as was witnessed at Powell, and at most other recent fields, would of course have been impossible but for the remarkable efficiency in drilling technique that has been attained in recent years. At the present time, with modern rotary drills, it is possible to reach a depth in two weeks which would have required perhaps three months with old methods of drilling.

The Powell field was already far gone in her decline when another Texas field suddenly loomed into prominence. The Wortham field, in Freestone County, not far from Powell, was opened on November 22, 1924, with a well of modest flow. Exploitation was even more rapid than at Powell. In less than a fortnight 265 drilling permits had been issued and there were 200 derricks on the ground. The astounding mechanical efficiency of present-day oil operations was well shown in the exploitation of this field. Crews putting up derricks delivered completed jobs in seven hours from the time the sills were laid. On one well, with rotary drill, the drillers made 900 feet in 24 hours, and wells were completed to the oil sand in as little as twelve days.

Production climbed with extreme rapidity, reaching a peak of 166,000 barrels daily on January 15. Then came a decline which was even more rapid, from 166,000 barrels on January 15, from 73 wells, to 100,000 barrels on January 26, from 138 wells, and to 84,000 barrels a week later, from 176 wells. In July the total production had declined to 32,-000 barrels.[10]

LOUISIANA

The oil regions of Louisiana are similar to those of Texas, and, as in Texas, indications of oil and gas were noticed very early. It was not until the spring of 1902, however, that the Jennings field was opened. In 1902 a half million barrels of oil were produced, and in 1903 twice as much. In August, 1904, the first of the great Jennings gushers was struck. Other large wells came in rapidly, and the yield soon rose to 50,000 barrels a day, almost all of it spontaneous flow. Among the remarkable wells of this region were the Bass and Beckenstein No. 1, which yielded 1,500,000 barrels in five months, or at the rate of over 3,500,000 barrels a year—one of the highest records in the oil history of the United States. The Wilkins No. 2 yielded nearly a million barrels in eighty-two days.[11] In 1904 the development of the Caddo field in Louisiana started, the product being at first largely gas. Here were encountered some of the most disastrous fires, and most unfortunate waste of gas. This waste became so great that the State Conservation Commission had to interfere.

The oil production of Louisiana decreased almost 50 per cent in 1907, and almost another 50 per cent in 1909, when salt water began to invade the Jennings pool. In 1910, however, three gushers were struck in the Vinton region, in the southwestern part of the state, each yielding 3,000 barrels or more per day. About this time, the output of the Caddo field began to increase rapidly. In October, the Standard Oil Company purchased over 100,000 acres of proved oil territory in the James Bayou region, and began active development work.

In 1911 the production of the state showed a marked increase, but the next year it decreased again, in spite of the drilling of several good wells in the bed of Ferry Lake, and in spite of deep drilling in the Vinton region. In 1913 over 10,000,000 barrels of Louisiana's total came from the Caddo field, and from the new Mansfield pool in De Soto Parish, both in the northwestern part of the state. The following year the Abington district, and the Crichton pool were opened, and another increase of 2,000,000 barrels was recorded, the gain coming entirely from the northern part of the new De Soto field, where there was much wildcatting, some of it successful. In 1915 the central region continued its decline, but an increase in the northern region more than offset this. In 1917, however, there was a sharp decline of over 3,000,000 barrels in the production of northern Louisiana and northeast Texas, in spite of the development of the Pine Island pool, an important extension of the Caddo district. The discovery of oil in the latter region caused considerable excitement and led to much wildcatting, most of which was fruitless, except in the production of gas, in which Louisiana took high rank among the states. At the close of 1917 a 10,000-barrel gusher on Pine Island was brought in, but the pressure was so high as to defy control for weeks. The excitement provoked the usual drilling boom, and over 300 wells were drilled during the year, with a small proportion of dry holes.

The Bull Bayou field, discovered in 1918, developed a modest production, and early in 1919 the important Homer field, in Claiborne Parish, was opened. Some of the wells in this field came in with initial production of from 20,000 to 30,000 barrels.[12] Two years later the Smitherman well opened the Haynesville field, and started a great wave of speculation. This field was in a poor agricultural section of Louisiana, and some of the poor farmers were soon the recipients of extraordinary incomes. One negro landowner sold his royalty on 80 acres for $90,000. Another poor farmer, the father of twenty children, according to reports, found himself worth a million. He owned 75 acres of land within the limits of Haynesville and 400 acres outside. But this is only a repetition of the story of most other oil pools. During the year 1922 683 wells were completed in the Haynesville field, but in 1923 drilling almost ceased and the production declined 50 per cent.[13] In 1922 a smaller Louisiana field was discovered—the Cotton Valley field, in Webster Parish, in the northwestern part. This field developed a modest production, which it still maintains.

ARKANSAS

Arkansas was for many years suspected of having oil deposits, and in 1917 the United States government sent geologists to examine some of the asphalt showings in the southwestern part of the state. No oil was found, however, until January, 1921, when the El Dorado field was discovered. Immediately the usual scramble for leases began, leases of from one to three acres and up being sought by promoters, who were anxious to inaugurate a town-lot drilling campaign in the town of El Dorado. The mayor of the city wanted to limit town leases to a minimum of two acres, and the city council passed ordinances to regulate the drilling of wells; but it is always hard to tame an oil boom. One enterprising Shreveport speculator told how he hitched up a buggy and team and drove a hundred miles through the surrounding country, seeking leases from the farmers, but "didn't find anybody at home but women and children. All the men had come to town." One correspondent described the town a week or so after the strike:

It looked as if every negro within forty miles of Eldorado had come to town Saturday. It was the biggest and doubtless the most interesting holiday they had ever enjoyed. One had sold a lease for $5,000, had the "cash money." Another had sold for $2,000. Nine children of a negro widower with a 10-acre tract out toward the well, were the envy of the colored population, so much interviewed were they by lease buyers. A negro preacher busied himself with collections. You never saw a busier negro. Boys ran hither and thither with much noise and elbowing, distributing thousands of circulars announcing that a preacher would deliver a sermon on Sunday relating to the blessedness of oil.[14]

So great was the pressure of the population seeking accommodations in El Dorado that the railroad had to declare an embargo on all traffic into the town, unless those wishing to go there could convince the railroad officials that they had successfully made arrangements for shelter. Officers of the railroad declared that they were being swamped with the tons of household goods offered for transportation. The population of the town doubled and quadrupled, and real estate values soared to unheard-of figures.

Drillers by the score were immediately busy, and new wells were brought in so rapidly that before November the new field had produced 10,000,000 barrels of oil, and was going strong. Some of the wells in this section were large producers, and a few were hard to control. One well, which came in with a gas production of 50,000,000 feet per day, blew the casing out, and presently began to spout gas at various points, some of them nearly a half mile from the well. This well soon became a crater 400 feet across and 100 feet deep, with the rig and engine and tools lost in the boiling sand and rock and oil at the bottom, and set itself afire several times.[15]

The next year a still more sensational discovery was made, when the Smackover pool, just north of El Dorado, was opened with a 7,000-barrel well. The first well caused little excitement, but as other large wells were opened, a sizable oil boom developed, and before the end of the year leases were selling for as high as $2,500 per acre. Within three months at least 40 derricks were up, and the sandy roads were jammed with mule teams and automobiles, hauling pipes and timbers, and carrying all classes of hurrying humanity to the new land of fortune—a typical oil boom, with its attendant excitement. The heavy flow of the first wells caught operators unprepared, and the only storage available was open sumps constructed by throwing earthen levees across the ravines. A heavy rain one day caused the breaking of one of these levees, and between 300,000 and 400,000 barrels of oil flowed down the creek. But there was "plenty more" oil for such storage, and great lakes of oil soon covered the ground, while steel tanks were erected as soon as they could be hauled in over the execrable roads. In the meantime it presently appeared that the oil which was stored in sumps was assimilating rainfall better than ordinary crude. Some of it which was shipped to eastern consumers—it was fit only for fuel purposes—would not burn when it reached its destination, because of the water content, and it was necessary to construct plants to dehydrate it. As the production climbed to 100,000 barrels a day, the price declined rapidly. The Standard Oil Company at first posted a price of 75 cents a barrel, but the growing flood of oil was too great to be handled easily, and this price was soon reduced to 50 cents, and, before the close of the year, to 30 cents a barrel. The entire fuel oil market felt the effects of this flow of cheap oil, but enterprising drillers continued their work of getting the oil into the earthen ponds as rapidly as possible. During the year 1923 this field produced about 36,000,000 barrels of oil, of which 17,000,000 barrels were in local earthen or steel storage at the end of the year.

The population of the town grew tremendously, from a few hundred in July to 10,000 or 20,000 in December,—lies regarding the population of oil towns vary in their dimensions and plausibility,—living mainly in hastily constructed rooming houses, and even in tents. There was no water system for the town, and no water except from a few wells, some of them not even equipped with pumps. The fire hazard was very great.

Smackover made fewer local landowners rich than most oil fields. The oil companies that did the drilling leased most of the territory at a small rental before oil was struck, and when the discovery well came in, landowners were unable to get high bonuses for their leases. Of course, the royalties presently began to pour in, and then the "money began to circulate."[16]

The Smackover field has recently established the highest record of daily production yet found in the oil history of the United States. For months the daily production hung around 100,000 barrels, from the shallow sands. In June, 1924, the output rose to 152,000 barrels, and then declined again below 100,000 barrels. In February, 1925, however, a deeper sand was opened, and, in spite of drastic price reductions by the purchasing companies, a succession of great wells drove the production to a peak of 473,000 barrels on May 26. This was more than 100,000 barrels above the previous high records of Cushing, Santa Fé Springs, and Powell. A rapid decline followed, and in June the daily record fell below 250,000 barrels. Early in July, 1925, the Smackover field had produced more than 110,000,000 barrels of oil, and was still making more than 200,000 barrels daily. The amount of oil in storage was estimated at 26,000,000 barrels.[17]

NOTES

1. *Sci. Am. Supp.*, Vol. 69, 229.

2. *Mineral Resources, 1897-1898*, Pt. II, 103.

3. *Eng. and Min. Jour.*, Jan. 19, 1901, 94. It has been claimed that it was not really Lucas, but a man named Sharp, who discovered Spindletop. (*Oil and Gas Jour.*, Oct. 14, 1921, 84.)

4. See below, p. 211.

5. Kansas City *Star*, Feb. 3, 1922.

6. *Petroleum Age*, Sept. 15, 1922, 17.

7. *Petroleum Age*, July 15, 1923, 35; *Oil Weekly*, Sept. 22, 1923, 9, 34; *Natl. Petroleum News*, July 11, 1923, 28.

8. *Oil Weekly*, Oct. 27, 1923, 36.

9. *Oil Weekly*, Oct. 20, 1923, 34.

10. *The Lamp*, Dec., 1924, 10.

11. The Lucas well at Spindletop had started with a higher production, but it had a very short life.

12. It is said that the company which opened the Homer pool did not profit greatly through the discovery, because of poor management. It was a stock promotion company, managed by a New York lawyer, and had leases on practically all of the Homer field, but permitted these leases to lapse just a few days before discovery. Even after discovery, the company was slow to make an effort to get back these leases, and, as a result, many of them got into the hands of other companies, and others were bought at high prices. (*Wall St. Jour.*, Sept. 4, 1920, 11.)

13. *Oil and Gas News*, May 19, 1921, 31; *Oil Weekly*, Jan. 19, 1924, 75.

14. *Oil and Gas Jour.*, Jan. 21, 1921, 81.

15. *Oil and Gas Jour.*, Jan. 21, 1921, 81; Jan. 28, 1921, 3; July 6, 1922, 117; *The Lamp*, June, 1922, 14; Kansas City *Times*, Jan. 2, 1923.

16. *Oil Weekly*, Dec. 23, 1922, 11; *Petroleum Age*, Oct. 15, 1922, 20; *Oil and Gas Jour.*, Nov. 21, 1922, 11.

17. *The Lamp*, June, 1925, 7; *Oil Weekly*, July 10, 1925, 25.

CHAPTER VIII

OIL EXPLOITATION IN THE MID-CONTINENT FIELD

KANSAS

IN the Mid-Continent field, including Oklahoma, Kansas, and northern Texas (northern Texas is treated in connection with the other Texas production, although geologically it belongs rather to the Mid-Continent field), oil has been known to exist almost since the earliest settlement of that country. The earliest settlers in Kansas discovered petroleum springs in Miami County, near Paola, in the eastern part of the state, and used the oil to some extent for lubricating purposes. As early as 1858, leases were obtained, and there was considerable talk of development. One pioneer began prospecting for petroleum near Paola, Kansas, as early as 1860; but the political unrest of the times prevented further exploration until after the Civil War.[1]

In a report made by Professor Mudge in 1864, the existence of oil and bitumen in the eastern counties of the state is referred to. The following year, Professor Swallow referred to nineteen different tar springs in Miami County, and added, "Scarcely a well has been dug without finding petroleum in some of its forms." He thought the facts were "very strong evidence of the existence of large reservoirs in these localities." In 1865 a St. Louis company drilled two wells near Paola, and in 1873 another company struck gas there. Considerable quantities of gas were found near Iola in 1873, and Paola was heated and lighted by natural gas in 1882.

During the eighties, with the opening of the rich Ohio and Indiana oil fields, there was considerable interest in oil exploitation in Kansas, and a great many wells were drilled in the eastern part of the state, but no important results accrued. A little oil was found at Paola, but the real exploitation of the state dates from 1890, when several Pennsylvania operators entered the field. In 1891 a Pennsylvanian by the name of Mills came to Neodesha, where he was so impressed with the chances for oil that he went back to Pittsburg and persuaded the firm of Guffey and Galey to come west and begin a series of developments on a large scale. He chose for his field of action the vicinity of Osawatomie, while Guffey and Galey made their headquarters at Neodesha. Success followed them in sufficient measure so that in 1894 they were able to sell their entire holdings to the Standard Oil Company. The Standard Oil

Company immediately began an extensive development program, and built a refinery at Neodesha, which was gradually enlarged to become one of the largest in the country.

Drilling was prosecuted with some vigor in other fields—Peru, Humboldt, Chanute, Cherryvale, Coffeyville, and Independence. Fifty wells were drilled within the city limits of Peru before the close of 1903, and at that time 200 drills were said to be running in the territory around Chanute. A number of small fields were opened during the next few years: Cherryvale, Neodesha, Erie, Chanute, Coffeyville, Moran, Independence, Bolton, Humboldt, Wayside, Tyro, Caney, Mound Valley, Spurlock, Blundell, Hoffman, Rantoul-Wellsville, and other lesser fields; while the Paola and Peru fields were further developed. The Kansas production was secured by the drilling of hundreds of new wells each year. In many of the fields gas was a more important product than oil.

The excessive drilling activity in Kansas during the years 1904 and 1905 resulted in a rapid increase in production. The total product of the state in 1905 was over 6,000,000 barrels. This swelling production came just at the time when the Oklahoma fields to the south were pouring out great quantities of oil, and the result of it all was a completely demoralized market. The low prices paid stirred up a bitter quarrel between the producers and the Standard Oil Company. This episode will, however, be treated in another connection.[2] Another result of the low prices, even more unfortunate, was the closing of hundreds of small wells. Those of small production could not be profitably operated at the prices prevailing, and many of them were closed and ruined, for when oil prices recovered, and they were opened, many of them never flowed oil again.[3] Much oil was wasted in various ways during this period of over-production. Much was burned as fuel. The producers spent a great deal of time and energy trying to persuade the railroads and industrial plants to burn their oil.

In 1905 Kansas produced over 6,000,000 barrels of oil, but for several years thereafter no important discoveries were made, aside from some developments near Fredonia in 1907; and in 1910 the production of the state had declined to a little over 1,000,000 barrels. This rapid decline caused an increase in the price of crude, which stimulated more intensive drilling of the old fields and led to the discovery of small pools in proved territory. As a result of such small discoveries, the output of the state increased somewhat, even before the discovery of the Butler County fields.

In 1914 the great Butler County fields were opened, including Augusta, Eldorado, and Towanda. The Augusta field had produced gas for some years,[4] a few wells dating from as early as 1905, but oil exploitation began in 1914, and before the end of 1915, a dozen wells were in operation, most of them larger producers than the wells which had hitherto been found in Kansas. Some of the unfortunate aspects of new oil field development were to some extent absent, because the field was largely controlled by a single company. It was not developed as rapidly as most new fields have been, although drilling was pretty rapid in 1916, and later. The speculative excitement, while not as intense as in some new oil fields, was fairly wild.[5]

The discovery of oil at Augusta naturally led to considerable excitement in nearby towns, and the town of Eldorado, as a municipal enterprise, presently drilled a well near the city. After the expenditure of thousands of dollars, the town abandoned its well at a depth of about 2,650 feet, and later sold the well and leases to the Wichita Natural Gas Company, later the Empire Gas and Fuel Company. This company promptly drilled in another part of the lease, and opened the Eldorado field with a 150-barrel well. Other wells, some of them 1,000-barrel wells, followed. Exploitation here was more rapid than in Augusta, because there were in the field a number of scattered tracts not owned by the dominating company, and much offset drilling was necessary. During 1917 the Towanda district furnished additional excitement, with the largest wells yet discovered in the Kansas fields, some of them producing from 12,000 to 20,000 barrels a day. These three fields absorbed millions of dollars of Kansas money, and turned out a fresh batch of oil millionaires, retired farmers, and disgruntled "investors," but it greatly increased the oil output of the state; in fact, these three pools, in five years, yielded nearly 100,000,000 barrels of oil. In April, 1918, the output of these fields was over 100,000 barrels a day. Within five years it had declined to 20,000 barrels.

A number of minor fields have been discovered in Butler County since the Augusta-Eldorado-Towanda boom. Among these new fields are the Douglass, Smock, Sluss, Potwin, and the Peabody-Elbing, which flashed into prominence in 1919. Still more recently, the Greenwood County field has been opened, and is now one of the most important in the state.

OKLAHOMA

There is no authentic record of the earliest discovery of petroleum and natural gas in Oklahoma. Oil springs near Mounds and Adair were well known to the early settlers in the vicinity. These springs yielded small quantities of crude oil, while the "burning spring" northeast of McAlester—a typical natural gas escapement—was noticed early in the history of the region. Shallow drilling near these so-called oil and gas springs failed to reveal economic deposits of either oil or gas, and they were considered of no consequence.

The discovery of oil and gas in Kansas in the early eighties excited the interest of the Five Civilized Tribes in Indian Territory, and in

1884 the Choctaw council passed an act forming the Choctaw Oil and Refining Company. The Cherokees followed this example almost immediately, and passed a similar act, and the two companies thus formed started drilling in the Choctaw nation, on Clear Boggy Creek, and in the Cherokee nation, on the Illinois River north of Tahlequah. The Cherokee council presently repealed the charter of 1884, and the Choctaw company failed to find oil. Nothing of importance was done until 1894, when the Cudahy Oil Company secured a blanket lease on the lands of the Creek nation and drilled two wells at Muskogee. Both these wells struck oil in small quantities, but there was no further development here until 1904, because of difficulty in securing title to the land. The Cudahy Oil Company also had leases covering about 200,000 acres near Bartlesville and started operations there, but the passage of the Curtis Act in 1898 forced them to give up all unproved lands, leaving them only the section on which Bartlesville now stands. Some development had been accomplished at Chelsea prior to 1893, and the Cherokee Oil and Gas Company had a large acreage leased, but the Curtis Act forced the surrender of these leases also. Little was done in the Cherokee nation until 1904, when it became possible to get allottee's leases approved by the Department of the Interior.

Tests had been made in the Osage lands as early as 1896. At that time Edwin B. Foster, who had secured a lease on all the lands belonging to the Osage nation, brought in a five-barrel well a few miles south of Chautauqua Springs, Kansas, but no important results followed for several years. A well drilled at Red Fork in 1901 opened the Red Fork-Tulsa district, and in the same year a well was drilled in the Granite field, but it brought no oil.

Important oil production in Oklahoma dates from 1904, when several fields were opened. The Lawton field was discovered, although it never became a sensational producer. The Cleveland field furnished more oil and more excitement. This field was in Pawnee County, Oklahoma, and a great rush followed, not only to the Pawnee lands, but to the Osage lands across the river. The Muskogee field had been discovered in 1894, but very little oil was obtained until 1904, when 40 wells were drilled on the present townsite. Hundreds of other wells followed during the next few years. Later in the year 1904 the Secretary of the Interior began to confirm leases in the Cherokee nation, and a tremendous drilling campaign was inaugurated at Bartlesville, Dewey, Chelsea, Alluwe, and Lenapah. The Bartlesville-Dewey field was the scene of great excitement, and before the close of the year 100 wells had been drilled there. Within ten years nearly 5,000 wells were drilled in this field, and the average product per well had declined from 73 barrels to 10 barrels per day. Chelsea was equally active, with 100 producing wells at the close of the year 1904.

The following year the Coody's Bluff-Alluwe field came into prominence, and developed very rapidly, reaching an output of 53,000 barrels a day within less than two years. The Bartlesville-Dewey field and the Cleveland field were extended in several directions. Production for Oklahoma increased so rapidly that the price of oil went as low as 50 cents a barrel; but drills by the hundred were employed to swell the over-production, and the next year several new fields were discovered. The Madill field proved small, but the Morris field responded with a number of fair wells, and the Glenn field, a few miles southeast of Red Fork, opened in the early part of the year, soon proved to be one of the world's greatest oil pools. Exploitation of this pool proceeded with great rapidity, and there were more than 100 wells in the field before the end of the year, with an average production of 400 barrels per well. Such a flood of oil, of course, resulted in serious demoralization of the oil market, and some oil was sold for 31 cents per barrel, but even this price did not discourage the drillers greatly, and hundreds of additional wells were drilled in the Glenn field each year following. It has been estimated that 1,000,000 barrels of oil were wasted outright in this field in two years of "development."[6]

In 1907 the Glenn pool maintained its remarkable output, reaching a maximum of 80,000 barrels per day in October, and then declined. The Morris pool was also being exploited in 1907, while drilling was begun in the Bald Hill field, and oil was found in the Canary-Copan field, and in the Hogshooter field. The Hogshooter field was one of the most important gas fields in the country, but it produced a great deal of oil also. In this one field 2,500 wells were drilled within the next seven years. The production of all these fields kept prices very low. High-grade light oil sold in 1907 for 40 cents, and heavy oils as low as 26 to 28 cents a barrel.

Yet the feverish exploitation continued, and hundreds of drillers were able to raise the 1908 production above that of 1907. The decrease in production of the Glenn field was checked by the drilling of new wells, by the cleaning out and shooting of old wells, and by drilling into deeper sands. A 1,400-barrel well in the Morris field led to active drilling there. A new pool of exceptionally high-grade oil was found at Muskogee. The Delaware-Childers field in Nowata County was discovered and, within four years, was punctured with over two thousand holes. In 1909 the Preston or Hamilton-Switch pool was the only new pool opened, but prices remained so low that a strong effort was made by the Producers' Association to curtail the production. This effort was not successful, and the following year the production increased further. The Henryetta pool in Okmulgee County was ushered in with a 600-barrel oil well, followed by a phenomenal gas well, with a capacity of 80 million feet. This field was promptly covered with several hundred wells, producing thousands of barrels of 30-cent and 40-cent oil. Another new pool at Osage Junction, across the Arkansas from the Cleveland pool, helped to swell the already too large production. During 1911 the prolonged drouth hindered both drilling and manufacturing enterprises, and no important fields were discovered, but the production increased several million barrels.

The year 1912 was marked by unprecedented drilling activity. The old fields were extended and filled in, and a great deal of wildcatting was done, some of it successful. The most striking feature of the year was the discovery of the great Cushing field, in eastern Payne County. This field was one of the most phenomenal producers in oil history. The first well was completed in March, and before the end of the year over 75 wells had been completed, and over 100 drilling rigs were up. The initial production of these wells was very high, and the oil of good quality. About this time came the discovery of deeper sands in the Cleveland field, the eastward extension of the Glenn field, the opening and rapid exploitation of the Adair field, west of Nowata, the extension westward of the Ponca City field in Kay County, and great activity in Okmulgee County. All this activity was not enough to increase the production of the state in 1912, but the following year the production increased enormously. Several new fields were discovered. The first wells in the Healdton field, in western Carter County, aroused great excitement, and 270 producing wells were drilled here before the end of the year. Several smaller fields were found: the Tiger Flats field in Okmulgee County, the Mervine field in Kay County, the Inola field in Rogers County, and the Owasso field near Tulsa; while extension of the Hogshooter field, further development of the Wicky field near Mounds, and of the Bald Hill field, the opening of the Booch Sand field in Okmulgee County, and, more important still, the further rapid development of the Cushing field, all helped to swell the production of the state to a new record—63,579,-384 barrels. Oklahoma was enjoying her "third drink." The year 1914 marked a new record for the state.

During 1914 Cushing remained the center of interest, with over 750 wells drilled during the year, over 100 of them gushers of over 1,000 barrels production. Several produced 10,000 barrels a day or more, and one came in with a production of 20,000 barrels. At one time the production from Cushing rose above 300,000 barrels a day—one-third of the production of the entire United States. The Prairie Oil and Gas Company issued a warning that the new field might flood the market if the operators persisted in drilling so many wells, but no attention was paid to this warning, nor to a second warning.[7] Prices of oil tumbled—declined below the price of good drinking water. At one time the lower grades of gasoline sold for 5 cents per gallon, and high-gravity crude oil for 40 cents a barrel. The extraordinary cheapness of the product, together with inadequate transportation facilities, made it necessary to store great quantities of oil, and the waste was very great. But the cheapness and waste of oil seemed to have little effect on the amount of drilling. At some points in the field the leases were so small that the derricks crowded each other uncomfortably, and such a situation has always necessitated hasty drilling. Even wildcatting was stimulated by the extraordinary productivity of the wells in this wonderful field, and by the general enthusiasm accompanying its exploitation.[8]

Finally, in accordance with the wishes of the more conservative operators in the state, the Oklahoma Corporation Commission complied with the resolutions adopted at a convention of more than 500 of the experienced operators, and prohibited any more drilling, except where necessary for offsetting other producers or protecting the terms of a lease. The final failure of this piece of radical legislation will be noted in another connection, but it shows the extent of the evil of over-production in the Cushing field; and for a while it helped to check the growing production.

The drilling craze at Cushing spread to other regions in Osage, Washington, Muskogee, and Creek counties, while the rapid exploitation of the Healdton field, the discovery of the Boynton field in Muskogee County, the Bixby field in Tulsa County, the Coweta field in Wagoner County, the Yale pool in Payne County, the Blackwell field in Kay County, and some development of the Booch Sand field in Okmulgee County, all helped in the wastage of oil, and in the general demoralization of the industry. The outbreak of the World War in August contributed to the same result.

The story of the Healdton field, in Carter County, the first oil pool of consequence to be developed in the southern part of the state, resembles that of Cushing. This field was opened in August, 1913; and before the end of the year, 275 productive wells had been drilled, some of them producing as high as 5,000 barrels a day. Operations were severely hampered by lack of adequate storage, and, as at Cushing, large amounts of oil were lost through leakage, through evaporation from open sumps, through washing away of the dikes, and through fires kindled by lightning.[9]

The over-production of 1914 continued into 1915. Healdton still produced heavily, the Boynton and Blackwell fields were extended somewhat, and two small fields were discovered, the Vera field in Washington County, and the Stone Bluff field in Wagoner County. A decline in the Cushing production, however, presently outweighed all other factors in the situation. The production of this field rose as high as 300,000 barrels per day in February, and again in April, but the decline set in shortly afterward, and in December the output was down to 100,000 barrels, while the price rose before the end of the year from 40 cents to $1.20.

During the next two years there were no very important discoveries. The Blackwell field was the scene of much activity in 1916, with 75 drilling rigs in operation and many good wells coming in. The Garber field in Garfield County came in with a 1,100-barrel well. The Ponca City field, which had yielded a small amount of oil in 1911, increased its output somewhat. The Beggs field, in Okmulgee County, which had been exploited somewhat in 1910 and 1911, furnished a number of thousand-barrel wells. The Mounds field, in Creek County, the Fox field in Carter County, and the Cement field in Caddo County, revealed several much smaller wells. The Billings field, in Noble County, assumed a slight importance in 1917, and the Yale field, north of Cushing, developed rapidly. The Shamrock extension of the Cushing pool offered some very productive wells, and some wells were drilled in the Hominy field in Osage County. The Healdton annual production increased from 5,000,-000 barrels to 22,000,000 barrels.

The tremendous demands of the war, and the high price of crude—from $1.00 in November, 1916, to $2.00 in August, 1917—stimulated heavy drilling of the older fields, but, in spite of that, the production of the state declined over 8,000,000 barrels from the 1916 production. In 1918 the production declined further, with a slump in the Healdton, Cushing, and Osage production, as well as that of some of the older fields. This decline came in spite of the discovery of a number of new fields: the Youngstown field in Okmulgee County, the Walters field in Cotton County, and the Barnes field in Garfield County, and in spite of an increased production from the Yale, Billings, Garber, and Blackwell fields. During the next year the decline was checked and the production was later greatly increased again, through developments in Osage, Okmulgee, and Cotton Counties, and by deeper drilling in various fields.

A number of less important fields have been opened within recent years: the Boston, Wynona, Haskell, Two-Four, Velma, Hewitt, Lyons, and Bristow fields; but the stories of these fields resemble those of other oil fields and need not be told in detail here.

In 1920 another great field was discovered in Oklahoma, the Burbank field, in Osage territory. The discovery well was completed by the Marland Oil Company in May, 1920, and soon dozens of drillers were on the ground, vying with each other in their efforts to reach the oil sands first. The exploitation of the field was far more orderly than that of most oil fields, yet, within three years, nearly a thousand wells had been drilled and production had risen above 120,000 barrels a day.[10]

The Tonkawa field, in Kay and Noble counties, was opened in June,

1921, likewise by the Marland Oil Company. The production of this field increased more rapidly than that of Burbank. In March, 1923, 90 wells in this district were producing over 50,000 barrels a day, and 125 wells were being drilled. Two months later, the output reached 112,000 barrels, and, in the next five months, declined 50 per cent.[11]

When the Burbank field was first brought in, oil prices were fairly remunerative, but as the production of the Mid-Continent field rose, and the industrial depression of 1921 came on, prices dropped more than 50 per cent. When, in the summer of 1923, oil from the Los Angeles basin began to displace the Mid-Continent oil in the eastern markets, a situation developed in Oklahoma which closely verged on chaos.[12]

Two important Oklahoma fields have been developed since Burbank and Tonkawa: the Cromwell field, and the Papoose field, in central Oklahoma, the two fields only a few miles apart.

One aspect of Kansas and Oklahoma history which will be treated in a later chapter is the dominant position of the Standard Oil Company, and the efforts of these states to secure what they considered fair treatment.[13]

NOTES

1. The following story has been told regarding this oil venture:

Dr. G. W. Brown, a former resident of Titusville, Pennsylvania, had been publishing the *Herald of Freedom* in Lawrence in 1855, and had published an account of oil springs on the Wea, in Miami County. During the winter of 1859-1860 he visited his old home in Pennsylvania, where he caught the oil fever which was raging throughout western Pennsylvania. Recalling the oil in Miami County, he hastened back to Kansas, and, after some prospecting around Paola, organized a company of eight men to explore for oil, got leases on 40,000 acres, and began to drill. The next year, however, the Civil War broke out and the bushwhackers stopped his work. After the war, two members of his company were dead, one was ruined by the Quantrell raid, and one had gone south, so the company was never reorganized. Here again is seen the influence of Pennsylvania in the development of western fields. (*History of the Mid-Continent Oil and Gas Field*, published by the *Independence* (Kans.) *Daily Reporter*, 1905.)

2. See below, pp. 241-245.

3. *Petroleum Age,* July 15, 1923, 16.

4. It is recorded that the Augusta field was originally discovered by a lawyer named Yeager, who had studied the geology of oil and gas at odd times, and who located his wells by using a carpenter's spirit level to determine the dip of the strata.

5. Thompson, Geo. F., *Fields of Fortune; A Souvenir History of Wichita's Oil and Gas Fields.*

6. U. S. Geol. Survey, *Bul. 394,* 31.

7. *Oil and Gas Jour.,* Dec. 31, 1914, 24.

8. *Oil and Gas Jour.,* July 2, 1914, 12; June 7, 1917, 2.

9. U. S. Geol. Survey, *Bul. 621,* 13; *Mineral Resources,* 1914, 1007.

10. *Natl. Petroleum News,* March 1, 1922, 34.

11. *Oil Weekly,* Dec. 16, 1922, 25; Jan. 13, 1923, 16; Feb. 24, 1923, 84; *Petroleum Age,* March 1, 1923, 19; *Oildom,* March, 1923, 15.

12. For further discussion of this phase of Oklahoma oil history, see below, pp. 111-120.

13. See below, pp. 241-245.

CHAPTER IX

OIL EXPLOITATION IN CALIFORNIA

THE existence of asphaltum and petroleum in California was familiar to the first settlers of the state, and semi-solid bitumen from seepages in the southern part of the state was used, from very early times, as fuel and as a cement. It is said that the early Catholic fathers of California used this bitumen on the roofs of their missions. In 1856 or 1857 a San Francisco druggist tried to collect some of it and refine it, and, following him, others made attempts at refining, but none of them succeeded financially.[1]

The oil industry in California is often dated from 1865, when Professor Benjamin Silliman was called upon, by men having interests in Ventura County, to examine samples of petroleum seepages from the Ventura district. It appears that Professor Silliman was grossly imposed upon in the matter. The samples submitted to him for examination were not taken from the local production at all, and, as a result, he issued a report glowing with enthusiasm over the quality of the California oil. His report was issued just at the time when the Pennsylvania oil excitement was at its height, and it aroused great enthusiasm.[2]

The oil craze in California in the sixties was about as wild as that of Pennsylvania, or as wild as later similar epidemics in California, in fact, one writer declares, "Everything in these lines in these later days pales to stupidity in the lurid light of the oil craze that attended the birth of the petroleum industry." The Pennsylvania discoveries naturally drew attention to the oil indications in California, and the wild excitement of Titusville and Oil City and Pithole found a ready echo in the Far West. Little was known about petroleum, and very extravagant ideas on the subject led to some of the most extravagant financial operations imaginable. Not only were the western cities—San Francisco and others farther south—stricken with oil dementia, but eastern cities caught the infection. In Philadelphia, according to one account, oil shares representing lands in the Los Angeles district were sold at $1,000 a share, and advanced to $1,500 two days after the subscription list was filled. One of the letters describing the lands declared, "The rancheros have always looked upon them as objectionable on account of their losing many cattle by their becoming mired in the oil that overflows hundreds of acres around the mouths of the wells;" and such a description seems to have been given credence. New York papers glowed with accounts of the wonderful California oil discoveries. One company was organized in New York with a capital of $5,000,000. A report in 1865 stated: "Professor J. T. Hodges, with a full corps of assistants, machinery, etc., is to leave New York soon for the California oil fields. The company expects to place cargoes of oil in New York in a few months."

Promotion schemes of the rankest variety were organized. One company, in its prospectus, announced its capital at $10,000,000. Its property was a Spanish grant in Santa Barbara County, of about 10,000 acres, which someone had bought for $22,000 in greenbacks, or about $10,000 in gold. The purchase of this "property" was telegraphed to eastern associates, who sold one-half for $50,000. Then it was sold for $450,000, and finally it went into the assets of the California Petroleum Company at $10,000,000.[3]

It is estimated that in 1865 and 1866 at least seventy oil companies were incorporated in California, and hundreds of thousands, perhaps millons of dollars, were spent in boring wells in various parts of the state—Humboldt, Mendocino, Santa Clara, Santa Cruz, Contra Costa, Colusa, Santa Barbara, Ventura, and perhaps in other counties. "The oil country is still full of people locating claims," wrote a newspaper correspondent in April, 1865, referring to Colusa County, where a number of operators were putting down prospect wells.

Enthusiasts over the California oil prospects even made pretty definite plans for the utilization of the oil that should be found. They persuaded the president of the California Steam Navigation Company to change the furnace of his steamboat "Amelia" to burn oil. A trial trip was made in May, 1866, and every prominent stockholder in the company was aboard for a voyage around Mare Island. According to one story, the boat caught fire four times in three hours, and "there were several millionaires on board of her who were glad to get ashore alive."

A few years found most oil companies with depleted treasuries and little or no oil, and with large amounts of material on hand; and, as a result, the search for oil was largely abandoned for a decade, although some desultory drilling was done, and a small amount of oil was found. The Central Pacific Railroad for a while took 200 barrels per month for lubricating purposes.

The opinion of Professor Whitney, State Geologist of California, Professor S. F. Peckham, and other eminent geologists, to the effect that California contained no such oil deposits as those of Pennsylvania, an opinion based on the absence of any extensive coal beds in California, was one of the causes of general discouragement.[4] In 1868 a government geologist took a different position on this question, but by this time most of the oil excitement had subsided. As Bowles said in 1869:

CROWDED DERRICKS IN THE LOS ANGELES FIELD

The state has gone through her petroleum fever; at one time she thot herself supremely rich in deposits of oil; but hardly in a single case has the boring for the luscious liquid been attended with profitable results.[5]

There was a definite effort at development in the Ojai Valley, Ventura County, in 1867, when a shallow well was drilled near one of the numerous brea or asphaltum deposits of that region. Owing to lack of proper tools and insufficient knowledge concerning the handling of the heavy oil, this was not a success. In 1874 there were several localities in Ventura County where oil was being produced from springs.

About 1875 the interest in oil began to revive. Three wells were put down at Pico Cañon, in Los Angeles County, which yielded 15 or 20 barrels each per day. In 1877 the Ventura wells produced about 80 barrels daily, and Pico Cañon produced half as much. The next year 60 barrels of crude oil were taken daily from the Boyer well, in the Santa Cruz Mountains. The oil here, as generally in California, was brought to the surface by pumps, no flowing wells having as yet been struck in the state.

From this time the production of oil grew somewhat, although California was for many years afterward a heavy importer of petroleum products. Already in 1885 one oil company—the Pacific Coast Oil Company—was reported as controlling 158,000 acres of oil territory in various sections of the state, and was said to have expended $3,000,000 in oil prospecting and development, for a production of 600 barrels of crude oil per day. At this time, oil was reported to have been found in 18 of the 52 counties of the state, and was being used to some extent as fuel—in oil stoves and in steam boilers for various types of mills. In 1887 there were 31 wells at Pico Cañon, producing about 500 barrels a day. In the vicinity of Sulphur Mountain, there were several wells producing a total of 142 barrels a day.

In the early nineties important events transpired. Oil was discovered near Coalinga, in Fresno County, in 1890, and, two years later, E. L. Doheny drilled the first successful well in Los Angeles.[6] This brought on a wild epidemic of town-lot drilling in Los Angeles. Hundreds of wells were drilled within the next two years, some of them so close together that the derricks almost touched. As exploitation proceeded three different fields were outlined: the Central or Old field, where there were 308 producing wells in 1895; the East field, with its first well in November, 1896, and with 200 or more wells a year later; and the West field, which first registered important production in 1899, and included about 200 wells a year later. In 1913, over 1,300 wells were reported to have been drilled in the Los Angeles field, and nearly 1,000 of them had been abandoned. Close drilling of the field had drained the oil deposits rapidly, while the great over-production and low price of oil—18 to 20 cents a barrel, from 1901 to 1905—made it difficult to operate profitably any but fairly productive wells.[7]

The Los Angeles episode led to extensive drilling in other fields. Systematic development of the Sunset-Midway field was begun about 1894, and, more novel and spectacular, if not more important, the Summerland field in Santa Barbara County was discovered. A well had been drilled at Summerland as early as 1877, and some oil had been found, and another well in 1887 had struck heavy oil; but nothing of commercial importance was found until 1894. The first wells were drilled along the bluffs near the ocean beach, but within a year or two the drillers worked down to the beach, and out into the ocean. The beach itself was government property, and was grabbed by the first comers, so that there were a large number of very small holdings. By means of wharves running out into the ocean, operators drilled about a hundred wells out in the water. But all this zeal overreached itself. There were over a hundred operators in this small field at one time, and in 1900 there were 305 producing wells, but ten years later, 265 wells had been abandoned. In January, 1912, the active wells averaged only about a barrel and a half a day.[8]

Summerland was only well started on her oil "development" when a number of new fields were discovered elsewhere in the state. In the late nineties and in 1900 several great California fields were opened. Coalinga came into importance in 1897. Oil had been discovered here in 1890, but it was not until five years later that some persistent operators brought in a small well, producing the lightest oil that had been found in the state. Some interest was aroused, several companies began drilling on neighboring tracts, and in 1897 the famous "Blue Goose" well was brought in, a flowing well of respectable proportions.[9] About the same time, the Whittier field, near Los Angeles, and the Ventura County fields were also being exploited, and oil was discovered on the famous Means ranch, near Bakersfield. The result of all this was one of the greatest oil booms in the history of California. Within three years about 2,400 oil companies filed incorporation papers in the state; most of them sold more or less stock; and at least 1,200 companies did some actual drilling. Again the state was punctured with hundreds of wells, and this time with some success. The northern counties again failed to meet expectations, but the operators in Kern River, Coalinga, Sunset-Midway, McKittrick, and Fullerton were successful—entirely too successful, if the resulting price of oil is any criterion. In the meantime the Los Angeles oil exchange was established, in 1899, and the business transacted there soon exceeded that of the mining stock exchange.

Oil was discovered in the Kern River field, near Bakersfield, in May, 1899. Within a few weeks oil men from every part of California were

THE SUMMERLAND FIELD

in Bakersfield to inspect the new discovery, and locators were scattered over the valley to a distance of twenty or thirty miles in every direction.[10] In September, 1900, there were 134 completed wells in the field, and within another decade nearly 1,500 wells had been drilled, and water was beginning to make inroads in the production. While the number of wells was increasing prodigiously, the average product per well was decreasing about as fast—from 90 barrels in 1903 to 17 barrels in 1913. Much of the oil secured by this reckless drilling was sold for 10 or 15 cents a barrel, and some as low as 8 cents.

The low price of oil, while it caused the ruin of many producing companies, did not seem to have much effect on the amount of drilling, for the next year after the Kern River field was opened, extensive developments were begun in the McKittrick field, not far from Bakersfield, and in the Fullerton field in Los Angeles County. Oil had been found in the McKittrick field in the sixties, and a refinery had even been erected, but not until about 1900 was any considerable exploitation undertaken. At this time also, the East Side field in the Coalinga district was being exploited, and there was some production in the Sunset-Midway field. The next year (1901) oil was found in the Santa Maria field, in Santa Barbara County, and the next year the Salt Lake field, in Los Angeles, was opened. In 1903 the Sunset field was producing steadily, the Santa Maria field was being developed as fast as the drillers could get their wells down, the Los Angeles output was increasing rapidly, and the Kern River field was still spouting its flood of oil. The rapid exploitation of all these fields raised the production of the state from a little over two and a half million barrels in 1899 to over 24 million barrels in 1903—nearly 1,000 per cent in four years. Such a prodigious increase was disastrous in many ways. The price of oil declined, before the close of 1903, to 20 cents a barrel, and even as low as 10 cents. This unfortunate cheapness of oil, of course, led to very great waste, for oil was hardly worth saving. At the close of 1904, 3,500,000 barrels were stored in earthen reservoirs or sumps, and the loss through seepage and evaporation was very heavy. Many producing companies that had started operations on the basis of $1.00 oil found that they could not make expenses at 10 cents, and were forced either to suspend operations or sell out to the stronger companies.[11] Thus a strong movement in the direction of consolidation of interests developed. At the same time the independent companies left in the field found themselves forced to coöperate to protect their interests, and thus was inaugurated a coöperative movement which played an important part in California developments for many years. The cheapness of oil, and the financial embarrassment of the producers, led also to the inauguration of an extensive campaign to stimulate consumption—a most unfortunate result of an unfortunate situation, since oil, an exhaustible and irreplaceable resource, should have been conserved until it was really needed instead of being forced upon the market.[12]

For several years California suffered greatly from over-production. The appearance of water in the Kern River field, and the decline of several of the other fields, were not sufficient to balance the discovery of new fields—the Cat Cañon field, in 1907, and the Coyote Hills field in 1908—and the growing production of other fields. In 1909 President Taft withdrew some of the public oil lands from entry, in spite of the opposition of many of the oil producers, who enjoyed the sight of 20-cent oil running down the creeks and cañons. Taft's wise action helped to stem the growing flood of oil, but on March 15 of the following year the famous Lakeview gusher was tapped in the Sunset-Midway field, with an initial flow of 18,000 barrels, which increased to a maximum of 65,-000 barrels a day.

For weeks it was impossible to cap this well. The owners tried to fasten a roof of 16- by 16-inch timbers over it, but the force of the stream tore such timbers to splinters. The fountain of oil rose 200 feet above the top of an 84-foot derrick, and sprayed the sage brush for half a mile in every direction, while the roar of the escaping oil could be heard a mile away. Until it could be capped the oil was run into open sumps, from which it was pumped into the pipe lines. Even after it was capped its flow was so great that the pipe lines leading from the field were inadequate, and much of the oil was stored in sumps hastily constructed by throwing dams across small cañons in the foothills, where the oil stood for weeks or months exposed to the sun and elements. This marvellous well produced 8,000,000 barrels of oil in 18 months, and at least 2,000,000 barrels of this was lost—one writer estimates 4,000,000 barrels lost.[13]

Such a spectacular well of course produced intense excitement. Land was bought in the vicinity for $1,000 per acre, and one tract of 20 acres was reported to have been sold for $200,000—an average of $10,-000 per acre. New companies by the dozen were organized. Some of them only sold stock, but some of them bored wells, and a great number of other gushers were brought in, with initial production of 1,000 barrels or more. Over 25 such wells were drilled in 1910, and over 100 such wells were opened in the Sunset-Midway field alone between 1910 and 1916. In 1914 another well in this field got beyond control, from May 10 to October 25, flowing at times as much as 50,000 barrels a day, the greater part of the product being stored in the old sumps of its world-famous predecessor, Lakeview No. 1.[14]

But the Sunset-Midway field was not the only performer on the stage. About the time that the Lakeview gusher was opened, oil was discovered in the Brea Cañon field, near Los Angeles, and in the Elk Hills and Lost Hills fields, a few miles north of the Sunset-Midway. The following year the Belridge field was discovered, in the same vicinity, followed by the Manel Minor or North Belridge and North McKittrick, and by a smaller field in the Simi Valley, in Ventura County.[15]

During all these years—from 1900 to 1912 or later—over-production was chronic, and prices never rose much higher than 50 cents a barrel. Organizations of producers spent hundreds of thousands of dollars trying to stimulate consumption, and succeeded in selling millions of barrels for all kind of low uses that no intelligent public would have permitted. Millions of barrels were wasted every year, wasted outright. Estimates of the waste from the inadequate storage available run from 4,000,000 barrels to 8,000,000 barrels per year. Mark Requa estimated the loss in 1910 at 4,000,000 barrels, and "possibly double that amount."[16] It was not until 1915 that operators succeeded in getting together effectively enough to secure any restriction of drilling.

The oil production of California reached a maximum of nearly 100,-000,000 barrels in 1914, but it was several years before this was again reached. Water had been creeping into the Kern River field for over a decade, and for several years had been more or less of a nuisance in several other fields, including the Sunset-Midway.[17] Operators were gradually learning how essential it was to the prosperity of all that water should be kept out of the oil sands, and urged legislation and state assistance in meeting the danger. In the meantime the production of some of the fields was greatly curtailed, and for several years there were no great discoveries to offset this. During the Great War, in spite of important discoveries in the Montebello and Richfield districts, and increasing production from the Elk Hills and a few other fields, there was much talk of scarcity, and much concerted effort to stimulate production. Even the geologists' predictions of approaching exhaustion were received with some respect. It hardly seemed likely that disastrous over-production would again disturb the California industry.

In 1920 and 1921, however, three great fields were discovered near Los Angeles, and there followed one of the most disastrous periods of over-production in the entire history of the oil industry. For a year or two, these great fields so dominated the oil industry that their history is almost the history of the oil industry of the entire country. This dramatic episode will be treated below.[18]

Many of the iniquities of our national oil policy have been illustrated in California history. First, there was the tendency to over-production, where offset wells are necessary to protect holders of leases, regardless of the demands for oil, or the price at which it could be sold. This was conspicuous in railroad lands, where each alternate section was controlled by the railroad, and where it was difficult to get a large enough compact holding to make conservative handling possible. In the second place, there was the consumption of oil for low uses, particularly for fuel. In the third place, there was the immense waste of oil; and, finally, there was the constant struggle of independents against monopoly control. Speculation was about the same as found in most oil regions. The difficulties encountered in handling the oil lands on the public domain will be treated in later chapters.[19]

NOTES

1. See Reports of the Cal. State Mineralogist, 1884 and 1887; *Bancroft's Works,* Vol. 24, 661; and *Overland Monthly,* 2d Series, Vol. 66, 352.

2. Cal. State Mining Bu., *Bul. 32,* 9.

3. Redpath, *Petroleum in California,* Los Angeles, 1900.

4. *Eng. and Min. Jour.,* June 20, 1896, 588.

5. Bowles, *Our New West,* 1869.

6. For an interesting account of the life of Doheny, see *Sunset,* Aug., 1910, 173.

7. Cal. State Mining Bu., *Bul. 32; Bul. 63,* 195-; *Sunset,* Aug., 1910, 173; *Good Words,* May, 1903, 327; U. S. Geol. Survey, *Bul. 309.*

8. U. S. Geol. Survey, *Bul. 321,* 16-; *Good Words,* May, 1903, 327.

9. U. S. Geol. Survey, *Bul. 398.*

10. *Pac. Oil Reporter,* Nov. 21, 1903, 5.

11. *Sunset,* Aug., 1910, 173.

12. *Mineral Resources,* 1906, 877; 1911, 335, 426; Cal. State Mining Bu., *Bul. 32.*

13. *Sci. Am. N. S.,* May 21, 1910, 419.

14. *Collier's,* May 14, 1910, 20.

15. U. S. Geol. Survey, *Bul. 721,* 35.

16. *Mineral Industry,* 1911, 561.

17. For a discussion of the water invasion of California oil fields, see the recent bulletins of the California State Mining Bureau.

18. See Ch. XII. Among the important fields recently developed in California are the Torrance, Inglewood, Dominguez, and Rosecrans fields, all in the Los Angeles Basin. None of these have attained the importance of the three great major fields mentioned, yet they have been important factors in the California oil situation. The Inglewood field alone was recently (July, 1925) producing nearly 100,000 barrels daily.

19. See Chs. XXI-XXV.

CHAPTER X

OIL EXPLOITATION IN WYOMING AND THE ROCKY MOUNTAIN STATES

WYOMING

WYOMING is so largely public land that her oil history is closely related to the history of the public lands. The history of Wyoming oil development is a rather long story. Oil seeps were noticed here early in the nineteenth century. The Indians are said to have used some of this petroleum for medicinal purposes, and some oil was later sold to the early travellers passing through. Bonneville discovered oil along the Little Popo Agie River in 1833, but previous to 1867 the oil spring was unknown except to the hunters and trappers who frequented the locality to get oil for medicinal purposes. In 1834 Zenas Leonard arrived at Popoasia Creek, where he found "an oil spring, rising out of the earth, similar to any other spring." The oil spring near Hilliard was doubtless well known to trappers who first built the trading post of Fort Bridger, but perhaps the first published account was the result of an examination made by the Mormons in their pioneer journey to Salt Lake in 1847. W. Clayton, who accompanied this expedition, published in 1848 a little book, the *Mormons' Guide Book*, containing the following:

About a mile from this place (the crossing of the road over Sulphur Creek) in a southwest course is a "Tar" or "Oil" spring, covering a surface of several rods of ground. A wagon trail runs within a short distance of it. It is situated in a small hollow, on the left of the wagon trail, at a point where the trail rises to a higher bench of land. When the oil can be obtained free from sand, it is useful to oil wagons. It gives a nice polish to gun stocks, and has proved to be highly beneficial when applied to sores on horses, cattle, etc.

The spring is mentioned by several observers, and it is reported that Brigham Young caused a shallow well to be dug at this point; in fact, the locality is now known as the Brigham Young oil well or spring. The oil was skimmed off the water in this well and sold to emigrants and carried in small quantities to Salt Lake City.

In 1859 oil was reported in Lincoln County. As early as 1863 oil was taken from a spring near Poison Spider Creek and sold for wagon grease to emigrants on the Mormon trail. In 1864 an oil spring was found in the Sage Creek field by a United States cavalryman. In 1867 Judge C. M. White began operations at a spring now known as the

White oil spring. He dug a large hole and skimmed off the oil, which he sent to Salt Lake and sold to tanners. He even did some drilling, but without results. The Carter oil spring locality became important in 1868, when workers driving a tunnel for coal found a slight flow of petroleum (8 to 10 gallons a day). This oil was sold to the Union Pacific Railroad and to neighboring coal mines for lubricating purposes. It is said that Judge Carter, then sutler at Fort Bridger, attempted to drill a well here, but finally abandoned the attempt. Several early writers mentioned various oil springs in Wyoming. The United States Geological Survey reported oil along the Little Popo Agie and Little Muddy rivers in the late seventies, and some borings were made along the latter stream in 1877.[1]

There was considerable activity in Wyoming in the eighties. The first wells in the Shannon pool, at the north end of the Salt Creek field, were drilled in the early eighties. The first gushers in the Dallas field were drilled in 1883, and about the same time oil was found along Powder River. The Bonanza oil seep, in the Big Horn Basin, and the Pilot Butte field were discovered not long afterward, but never attained any importance. Dr. George B. Graff, of Omaha, had become interested in the oil possibilities of Wyoming in 1876, and a few years later he and his associates purchased a number of claims and did some surface work. In 1883 and 1884 they drilled several wells in the Lander field, and secured a small amount of oil, which was hauled in iron barrels to the Union Pacific Railroad for shipment. The competition of eastern oil presently stopped this traffic.

About this time many operators became enthusiastic about the oil possibilities of Wyoming, and wells were sunk in various regions: along the Powder River, along Poison Spider Creek, in the Big Horn Basin, and near Moorcroft, in northeastern Wyoming. Most of this drilling was unsuccessful, although a small amount of oil was found near Moorcroft and was transported to Black Hills towns, where it was sold as lubricating oil for $28 per barrel.

In 1889 D. M. Shannon, who had been interested in Wyoming oil possibilities for several years, drilled a producing well in what came to be known as the Shannon pool, in the northern part of the Salt Creek field (the Salt Creek field includes three pools: the Shannon, Salt Creek, and Teapot Dome pools). Shannon acquired locations on a total of about 100,000 acres of land, and drilled two wells not far from Casper, but secured little oil.

During the late eighties Wyoming was getting a very bad reputation among oil men, for many fortunes had been lost and still there was no important production. Some European capital—English, Belgian, Dutch, and French—was coming into the state, however, and it is to this foreign capital that the development of the Shannon pool was due.[2] In the nineties there was considerable drilling in what are sometimes called the Douglas, Brenning, Belle Fourche, and Baxter Basin "fields," but the smallness of the production may be judged by the fact that the price of oil in Casper in 1897 was $8 per barrel.[3]

It has often been said that the oil industry in Wyoming, previous to the year 1900, and to some extent even later, was largely in the hands of inexperienced oil men, who knew little about the business; and many of the men in the field were promoters, who organized companies, sold stock, filed oil claims on the public lands, and perhaps did a little shallow drilling. Professor W. C. Knight wrote in 1901:

Over a quarter of a century has passed since the oil fields of Wyoming commenced to attract attention. Companies have been organized by the score, and, had their represented capital been sold at par for cash, there would have been sufficient money realized to purchase the state of Wyoming, and part of Colorado.[4]

Perhaps the year 1900 may be said to mark the beginning of important oil production in Wyoming. The Union Pacific Railroad, in drilling a water well at Spring Valley, found traces of oil. A great deal of excitement ensued, with the filing of claims on thousands of acres of land in the vicinity. In some cases promising land was filed upon as many as three times, and prominent oil men entered the field, but no important developments resulted. Discoveries reported in several other parts of the state proved of no importance.

The early history of the Salt Creek and Shannon pools has been noted. An English syndicate reported some production from the Shannon pool in 1902. Not long afterward a man by the name of Lobdell purchased the Shannon claims of some 200,000 acres, and the refinery at Casper, for the Société Belge-Américaine des Pétroles du Wyoming —a Wyoming company backed by French and Belgian capital. Lobdell afterward purchased what had been known as the Cy-Iba claims on the Salt Creek dome for a Dutch corporation known as the Petroleum Maatschappij Salt Creek of Wyoming. These claims were later contested by the government, and were the subject of extended litigation.[5] In the development of this field a great many gross frauds were perpetrated on investors, particularly on foreign investors, who were unfamiliar with our land laws.[6]

In 1909 all public lands in Wyoming thought to be valuable for oil were withdrawn. This stopped some of the frauds, but it also stopped some development, and during the next decade, much of the energy of Wyoming oil operators was devoted to violent criticism, not always disinterested, of the policy of reservation, and to efforts to get the oil lands opened again. For a year or two before the government withdrawal order was issued, there had been considerable activity in Wyoming, and discoveries had been made in the Byron field, northwest of Greybull, in the Dallas field, in Fremont County, and in the Salt Creek field. The government withdrawal order retarded the exploitation of the Wyoming fields, but it did not stop it, and in 1912 a gusher in the Salt Creek field ushered in a period of rapid growth. Before the close of the year, 59 producing wells and 30 dry holes were drilled in the various fields of the state, and the small annual production of 186,000 barrels in 1911 rose to 1,500,000 barrels in 1912. The following years saw a steadily increasing production, to a' total of nearly 9,000,000 barrels in 1917, and over 12,500,000 barrels in 1918; in fact, the production increased so rapidly that the operators and refineries found great difficulty in handling the product, in spite of the building of a number of new pipe lines. Over-production became chronic, with prices running as low as 47 cents a barrel. At the end of 1922 the Salt Creek field had about 500 wells, with a daily production of about 80,000 barrels, and a potential production of more than twice as much.

This rapid increase in Wyoming production was largely the result of the exploitation of the great Salt Creek field, but many other fields were opened up, in various parts of the state. In 1913 there was extensive drilling near Basin, in Big Horn County. In 1914 the discovery well in the Grass Creek field, in Hot Springs County, was opened, and the next year this field took rank next to Salt Creek. In 1915 the Elk Basin field, near the Montana line, was opened with a 400-barrel well, and a large number of wells were soon drilled here, and across the line in Montana. The following year the Lost Soldier field was discovered, in the extreme northeast corner of Sweetwater County, and the Big Muddy field, 15 miles east of Casper. About this time also the Lance Creek field, in Niobrara County, was opened with several large wells, one of them a 2,000-barrel producer. Within a year 30 wells were completed or drilling in the Big Muddy field, and within six years 330 producing wells had been drilled here. Such discoveries as Big Muddy and Lance Creek created great excitement, coming at the time of war prices for crude, and led to much speculation and much activity in the acquisition of surrounding public lands.

During 1917 the geologists of many large oil companies investigated Wyoming carefully, and immense sums of money were invested in oil lands here. Every acre of land that was not withdrawn and showed reasonable promise of oil was leased or bought, and within the next few years several new fields were discovered, including the Maverick Springs field, in Fremont County. The next year the Rock Creek field was discovered, in the Laramie Plains, in the extreme eastern part of Carbon County, and this field was exploited rapidly. Within the next few years a number of fields and "prospects" were discovered in rapid succession: Poison Spider, Ferris, Pine Mountain, Hamilton Dome, Mule Creek,

Bolton Creek, Osage, Sheldon Dome, and the much-discussed Teapot Dome. Poison Spider, Ferris, Hamilton Dome, Mule Creek, Osage, and Teapot Dome have developed into producing fields.[7]

The depression of 1920 and 1921 brought a very severe reaction in the price of oil, and for some time the burning question was how to keep down the production. The situation never became so acute here as in Oklahoma and California, partly because the Wyoming oil industry is largely in the hands of strong companies, partly because the government withdrawal order for over a decade put a decided check upon drilling, and partly because fewer great fields have been discovered here than in Oklahoma or California. Most of the Wyoming fields were for many years very disappointing to operators, and many fortunes have been lost, trying to find oil that never existed. For instance, in the Powder River field, known as a possible oil field since the early eighties, a well was drilled in 1886, several in 1893, one in 1896, 1,200 feet deep, two in 1902, and a big well seven feet square in 1903, and others in 1904 and 1907, and none of them successful.[8]

MONTANA

Montana has been the field of considerable prospecting, although the oil production of the state is still small. As early as 1893 some gas was obtained in the Havre field, and a little oil prospecting was done in the Stillwater Basin previous to 1908, but only losses were recorded in these attempts. The discovery of oil in Montana should probably be dated in 1915, when a Thermopolis syndicate found oil in the Elk Basin district, which lies partly in Wyoming and partly in Montana. There was some drilling along the Sun River about this time, and in 1916 Montana produced a few thousand barrels of oil. A considerable number of wells were drilled during the next few years. Five deep wells were drilled at Havre and Chinook, and one at Fort Assiniboine, several deep wells between Great Falls and the Sweetgrass Hills, several in the vicinity of the West Butte of the Sweetgrass Hills, several in Chouteau County, and a number of wells in other regions, but almost all of them were dry or gave only "showings" of oil.

The oil development of Montana really had its inception in 1919, in the strike of the Van Duzen Oil Company in the Devil's Basin territory, about 15 miles north of Roundup. This well flowed for a short time, but, upon subsequent deeper drilling, water replaced the oil. Other operations started immediately in the Devil's Basin field, and a few wells developed commercial production, but the oil was low in gravity and gasoline content and was used mainly for fuel.

In February, 1920, the Franz Corporation struck some oil near the

Musselshell River, and about the same time a high quality of oil was found in the Elk Basin, along the Wyoming line. The most important discovery, however, was in the Cat Creek field, where several large wells were brought in, wells of 2,000 barrels production. Of course such wells attracted wide attention and led to a characteristic oil fever epidemic. Nearly 100 wells were drilled here in the course of a year or two. Some drillers in the Cat Creek field were said to have paid $1 a barrel for water with which to drill wells for the production of $1 oil.[9]

About the time of the Cat Creek discovery a 600-barrel well was brought in on the Soap Creek Structure of the Big Horn Indian Reservation, but it was a low grade of oil, suitable only for fuel use. In March, 1922, the Kevin-Sunburst field, in Toole County, northern Montana, was discovered, and, within a little more than a year, over a hundred wells had been drilled and fifty-eight more were going down. In July, 1923, the field was producing 3,000 barrels a day. One geologist estimated that 500 test wells had been drilled in Montana, previous to August, 1923.[10]

While Montana is producing only modest quantities of oil, the state may some day become a large producer, for it is largely untested. There is a possibility that it may hold valuable deposits of oil shale, although the shales heretofore found and tested have not been valuable.

COLORADO

Colorado has never been an important oil producer, but the state has at times been the scene of considerable excitement and activity. Zebulon Pike, in his expedition in 1806, wrote of "Oil Creek" in Colorado, a branch of the Colorado, which must have shown signs of oil.[11] As early as 1860—only a year after the discovery of the Drake well in Pennsylvania—prospecting was being carried on in Oil Creek Valley, near what is now Cañon City. Between 1862 and 1865 the owner of an oil spring in Oil Creek Cañon hauled small quantities of oil to Pueblo, Denver, and Santa Fé, where he realized as high as $5 a gallon for refined oil. A few years later several shafts were sunk, and a small amount of oil was secured, which was hauled in wagons to Denver, and sold for a high price. The industry soon died out, however, and no further steps were taken toward development until the early eighties, when the attention of Pennsylvania oil men was directed to Colorado. The Land Investment Coal & Oil Mining Company purchased about 14,000 acres in the vicinity of Cañon City, bored two wells, and got from them both about three barrels of oil a day. This company spent about $20,000 in drilling, and later failed and was merged into another company. Still another company bored near Cañon City and got about a barrel a day. This success led the company to contract for the drilling of ten more wells, the first of which produced about five barrels a day. There was some other oil prospecting in the state about this time, but no significant results accrued.

In 1888 the field at Florence, near Cañon City, was assuming some importance in the local market, the production running above 400 barrels a day. One peculiarity that was early noticed in Colorado was that the wells, instead of decreasing in flow after a while, often actually increased with age, in some instances even after a life of several years. The life of the Florence wells was notably long at first, perhaps because there were so few wells tapping the reserves.

In 1902 oil was discovered near DeBeque, and considerable excitement and some drilling ensued, but very little oil was secured. The Rangely field, in Rio Blanco County, in the northwestern part of Colorado, was exploited early in the history of the state, but at the present time this field is yielding only about 100 barrels a week. In 1913 the successful oil exploration in Wyoming led to considerable interest in Colorado prospects, but no oil was secured, and the production of the state has dwindled since 1905 to insignificant proportions.

The Florence field, the only important oil field in Colorado, has always been in the hands of a very few companies. In 1911 it was said to be almost entirely controlled by the United Oil Company of Denver, a Standard subsidiary, which helped independent drillers by lending them tools and rigs and then paying them $1 a barrel for the oil produced.

The Boulder oil field attracted considerable interest in 1905 and for several years thereafter, and in 1908 a 250-barrel gusher was brought in. This, of course, aroused a great deal of enthusiasm, but no great further discoveries resulted. The total production of this field in 1917 was only 6,000 barrels, and in January, 1923, less than a dozen wells were producing only about 15 barrels per day.

In recent years there has been considerable drilling in southwestern Colorado. Thousands of acres of oil claims were taken up at the Durango land office in 1921, and a number of wells were drilled. Much of this drilling has been very expensive, since it was done far from any railroad, and supplies had to be hauled many miles over execrable roads, but it takes worse obstacles to stop the American oil driller. The recent discovery at Farmington, New Mexico, just across the line from Durango, led to considerable speculative excitement, even to an ambitious town-site proposition fathered by Durango men.

More recently, in January, 1924, there has been an oil strike in Moffat County, in northwestern Colorado, but it is not yet certain how important this will prove. The discovery well was reported at 400 barrels a day, but has more recently been said to be producing 4,000 barrels.

UTAH

Utah, like Colorado, presents a story of persistent efforts with small results. The early prospectors found oil seeps and recognized their meaning, but did not turn from their quest for gold. Signs of the existence of oil were reported along the line of the newly completed Union Pacific Railroad, in Summit County, Utah, in 1869; and it was claimed that samples of the oil compared favorably with the Pennsylvania product.[12] In 1882 E. L. Goodridge, attracted by indications of oil, located the first oil claim in the San Juan field, in southeastern Utah, but did nothing further. The Census report of 1885 mentioned "tar springs" in southern Utah. In 1888 considerable interest was evinced in the oil indications found near Green River, in Grand County, and plans were made for development here, but nothing was done until 1891, when a dry well was drilled. About ten years later increasing demand for oil led to further tests in the Green River region, but only traces of oil were discovered. Ten years later, in 1910, another spurt of activity led to the drilling of seven more dry holes.

In 1907 a small oil production was secured from a well near Virgin City, in southwestern Utah, and a great rush of hopeful speculators ensued, but in January, 1923, there were only about 10 producing wells in the field. In 1907 drilling also began in the San Juan field, in another southern county of the state; and the next year high-grade oil under artesian pressure was struck at 225 feet. Considerable excitement was aroused, and operators hurried into the new field. About 27 wells were driven, and nine of them proved moderately productive. The oil was of good quality, but the yield was so small that drilling presently ceased. Many of the holes drilled here during this period were only prospect holes, put down to validate claims and not with any serious intention of determining the resources of the field. In 1916 oil leases were reported much in demand in the vicinity of Manti. The Arcola well, in the San Juan field, is claimed to have produced 17,000 barrels of oil in 1917, but it was practically all used as fuel for drilling operations. After the close of the Great War, there were a number of oil booms in various parts of the state, and in August, 1921, the Land Office at Washington reported more applications for prospecting permits in Utah than in any other state. Permits covering more than 200,000 acres have been granted in eastern and southeastern Utah, and several companies were operating in the San Juan field in 1921.

In general, the oil history of Utah has been one of disappointed hopes. One reason for the difficulty in exploration was the fact that the oil fields were arid and isolated, and prices exorbitantly high. In the San Juan field, hay was said to cost $70 a ton, and oats $70 to

$100 a ton, and other things in proportion. With prices and wages on such a level, prospecting was an expensive matter.

NEW MEXICO

"Tar springs" in New Mexico were mentioned by the United States Census report in 1885. Somewhat later, traces of oil were found in a number of water wells near Dayton, New Mexico, and in 1909 and 1910 there was much excitement in this district. About this time oil was discovered in the Seven Lakes field, by a man who was drilling for water. Great excitement prevailed for a while and some 3,000 land claims were filed in 20 nearby townships. Drilling proceeded with vigor for a short time and oil was found, but not enough to make development profitable. In 1913 the field had been practically abandoned.

Little was done in the state for nearly a decade, although the United States Geological Survey reported in 1920 that about fifty wells had been started in the state. Many of them were abandoned before they reached any considerable depth. In October, 1922, however, the Midwest Oil Company brought in a 350-barrel well near Farmington, in the San Juan region in the northwestern part of the state, in the Navajo Indian Reservation. A great deal of excitement followed, with some wild rumors as to the production of the discovery well, and as to vast reserves of the new region. Some 200,000 acres of land were reported to have been leased in New Mexico within a few months after this discovery. In August, 1923, another well in this field was reported to have started at between 400 and 800 barrels a day of extremely light oil—61 degree gravity.

More recently a field of uncertain proportions has been discovered near Artesia, Eddy County, in the southern part of the state. A pipe line has been constructed to the Santa Fé Railroad, 17 miles distant, and plans have been made for refining the oil at Roswell and El Paso. Reports indicate considerable activity in the acquisition of leases in this field.

NOTES

1. *U. S. Geol. Survey of the Territories of Idaho and Wyoming*, Hayden, 1877, p. 151.

2. *Eng. and Min. Jour.*, May 3, 1890, 505.

3. For an interesting description of the early transportation of oil in the Salt Creek region, see the *Inland Oil Index*, Sept. 8, 1923, 3.

4. *Bancroft's Works* (1889), Vol. 25, 804; U. S. Geol. Survey, *Bul. 285, 342; 541, 49*; Univ. of Wyo., School of Mines, *Bul. 4* (1901), p. 4; *Eng. and Min. Jour.*, May 24, 1902, 720.

5. U. S. Geol. Survey, *Bul. 670*.

6. *Eng. and Min. Jour.*, June 22, 1912, 1229; *Oil and Gas Jour.*, Nov. 16, 1916, 26.

7. U. S. Geol. Survey, *Bulletins 711 D, 711 H, 716 B; Oil Weekly*, July 24, 1925, 31.

8. U. S. Geol. Survey, *Bul. 471*, 56.

9. *Oil and Gas Jour.*, March 26, 1920, 46; April 16, 1920, 70; *Wyo. Oil News*, Aug. 6, 1921.

10. *Oil Weekly*, Jan. 27, 1923, 9; *Natl. Petroleum News*, Aug. 22, 1923, 34, 78.

11. Coues, *The Expedition of Zebulon Pike*, Vol. II, 464, 465.

12. Derrick, *Handbook of Petroleum*, June 24, 1869, 114.

CHAPTER XI

DISAPPOINTING FIELDS OF EXPLORATION

SOUTH DAKOTA

SOUTH DAKOTA has long been suspected of having oil, because of its proximity to Wyoming and Montana, but no oil has been discovered in paying quantities. Late in the year 1920 considerable excitement arose in Lemmon, with the drilling of a "mystery well," which many people believed a successful well. This well was drilled by a group of farmers in the Lemmon Basin, and when down about 1,400 feet, was fenced in and guarded, but whether successful or not was not divulged.

WASHINGTON

Washington, like New Mexico and Utah, has been a disappointing field for oil prospectors. High-grade paraffin oil is reported to have been discovered in the western part of the Olympic Peninsula as early as 1881. Over thirty years ago oil was reported to have been discovered on the beach in the vicinity of Copalis Rock, which is about three miles north of Copalis. Persons walking along the beach had noticed an offensive odor, and attributed it to escaping gas. In 1901 the Olympic Oil Company was organized and began drilling, struck gas and a small amount of oil, but nothing worth while. Other wells were drilled in various places, but "indications" only were found, and of course there is no market for "indications." In 1916 there was a considerable oil boom in western Washington, but no results accrued.

ARIZONA

The Census report of 1885 mentions "tar springs" in Arizona, as in many other western states, and almost every part of the state has had an oil boom at some time. As early as 1900 oil wells were drilled near Phoenix, at Riverside, in the Gila Valley, in the Verde Valley near Camp Verde, and on the plain north of Agua Caliente. In 1919 there were a number of centers of oil fever: Holbrook, Snowflake, St. Johns, and Tucson. At Tucson thousands of acres of oil claims were entered, from the Cienega section and eastward. There was also much activity in the Rialto Valley and in the Santa Cruz Valley, but thus far all Arizona activity has been fruitless.

NEVADA

Even Nevada has been tested out somewhat. In the region around Reno, a number of claims were taken up in 1907 and in 1908, and a number of oil companies were formed. Some drilling was done, but results were negative and Nevada still remains unproved.

ALABAMA

There was considerable prospecting and even some drilling in Alabama a few years after the Drake well was opened in Pennsylvania, and a little oil was found. In 1884 or 1885 a well drilled near Bladon Springs brought a small flow of gas. In 1890 a well completed in Lawrence County was said to have produced 25 barrels of oil per day. After the spectacular opening of the Spindletop pool in Texas in 1901, there was considerable prospecting of similar geological formations—"salt domes," they were called—all along the Gulf coast, and several wells were drilled in Alabama, but no oil was found.

MISSOURI

Oil and gas discoveries in the eastern part of Kansas led to considerable wildcatting in Missouri, particularly along the Kansas border, but the discoveries were unimportant. A small amount was secured in Bates County in the early nineties, and was sold for lubricating oil at a very high price; and some commercial production was secured in Jackson County, but the total production of the state is relatively unimportant, and is not recorded separately in the annual reports of the United States Geological Survey.

OTHER STATES

It is unnecessary, however, to go into the details of unsuccessful exploration in all the various states. Not only the states mentioned above, but many other states as well have been the scene of more or less wildcatting: Nebraska, Iowa, Wisconsin, Mississippi, Georgia, Florida, the Carolinas, and Virginia. It is likely, in fact, that if records could be obtained covering all the oil explorations that have been carried on, even in states that are never thought of in connection with oil, those records would make a surprisingly long story. The wildcatter has gone everywhere that a hope of discovery was offered, undaunted by failures and disappointments, heedless of the heavy odds against him. If this were a story of human pertinacity, instead of a history of oil development, the story would have to be much longer.

CHAPTER XII

THE OVER-PRODUCTION SPREE
OF 1921-1923

THE history of the oil industry during the troubled years from 1921 to 1923, with the rise of the three great oil fields of southern California to an unprecedented flush production, and with the consequent chaotic conditions in the oil industry all over the United States, and for that matter all over the world, constitutes an interesting and dramatic story, yet a story that thoughtful Americans will follow with little pride.

During the Great War, and for a year or two afterward, it seemed unlikely that over-production could ever appear again. Oil was scarce and prices were high, and the impending "oil famine" was a common topic of conversation. Late in the year 1920, however, a new field was opened, the first of several great fields that were destined to mark out an epoch in oil history. The first well in the Huntington Beach field was completed in May (some accounts state that it was in October), 1920, and a few months later two good wells came in, one of them producing 6,000 barrels a day. Very soon one of the most unfortunate speculative booms in the history of the state was on foot. Every available lot was leased, and even the beach was taken up under the Placer Act, for miles up and down the coast. The leases drawn during the cheerful days of the boom generally called for prompt drilling of offset wells, and thus a tremendous campaign of drilling was inaugurated. In January, 1923, there were 200 wells in the field, with a production of over 80,000 barrels a day.[1] Many of the oil companies were poorly financed, and of course many were organized only to sell stock. This situation later improved, as the early promotion companies were replaced by substantial interests.[2]

The Huntington Beach stampede was just about well started when another, a more remarkable field, was discovered not far away. The geological structure at Signal Hill had long been known as an oil possibility, and oil had been noticed in some of the wells in Long Beach at least as early as 1912. Lands on Signal Hill had been leased for speculative purposes from time to time, but no deep drilling was done until 1916. The first well was dry, and this gave the field a bad name. The fact that the area was divided into small tracts and town lots made it unattractive to wildcatters anyhow. In March, 1921, however, the Shell Company spudded another well, and in June struck oil at 3,000 feet. This well produced 500 barrels a day, of 21-degree gravity oil.

This discovery immediately started a wild scramble for leases and for drilling apparatus—for what is popularly called "development" in the oil fields. Drilling rigs were quickly obtained from Huntington Beach, and many small operators who had gone into the Huntington Beach field too late to secure satisfactory acreage moved to Signal Hill and began drilling there. One of the maddest drilling campaigns in the history of the state resulted. Before the end of the year four more wells were producing, and the next May, 21 wells were producing 11,000 barrels a day. For some time thereafter an average of four wells per week were completed, and production leaped to 95,000 barrels a day in September—from only 205 acres of land. Signal Hill was covered thick with derricks, and new wells were crowding down farther into the city. A production of 105,000 barrels daily in January, 1923, was raised to 140,000 barrels a month later, 210,000 barrels six months later, and 258,000 barrels in October. At the end of the second year of "development," in June, 1923, there were 200 operators in the field, with 600 rigs, and 350 wells drilling. For several months the daily production hung above 250,000 barrels.

The city of Long Beach grew with astonishing rapidity. From a population of 55,000 in 1920, it increased to 65,000—the natives claimed 100,000—in 1922, with 1,600 real estate agents, an unnumbered horde of "unit" oil promoters, sharks, lawyers, and the necessary complement of dupes and suckers. For rapidity of exploitation, Long Beach, or Signal Hill, stands almost alone among the oil fields of the country.[3]

Still another great field was coming to the front, however. Santa Fé Springs history runs closely parallel to that of Signal Hill, although exploitation was somewhat less spectacular, and less rapid. This field, in some respects the greatest field that the United States has produced, was prospected more or less at least 25 years ago, but the first well resulted in a disastrous blowout, and for years the whole district was taboo. In 1907 the Union Oil Company started a well, and later started another one, but after several years abandoned both of them. In 1917 another was started and, after many difficulties, finished in 1919. This produced a small amount of oil, but attracted little attention until a second producer was brought in two years later. Many wells were started immediately, and in January, 1923, the field was producing 97,000 barrels a day. Four months later the output had doubled, and in July rose above 300,000 barrels, rose above the record established by the famous Cushing field of Oklahoma. In August it reached nearly 350,000 barrels a day. Only Cushing, Powell, and Smackover could compare with this, and these fields declined far more rapidly than Santa Fé Springs.

A most extraordinary speculative fever accompanied this exploita-

A GUSHER AT LONG BEACH

tion. Promoters ran free stages to the oil field, from points as far away as San Diego, with free chicken dinners to prospective buyers of stock who would make the trip. Leases were sold at fabulous prices—as high as $50,000 per acre, with royalties running from one-sixth to two-fifths.[4]

Just about the time that Huntington Beach, Long Beach, and Santa Fé Springs showed signs of having reached their peak of production, along came two more fields, with a threat of furnishing additional surplus oil, and so of increasing and prolonging the general disorganization. The Torrance field, in the Los Angeles Basin, was discovered early in August, 1922, and in spite of the deluge of oil from the three great neighboring fields, was exploited with such energy that at the end of a year it had 36 producing wells, with 47 drilling. Two months later there were 90 derricks in the field, and Torrance promised to become an additional argument against private ownership of oil deposits; but it never reached the importance predicted for it. Much the same must be said of the Compton field, in the same region, which failed to register important production, in spite of the energetic efforts of many enterprising drillers.[5]

The feverish exploitation of the three great California fields, with the resulting addition of several hundred thousand barrels per day to the previous production of the country, was disastrous. It broke the market repeatedly, ruined hundreds of producers in all parts of the country, forced the abandonment of thousands of wells, and threw the entire oil industry into such a state of utter distress as had not been seen in years.

During the year 1921 the California fields created no great alarm. Geologists were predicting that the Mexican fields would soon decline, and when, in 1922, this decline actually set in, there seemed to be a prospect for scarcity of oil, with high prices and prosperity for producers. California production, however, increased so rapidly that it completely outweighed the Mexican decline, and upset the domestic market in addition. In the month of January, 1922, California produced less than 10,000,000 barrels. In December the state produced 15,000,000 barrels, the increase coming largely from the three great southern fields. The total production of the state, which had reached a new high record at 114,000,000 barrels in 1921, rose 24,000,000 barrels higher, to 138,000,000 barrels, in 1922, and climbed more than 100,000,000 barrels, to 258,000,000 barrels in 1923.

With this vast increase in production, there was also a rise in the quality of the oil, for the new production in southern California was high-gravity oil, with a far higher gasoline content than most other California oils.

The wells drilled in the Los Angeles Basin were extraordinarily prolific, many of them running 10,000 barrels a day, and over. The average production per well was far higher than in any other section of the country. New wells in Santa Fé Springs at one time flowed an average of over 3,000 barrels a day, and those at Long Beach and Huntington Beach did nearly as well. A few new wells of that caliber raised production at a very rapid rate.

The gusher character of the wells, and the close town-lot drilling led many to believe that these fields would be short-lived, like the Burkburnett field of northern Texas; and predictions of approaching decline in production were made every week. Late in 1922 the *Standard Oil Bulletin* predicted that Long Beach would begin to decline soon, and in January there was considerable discussion of the appearance of salt water, the encroachment of edge water, the decrease of gas pressure, and the impending decline of the field; but in February production had risen to 140,000 barrels. It presently appeared that these oil fields were not like any others that had ever been known in the United States. There were several oil-bearing strata here—as many as five in some places—with a total thickness in some instances of from 2,000 to 3,000 feet. Such oil sands had never been known in the United States before, and the oil recovery per acre reached unprecedented figures. When wells ceased to produce, it was only necessary to drill into a deeper horizon, and in many instances the flow from the deeper strata was greater than from the shallower sands.

Here was the unfortunate spectacle of a vast over-production of oil, prices ruinously low, and yet thousands of wells being drilled all the time to help in the general waste and disorganization. At Long Beach, for instance, in August, 1922, after one of the most serious price reductions, and at a time when oil producers were talking about ruin and bankruptcy, new wells were being brought in at the average rate of two each day. At this time there were 612 wells going down in California, and there were 109 new rigs up and starting operations.[6]

Stocks of crude oil held in California grew by the millions of barrels, from 36,000,000 barrels in January, 1922, to 49,000,000 barrels in December. In the following April, stocks reached 66,000,000 barrels, and three months later 72,000,000 barrels. One writer summarized the situation in the Los Angeles Basin in March as follows: "14,000,000 barrels steel storage completed; 2,500,000 barrels steel storage building; 5,390,000 barrels steel storage proposed immediately; 9,280,000 barrels concrete reservoir storage completed; 6,950,000 barrels concrete reservoir storage building."[7] Some of the oil was shipped north to San Francisco, and piped into the old storage reservoirs of the San Joaquin Valley.

The larger oil companies threw millions of dollars into a vast storage campaign. In the latter part of July, The Pan American Petroleum and Transport Company was reported ready to open a new $18,000,000 refinery at Watson Junction, near Los Angeles harbor, with immense storage capacity for crude oil and for gasoline. One reservoir belonging to this company was calculated to hold 4,000,000 barrels of fuel oil. The Standard Oil Company was starting work on storage facilities for several million barrels, the General Petroleum Corporation was working on two $2,000,000 storage reservoirs, and the Union Oil Company and several smaller companies were working on storage as rapidly as possible. All this was to take the place of the storage in the oil sands, which was perfect in every mechanical detail, and cost nothing.[8]

Stocks of gasoline were also reaching unprecedented figures. In June, 1922, stocks in California had been 46,000,000 gallons. One year later they reached 142,000,000 gallons—200 per cent increase in one year.

With reservoirs full and production increasing, it seemed inevitable that prices must be cut again, but other means of keeping down production were tried first. In March and April, 1923, there was some talk of a curtailment campaign or some sort of proration, and a meeting of the producers in the southern fields was called in Los Angeles on April 27 to consider the question. There were over 300 producers at this meeting, and they voted, almost unanimously, to ask the marketing companies to cut all runs 30 per cent, regardless of existing contracts. Most of the marketing companies complied with this request, and some of the larger companies shut in as much as half of their own production. The Standard Oil Company, while refusing to be bound by any agreement, announced its sympathy with the movement, and stated that it was shutting in more than 40 per cent of its own production. Several hundred workmen were immediately dismissed in the southern fields, and it was estimated in April that over 200,000 barrels of daily production were shut in, mostly in the southern fields.[9] While these wells were shut in, hundreds of new wells were being drilled in the same territory.

The California situation in the early part of 1923 constituted a strange chapter in the history of a people who venture to call themselves not only civilized, but enlightened and intelligent. The spectacle of a vast over-production of this limited natural resource, growing stocks, overflowing tanks, and declining prices, frantic efforts to stimulate more low and unimportant uses, or to sell for next to nothing, passed dividends and bankruptcy for helpless producing companies; and yet dozens of new wells, and more oil, more oil, and more cries that it was bringing ruin to all in the industry! It was a case of "being choked, and strangled, and gagged, by the very thing most wanted—oil," as one of the oil journals expressed it; while an energetic and resourceful people stood by, as helpless as so many medieval peasants in the presence of the bubonic plague.

Before the end of January, 1923, Long Beach was doing 130,000 barrels per day, Huntington Beach about 90,000, and Santa Fé Springs over 90,000 barrels—a total for these three fields of over 300,000 barrels a day, and for the entire state a total of over 500,000 barrels a day, with 72,000 barrels shut in. The refineries of the state were capable of handling only about 365,000 barrels, and, while the larger companies were building storage on a vast scale, they were not building fast enough to care for the mounting production.[10]

Production continued to climb. The week ending May 5 showed, for the state, an increase of 15,000 barrels a day. The next week production declined 15,000 barrels, and the following week fell 10,000 barrels more, but the next week it filliped back to the record production of 710,000 barrels a day, during June climbed 120,000 barrels higher, and in July rose still higher. In July, Santa Fé Springs alone reached a daily production of over 300,000 barrels, and Long Beach over 200,000 barrels, while Huntington Beach was still producing over 100,000 barrels. These three fields alone, during July and August, 1923, produced more oil per day than Pennsylvania and New York and all the fields east of Oklahoma had ever produced at any time in their history—several times as much; and in the first week of September, Santa Fé Springs rose to 338,000 barrels, and Long Beach to 254,000 barrels.

Thus proration was not sufficiently effective. Near the end of May, the chairman of the executive committee having charge of the movement sent out a warning that available storage was almost completely filled, and that the runs would have to be cut more that 30 per cent; and early in July the committee called for a 45 per cent cut, which was promptly put into operation. The following week California production was up 30,000 barrels a day, and new wells were being started at the rate of three each day.

Early in August the proration committee called for a more drastic cut in runs. A few days afterward California production reached a new high record of 872,000 barrels daily. In the last week of August, 17 new wells were completed in the three big southern fields, with an initial production of 65,000 barrels. One oil journal described the situation briefly:

In the three pools—Santa Fe Springs, Long Beach, and Huntington Beach—production of 682,000 barrels a day is being obtained from 520 producing wells. There are 592 wells drilling in those three pools in areas that are regarded as certain of production.[11]

The flood of oil was more than could be handled properly, and more than the market could absorb at prevailing prices. In August, 1921, California prices had ranged from $1.10 to $2.45 per barrel, according to gravity. On July 15, 1922, prices were cut to a new schedule, ranging from 85 cents to \$2.20, and, on July 25, the higher grades were reduced again. On January 6 of the next year, and again on April 10, the higher grades were still further reduced, finally bringing prices down to from 60 cents to \$1.04 a barrel. It seemed that the bottom was finally reached, and when in September the production seemed to be declining, operators felt that they were at last on solid footing, when, on October 9, the Standard Oil Company announced that there was still a serious over-production of oil, necessitating another cut in the prices of high-gravity oil, to 76 cents for the highest grade. This new price schedule did not seriously affect the operators in the San Joaquin Valley, who produced mainly the lower grades of oil, and who had kept many of their wells shut in for months.

The price of gasoline was falling, along with the price of crude oil. The year 1923 opened with the Standard retail price 22 cents a gallon. On January 5, this was cut to 20 cents, in April to 19 cents, on August 1 to 17 cents, on August 23 to 15 cents, and in September to 13 cents. "Bootleg" gasoline was sold by some small retailers as low as 10 cents a gallon, and sales were even reported at 6 cents a gallon—about the price of distilled water.[12]

These reductions seemed for a long time powerless to check production. Many operators were bound by the terms of their leases to drill, regardless of conditions, and, in town-lot "development," operators usually do that in any event, so that price reductions had little effect, at least in the southern California fields. To the north, in the San Joaquin Valley, where wells were smaller and where holdings were larger, many wells were shut in during the summer of 1923. Early in October the Associated Oil Company announced that it was closing 479 wells in the Kern River field, for a period of from four to six months.

In spite of this, there seemed to be no limit to California oil production, and all means employed to keep it down had proved almost equally impotent. One of the favorite pastimes of operators, not only in California, but all over the United States, and all over the world, was to guess at the time when California should reach her "peak." Every week during 1923 it seemed that the peak must be near, but production continued to climb. New extensions were added, and deeper sands were found, which constantly staved off the decline that all were hoping for. Finally, a committee of producers was appointed to study the question of probable future of production, and early in September this committee reported that the peak of potential production had been passed, and that the actual production would probably soon begin to decline.

Unfortunately, California operators were not the only ones suffering from the flood of oil. Previous to 1923 California oil had had little influence on eastern or Mid-Continent conditions, because the California product had been mainly low-gravity fuel oil, and because the expense of shipment to the eastern markets was generally prohibitive. In spite of her great production of oil, California had generally been an importer of gasoline. In the latter months of 1922, however, facilities for transporting and storing the California oil were becoming hopelessly inadequate, and operators began to look earnestly for some outside market. Harbor facilities were enlarged, and a huge fleet of tankers was brought into service to haul the oil around to Gulf and Atlantic Coast markets. When eastern refiners could get high-gravity oil from Los Angeles for $1 per barrel, they bought it in preference to Mid-Continent crude at twice the price. In February, 1923, over 1,000,000 barrels of oil went eastward through the Panama Canal; in March nearly 3,000,000 barrels were sent, and in June over 6,000,000 barrels. In August 117 tankers were engaged in hauling oil to the eastern markets and to Europe. These eastern markets had for years taken much of the Mid-Continent production, and Mid-Continent producers suddenly awoke to find that their market was being taken by California oil.

Mid-Continent producers had for several years been suffering from an almost chronic over-production of oil anyhow. The industrial depression of 1921 hit the Mid-Continent oil industry a hard blow, for the price of crude oil tumbled from $3.50 per barrel in January to $1 per barrel in midsummer. Repeated cuts in the price of crude oil seemed to have no effect on the swelling production. While the price of oil declined from $3.50 a barrel to $1 a barrel, production increased over 1,000,000 barrels a month in Oklahoma alone, and nearly as much in Kansas, although North Texas production declined considerably. In the later months of the year, however, general business conditions began to improve, and oil prices rose somewhat, in spite of the large production. The year 1922 opened with Oklahoma-Kansas oil selling at $2 a barrel. In July, however, the price was cut to $1.50 a barrel, and in August to $1.25. In November, prices of Mid-Continent crude oil were put on a gravity basis, but the $1.25 (maintained for oil having a range in specific gravity of from 33 to 34.9 degrees) remained unchanged to the end of the year. Producers insisted that this price was far below cost of production, and many of them proved it by going into bankruptcy; but the old law of supply and demand remained in operation.

Meanwhile operators were casting about for some way to avert the general disaster their own enterprise had brought upon them. In the first place, a "shutdown" campaign was urged, and even initiated with some success in one or two fields, but in general the operators were unwilling to be bound or seriously handicapped in their zeal for "development"—development in this case meaning the extraction of dollar-and-a-quarter oil from the natural reservoirs in the earth, where it was perfectly safe, and where it could be stored without risk or expense, and running it into earthen or steel tanks, where all the waste and risk and cost of storage would begin to run against it. Stocks of crude oil increased rapidly, from 68,000,000 barrels in January to 90,000,000 barrels in September; and purchasing pipe lines and refineries became more and more reluctant to buy the current production, as they might well be, for the decline in the value of the crude stored in tanks throughout the country had cost the Standard companies alone nearly $100,-000,000. There was even talk of prorating the pipe-line runs, after the fashion made necessary in the period of Cushing over-production.[13] When these proposals proved futile, a "buy-a-tank" movement was inaugurated, several of the oil journals urging oil men to invest, either individually or through coöperative associations or private corporations, in storage tanks to hold the surplus of oil.[14] Finally, the operators in Oklahoma assembled and petitioned the Secretary of the Interior to stop operations on Indian lands. This petition was granted, but not long afterward the operators asked for the reversal of this order, and the Secretary again acceded to their wishes. In December, when oil production was higher than it had ever been, he announced another sale of Osage leases, as required by law. There was some talk of asking the federal government to intervene directly, as the government of Oklahoma had done in 1914, but this plan was not generally favored among operators.[15]

The federal government should at least have stopped the wildcatting on the public lands, but the Leasing Act did not provide for such action, and hundreds of permits were issued to oil seekers in the West; in fact, it was reported in September, 1922, that more permits were being issued than ever before. Thus the government was doing what it could to increase the general demoralization, by opening Indian lands and the public domain to prospecting and exploitation. And some cities were doing all they could to stimulate prospecting. One Kansas town offered $50,000 to any person who would bring in a 200-barrel well within ten miles of the town, and a New Mexico town offered half as much.[16]

Thus the over-production in the Mid-Continent, the "crime against the oil industry," continued, in spite of low prices, and in spite of all efforts at reduction. Oil journals and leaders in the industry called for some effective remedy for the situation, but none was forthcoming. One oil journal declared that the oil producers were committing "hari-kari" by producing so much oil. All saw the remedy, but would not adopt it. The remedy was, of course, a reduction in the production and restriction of drilling, but under our system of private ownership this was impossible.

Toward the end of the year 1922 conditions gradually changed for the better. In November, stocks of petroleum, which had been accumulating since May, 1920, began to decline, and Mid-Continent production even began to decline slightly, with the result that refiners and purchasing companies began to show signs of apprehension regarding their future supply of crude oil. The posted price was not raised during 1922, but late in the year various purchasers began to offer premiums for the higher grades. In December North Texas refiners were offering premiums of 25 cents to 30 cents a barrel; and on December 14 the Texas Pipe Line Company advanced the posted price of Mexia and Currie crude 10 cents a barrel. On the same day the Texas Company raised the price of various crudes from 25 cents to 35 cents a barrel. The next day the Magnolia Petroleum Company raised the price in southern Oklahoma, and the following day the Sinclair Crude Oil Purchasing Company met the Magnolia price. On December 23 the Ohio Oil Company advanced the price of Wyoming oil 30 to 35 cents, and, five days later the Humble Oil and Refining Company raised the price of Mexia and Currie crude 20 cents a barrel. Pennsylvania oils got a 25-cent increase on December 30; and this was always supposed to augur a coming increase in Mid-Continent prices.

Thus the year 1923 opened auspiciously for Mid-Continent oil producers. It is true that there were several hundred drillers in the field, doing their best to ward off any such thing as a reasonable price for oil, by drilling hundreds of new wells. Nevertheless, the outlook was favorable, and during January, February, and March there was a scramble for the light oil production, with numerous price advances. In January there were several advances in the price of Mid-Continent, Pennsylvania, Louisiana, Arkansas, and Wyoming oils (Wyoming production is generally of the same high gravity as that of the Mid-Continent fields, and is, therefore, valuable for its gasoline content); and in early February scarcely a day passed without a crude advance somewhere. There were six advances in the price of Louisiana crude within a month. The large purchasing and refining companies evidently feared that the long-heralded oil famine was approaching, for they bought and stored great quantities of crude oil, at constantly rising prices. In February the Prairie Oil and Gas Company was said to have in storage about 54,000,000 barrels, and was buying at the rate of 110,000 barrels daily. Sinclair companies were reported to have 30,000,000 barrels in storage.[17]

This general situation carried over into March, with several advances in the first half of the month, but suddenly the situation changed completely. During March Smackover production ran so high that on March 28 the Standard Oil Company cut the price 20 cents, but that was not very significant, because Smackover oil was mainly fuel oil. Something more significant was happening, and, almost before they knew what it was, the Mid-Continent producers and refiners were facing the most serious situation that they had faced in many years. California production was taking their eastern markets.

It was not California oil alone that caused grief to the Mid-Continent operators, for Oklahoma production presently began to increase at a tremendous rate, and rose rapidly during the very time when the California output was taking such prodigious leaps. In January Oklahoma production was 311,000 barrels a day, and in May this rose above 500,000 barrels a day. The Burbank and Tonkawa fields were being exploited rapidly, and were together producing over 200,000 barrels a day, while Mexia and Bristow were producing over half that much. The Mid-Continent fields were breaking all records, while Salt Creek was producing 100,000 barrels a day, and Smackover flowed more than that. To make matters worse, on April 5, the government opened up a new tract of Osage lands, in the vicinity of the Burbank field.

Operators began to talk again about curtailment, and even made some rather ineffectual efforts at curtailment. The Prairie Oil and Gas Company, with unprecedented pipe-line runs, and with storage almost full, issued a warning that unless production was reduced, a proration of runs must come, but this warning seemed wholly without effect, for in June, Burbank production was up to 115,000 barrels, and Tonkawa was running at about the same rate; and drillers were drilling wells by the hundred.

Storage for millions of barrels was erected. In July more than 10,-000,000 barrels of steel tankage was reported to be under construction or just completed in the Mid-Continent field. In other fields a similar movement was under way. In Wyoming the Sinclair Crude Oil Purchasing Company, early in August, reported the purchase of 60 tanks of 80,000 barrels capacity each, the Texas Company reported four of the same size, and the Producers and Refiners Corporation reported eight 55,000-barrel tanks.

This vast production was of course accompanied by a great deal of waste, especially in the Burbank field. For instance, a 2,800-barrel well belonging to Waite Phillips was completed before any proper storage was available, and the oil was run into earthen storage. When several thousand barrels had been stored, a heavy rain so weakened the retaining walls that the entire pool went coursing down a nearby creek. Instances similar to this were common in the Burbank and Tonkawa fields at this time.[18]

When all efforts at curtailment proved fruitless, and the warning issued by the Prairie Oil and Gas Company failed to bring any cessation of drilling, the Prairie Company, on June 26, put into effect a proration which for some producers ran as high as 50 per cent. Several other large purchasing companies announced a similar policy at the same time. The Prairie announced that it would connect with no new wells until production had declined to a point where the refineries could take the oil offered. The over-production situation in the Mid-Continent fields was so serious at this time that when a succession of heavy rains and floods in Oklahoma and Kansas fields in June destroyed millions of dollars worth of field equipment, refineries, etc., one of the prominent oil men, E. W. Marland, declared the floods a blessing just because they kept drillers out of the fields, and so helped to keep production down.[19]

The proration scheme of the Prairie and other companies seemed sufficiently drastic to reduce production somewhat, and during July there was a slight decline in Oklahoma and Kansas production, but during the same month a great increase in northern Texas production more than offset this, with a resulting increase of over 40,000 barrels a day in the total Mid-Continent figures. There seemed to be no way of stopping the enterprising drillers of the Mid-Continent fields.

The proration policy of the purchasing companies was said to have caused great damage to the oil fields in northern, central, and western Texas districts, since the wells could not be pumped sufficiently to keep the water out of the oil horizons. The owners of many of the wells were not financially able to erect storage and hold the oil until it could be sold.[20]

California oil was cutting deeper into the eastern markets all the time, and in July some of the pipe-line companies in the eastern fields started to prorate runs. Many of the smaller wells in the east were abandoned, because of the low price paid for oil, yet new wells were going down all the time. In the last week of July, 136 wells were completed in the eastern fields—an average of nearly 20 a day.

A peculiar situation gradually developed in the Mid-Continent fields. The large refining companies, finding it impossible to keep down production, made a strong effort to keep the price of gasoline up, until they should be able to get rid of their large accumulations of crude oil and gasoline. They had been storing immense quantities of oil, in the hope that production would presently decline and that prices would again go up. When their storage was full and production continued to increase, it was necessary for them either to cut crude prices drastically and write off a heavy inventory loss on the oil and gasoline they held in storage, or hold crude prices firm, and keep the price of gasoline as high as possible until they could work off their accumulations at a profit, or at any rate at the least possible loss. For some time many of them adhered to the second policy. The Prairie Oil and Gas Company consistently refused to cut crude prices, and, apparently by some sort of gentlemen's agreement which has not been made public, the refiners were for a while able to keep the price of gasoline at a remunerative level, or even, for some of them, a very profitable level.

The situation was quite too artificial however, and late in July the whole scheme blew up. The price of crude oil, although shamefully low, was too high to depress production, and high gasoline prices were causing much dissatisfaction among the jobbers and marketers, who thought their margin too small, and who found that high prices do not stimulate sales. At a meeting of the National Petroleum Marketers' Association in Chicago on July 26, there was a heated discussion of the question of crude oil and gasoline prices, the marketers demanding a reduction in both, while producers and refiners who were present at the meeting called attention to the situation they were in, and declared that a reduction of gasoline prices would cause serious loss to many refiners and purchasing companies, while a reduction in the prices of crude would ruin many producers and cause the abandonment of hundreds of small wells, thus inflicting irreparable injury on the oil industry. L. V. Nicholas, president of the Marketers' Association, had previously demanded of the Prairie Oil and Gas Company that it lower the price of crude oil and gasoline, and at the Chicago meeting he repeated this demand, and incidentally "spilled the beans" regarding some of the refiners' tactics in keeping prices up. In the meeting there was considerable discussion that was not meant for the public.

One of the results of the Chicago meeting was an agreement among many of the Mid-Continent refiners to shut down their plants for a while. Of course, this brought threats of prosecution from the attorneys-general of several states, as well as from the Federal District Attorney of Chicago.[21] Investigations were started by several different officials. The refiners were really having troubles enough without any government prosecution, because crude prices had not been reduced to correspond with the numerous cuts in gasoline prices, the Prairie Oil and Gas Company having consistently refused to reduce the price of crude oil; and gasoline was becoming cheap, and increasingly difficult to sell.

For weeks the Prairie Oil and Gas Company steadfastly refused to reduce the price of the crude oil. When it adopted a policy of prorating purchases, the producers who were unable to sell all their oil began to "bootleg" it around to independent refiners for about anything it would bring, in some instances 50 cents below the posted price. The law of supply and demand was out of joint. A cut in crude prices seemed certain to come. There seemed to be no end to the increase in California production, and California oil was becoming more and more firmly intrenched in the eastern markets.

Crude prices were not the first to be cut, however. The Chicago meeting of marketers had made the public very suspicious of the gasoline price structure, and early in August Governor McMaster of South Dakota announced that he had asked the State Supply Depot to begin selling gasoline to the public at 16 cents a gallon—a price about 6 cents lower than the regular service station price. The Standard Oil Company of Indiana promptly met this cut with a reduction of 6.6 cents, effective throughout its entire territory of some ten or a dozen states. This reduction was necessary, under the circumstances, but it was a very severe blow to many smaller independent companies that were unable to produce as cheaply as the Standard.[22]

Thus the Mid-Continent refiners found themselves in a worse position than ever. They must still buy at least a part of their crude oil at an artificially high price, while their eastern markets were partly gone, and the price of gasoline had fallen to a ruinous level. The extent of their demoralization may be judged by the fact that the stocks of some of them, even the best companies, fell 50 per cent or more in the New York exchanges, and many of the smaller ones were forced into bankruptcy, or into the outstretched hands of "Uncle John." The producers were in little better plight, for they could sell only a part of their production, and few of them were financially able to enter any extensive campaign of storage.

The Prairie Oil and Gas Company persistently refused to reduce crude prices, but, instead, narrowed its proration schedule, as did also several of the other large buyers in the Mid-Continent field; and for a while it seemed that proration was effective. During August, Oklahoma production steadily declined.[23]

Just at the time when Oklahoma production seemed well started on its decline, along came the Powell field in central Texas, with an increase to over 100,000 barrels a day, and within a few weeks to 300,000 barrels a day. Part of the Powell field was divided up into a large number of plots of from 10 acres to 40 acres each. For a while it had been possible to avoid starting any drilling campaign, but one of the landowners presently demanded that the company which had leased his land comply with the provisions requiring the drilling of a certain number of wells. The company was unwilling to start a disastrous drilling campaign at such a time of general demoralization, and offered a handsome bonus to the landlord to be relieved of its drilling requirements, but the landowner insisted and forced the drilling of the first well. Soon a number of offset wells were going down, and within a short time a drilling campaign was on which brought further demoralization to the oil industry.

Thus over-production, declining prices, proration, even bankruptcy of producing companies, all seemed impotent to stop the drill. The Oil Weekly of August 25 reported 604 wells shut down in Oklahoma, and 640 wells drilling.

As burdensome stocks of crude oil in storage increased, and "boot-legging" became more common, it became increasingly clear that the price of crude oil must be reduced. On September 1, the Magnolia Petroleum Company (a Standard subsidiary) announced a cut on all grades of oil. The following day the Humble Oil and Refining Company (also a Standard subsidiary) met the cut. The Standard Oil Company of Louisiana also cut prices of most grades of crude oil. Other purchasing companies followed, not only in the Mid-Continent, but in Wyoming and elsewhere, but the Prairie Oil and Gas only cut runs a little lower.

The unending flood of California oil continued, however, and finally, in the latter part of September, the Prairie Oil and Gas Company announced a severe cut in prices, bringing Mid-Continent oil down to from 90 cents to $1.75, according to gravity. This price was below the prices already fixed by other purchasers, and the other large companies immediately lowered their prices to meet it. This last reduction almost brought a sigh of relief from the Mid-Continent producers, for they at last felt that the bottom was reached and that they finally knew "how they stood." Many felt that they would really benefit from the reduction, because it would bring Mid-Continent oil more nearly on a level with California oil in eastern markets, and so would enable them to sell more of their crude, even if they had to sell it at a low price. It is true that the price cuts did not alter the proration schedule of most of the large companies, who still took only a certain percentage of the oil offered, but California production was finally declining, and there seemed a fair probability that the proration of runs might be discontinued. Mid-Continent operators began to feel more cheerful—so cheerful, in fact, that a considerable number of them began drilling for more oil.[24]

The relief of the Mid-Continent oil men was only temporary, for the Prairie Oil and Gas cut was answered on October 9 by a cut in California oil prices, which brought the highest grade of California oil down to 76 cents a barrel. It almost looked like the days of the Sunset-Midway No. 1; but California producers were in no position to haggle about prices. Again Mid-Continent operators found their hopes of an eastern market glimmering.

For a time a spirited contest was staged among the producers of the various sections of the country, especially between the California and the Mid-Continent producers, to see which could sell oil the cheapest, on an already over-sold market. The Prairie, Magnolia, Sinclair, Texas, Empire, Cosden, and Gypsy companies announced severe reductions on Mid-Continent prices on September 19. The next day cuts were announced in Pennsylvania, Ohio, Illinois, and Indiana prices. The next morning reductions of from 20 cents to 50 cents a barrel were announced by the Midwest, for Salt Creek oils. Four days later, the

Atlantic Producing Company posted a new schedule of lower prices for Stephens County, Texas; and on October 8 the Humble Oil and Refining Company announced lower prices for north-central Texas crudes. The next morning the Texas Company announced lower prices in the same region; and the Standard Oil Company cut prices from two to 28 cents in California. Three days later, Magnolia pared from 20 cents to 50 cents off north Texas and Oklahoma prices, and four days later the Gulf Pipe Line Company announced a reduction of 25 cents in Gulf Coast oils. Three days later, on October 19, the Champlin Refining Company posted a new low price on high-gravity crudes in northern and western Oklahoma, and the next day the Humble company announced further reductions for north Texas and central Texas crude oils. Two days later the Midwest Refining Company reduced Osage, Wyoming, prices. About a week later, on October 30, the Magnolia Company announced cuts of from 20 cents to 50 cents a barrel on the higher grades of Mid-Continent oils, and reduced Powell prices 25 cents, bringing the best grades of Powell oil down to 75 cents a barrel. On November 8 the Ohio Oil Company reduced Illinois and Princeton oils 25 cents a barrel, and on the same day the Prairie Oil and Gas Company cut Mid-Continent oils again. A week later reductions were announced in Pennsylvania, Louisiana, and Arkansas.

Just when conditions seemed to be at their worst, and prices were being cut somewhere almost every day in the week, and when, in spite of price cuts, proration, and all efforts at curtailment, productions ranged far above the demands of the market, the Secretary of the Interior added a final touch by announcing the sale of leases in the Navajo Indian lands in New Mexico, the sale to be held October 15.

In October and early November the over-production of oil seemed to have brought the oil industry to about the lowest depths of distress possible. Scores of producers and small refiners had been forced into bankruptcy, and scores of others were grimly hanging on, "waiting for the undertaker," as one of them expressed it. Oil was so cheap that most producers could not produce at a profit, and gasoline was so plentiful and so cheap that most refineries were operating at a loss. Gasoline was selling as low as 6 cents a gallon in bulk in California, and at about the same price in some places in the Mid-Continent. For a while the tank wagon price in some Texas cities was only 7 cents, and the service station price was only 10 cents.[25]

The cheapness of the oil products, and the energetic efforts of producers and refiners to find new markets for their products, stimulated a vast consumption for all sorts of low and unessential purposes. Automobile sales increased prodigiously, even in the season when they should normally have declined, and gasoline consumption remained high far into the winter. Street-car systems were scrapped in some small towns, and motor busses substituted. Oil was substituted for coal in furnaces in thousands of homes, and even in many industrial plants and large institutions, while the railroad and shipping consumption of oil reached new high levels. There were not enough reasonably important uses to absorb over 2,000,000 barrels of crude oil each day, and much of it was wasted in various ways.

But just about the time the situation seemed most desperate, underlying conditions were improving. California production, which had been the cause of much of the distress, was declining. Already in September, California records began to decline somewhat, and during the next two months it became evident that this was not a mere temporary fall, but the beginning of the long-looked-for decline of the great fields of southern California. Millions of barrels of storage in California had not been filled, and suddenly owners of this began to fear that they would not be able to secure enough oil to fill their tanks. In March, 1924, the production of the three great fields of southern California had declined nearly 50 per cent.

The decline in California production was hardly well established before refiners began to get apprehensive about their supplies of crude oil, and price increases began to come as fast as the cuts had come previously. On November 19, four days after the last important price cut had been announced, the Magnolia Petroleum Company raised the price of Powell crude 25 cents a barrel. The next day the Midwest Refining Company raised Osage, Wyoming, crude 10 cents, and the Ohio Oil Company raised Lance Creek crude 10 cents. A few days later premiums appeared in California, and a week or so later appeared in the Mid-Continent. On November 28 the Empire company announced its willingness to purchase all crude oil it was holding in storage for other companies. Proration restrictions were ended almost overnight. During December and January numerous price increases were announced in every part of the United States, not only in crude oil but in gasoline as well. Within two months after operators had been complaining of the impossibility of stopping the flood of oil, they were discussing the difficulty of securing enough crude oil to keep their refineries going, and crude prices had risen in some instances nearly 100 per cent, while advances in gasoline prices had been so conspicuous that the federal government and a half dozen state governments were ordering investigations.

In the meantime, the new prosperity was quite too much for the producers, who invaded the fields with hundreds of drills, in a determined effort to stave off any such thing as a really remunerative price for oil as long as possible. In December 60 wildcat wells were drilling in the Los Angeles Basin in southern California. There were 340 wells going down in the Long Beach and Santa Fé Springs fields. Early in December over 200 new rigs were up in Oklahoma, and new wells were coming in at the rate of a half dozen a day. In California 134 wells were completed in December, an average of four or five a day. Early in February Oklahoma reported 240 new rigs, and wells coming in at the rate of more than ten each day.

During the summer of 1924 oil production climbed again to a point where it seemed possible, if not probable, that the country was to indulge in another over-production debauch; but Mid-Continent prices were reduced in June and July, and production again started downward. At the present time, July, 1925, production is considerably below the peak of 1923, and it seems probable that the long decline has begun.

NOTES

1. *Cal. Oil Fields,* Nov., 1921, 6.
2. *Oil and Gas Jour.,* Dec. 17, 1920, 36; Jan. 20, 1922, 42.
3. *Standard Oil Bulletin,* Oct., 1922, 1; Cal. State Mining Bu., *Bul. 63,* 329; *Petroleum Age,* Dec. 15, 1921, 78; Sept. 15, 1922, 20; *Oildom,* Aug., 1923, 14.
4. *Cal. Oil World,* Dec. 21, 1922, 1; *The Lamp,* Aug., 1923, 15.
5. *Natl. Petroleum News,* Aug. 22, 1923, 93; *Oil Weekly,* Oct. 20, 1923, 29, 103; Jan. 19, 1923, 103.
6. *Oil Weekly,* Aug. 19, 1922, 11.
7. *Oil Weekly,* March 24, 1923, 55.
8. *Natl. Petroleum News,* Aug. 1, 1923, 73; *The Lamp,* Aug., 1923, 11.
9. *Oil Weekly,* April 28, 1923, 23.
10. *Oil Weekly,* Jan. 27, 1923, 54.
11. *Oildom,* Sept., 1923, 24.
12. *Cal. Oil World,* Aug. 30, 1923, 2; *Oildom,* Sept., 1923, 27.
13. *Oil and Gas Jour.,* July 1, 1921, 3.
14. *Oil Weekly,* Aug. 12, 1922, 11.
15. *Oil Weekly,* Dec. 9, 1922, 15.
16. *Oil and Gas Jour.,* Feb. 17, 1922, 7; *Petroleum Age,* Sept. 1, 1922, 21; *Natl. Petroleum News,* Aug. 22, 1923, 48.
17. *Oil Weekly,* Feb. 24, 1923, 15.
18. *Oil Weekly,* June 9, 1923, 31.
19. *Petroleum Age,* July 1, 1923, 18.
20. *Oil Weekly,* Sept. 15, 1923, 36.
21. *Oil Weekly,* July 28, 1923, 23; Aug. 4, 1923, 23.
22. *Petroleum Age,* Aug. 15, 1923, 13.
23. *Oil Weekly,* Aug. 18, 1923, 31.
24. *Petroleum Age,* Oct. 1, 1923, 11; *Oil Weekly,* Sept. 22, 1923, 12.
25. *Oildom,* Nov., 1923, 33.

CHAPTER XIII

OVER-PRODUCTION AND CURTAIL-
MENT AGREEMENTS

THE unfortunate features of the oil history of Pennsylvania have been repeated in the later history of almost every other region. There has been the same instability in the industry, the same recurrent or chronic over-production, the same wide fluctuations in prices, with consequent curtailment agreements, the same waste of oil, of capital, and of energy, the same fraudulent promotion, wild and senseless speculation, aimless drilling by green hands, the same unearned fortunes, the same "newly rich," with their ostentatious display of wealth, the same monopolistic tendencies, the same unhealthy conditions, economic and social. Not all regions showed all these characteristics, but they all showed most of them, and in the history of some could be found all the features that make Pennsylvania oil history depressing to read.[1]

Over-production has been chronic. There has hardly been a time since 1860 when too much oil was not being produced; although this over-production has varied in degree from time to time, and has been greater in some fields than in others. From the time of the first Pennsylvania well, there has been an almost uninterrupted increase in the production, an increase so rapid that consumption could be maintained only at extremely low prices, and by concerted effort of the producers to stimulate the public appetite for oil. In other words, our oil producers have for over half a century been constantly engaged in producing more oil than was needed or than was demanded at any reasonable price; have been forcing upon the consuming public, sometimes at almost no price at all, millions of barrels of an irreplaceable natural resource, a resource for which future generations will almost certainly be glad to pay ten—or even a hundred—times as much, and perhaps will hardly be able to secure at any price. That there was no real need or use for the oil was not an important matter to those who were seeking only personal gain. It was the spirit of "American enterprise" to "develop," "development" being generally synonymous with wasteful exploitation and despoliation, and that man being most "enterprising" who could seize and waste the most oil in the shortest time. Greedy operators have found the pools of oil here and there, have rapidly and wastefully appropriated the most accessible deposits and passed on to other fields to despoil them in like fashion.

The total production of the United States has mounted with astonishing rapidity, has doubled about every ten years. When production reached 2,000,000 barrels in 1861, it was regarded as a stupendous total, but eight years later this was doubled, and four years later doubled again, and six years later, in 1879, doubled yet again, and ten years later doubled once more, with the imposing figure of 35,000,000 barrels. This seemed the limit to possible production, and for a few years, in spite of the growing flow of Ohio oil, the production did not make any spectacular advance, but in the late nineties the West Virginia production began to swell the total, and in the early years of the next decade, the vast floods of Texas and California oil brought the total production up to 134,000,000 barrels in 1905. This seemed the utmost that could be expected, but about this time the Kansas and Oklahoma production began to grow with astounding rapidity; and a year or two later Illinois entered the stage, with the result that the production more than doubled between 1905 and 1915, with a total flow of 281,-000,000 barrels for the latter year. This seemed the very maximum possible, and with the insatiable war demands, and a temporary lack of success in wildcatting, there was much talk, even among practical oil men, of an impending oil famine. Predictions were freely made each year that the crest had been reached and that figures of output were headed downward. Oil prices advanced, and it began to seem that our half-century spree of over-production and cheap oil was winding up in the cold gray dawn of the morning after. But one more drink was still in the jug, and the country was to have one more grand, general, all-round debauch. In 1921, 1922, and 1923, several new fields were opened: Mexia, Huntington Beach, Signal Hill, Santa Fé Springs, Burbank, Smackover, Tonkawa, Powell. Each in turn added its thousands of barrels to the swelling production, with the result that the total production of the country almost doubled again, reaching a total of 556,000,-000 barrels for 1922.

It seemed impossible that such a tremendous production could be maintained, and, each month during the year 1923, producers in a demoralized market looked anxiously for the "peak" and for the decline that they believed must surely follow; but production rose steadily and rapidly until August, and did not decline noticeably until November. The production for the year, 735,000,000 barrels, nearly 2,000,-000 barrels a day, represented an increase of 169,000,000 barrels over the production of the preceding year. The increase in one year was more than the total production of the country in 1907.

Some idea of the tremendous volume of our oil production may be gained from the picture drawn by George Otis Smith, who figured that the 1918 consumption of oil—413,000,000 barrels—would make a stream as large as the Niagara River as it flows over Niagara Falls, for over three hours, or a stream as large as the Potomac River at Washington flowing at its summer rate for nearly four days and a half; and the production during 1923 was nearly double the figure which was used in these computations.[2]

The causes of over-production of oil have been touched upon at various points in the preceding pages, and need be only briefly recounted here. The small size of the leases in many fields has forced the drilling of too many wells, in the greatest haste, and without regard to the conditions of the market. Furthermore, lease contracts have generally required the drilling of a certain number of wells, and have been so drawn that it was difficult or impossible for operators to delay drilling or to shut in producing wells. Shutting in of wells has been opposed with considerable reason, because in some fields it has proved ruinous to shut in flowing wells, or to stop pumping when it has once been started.

To some extent, state and national governments have contributed to this over-production by refusing to sanction agreements among operators to limit production. The theory of our law has been that such agreements are in restraint of trade, and therefore contrary to the anti-trust laws; and, not a few times, state and federal officials have taken alarm at the laudable efforts of oil producers to limit production by agreement. As pointed out in a later connection, there is no possible harm that could come to the public from limitation of oil production, through organization or otherwise.[3]

There has even been some doubt whether the collection of statistics of production, shipments, stocks, etc., and the distribution of these statistics to the operators, were permissible under the present ruling of the federal Department of Justice. Attorney-General Daugherty not long ago submitted an opinion to Secretary Hoover, of the Department of Commerce, in which he apparently held activities of this sort to be unlawful. The American Petroleum Institute has for several years been compiling statistics of various kinds for the trade, statistics which are valuable from every point of view. It is only unfortunate that they cannot be made more complete and more accurate. Yet Daugherty's decision raised some doubt as to the legality of this work.[4]

Over-production was generally worst in those fields where many small companies were operating. The oil-producing industry has in most fields been burdened with a multitude of small companies, often controlled by men utterly unfamiliar with the business, men whose "sole aim," as one man expressed it, "has been to produce oil, regardless of the requirements of the market." But even the big companies were often helpless. In California, when prices were tumbling most rapidly, many of the leading producers would have been glad to curtail, but were forced to drill, by their lease obligations; and, of course, when the smaller companies saw the large ones drilling, they got into the game too, in the face of a practical certainty that their activity was going to bring lower prices and even ruin to many. Generally, the larger companies were more conservative in their operations at times when oil was too cheap. The big companies have wanted steady development, because they have refineries and pipe lines and other equipment which they would like to keep employed for a reasonable number of years. The town-lot promoters, however, and perhaps even more particularly the owners of land, through the leases exacted of operators, have often forced the orgies of rapid drilling.

"The landowner and [the] promoter," as one writer has said, in describing the situation in the southern California fields, "are not worried about the expense of building refineries, tank farms, steamships, and the like. That is done by the big companies to which they sell the oil. Nor are they worried about oil reserves twenty years hence. Nor do they worry even about the fact that the tremendous flush production is a temporary peak at most. . . . The promoter and landowner have but one object in mind—to make a fortune, make it quick, and cash in."[5]

In most periods of over-production and low prices, there has been a great deal of talk about the "big fellows" crowding out the "little fellows" by reducing prices. The big purchasing companies often found it unprofitable, or even impossible, to buy and store all the oil that enthusiastic producers wanted to sell. In some instances they lost heavily by holding prices up in these deluges, by buying oil and then selling refined products later on a lower market; while, on the other hand, they often established the basis for immense profits by reducing prices very low in these periods of over-production and selling their products later on a better market. The Standard Oil companies owed much of their power to just such situations as these. As indicated in a later chapter, they could take advantage of them in two ways. In the first place, they were able to buy the oil at attractive prices, and in the second place, they could buy many small companies outright, because some of the small and financially weak companies were always in distress in periods of over-production and low prices, and the Standard companies almost always managed to have immense sums of cash on hand at such times. Other large companies also took advantage of these periods of over-production in the same way. The smaller producers were hardly fair, however, in blaming the larger companies for their distress, for the larger companies were simply protecting themselves, taking advantage of an evil situation which all producers had helped to bring on; and doubtless even the officials of the Standard Oil companies would have been glad to see over-production eliminated and the oil industry put on a saner and more stable basis.

It has been suggested that there is a close analogy between the condition of chronic over-drilling and over-production in the oil industry and the present over-development in the coal industry.[6] There can be no doubt that both industries have been over-developed; but there is an essential difference between the situation in the two industries. The demand for oil is practically unlimited, and, no matter how much is produced, it will always be used for some purpose. If oil producers were able to guarantee a twenty-five-year supply of fuel oil at 50 cents a barrel, or even more, a billion barrels per year would be demanded as soon as fuel-burning apparatus could be changed to handle the new fuel. One billion barrels would not be enough. Many billions of barrels would be used annually, if oil were cheap enough. It is the vast potential demand for oil and its products that has made possible the meteoric rise of the oil industry during the past fifteen or twenty years; and this demand will always absorb any possible amount of production. This is not saying that there has not been over-production, or that there will not be over-production in the future; but the demand always increases very rapidly, and unlimited amounts of oil could always be used for some purpose. The demand for coal is more nearly fixed, and when a certain amount is produced, little more will be taken, no matter how low the price.

In the conditions of supply, perhaps oil and coal differ even more. Within a few months after discovery, the production of almost every oil field begins to decline rapidly, and the decline continues to the exhaustion of the field; while the production of every coal mine increases year by year as new machinery is installed, as the number of working rooms increases, and as new shafts and tunnels are driven. Since the production of oil is maintained only by the constant opening of new fields, and the number of new fields is necessarily limited, it is clear that the general situation in the future will not be that of over-expansion and surplus production, but rather that of inability to meet the vast demands that have been built up. Just as the situation has in the past been one of over-production, with demands rapidly built up to absorb the production, it seems certain that at no distant date the condition will be reversed, with production dwindling, prices rising, and consumption slowly falling, with the rise in prices. In other words, the prospect for the future, and probably at no distant date, is for oil famine and not over-development.

Just as in Pennsylvania, in all later fields the recurrent periods of over-production have meant a very unstable industry. The story of the "development" of almost every field has been a story of sudden discovery, feverish and wasteful drilling, a short period of flush production, then rapid decline of the field, showing itself in a rapid decline in the average production per well as the number of wells increased, and then a long period of decadence before being finally abandoned. Some fields have lasted longer than others, but a decade has been enough to include the important production of many of them; and some, as for instance Spindletop, have passed into decadent senility in less than half that time.[7]

Not only for individual fields and regions, but for the country as a whole, the oil industry has always been subject to wide fluctuations. Some even claim that there is a certain inevitable sequence in these fluctuations, a period of over-production and low prices inevitably following a period of smaller production and higher prices. Amos Beaty, of the Texas Company, in his testimony before a senate investigating committee, declared that psychology had much to do with this, that, whenever there was the slightest evidence of a shortage of crude oil, refiners became apprehensive about the future supply, and bid up prices far higher than conditions justified, and that these high prices stimulated production to a point where the market broke to low levels again. In this way Beaty explained the fact that, in the early spring of 1923, the price of Mid-Continent oil rose to $3.50 a barrel, and within a few months again declined almost to $1. Many writers have attributed the tremendous over-production of 1922 and 1923 to the high prices of the war period and of 1919 and 1920, which stimulated drilling and resulted in the finding of several great pools. Pools, once discovered, are usually exhausted so quickly that refiners dependent on them for crude are haunted by a fear that they may have to scrap their investment in a short time, and many of them become very apprehensive when any shortage appears. On the other hand, when over-production brings declining prices, refiners have to mark off inventory losses which are sometimes disastrous.

In such periods of over-production and reaction, prices of oil have of course varied widely. Only a few illustrations need be given. In the Beaumont field of Texas, the price dropped from 60 cents to 10 cents a barrel within a few months in 1903, and then started up again. Kansas oil declined from $1.38 early in 1904 to 50 cents in 1905, and to 30 cents in the next year. With the tremendous production of the Kern River field, California oil declined from $1 a barrel in 1900 to as low as 20 cents, and even less, in 1903. In the Cushing field, the price of oil was 40 cents a barrel or even less in 1915, when the development of the field was at its height, but by January, 1916, the posted price was $1.20, and in September, 1917, was $2, with premiums on Cushing crude of 30 cents to 50 cents a barrel. In two years' time the price of oil increased 500 or 600 per cent. Fluctuations of 100 per cent of the lower price have been very common within a few months, in many of the oil regions. Such rapidly shifting values make a stable industry impossible.

Not only have prices of petroleum products been unstable, but they have always been far too low. Appalachian oil has often sold for less than $1 a barrel; in fact, the yearly average price scarcely went above $1 between 1879 and 1895, and at one time went as low as 55 cents a barrel. Since that time the price has generally hung between $1 and $2 a barrel, although it rose to $3 and $4 for several years. The yearly average price of Lima-Indiana oil, as low as 15 cents a barrel for two years successively in 1888 and 1889, never rose above $1 between 1886 and 1902, and was above that figure only eight years of the thirty-two years between 1886 and 1918. Illinois oil sold for 60 cents to 85 cents (yearly average price) between 1905 and 1912. Kansas and Oklahoma oil sold for an average price of 36 cents in 1909, and did not rise to $1 until the era of war prices came in 1916; and northern Texas and northern Louisiana oil was nearly as cheap. The 40-cent oil of Cushing days has often been mentioned. The yearly average price of Gulf Coast oil in 1901, the year of the Lucas gusher, was about 17 cents, and it never reached 50 cents until 1907, or $1 until 1917. Rocky Mountain oil, some of it very high-grade oil, similar to the Pennsylvania product, never sold for a yearly average price of $1 between 1888 and 1898, and sold for that price only once between 1906 and 1916. California petroleum, generally heavier and less valuable, never rose above a yearly average price of $1 between 1895 and 1917, and much of that time it was below 50 cents a barrel. Between 1876 and 1918, California produced over one billion barrels of oil, which was sold at an average price of 57 cents a barrel. In recent years, prices have everywhere been higher, yet in 1923 Smackover oil sold for 30 cents a barrel.[8]

Glutted markets, overflowing storage, and low prices have often resulted in curtailment campaigns among producers, such efforts being particularly common in Oklahoma and California. In Oklahoma, such a movement has, on one occasion, been encouraged and even directly engineered by the state government, while in California the operators never secured state assistance. One of the most remarkable curtailment campaigns was that in Oklahoma in 1914, when the tremendous flow from the Cushing field had brought prices down to almost nothing, and many of the producing companies faced ruin. Even the most zealous exponents of "American enterprise" and "free development" were able to see that American enterprise was wasting millions of barrels of oil, and ruining the American enterprisers in the process. In December, President James E. O'Neil, of the Prairie Oil and Gas Company, sent a personal letter to all producers served by the lines of his company, calling attention to the fact that production was increasing at an alarming rate, and declaring that unless some action was taken to curtail operations, there would certainly be a reduction in price. This warning was generally disregarded by producers, who saw in it only part of a scheme to reduce prices. Operators continued drilling, and the flood of oil grew until the threatened price cut was announced.

Then the operators began to see the true situation. Meetings were held, the Standard Oil Company—the Prairie Oil and Gas Company, in this field—was roundly denounced, although there was some talk of getting at the root of the trouble by curtailing the production. In May, 1914, at the urgent solicitation of many of the producers, the Oklahoma Corporation Commission issued an order forbidding any further drilling, except where necessary to offset wells already drilled or to comply with lease requirements, prorating runs from the Cushing district, and fixing a minimum price at which oil might be sold. This step seemed a strange one for the operators to be urging, but it seemed absolutely necessary under the conditions existing at the time. As one correspondent expressed it:

The average producer of oil here, as elsewhere, has a horror of the idea that anybody can interfere with him in the conduct of his business, or that anybody should be permitted to control the conduct of his business, but in this case it was the assertion of the old first law of self-preservation, and, much as he disliked it, the producer swallowed the implied acknowledgment of the affirmation of the doctrine of modified state socialism with what grace he could, knowing that it was the best that could be done under the circumstances.[9]

On June 24, only a few weeks after the Corporation Commission issued its order, 500 oil producers assembled in a mass meeting in Tulsa to discuss the situation. It was a wild and stormy session, with much heated and acrimonious debate, much questioning of motives, political maneuvering, criticism of the pipe-line companies, and a moderate amount of robust Oklahoma cursing and swearing; but late at night the assembled producers finally agreed upon resolutions, approving the shutdown order of the State Corporation Commission, and providing means for assisting in carrying out that order.

For a while it appeared that the curtailment campaign was going to be a success; but presently all sorts of difficulties arose. In the first place, on the "morning after" the Tulsa meeting, a report was circulated that the meeting had been controlled by the pipe-line companies, and this report tended to discredit the new policy. In the second place, a quarrel soon arose among producers as to the details of the shutdown order, one important point of disagreement being as to whether producers should be permitted to drill all the wells for which rigs were up or contracted. One faction claimed this right, while the opposing faction pointed out that there were 216 rigs and wells drilling at Cushing, and that the completion of so many wells in a pool like Cushing would actually increase the production instead of reducing it. In the third place, the question of unemployment of drillers was soon raised, some claiming that it was unjust to deprive these men of their means of livelihood. As a means of affording some employment, some operators drilled wells to, but not into, the oil sands, thus putting themselves in readiness to secure production quickly when the order should be rescinded, and at the same time affording employment to drilling crews. In the fourth place, the State Corporation Commission was always doubtful as to its constitutional power to follow out such a policy as it had adopted, and many of the operators and pipe-line officials shared this doubt.

The policy of curtailment was not sufficiently rigorous, and it was soon evident that the production was still too great, and that something further must be done if price cuts were to be avoided. On the question of further reduction there were two opposing factions, one favoring a severe restriction of production, and the other holding that the best way out of the difficulty was not to restrict production, but to proceed until the Cushing pool was fully "developed" and relegated to the class of producing districts that have no power to influence the market.[10]

Economic law was more powerful than the edict of the Corporation Commission, and when the over-production continued and tanks were full, the Prairie Oil and Gas Company announced a reduction in price to 55 cents a barrel, contrary to the order of the Corporation Commission. Upon this, the Corporation Commission called in operators and pipe-line representatives for an investigation, in the meantime forbidding any sales at less than 65 cents a barrel until the investigation could be completed. The Prairie Oil and Gas Company immediately responded by refusing to take any oil from leases in Oklahoma. In the course of an investigation, it was ascertained that "too much oil was being produced," and the Commission presently receded from its 65-cent price and allowed purchase at 55 cents, in spite of the energetic opposition of Attorney-General West, who proposed to handle the pipe-line companies in masterly fashion and compel them to buy oil at the regulation price. The adoption of the new price established the business on a firmer basis for awhile, but in October the Magnolia Petroleum Company—another Standard Company—applied to the Corporation Commission for permission to reduce the price of Healdton crude from 50 cents to 40 cents a barrel, and this stirred the brew once more. In response to the Magnolia petition, Commissioner Henshaw stated that the price of oil should be going up and not down, and presently there was a well-

developed quarrel between the Corporation Commission and the pipe-
line companies, the latter generally refusing to concede that the Commission had the power either to fix prices or enforce a shutdown. Finally,
two of the oil companies, the Silurian Oil Company and the Quaker Oil
Company, asked the federal court at Muskogee for an injunction to
restrain the Corporation Commission from enforcing any of its orders,
thus throwing into the courts the question of the constitutionality of the
acts of the Commission.

Economic law finally decided the matter, however, for some of the
producers presently found themselves with so much oil on their hands
that they could not provide storage for it all, and were forced to dispose
of it at any price they could secure. As soon as it became evident that
price-cutting was the order of the day, the bottom dropped out of the
artificial price structure, and the Commission receded from its position.
Its action had been founded on a false notion of the operation of economic law, but nevertheless it had for a time been a benefit to the industry. If the state of Oklahoma had forbidden all drilling ten years
ago, and had not drilled a single well since then, it would be in a far
better position today. It would have fewer new millionaires, but a great
deal more oil, and the oil will presently be worth far more than it ever
sold for in the days of Cushing gushers.[11]

In California there have been numerous curtailment campaigns. In
1911, an "earnest effort to limit the production" was reported, but no
success attended this effort, and the next step was an organized effort
on the part of producers to stimulate the consumption of oil. In 1915,
the efforts to stop drilling were reported as the "all-important feature
of the oil situation in California."[12] Many of the important producers
were barely covering costs of production, and yet any considerable increase in price promptly resulted in more drilling. Some conservative
men—McQuigg, Requa, St. Clair, O'Donnell, and others—worked consistently for reduction, and were successful in reducing the output
several million barrels. There is no record of oath-bound agreements,
such as were often used in Pennsylvania.

With the over-production of 1921, 1922, and 1923, came another
series of curtailment campaigns, this time in several of the important
fields. The flood of oil from Oklahoma, Texas, and California forced
prices lower and lower, producers began to discuss the possibility of a
concerted curtailment, and in February, 1921, 300 operators in the
Mid-Continent field met at Tulsa and passed a resolution, favoring the
restriction of drilling, referring the details of such a restriction to a
committee of the Mid-Continent Oil and Gas Association. The committee entrusted with formulating the plans, representing production
in Kansas, Oklahoma, Texas, and Louisiana, met and adopted a resolu-

tion against the drilling of any "new or unnecessary wells," "as long as the present economic waste continues." A committee of nine producers was appointed to formulate and administer some plan of curtailment, and this committee of nine drew up in very vague and indefinite language four principles that were to govern their plan. Within a very short time a considerable number of wells were reported closed. Early in March, the *Oil and Gas Journal* reported that the movement was a success, but in April the committee in charge of the curtailment campaign reported that pipe-line companies were again taking all the oil offered, and that the market situation was sufficiently improved to justify calling off the movement. The Mid-Continent Oil and Gas Association declared the curtailment campaign at an end, but warned producers that unless consumption increased, the industry would suffer further demoralization. This warning was timely, for drillers and operators went back to their work with quite too much energy, and on May 2, the Sinclair Oil Company announced a 25-cent reduction in crude prices. The Prairie Oil and Gas and other purchasers followed with similar reductions.

Again there was much talk of curtailment, and the state was once more invoked to help the oil producers out of their difficulties. In 1915 the state legislature had passed a law specifically granting the Corporation Commission the authority to close down any industry when there was danger of waste through over-production, and a large number of operators met at Tulsa early in May, and adopted resolutions appealing to the Corporation Commission for relief. The Commission called upon all operators to present their views on the question of state intervention at a hearing at Oklahoma City, and pending this hearing, a hot debate raged in the ranks of the oil producers. Some of them argued that unless a general and very considerable reduction in oil production could be secured, further price cuts were sure to come, and that they might reduce the industry to a state of general demoralization. They pointed out that the voluntary curtailment movement earlier in the year had failed to achieve its purpose, and that the only agency which could insure success in such a movement was the state. Some also pointed to the elements of unfairness in a voluntary shutdown, arising from the fact that there were always some selfish individuals who would take advantage of the situation, and perhaps sell more than their share on a market supported at the expense of the more loyal operators.

On the other hand, men of a more conservative frame of mind preferred to keep the control of the industry in the hands of the operators. Many of them, like business men generally, had a deep distrust of state interference. Probably some of them really feared that state intervention in their behalf in this crisis might lead to state intervention and control on other pretexts, and eventually to state control as a permanent policy; and, of course, if state intervention in this instance proved successful, it would, from this point of view, be only the more dangerous to that "American enterprise" which business men always speak of with such pride and solicitude. This view finally prevailed at the hearings, and the Corporation Commission declined to intervene.[13]

But the situation did not improve, and the following year, early in August, the Mid-Continent producers began to talk of curtailment again. On August 12, a meeting was held in Tulsa, where various schemes were discussed. One plan suggested was to ask the state governments to provide oil storage for the independent operators who were unable to provide their own storage, but the Texas operators immediately declined to have anything to do with such a plan. "We men from Texas," as one of them stated, "hope you will try that scheme in Oklahoma first. We have enough of legislation in the oil business down our way." Another scheme more generally favored was that of coöperative storage tanks. One operator expressed himself as follows:

It is really the most singular thing, . . . that the independent oil producer has never thought to provide himself with any facilities for storing his product. In no other business of which I know is there such implicit reliance by the producer of a commodity upon the purchaser of it. Under present conditions, the buyer fixes the price at which the seller will sell and the seller sells at the buyers' price and has no practical alternative whatever. The producer sells always on a buyer's market.[14]

Increased storage facilities, necessary though they were, did not strike at the root of the difficulty, and the proposal of an organized cut on production was received with strong approval. Committees were appointed to arrange for a concerted plan of tank-building and curtailment.

In the Oklahoma fields the curtailment campaign was not very successful. Some oil statisticians were able to show a reduction in the output, but it was generally admitted that the operators were not curtailing much. There was considerable dispute all the time as to whether the large or the small operators were responsible for the failure, some claiming that the large producers were doing all the curtailing, and others claiming that they were just as busy as the "small fry." It was relatively harder for the small operators to close down, to be sure, because, since they were only producers, their business depended entirely upon their regular sales of crude oil. Some of the large producers were also engaged in the refining and transporting of oil, and they could cut their crude production without reducing their income so seriously. There was also considerable discussion of the Standard companies and their attitude in the curtailment campaign, some writers claiming that they were doing more than their share, and others claiming that they were not doing anything. Thus, with a great deal of discussion and dispute and selfish contention, the curtailment movement did not amount to much.

For a time it seemed that aid was coming from the government. On September 25 the shutdown and storage committee of the Mid-Continent Oil and Gas Association adopted a resolution calling upon the Secretary of the Interior to order a stop to all unnecessary development work on lands under his control. On receipt of this resolution, Secretary Fall promptly ordered a cessation of all development work on Osage lands, except necessary offset wells, and offered to release operators on other Indian lands of all obligation to drill under the terms of their leases. Osage lands were of particular importance at this time, because of the rapidly developing Burbank pool, and so the order was mandatory only as to these lands.

The new order immediately precipitated a fresh quarrel in oildom. Many of the operators outside the Osage lands were of course much pleased, because Burbank production was seriously menacing crude prices, and every operator in Oklahoma would have been pleased if every other operator could have been compelled to close down. Also, from the point of view of the conservationist, there was much to be said for the order, because the government should long before have stopped all production from its lands. There was no excuse for permitting production from government lands, even from Indian lands, to swell the vast over-production. But in the application of the order, some vexing questions arose. In the first place, there was the question of dealing fairly with the companies operating on Osage lands. The Gypsy Oil Company, for instance, had just paid $1,000,000 for a lease in the Burbank field, which was offset on one side by a Prairie Oil and Gas lease producing from 16 wells, and on another side by a Sinclair lease with a similar number of wells. The Gypsy Company had several leases similarly situated, and was interested in the question as to what were to be regarded "necessary offset wells." There was no way of eliminating all development on the Osage lands without doing injustice to some of the companies operating there.

In the second place the question immediately arose as to why the government had not also ordered a shutdown of government lands in the West, for instance, Salt Creek, and Teapot Dome; and a great deal of criticism was aimed at Secretary Fall because of the "discrimination" shown in his order. There was some justification for his action on this point. The Salt Creek production was in stronger hands, and producers there were already curtailing effectually, and Teapot Dome was in the hands of a single company. But there was much criticism of the action of Secretary Fall, and within less than a month after the order was issued, the committee which had asked for it requested that it be rescinded, and again the Secretary complied. Perhaps a fair explanation of the sudden change of mind on the part of the shutdown committee is found in the following quotation from one of the oil journals: "What is the use of an enforced shut-down in the Osage reservation when producers in every state drill their heads off?"[15] About this time the situation in the Mid-Continent fields began to improve slightly, anyhow. It was generally admitted that the curtailment campaign in Oklahoma had been of little effect.

In Kansas curtailment was distinctly more effective, because the fields were in the hands of fewer and stronger companies. In August, 1922, Governor Allen threatened to order the Industrial Court to investigate the curtailment movement, on the ground that it was an effort to increase the price of an essential commodity; but a month later the "lid" was reported to be "on tight."[16]

The price increases of the early months of 1923 suggested that there was no further need of curtailment, but when the California fields, and the Burbank and Tonkawa fields, again broke the market, and prices began to fall in March and April, producers were soon discussing curtailment again. Even in May meetings were called to consider such a move in the Burbank field, and early in June, at a meeting in Tulsa, the principal operators in the Burbank and Tonkawa fields agreed to stop all unnecessary operations. The Secretary of the Interior authorized a shutdown on Indian leases. This move was said to affect some 300 wells, and it was claimed that it would cut off more than 300,000 barrels of flush production, since the wells in Burbank and Tonkawa averaged about 1,000 barrels each.

This movement was carried on in the face of considerable difficulties. It presently appeared that operators in several other Mid-Continent fields were increasing their production, while the Burbank and Tonkawa producers, who were loyal to the curtailment agreement, were "holding the sack," and of course this caused dissatisfaction. Yet, at least as far as the Burbank field is concerned, the curtailment agreement was apparently effective until the next January, when, with the partial disappearance of the over-production menace, the agreement was in part rescinded.

In California, as in the Mid-Continent fields, during these years, there was much talk of curtailment, but very little curtailment. Much of the California production was coming from town-lot drilling, and the operators were helpless in the face of the growing production. Committees were appointed at Long Beach to "study the situation," but the only information such committees seemed able to get was that there was too much oil, and that if there were less of it the price would be higher. In 1923, the curtailment movement in California took the form mainly of proration of runs by the purchasing companies, at the request of committees of producers. This has been discussed in a previous connection. Proration was naturally more effective than ordinary curtailment agreements.[17]

In several other Mid-Continent fields, curtailment was said to be effective. In the Gulf Coast field of Texas, it was claimed that curtailment was to some extent effective, because the production was largely in the hands of a few large companies, although wildcatters were about as busy as ever in their search for new fields. So likewise in the Salt Creek field of Wyoming, curtailment was undoubtedly effective. About fifty representatives of the more important producing companies operating in the Salt Creek field met at Casper on September 20, 1922, to consider the situation, and, without wasting any time on the appointment of committees, agreed to suspend all drilling operations on December 1, 1922, and to do no further drilling until conditions should change.

NOTES

1. For a classification of wastes in the oil industry, see *Oil Weekly,* June 12, 1925, 29.

2. *Natl. Geographic Mag.,* Feb., 1920, 181.

3. See below, pp. 512-516; *Oil Weekly,* Oct. 10, 1924, 28; *The Lamp,* Dec., 1924, 5.

4. *Cal. Oil World,* Jan. 24, 1924, 1; *Natl. Petroleum News,* Jan. 16, 1924, 25.

5. Albert Atwood, in *Sat. Evening Post,* July 7, 1923, 93.

6. *Natl. Petroleum News,* Nov. 22, 1922, 47; Dec. 27, 1922, 43.

7. Cushing production rose from 652,000 barrels in January, 1914, to 8,352,000 barrels fourteen months later, and declined over 50 per cent to 3,793,000 barrels within seven months thereafter. The production of the Glenn pool rose from 385,000 barrels in January, 1907, to 2,441,000 barrels in October; and seven years later had dwindled to 535,000 barrels a year. The Glenn pool reached its maximum production within thirteen months after discovery.

Figures of drilling also show the same instability of the industry, the same rapid exploitation and partial or complete abandonment. In Franklin and Miami counties, Kansas, 252 oil wells were completed in 1905, and only one well three years later; but eight years later, in 1916, a total of 526 wells were drilled, 451 of them oil wells. In Allen County 450 wells were drilled in 1904, only two wells in 1906, and 192 wells in 1908. In Montgomery County, 828 wells were completed in 1904, and three years later only 56—about one-sixteenth as many; and six years later the number had again risen to over 800. In Labette County, 82 wells were drilled in 1904, and only one the next year. In Chautauqua County 630 wells were drilled in 1904, and only 47 four years later. In Butler County the extraordinarily rapid development may be seen in the following table of wells completed: (Data from Snyder, *Oil and Gas in the Mid-Continent Fields.*)

1914	29
1915	60
1916	977
1917	1184
1918	1440

Figures for some of the Oklahoma fields are similar in their significance. The following tables show the number of wells completed each year in several of the Oklahoma fields:

WELLS COMPLETED

Year	Coody's Bluff	Alluwe
1905	280	165
1906	549	441
1907	72	246
1911	56	98

Year	Nowata District
1912	1231
1913	1417
1914	1558
1915	600
1916	1219
1917	757

Bartlesville field

Year	Wells completed	Average initial production
1906	790	...
1909	254	42.2
1910	251	43.9
1911	188	30.2
1912	584	23.5
1913	948	28.4
1914	454	10.1

Tulsa and Bartlesville districts

	Wells completed	Average initial production
1905	273	65.7
1906	790	73.2
1907	941	91.8
1908	690	60.4
1909	652	65.8
1910	802	46.1
1911	1074	37.4
1912	2906	31.1
1913	3429	24.8
1914	1789	15.0
1915	471	24.5
1916	1294	17.3
1917	763	23.2
1918	1330	25.4

Muskogee field

1909	. . . .	129	104.4
1910	. . . .	171	135.3
1911	. . . .	117	86.0
1912	. . . .	38	24.4
1913	. . . .	100	26.1

Carter County (Healdton pool)

1913	. . . .	23	56.2
1914	. . . .	392	312.3
1915	. . . .	318	295.2
1916	. . . .	685	126.6
1917	. . . .	507	100.3
1918	. . . .	269	51.3

Figures for some of the older California fields tell a similar story. (Data from *Bul. 69* of the Cal. State Mining Bu.)

Coalinga field

Year		Wells completed	Average production of producing wells (In January of each year)
1904	. . . .	65	95
1905	. . . .	77	126
1906	. . . .	52	107
1907	. . . .	44	92
1908	. . . .	152	87
1909	. . . .	170	84
1910	. . . .	218	78
1911	. . . .	139	71
1912	. . . .	107	68
1913	. . . .	64	55

Kern River field

Year		Wells completed	Average production per well (In January of each year)
1903	. . . .	102	73
1904	. . . .	106	80
1905	. . . .	86	65
1906	. . . .	50	57
1907	. . . .	148	43
1908	. . . .	194	35
1909	. . . .	195	32
1910	. . . .	301	30
1911	. . . .	136	25
1912	. . . .	90	22
1913	. . . .	33	19

It is true that these records do not include some fields in which exploitation has proceeded more evenly; and some figures are less significant than they might be if they did not cover the years of the Great War; but they show clearly the instability of the industry, and some of them are typical examples of feverish and over-rapid development, followed by conditions of stagnation and dwindling runs.

During the year 1922, to bring our figures down to date, there were extraordinary changes in the production of many oil fields. El Dorado, Arkansas, declined from 1,284,000 barrels in the month of January to 697,000 barrels in December. Smackover rose from 125,000 barrels in September to 2,442,000 barrels in December. Homer, Louisiana, declined from 646,000 barrels monthly production to 392,000 barrels during the year, and Bull Bayou, in the same state, declined from 257,000 to 186,000 barrels. Mexia declined from a monthly production of 4,200,000 barrels in January to less than half as much in December; Desdemona-Pioneer increased from 163,000 to 297,-000 barrels; Moran increased from 6,000 to 85,000 barrels; Burkburnett declined from 1,165,000 to 800,000 barrels; Electra rose 100 per cent, from 359,000 to 765,000 barrels; Goose Creek rose from 365,000 to 587,000 barrels; Hull rose from 272,000 to 751,000 barrels; Somerset rose from 54,-000 to 118,000 barrels; West Columbia declined from 1,076,000 to 823,000 barrels; and Pierce-Junction declined 70 per cent, from 150,000 to 41,000 barrels per month. (*Oil Weekly,* Jan. 20, 1923.)

8. See *Mineral Resources* for various years, on oil prices.

9. *Oil and Gas Jour.,* July 2, 1914, 33.

10. *Oil and Gas Jour.,* Sept. 24, 1914, 12.

11. *Mineral Resources,* 1914, 1006.

12. *Mineral Industry,* 1915, 528.

13. *Oil and Gas Jour.,* Feb. 25, 1921, 74; March 11, 52; April 15, 62; May 6, 56; May 13, 2.   •

14. *Oil Weekly,* Aug. 19, 1922, 11.

15. *Petroleum Age,* Nov. 1, 1922, 19.

16. *Oil Weekly,* Aug. 26, 1922, 24; *Petroleum Age,* Sept. 15, 1922, 50.

17. See above, pp. 109-111.

CHAPTER XIV

WASTE OF OIL

IN the waste of vast quantities of oil, later oil history is again a
second edition of Pennsylvania history. Great over-production
and low prices have everywhere led to waste of this precious resource, by hundreds, thousands, millions, and hundreds of millions of
barrels. One writer described the waste in the Indiana oil fields as
follows:

> One cannot pass through the oil field without noting that every pool and
> stream of water is covered with oil, thousands of barrels being allowed to
> go to waste through leakage and overflow of tanks, overflow of wells when
> first shot, or through the oil passing off in quantity with the salt water flow-
> ing from the tanks.

The Director of the Bureau of Mines estimated in 1914 that the total
waste or loss, in terms of dollars, in drilling and exploiting the oil
fields, and in the storage and transportation of oil, was not less than
$50,000,000.[1] In writing of the waste in 1916, Mark Requa declared
it was "difficult to speak dispassionately and judiciously." In 1916 the
same high authority estimated that 10,000,000 barrels of oil were
being wasted in California each year through seepage, evaporation,
and by burning.[2] Speaking before the American Petroleum Institute in
November, 1920, Requa declared: "I am prepared to make specific
charge that the consumption of fuel oil, natural gas and gasoline is
inexcusably wasteful."[3] Technologists in the Bureau of Mines, writing
of the waste in California fields in 1914, reported: "If past conditions
prevail throughout the life of the fields, it is no exaggeration to state
that of the supply of recoverable oil throughout the oil fields of California, not over 40 to 50 per cent will ever be marketed."[4] In the Cushing
and Healdton fields of Oklahoma "hundreds of thousands" of barrels
were reported lost in 1914.[5] One writer mentions "millions of barrels"
as having been wasted. Another observer once declared, "More waste
oil has run down the creeks from the famous Glenn pool than was ever
produced in Illinois," which was an exaggeration, yet far too near the
truth. The state legislature finally passed a law prohibiting unnecessary waste of oil.[6]

Concerning various wastes, the Director of the United States Bureau
of Mines has said:

> We have been wasteful, careless and recklessly ignorant. We have aban-
> doned oil-fields while a large part of the oil was still in the ground. We

have allowed tremendous quantities of gas to waste in the air. We have let water into the oil sands, ruining areas that should have produced hundreds of thousands of barrels of oil. We lacked the knowledge to properly produce one needed product without over-producing products for which we have little need. We have used the most valuable parts of the oil for purposes to which the cheapest should have been devoted.[7]

Waste of oil and oil products has arisen at all stages in the production of that resource. In the first place, too much oil has, for various reasons, been left in the ground. In the second place, there has often been great waste in bringing new wells under control. In the third place, there has been waste in the storage, transportation, and refining of oil; and in the fourth place, there has been immense waste in the consumption of all oil products.[8]

It would be impossible to estimate or even give an intelligent guess as to the amount of oil that has been rendered inaccessible in the oil sands by water flooding and otherwise, but there can be no doubt that it would run into immense figures.

"Water troubles" may be due either to natural conditions or to a combination of natural conditions and human carelessness or ignorance; and until recent years most such troubles were largely the result of inexcusable negligence. Water caused the final ruin of almost all the older fields; and it is the omnipresent and greatest menace of producing fields.[9]

Oil, in most fields of the United States, and, in fact, throughout the world, occurs in inclined or sloping beds of sand or other porous rock; and these oil zones or sands are sometimes separated from the water zones by impervious clay, shale, or other strata, although in some fields oil and water are found in the same sand, the oil above the water. Water zones above the oil sands are called "top" waters, and those below the oil sands are called "bottom" waters, while water in the same sand with the oil but farther down the dip is called "edge" water. In a properly finished well the "top" water is cased off or cemented off before the well is drilled into the oil sand. The "bottom" water is never drilled into except by accident, in which event it is plugged off. With the top water shut off and the bottom water untouched, the oil is produced practically free from water, unless there is water in the oil sands. Since water is heavier than oil, it replaces part or all of the oil wherever it is admitted to the oil sands, and so increases the cost of recovery. Many water troubles have been due to failure to shut off the top water in the process of drilling. Wells, holdings, and entire fields have been seriously damaged or entirely ruined by the water, sometimes from only a few offending wells. Every large oil field is exposed to great danger of flooding through careless or inefficient drilling and operation of wells.

Geologists who have investigated underground wastes with considerable care declare that the number of instances in which the drowning of wells, attributed to natural exhaustion, could have been prevented or delayed by proper methods, is much greater than is commonly supposed. It must be said, however, that in recent years there has been a great increase in the care and intelligence used by the larger oil companies in fighting water troubles.[10]

The damage caused by flooding has, of course, varied according to a number of local conditions—according to the volume and head of the water, the volume and pressure of the oil and gas, and the character of the oil and the oil sand. Under some conditions, water has been known to make the production of oil impossible, while under other conditions the producers have even considered small amounts of water with the oil to be desirable. In fields with high gas pressure, the water has sometimes "cut the oil," making an emulsion that is hard to treat. It has been estimated on reliable authority that at one time the Cushing field produced 25,000 barrels of "cut" oil daily, much of which was allowed to run into the Cimarron River. Single wells in California have produced thousands of barrels of "cut" oil daily.

In some California fields the water seems to "set" the sands, and it has sometimes been found difficult to get them to produce after water has once appeared. So marked has been this tendency that it has been the practice of some California operators to drill into the oil sand with oil instead of water in the hole.

Damage from water in the sands tends to increase in times of low prices. Wells which are suffering from water intrusion cannot usually be shut down, because the water would gain such headway as to ruin the wells. Operators have often been confronted with the alternative of either pumping at a loss to protect their property, or of abandoning their wells permanently. Not infrequently it has proved necessary to do the latter.

In the Cushing, or Glenn, or Burkburnett, or Kern River, or Sunset-Midway, or Signal Hill, or any other large field, where hundreds of wells were drilled, some of them by inexperienced men in poorly financed oil companies, it was inevitable that some wells should be improperly drilled, and carelessly operated later, in spite of all efforts of the state authorities to see that the laws were observed. In most fields, and particularly in the closely drilled fields, all operators are to some extent at the mercy of indifferent or careless drillers and operators. The Cushing field in Oklahoma had not been producing long before operators began to notice that some of their wells were producing a large amount of water mixed with the oil. The Bureau of Mines afterward claimed to have reduced the volume of water lifted from these wells from 7,520 barrels to 628 barrels per day, by proper cementing, at the same time greatly increasing the volume of oil produced.[11] In spite of the efforts of the Bureau of Mines, however, the Cushing field has been suffering a great deal from the inroads of underground water, even within the last year or two.

In several of the California fields, the water problem has been critical for many years. In the Kern River field, for instance, a serious water problem began to develop within a few years after the discovery of the field, in fact, considerable trouble was encountered as early as 1903, and conditions gradually got worse. In 1916, some wells were shut down, and others were being operated at great expense, because of the cost of lifting the increasing volume of water. About this time the State Oil and Gas Supervisor and the State Mining Bureau began work on the problem, and finally closed one of the wells, which was found to be letting water into the oil sands. This reduced the water production of the wells over a considerable area, but it did not shut out the water altogether, and the Oil and Gas Supervisor continued his investigation, finally, in 1919, issuing a report in which he listed 76 wells, or 23 per cent of those examined, as probable sources of water infiltration, and named 60 of these as needing immediate repairs. The repairing of these wells improved the situation somewhat, but water has always been a menace in this field. Some companies plugged certain of their wells and tried various suggestions of the supervisor, yet in places the sands are still badly flooded, even where only an exceedingly small portion of the oil has been extracted. In September, 1919, the wells in this field were producing daily over 43,000 barrels of water, the cost of lifting this water being doubtless more than $1,000 per month. One of the reasons given for the failure of operators to take greater care of their wells in the first place was that the over-production of oil had so reduced their incomes as to make it impossible for them to meet the expense. Thus, in another way, has over-production been the cause of waste of oil.[12]

It was not the Kern River field alone that suffered, however. In the Midway field a serious water situation developed at least as early as 1915, and at the close of 1917 the estimated daily production of water from two sections in this field was twice as great as the oil production from the same areas, and the water production from 4,000 acres investigated by the State Supervisor was 15,500 barrels per day. Two years later the average production of the wells in this area was about 39 barrels of water and only 14 barrels of oil. In the Huntington Beach field there has recently been considerable trouble with water.[13]

The intentional flooding of the older Pennsylvania fields, particularly the Bradford field, in an effort to increase recovery, has been discussed elsewhere, but it is worth noting here that some authorities are not convinced that this is wise practice. At least one authority on the subject, J. O. Lewis, of the United States Bureau of Mines, has distrusted this procedure, and has urged that it be used only as a last resort.[14]

Under conditions that have existed hitherto, a very large percentage of the oil has been left in the ground—from 20 per cent to 90 per cent, according to various authorities. The percentage left underground differs with conditions in the different districts, and even with different sections of the same pool. It differs also with the price of oil, and with the efficiency of the companies dominating the various fields. I. C. White has given the estimate of 25 per cent for West Virginia; Arnold and Garfias, 40 to 60 per cent for California; Ashburner in 1887, 90 per cent for Pennsylvania. As an average for the United States, perhaps the commonest estimate of the amount left underground is 50 per cent. I. L. Dunn made the estimate of 25 to 85 per cent; Chester Naramore, 90 per cent; C. W. Washburne, 36 to 60 per cent; Walter F. Rittman, 70 per cent (1916); and A. W. Ambrose, "possibly" 70 per cent (1920). J. O. Lewis, of the United States Bureau of Mines, after a careful consideration of the main authorities, came to the conclusion that perhaps between 80 and 90 per cent of the oil is not recovered under present methods of exploitation.[15]

There are various reasons for believing that a very large percentage of the oil is left in the ground. In the first place, wells are seldom entirely exhausted when abandoned, and in eastern fields, as the price of oil increased, many old wells were rejuvenated and made productive again by what is called the Smith-Dunn or Marietta process, or by flooding. New wells drilled among the old wells show that much oil still remains in the sands, and in some of the older Appalachian fields a second or even a third crop of wells has been drilled. In the second place, since the gas in the sands escapes much more rapidly than the oil, and since commercial oil production ceases when the gas is exhausted, it seems probable that only a minor part of the oil is brought to the surface. After the gas is gone, only the comparatively weak force of gravity remains to expel the remaining oil. This, of course, emphasizes the tremendous injury done by wantonly wasting the gas in the oil fields.[16] In the third place, figures of the average capacities of various sands, compared with figures of actual recovery, indicate that a very large proportion of the oil remains underground. The total production of the United States previous to 1914, with the estimated total future production of the 179,129 wells producing in 1914—a total of 4,700,-000,000 barrels of oil—from a total area of about 2,000,000 acres, would approximate 2,350 barrels per acre, or barely enough to saturate an oil sand of average porosity two feet thick; and the average oil sand is much thicker than that. Lewis figured also that the gas sands are no greater in capacity than the oil sands, and the 591,866,733,000 cubic feet of natural gas produced in 1914, not including an enormous wastage of perhaps an equal amount, would, at an average pressure of 250 pounds, occupy a space equivalent to 5,850,000,000 barrels of oil, nearly double the total oil production to that date, and 22 times the oil production of 1914.

In other words, the oil sands are probably as great in capacity as the gas sands, and in one year the gas sands produced gas which occupied as much space as nearly six billion barrels of oil. Following the same line of investigation, Lewis estimated that the recovery from the Appalachian fields had been about 1,000 barrels per acre, or only enough to saturate an oil sand one foot thick, with a porosity of 12½ per cent; and the average thickness of the pay streaks alone has been estimated at five feet in West Virginia, and at least that in the Appalachian fields, as a whole. The Bradford field of Pennsylvania has produced only 2,700 barrels per acre, or about enough to saturate 3 feet of sand, and the sand in this field is supposed to average 45 feet thick and to be oil bearing and comparatively uniform throughout. The past production of Illinois and the future estimated production of wells now producing amounts to about 2,750 barrels per acre, enough to saturate an oil sand of 17½ per cent porosity two feet thick, yet the "pay streaks" in the average well of Illinois have been estimated at 25 feet thick. The total average production, past and future, from the developed fields of Oklahoma will, it is estimated, approximate 3,400 barrels, which would not saturate three feet of sand of the estimated porosity, although the average thickness of the Oklahoma oil sands is many times greater. Figures of gas production in one section of the state tend to show sands which average 24 feet in thickness. The total past and estimated future production of the Kern River field in California amounts to over 40,000 barrels per acre, enough to saturate about 20 feet of sand, whereas a thickness of 200 to 300 feet is reported for that field.

Thus, the best estimates available agree that a very large part of the oil is never recovered, and one of the most recent and authoritative estimates puts the percentage left at from 80 to 90 per cent. We may make fair deductions for errors and over-estimates, but we cannot well doubt that the percentage of oil left in the ground is very high. Much of the oil would not be recoverable under any conceivable circumstances; some of it has been permanently lost which might have been recovered if we had followed a saner policy in exploiting our oil fields; and some of it will be recovered at much greater expense because of the manner in which we have "developed" our fields. The flooding of oil sands, already referred to, has almost ruined some fields, and will injure other fields, as they approach commercial exhaustion. The rapid and wasteful exploitation of natural gas has greatly reduced the recovery of oil in many fields, notably in the Cushing field.

The abandonment of wells in times of over-production and low prices has rendered much oil either entirely inaccessible or accessible at a very high cost. In the Allegany field, in New York, for instance, hundreds of wells about as good as those now producing were "pulled out" at the time of the excitement in other fields. Twenty or thirty years later many of these wells were redrilled, and made to produce small amounts of oil, but of course this involved a large duplication of expense and the permanent loss of much oil, for some oil which could have been secured in 1885 was no longer available after surface water and mud had for two or three decades percolated into the sands through these abandoned wells. In other fields, likewise, when prices were low, many wells were pulled out and plugged which had considerable amounts of oil in them. In the pulling, water often leaked into the oil sands, and when these wells are later redrilled, as many of them almost certainly will be, some of the oil will no longer be recoverable. Thousands of wells are abandoned in the United States each year, and some of these wells are producing, but perhaps not enough to pay expenses, with crude oil selling at 50 cents or $1 a barrel.

Within the past two or three years, a great many small wells have been abandoned because of the cheapness of oil. Many of the eastern wells produce only a few barrels a day, in fact it has been estimated that over 29 per cent of the gasoline-bearing oil produced east of the Rocky Mountains comes from fields in which the average daily production is ten barrels or less. There are many wells in the eastern fields which produce less than a barrel a day, and such wells have been abandoned by the hundred. The waste involved in this has been recognized by some public officials. Governor Donahey of Ohio recently declared that he did not wish gasoline prices to go too low, because that would throw hundreds of low-producing oil wells out of business for all time, in the state of Ohio; and Governor Walton of Oklahoma announced that the state, in order to conserve its resources, would act to prevent the sale of oil and gasoline at too cheap a price.

The custom of paying "flat" royalties to landowners is another cause of the abandonment of wells of small yield. The operator usually must pay a certain royalty, regardless of the production of the wells, and so is forced to abandon them whenever they cease to produce enough to cover his own expenses and the royalty too. It is unfortunate that the royalties cannot be graduated according to the flow of the well, for then operators could afford to pump them somewhat longer.

The outstanding reason for waste is always the same, the unfortunate cheapness of the oil. When crude oil is selling for 10 cents or 50 cents or even $1 a barrel, it may be profitable for the individual operator to drill carelessly, to take relatively few precautions against water infiltration, and to abandon wells long before they are entirely exhausted.[17]

While over-production and low prices force the abandonment of many small wells and so reduce oil recovery in one way, it is now generally conceded that rapid drilling of a given field, with close spacing of wells, may result in greater oil recovery from the oil sands than slower drilling with wider spacing of the wells. It is now generally believed that the recovery from rapidly and closely drilled fields, such as Burkburnett and Long Beach will be greater than it would have been if fewer wells had been drilled, and had the exploitation been more deliberate.[18]

Waste in the controlling of new gushers has been a spectacular feature of many fields. It has been so common in some fields as to cause little comment. About 400,000 barrels of oil were wasted from the Lucas well at Spindletop. Several million barrels of the production of the Lakeview gusher in the Sunset-Midway field were reported to have been lost, and a few years later, immense quantities were lost from another gusher in the same vicinity, a gusher reported at 50,000 barrels production per day. Some idea of the waste from the Kern River field in California may be gained from the fact that a company controlling the land near the mouth of a ravine running down from the oil field installed a catch-basin to gather the oil wasted by companies having wells along the ravine, and in the course of eight years recovered nearly 250,000 barrels of oil, or about 85 barrels per day.[19] It was estimated in 1916 that 10 to 15 per cent of the California oil was either used as fuel or wasted in production. In general, however, it is said that California gushers have been capped with less waste than those of Texas and Mexico.

Losses of this kind were very great in those fields of Texas and Louisiana where "wild wells" were common; but it would hardly be just to criticise operators for failing to control some of these eruptions. In Texas, in 1919, the State Oil and Gas Department made the rather significant ruling that the blowing of wells over the derricks, just to get pictures, was waste, and a violation of the conservation law.[20]

Some waste from new wells is of course inevitable. It is not to be expected that operators, every time they drill a well, should have at hand sufficient apparatus and storage to control and handle the flow of a Lakeview gusher, for some wells will be dry, and many will be small producers; and many of the important oil fields are so far from railroads and sources of supplies that it costs heavily to get storage tanks and pipes to the scene of operations. In wildcatting, this is particularly expensive, because the wells are often in isolated regions, are scattered,

THE LAKEVIEW GUSHER

Flowing between 40,000 and 65,000 barrels per day. One of the two greatest oil wells in the oil history of the United States. The other was the Lucas gusher at Spindletop.

Courtesy W. E. Herkemer, Photographer

A CRATER IN THE SMACKOVER FIELD, ARKANSAS

From the state-federal report on "Eastern Smackover, Arkansas, Oil and Gas Field."

and are naturally more likely to be dry than in the proved fields. With
all consideration for these facts, it is nevertheless true that there has
been immense unjustifiable waste of oil from new wells, in every field in
the country. Prospectors have often drilled when they were wholly unprepared to take care of the oil; and in some instances have opened
wells merely so they could sell their land to advantage. The total waste
in this way in the past thirty or forty years has certainly amounted
to many millions of barrels.

Investigations carried on by the United States Bureau of Mines
have been far from excusing much of the waste in capping wells:

With the present state of our knowledge regarding the situation of the
"gusher" and "gasser" strata in the developed fields, there is no excuse for
the general lack of precautions taken before the depth is reached at which
the flow is expected. Furthermore, if the well is being drilled in a new or
prospective field, adequate precautions should be taken so as not to jeopardize in advance the possible profits of such an uncertain and expensive undertaking. The additional cost of the safety devices is insignificant in comparison with the total cost of drilling such wells, and with the amount that
can usually be saved if their production is regulated. It may be safely stated
that practically all the great waste resulting from the unrestricted flow of
gassers and gushers can be prevented easily by known means which generally are within reach.[21]

In most new fields, the first production has been run into earthen
sumps and stored there until storage tanks could be erected, and in
some fields these open earthen sumps have for years been used regularly
for storing oil. In many fields, operators have drilled new wells so fast
that they were never able to get enough steel storage for all their production, and so have used the earthen sumps even after the first need
for them was past. A very large part of the oil in the Beaumont field, in
Texas, was at first stored in earthen reservoirs, where it deteriorated so
much that it was later sold for from 3 cents to 10 cents a barrel. It was
estimated in 1905 that the oil producers of Texas and Louisiana were
losing over $1,000,000 a year through evaporation and seepage from
earthen reservoirs. Perhaps waste of this sort has been worse in California than anywhere else. In 1910 the Standard Oil Company alone
had 15,000,000 barrels of earthen storage capacity in California. It
is easy to imagine the heavy loss that a lake of oil suffers, standing for
weeks or months under the scorching sun of interior California.[22]

Mark Requa described the situation prevailing in the Midway field
of California in 1910 and 1911 as follows:

The waste of oil is appalling. Brought to the surface, it is allowed to lie
for months in open earthen sumps. Storage tanks of steel, concrete, and
earth are full to overflowing, and yet the daily surplus of 31,000 to 51,000

barrels accumulates and is in part dissipated by evaporation. Probably not less than 4,000,000 barrels, and possibly double this amount, was lost last year by seepage and evaporation. This year will see quite as much similarly wasted. Much of this loss could be eliminated by agreements among the producers. . . . To improve prices and relieve surplus, the suggestion has been made that large quantities of oil be burned.[23]

In 1914 the Bureau of Mines estimated that 10 to 15 per cent of the total gross production of the state of California—between 10,000,000 and 15,000,000 barrels per year—was being wasted in this way; and some reports later than this have given a still higher figure. In the early days of the Glenn pool, in Oklahoma, one company alone had 1,000,000 barrels of oil in earthen storage. In the Healdton pool much oil stored in this way had to be sold at a greatly reduced price. In 1912, Raymond Blatchley was sent to investigate the waste of oil and gas in the Mid-Continent fields; and his report, issued the following year, discussed the waste in various Mid-Continent fields, described the manner in which vast quantities of oil were stored in open pits until the lighter constituents had evaporated and left a heavy residue, which was sometimes burned to get it out of the way. The Mid-Continent oil was of much lighter grade than California oil, and the loss from evaporation of the lighter constituents was much higher. In 1920 the Bureau of Mines estimated that 6.2 per cent of the Mid-Continent production— over 12,000,000 barrels, or 510,000,000 gallons—was lost by evaporation alone, and this of course represented the lightest and most valuable part of the crude oil—the gasoline content. One authority has estimated that 20 per cent of the gasoline in the crude oil is lost.

The following quotations from a bulletin of the United States Bureau of Mines give a graphic picture of the wasteful manner in which oil has been handled in Oklahoma:

The first pool visited was the Bartlesville, in the vicinity of Dewey. The oil in this area produces a heavy residue, which is allowed to collect in the receiving tanks, then drawn off, and collected in a waste pit. Those operators who are particular of the condition of their properties cause the collected residue to be burned periodically. In many cases, however, the residue is allowed to flow over the retaining pit and down the gullies to the nearest stream. It often is allowed to run out of the waste pit and over the ground adjoining the tanks, where it is absorbed by the soil.

The oil (from the Cleveland pool) . . . yields much residue, caused by the oil being churned and emulsified by gas and water. Under such conditions, the product can not be marketed until the oil is practically free of sediment. In several cases the flow pipes are turned into large pits, where the oil and residue separate by standing. The loss by evaporation is large in such instances. After the oil separates, the remaining residue is burned. In some cases where the gas is wasted close by, the oil waste is not burned for fear

THE STORAGE POOL OF THE ROXANA COMPANY, AT SMACKOVER, COVERING 34 ACRES, 50 FEET DEEP IN PLACES

S. C. Wilson, Photographer

of conflagration and therefore is allowed to stand permanently. Evaporation continues until the mass has the appearance of a vat of grease.

A large open pit, containing thick residue, was observed on the southwest edge of Cleveland. The pit is about 125 feet in diameter and contained about 25,000 barrels of the residue. The pit was built when the Glenn pool was opened and developed in 1906-1907. The immense flow of oil from the Glenn pool required the full capacity of the pipe lines then in use. As a result, the Cleveland oil had to be stored temporarily. No runs of oil could be made, and since many of the wells had to be pumped, the oil was stored in the best manner possible. For this reason, the Cleveland pit was built and about 250,000 barrels of oil stored in it. Evaporation, however, quickly reduced the oil to a thick residue. A large proportion of the mass was later shipped from Cleveland. The balance remained in the pit several years.[24]

Even in very recent times, there has been extensive use of earthen sumps, as in the Eldorado and Smackover fields in Arkansas, where hundreds of thousands of barrels were thus stored. Here the loss from evaporation was not so serious, because the oil was of low grade, most of it fit only for fuel.

Not all the evaporation losses have been from earthen reservoirs, of course. In some regions big open concrete reservoirs have been built, which represent an advance over earthen sumps, but allow a great deal of evaporation. In California some of these have been built to hold several million barrels of oil.

Experiments made by engineers in the Bureau of Mines in the Mid-Continent field indicate that the annual evaporation losses of crude oil from a 55,000-barrel tank amount in value to $9,900 as gasoline, or $2,140 as crude oil. The loss from seepage and evaporation in large, well-constructed earthen reservoirs, holding as much as 500,000 barrels, and covered with good roofs, which prevent air currents from sweeping over the surface of the oil, varies from 1,500 to 4,000 barrels per month for California crudes of fairly light grade. A loss of several thousand barrels per month has been known to occur.[25]

Fires have destroyed millions of barrels of oil, sometimes millions in a single year. Oil in storage and flowing oil wells are always in constant danger from this source. Ignition has been caused by fires under the boilers or in the forges, by unprotected lamps, burning matches, cigarettes, automobile sparks, by heat in the bearings of the bull wheel or band wheel or other machinery, and more frequently by lightning. It is claimed that the friction of pebbles, striking each other or striking the casing as they are expelled with gas or oil, has set on fire the more inflammable hydrocarbons, and some authorities believe that static electricity has caused fires. In the Spindletop field, in September, 1902, a fire destroyed 100 derricks, and at one time 50 of the largest wells were afire. This "ten-acre fire" lasted a week, destroyed hundreds of thousands of barrels of oil, and cost the lives of twenty men. In one week during the summer of 1914, fires in the Healdton and Cushing fields of Oklahoma destroyed 86 tanks of oil, ranging in size from 250 barrels to 55,000 barrels, besides numerous oil rigs and other property, with a total loss of $1,000,000. A fire swept a small portion of the Cushing field in 1916, destroying $2,000,000 worth of oil and other property, including 100 derricks, and scores of homes, pump houses, etc. At one time 20 wells were burning, some of them producing thousands of barrels of oil per day.

One of the most disastrous oil fires of recent years occurred at Greenpoint, New York, in September, 1919, when 25 large tanks were destroyed, with a total loss of about $5,000,000 worth of oil and other property. A fire at Casper, Wyoming, in June, 1921, destroyed nearly 350,000 barrels of oil before it could be put out. In the years 1908 to 1918 the Director of the Bureau of Mines reported that fire destroyed a total of 12,850,000 barrels of oil and over 5,000,000,000 feet of gas. In the same period over 500 oil fires were reported, of which 310 were caused by lightning and 193 by other causes. In some California fields the loss of oil from fire was estimated at 5 per cent in 1918.

It will perhaps be argued that fire losses are unavoidable in the handling of a highly combustible commodity. With respect to part of the loss, that is true, but very much of this loss has arisen from the wild haste and carelessness with which oil fields were exploited, from the production of more oil than could be properly handled, and from the fact that oil has been so cheap that it was not worth guarding carefully. Oil stored in open sumps, or in wooden tanks, or in steel tanks with wooden roofs, or in the many types of makeshift reservoirs or tanks common in most of the oil fields, is, of course, in great danger from fire, but it has been proved that the erection of properly designed steel tanks eliminates the fire hazard almost entirely, and if the oil had not been brought out in such insane haste, there would have been time to erect proper storage for it. It might also be pointed out that if there had been less over-production of oil, there would have been less oil in storage. Thus the fire waste has been one of the unfortunate results of over-production and cheap oil.[26]

It is fair to state that within recent years the more efficient oil companies have greatly reduced the loss of oil from uncontrolled wells, from evaporation, and from fire; and the charge of careless and wasteful methods is no longer so justly made as formerly. Most of the waste at the present time arises, not from carelessness of the individual operators, but from the unfortunate system under which they work, the system of private ownership and exploitation, under which it has been impossible to exercise any control over production.

500,000 BARRELS OF OIL GOING UP IN SMOKE, AT POINT FERMIN, CALIFORNIA

The petroleum refineries of the United States have certainly been the most efficient in the world, and the charge of wastefulness could hardly be laid upon them in general; yet "efficiency" in the refining of cheap crude oil has naturally meant more emphasis on economy of labor and apparatus than on economy of the raw material, and some of the smaller refineries have fallen far short of the standard set by the large Standard companies, and even by some of the large independent companies, such as Sinclair, Cosden, and Phillips. That some refineries are almost criminally inefficient is proved by the fact that in some cases Standard refineries have been built close to other refineries or "skimming plants," and have been able to make a handsome profit in refining the residue which the other plants leave after skimming off the lighter components. The Bureau of Mines has estimated that about 4 per cent of the oil run to stills is unrecovered in the form of products. Part of this loss is inevitable, but part of it could be avoided, and would be avoided, if oil were worth saving.[27]

While transportation of petroleum in the United States is likewise the most efficient in the world, yet there is considerable loss in transportation which could be avoided if oil were worth saving. According to the United States Bureau of Mines, the loss of oil through evaporation and leakage approximates 2 per cent in the pipe lines, 1 per cent in the gathering lines, and 1 per cent in the trunk lines. In shipping petroleum products in bulk, in tank cars, there is considerable loss. In the most carefully built but non-insulated tank car with dome cover and safety valve fitting tightly and accurately, gasoline sometimes evaporated to the extent of 5 or 6 per cent in a six-day trip in warm weather. Insulated tank cars are built which eliminate some of this loss, and they will probably be used more as gasoline becomes more expensive.[28]

The wastefulness of many of the refineries is shown by the large percentage of skimming plants and topping plants. Skimming plants, as the name suggests, remove only the lighter fractions from the crude oil, selling all the rest as fuel oil. They represent the simplest and cheapest type of refinery, and make only a rough separation of the raw material into a few products in ready demand. These plants produce a considerable proportion of the country's gasoline supply, but are wasteful of the lubricating values contained in the oil. They have sometimes been very profitable where accessible to cheap oil, but are not able to operate on expensive crude. In 1920 nearly 40 per cent of the refinery capacity of the United States was of the skimming type. Most of the plants were in the Mid-Continent region near the producing fields yielding oils rich in gasoline.[29]

The so-called "topping plant" is a variant of the skimming plant used for heavy crudes, such a those of California and Mexico, which contain small percentages of the lighter products. This type of refinery has been chiefly concerned with the producing of fuel oil, removing from the crude oil the volatile components which, if left in the fuel oil, would make it unsafe for general use. Like the skimming plants, the topping plant has been inefficient and wasteful of some of the most valuable products in the oil.

In various other ways, oil has been wasted in large quantities. Considerable amounts of oil have been lost through leaks in tanks and pipes and through oil spray being whipped away by the wind as it flows into the tanks. When oil is so cheap as to be hardly worth saving, there are many ways in which it is wasted.

NOTES

1. U. S. Bu. of Mines, *Rept., Director,* 1914, 23.

2. *S. Doc. 363,* 64 Cong., 1 sess., p. 16.

3. Am. Petroleum Inst., *Proceedings,* Annual Meeting, 1920, p. 57.

4. U. S. Bu. of Mines, *Tech. Paper 70,* 7.

5. Ok. Geol. Survey, *Bul. 22;* U. S. Geol. Survey, *Professional Paper 116,* 68.

6. U. S. Bu. of Mines, *Tech. Paper 45,* 24; *R. of Rs.,* Jan., 1904, 29, 58.

7. Pogue, *Economics of Petroleum,* 350.

8. U. S. Bu. of Mines, *Tech. Papers 45* and *70; Outlook,* March 13, 1918, 42; *R. of Rs.,* Jan., 1909, 49-.

9. Ralph Arnold, in *Economic Geology,* X, 704.

10. U. S. Bu. of Mines, *Tech. Paper 130,* 5.

11. U. S. Bu. of Mines, *Bul. 163;* U. S. Geol. Survey, *Bul. 705,* 16.

12. See annual reports and various bulletins of the Cal. State Oil and Gas Supervisor, particularly *Bul. 82;* and *Cal. Oil Fields,* particularly issues of Sept., 1919, and Oct., 1921; also *Min. and Sci. Press,* Apr. 13, 1918, 517.

13. *Cal. Oil Fields,* July, 1919; Dec., 1919, 7-.

14. U. S. Bu. of Mines, *Bul. 148,* 108-; *Natl. Petroleum News,* Feb. 15, 1922, 67; *Oil and Gas Jour.,* April 20, 1916, 29; May 3, 1917, 3; *Oil Weekly,* Oct. 10, 1924, 16, 17.

15. U. S. Bu. of Mines, *Bul. 148,* 25-32; *Oil and Gas Jour.,* Nov. 30, 1916, 2; *American Petroleum, Supply and Demand,* By the Committee of Eleven of the Am. Petroleum Institute, 97.

16. *Oil Weekly,* June 28, 1924, 22.

17. On the general question of the amount of oil left in the ground, see: U. S. Bu. of Mines, *Bulletins 148, 177, 194; Tech. Papers 51, 130; Min. and Sci. Press,* Dec. 9, 1911, 739; *Mineral Industry,* 1909, 555; *Oil News,* March 5, 1922, 40; *Oil and Gas Jour.,* Jan. 14, 1915, 3; Nov. 30, 1916, 2; Aug. 30, 1917, 2; April 9, 1920, 60.

18. *Oil Weekly,* March 6, 1925, 23; *Mineral Industry,* 1923, 477.

19. U. S. Bu. of Mines, *Tech. Paper 70,* 7.

WASTE OF OIL

20. *Oil and Gas Jour.,* Nov. 28, 1919, 81.

21. U. S. Bu. of Mines, *Tech. Paper 42,* p. 4 (1914).

22. *Oil and Gas Jour.,* Dec. 24, 1920, 86; July 29, 1921, 2; U. S. Bu. of Mines, *Tech. Papers 42* and *70;* U. S. Bu. of Mines, *Rept., Director,* 1913, 103; Cal. State Oil and Gas Supervisor, *Third Ann. Rept.,* p. 10; *Econ. Geol.,* Vol. 11, 323, 324; *Sci. Am. Supp.,* April 11, 1914, 229.

23. *Min. and Sci. Press,* Nov. 18, 1911, 644.

24. U. S. Bu. of Mines, *Tech. Paper 45,* pp. 21, 22.

25. *Oil and Gas Jour.,* Nov. 26, 1914, 2; Oct. 15, 1920, 86; *Oil Weekly,* June 26, 1925, 60-67; U. S. Geol. Survey, *Bul. 394,* 46; *623,* 20; *Econ. Geol.,* Vol. 11, 324; *Sci. Am., N. S.,* Dec. 10, 1910, 462; *Oil Weekly,* Aug. 19, 1922, 18; Nov. 3, 1923, 66; U. S. Bu. of Mines, *Tech. Papers 45, 70,* and *319; Bul. 200; Eng. and Min. Jour.,* Sept. 24, 1910, 591; Jan. 6, 1912, 94; *Petroleum Age,* Sept. 1, 1922, 87; Oct. 15, 1922, 78; July 15, 1923, 26; Aug. 15, 1924, 16; *Natl. Petroleum News,* Nov. 22, 1922, 30; *Oildom,* Nov., 1922, 34.

26. U. S. Bu. of Mines, *Bul. 170; Tech. Paper 42,* 7; *Natl. Petroleum News,* May 31, 1922, 21; Sept. 6, 1922, 39; *Petroleum Age,* Dec. 1, 1922, 24; U. S. Bu. of Mines, *Rept., Director,* 1918, 99; *Sci. Am. Supp.,* May 18, 1912, 315; Feb. 24, 1917, 120; *R. of Rs.,* Jan., 1904, 58; *Oildom,* Nov., 1922, 29.

27. Am. Petroleum Inst., *Proceedings,* Annual Meeting, 1920, p. 57; *Oil and Gas Jour.,* July 1, 1921, 80; Pogue, *Economics of Petroleum,* 345.

28. Pogue, *Economics of Petroleum,* 345.

29. Pogue, *Economics of Petroleum,* Ch. VI, VII.

CHAPTER XV

WASTE OF OIL (Continued)

THE greatest and most unpardonable waste of oil, however, has been its consumption for all kinds of low or unimportant uses, particularly its use as fuel, in place of coal, and its use for all sorts of unessential automotive purposes.

The amount of oil which has been devoted to unimportant uses is to some extent indicated by the vastly increasing per capita consumption. This is shown by the following table, from the United States Geological Survey:

Consumption per capita of crude petroleum in the United States, 1870-1921.

Year	Population (thousands)	Consumption of crude petroleum (thousands of barrels)	Annual per capita consumption (barrels)
1870	38,558	5,261	0.14
1880	50,156	26,286	.52
1890	62,948	45,824	.73
1900	75,955	63,620	.84
1910	91,972	191,483	2.08
1920	105,711	530,532	5.02

These figures do not cover the years since 1920, but there is no doubt that the per capita consumption has increased very greatly in the past four or five years. One estimate has put the per capita consumption in 1923 at 6½ barrels.[1]

Certainly one of the worst forms of waste is the use of oil as fuel. The use of oil for fuel purposes was tried as early as 1862 in Pennsylvania, and not long after this was claimed to be successful. In the early eighties, considerable oil was being burned in California, and its use for this purpose increased rapidly during that decade. During the late eighties and early nineties, its use as fuel was much discussed in the engineering and scientific journals. In 1888 Dr. Charles B. Dudley presented to the Franklin Institute of Philadelphia a comprehensive paper dealing with the subject of oil fuel for locomotives, based on a series of experiments on the Pennsylvania Railroad.

Of the total oil production of Ohio in 1889, no less than 12,153,188 barrels were used as fuel, only 317,037 barrels for illuminating and 1,740 barrels for lubrication. This waste was unavoidable at the time, since the refining methods of the time were inadequate to permit its utilization for any but fuel use, but in the light of present knowledge it seems unfortunate. In the early nineties the French government was conducting experiments in its use for torpedo boats, but it was said to have been found unsafe because easily ignited. A decade later, with the tremendous increase in the oil production of California and Texas, much attention was given to stimulating oil-burning, and its use was greatly increased. The Santa Fé Railroad had 180 oil-burning engines in 1901, and not long afterward the Southern Pacific began to convert its engines into oil-burners. In 1904, practically all the railroads in California were reported to be using oil fuel, as were also the street railways, stationary engines, machine shops, and manufacturing establishments.[2] Even before this, the Royal Shell Oil Company burned oil in one of its tank steamers, partly as a matter of experiment, and some of the big line ship companies were considering its use. The United States Geological Survey even tried to help in the movement, by working out comparisons of the fuel value of coal and petroleum. The chief market for the floods of Texas oil in 1901 and 1902 was as fuel, and in Louisiana the sugar refineries and other industries were turning to oil-burning about this time. The unfortunate cheapness of the oil was, of course, one of the necessary conditions to such a change. About 1910 considerable agitation arose for the use of oil by the United States Navy, and some types of war vessels were constructed to burn oil either exclusively or as auxiliary to coal, in fact, some battleships were authorized at this time to burn oil exclusively. Within a few years after this, all battleships and torpedo vessels built by the United States Navy were designed to use oil fuel. Other nations were following a similar policy.

In 1917 the total consumption of fuel oil in the United States was over 200,000,000 barrels, or about 61 per cent of the marketed production of petroleum in the entire country. The railroads used 45,000,-000 barrels, and the United States Navy over 2,000,000 barrels. In 1922 the railroads consumed 44,752,344 barrels, merchant vessels 51,-996,000 barrels, and the Navy 5,800,000 barrels. In 1923, the consumption by merchant vessels increased to 63,000,000 barrels.

In the report of the United States Shipping Board, oil was credited with the following advantages over coal, as a fuel in marine transportation:

1. Less bunker space required, a barrel of oil being equal to a ton of coal, and occupying four-sevenths as much space.
2. Oil can be carried in spaces not available for cargo, as, for instance, the double bottom, leaving for the cargo the space ordinarily occupied by the coal.

3. Greater dispatch in bunkering, a consideration of special advantage at a time of shortage of ships.
4. No labor or machinery required to handle ashes.
5. No stoking required, permitting reduction in labor costs.
6. Uniform pressure easily maintained, thus insuring a steady speed, and reducing boiler depreciation from uneven temperature.[3]

Not only is oil less bulky, but the heat value of a ton of oil is considerably greater than the heat value of a ton of coal. An oil-burning ship can go much farther with a given weight of fuel cargo, a consideration of great importance in naval strategy; or, to express it in another way, an oil-burning ship can carry more cargo in proportion to its displacement. It has been estimated that a ship which can sail 1,800 miles on a full cargo of coal, can sail 3,000 miles on oil, without reloading. The significance of this in time of war can easily be seen. At the same time, because of the ease with which the oil is handled, a much smaller crew is required, and of course the expense of "oiling" is very much less than the expense of coaling a ship. It is reported that when the White Star liner *Olympic* was converted to an oil-burner, the work of "oiling" the ship for each trip required only ten men working eight hours, whereas the coaling of the ship had required the labor of 300 men working from four to four and a half days. In other words, the labor hours were reduced from 9,600 to 80. At the same time, the number of firemen required to keep up steam was reduced from 246 to 60, and their work made much more pleasant, or at least more tolerable. Similarly, when the *Leviathan* was converted into an oil-burner, the number of firemen in the engine and firing rooms was reduced from 455 to 285 men.[4]

There are still other advantages, too many to be discussed in detail here: the absence of clinkers or ashes, the absence of soot, a great reduction in the amount of smoke, cleaner boiler flues, and reduced wear and tear on firedoors. All these advantages have led to a rapid increase in the amount of oil burned as fuel, particularly by steamships, and in 1920 it was estimated that 16.3 per cent of the world's tonnage had turned to oil fuel.

Recognizing the very great advantages of fuel oil for naval purposes, most officials of the Navy have proceeded on the theory that the Navy must and will have oil fuel for an indefinite future, because it is "absolutely necessary" to the national defense. It is reasonable to suppose that the Navy will get oil fuel as long as there is oil fuel to be had, for, as Redwood says, "This demand is not primarily governed by price," but is "irresistibly preferential in character," taking precedence over all industrial needs. During the World War we had some reason for rejoicing that we had an oil-burning navy, for it was, of course, one of the factors contributing to our success in the war, even though we might not put it as strongly as Lord Curzon, who said that we "floated to victory on a wave of oil." Another point of view, however, is worth considering. It is perhaps more important, from the point of view of national defense, that we should have a small amount of fuel oil for our submarines and plenty of lubricating oil and certain other products of petroleum for a long time in the future, than that we should have fuel oil for all sorts of war crafts for a decade or two. We should probably have been able to turn the scale in the Great War, even with a coal-burning navy, but the question of an adequate supply of lubricating oil and gasoline and other oil products for an indefinite future, not only for direct naval and military purposes, but for the industries that work behind the Navy and Army—that question is entitled to some very careful consideration. Perhaps an oil-burning navy is not a wise investment for national defense, in the long run. One thing seems almost certain. We, or our children, will some day, in a time of national peril, look back regretfully at the wanton waste of oil and oil products in these days of plenty, and we will perhaps squander vast sums of money in bootless efforts to secure satisfactory substitutes for material which we have wasted.

This general view has been taken by a number of writers on the subject. Sir Boverton Redwood, and a number of English authorities, have been rather cautious in their attitude toward oil-burning in the English navy, fearing that the supply of oil might prove inadequate; and many American authorities have been unwilling to urge oil-burning very strongly. The United States Naval "Liquid Fuel" Board, in its report of 1904, was modest in its demands, although it recommended that oil fuel be tried. Pogue has recently given his view that it is "questionable . . . if the world's resources in petroleum can support, for more than a transient period, widespread shipping operations on an oil-fuel basis. Ultimately automotive transportation on land may be expected to come into direct competition with marine transportation for petroleum supplies, and then the economic advantage now in favor of petroleum fuel oil for ocean shipping will be forced back to the side of coal in some form."[5] The recent decline in the production of the Mexican fields, on which both English and American navies have drawn heavily for fuel, lends some support to a cautious attitude.

Oil fuel has somewhat the same advantages in railway locomotives that it has in ships, and the railroads have consumed immense quantities of oil. Since 1910, the annual consumption has run from 23,000,000 barrels to over 40,000,000 barrels; reaching a total in 1922 of over

44,000,000 barrels. In 1923, the consumption was probably more than this. In addition to this, public utility power plants in the United States consumed 13,000,000 barrels of fuel oil in 1920, and the consumption for this purpose was growing. Almost half of this consumption was in California, and considerably more than half was in California and Texas.

A large proportion of our artificial gas is also made from petroleum, from what is known as "gas oil," the fraction just above fuel oil, notwithstanding the fact that coal can be used for this purpose. Over 18,-000,000 barrels of gas oil were consumed in this way in 1920.

This wasteful use of oil is an anachronism for which the public is largely to blame. About 1880 it was discovered that water gas, which could be made much more cheaply than coal gas, but which lacked luminosity, could be made luminous by enriching it with gas oil. At that time the gas used as an illuminant in cities was burned in an open flame, and so had to be luminous. Since then the old type of open burner has been practically discarded in favor of the Bunsen flame, in the use of which the luminosity of the gas is a matter of no importance. With this change, the reason for using gas oil practically disappeared, yet many municipalities still require the gas companies to continue to use it, as if the product were still used in the old-fashioned open-flame burners; and in 1920 some cities, including New York, even granted substantial increases in rates, in order that the gas companies might be able to afford the purchase of the costly hydrocarbons for this use. Thus millions of barrels of gas oil are consumed for a use for which other fuels would serve just as well, despite the fact that this gas oil has in it valuable lubricating elements which ought to be put to a higher use.[6]

Within very recent years there has been a very considerable use of fuel oil for domestic purposes. During 1922 and 1923, when oil was cheap, a great many people installed oil-burners, and the consumption of oil in this way has risen rapidly. During the late fall of 1923, Henry L. Doherty, of the Cities Service Company, put on a "drive," with 50 special salesmen, to stimulate the sale of oil-burners, and the results were claimed to be remunerative, and beneficial to the oil industry in general.

Oil burned as fuel has been of several different kinds. Some of it has been the residue left after refining, after the more valuable elements have been extracted. Some of it has been crude oil of the heavier grades, either "topped" for whatever valuable factors might be in it, or entirely untreated. Much of the California oil has been burned without any treatment or refining whatever. Some of the oil burned, finally, has been of high enough gravity to be valuable for its gasoline and lubricating factors, yet it has been burned with all these elements in it. In recent years this sort of criminal waste has been less common than it was formerly.

There are many who think that oil fuel will play an ever increasing part in our economic life, who believe that its use will expand until ships, railroads, industrial plants, and even home furnaces burn little else. The overthrow of "Old King Coal" is prophesied by numerous enthusiasts, some of them practical oil men, who believe that the supply of oil is unlimited, and that all uses may be supplied for an indefinite period. As one writer expressed the idea a decade ago: "The supremacy of oil in its ever broadening sphere is now so well established that nothing except a falling off of the supply could cause it to give ground, and this contingency, thanks to the incalculable promise of the North American fields, especially those of Mexico and California, is not one that need be considered for many decades to come."[7] More judicious spirits, however, see that the burning of oil is a low use, and that it will be impossible to supply such a use for an indefinite future.

Oil which is unfit for any other use may properly be burned, and some oil is said to be unfit for any higher use. Much of the Mexican, Gulf Coast, Louisiana, and Arkansas oil, and some of the California and Montana oil, has been classed as fuel oil, and millions of barrels have been burned without refining. Thomas O'Donnell testified in 1910 that approximately 75 per cent of the oil produced in California was burned in its crude state, and only 25 per cent went to the refineries.[8] In 1921, 82,000,000 barrels of crude oil were burned without refining.[9] There was immense waste involved in this, for later developments in refining and higher prices of crude and refined products have demonstrated that much of this "fuel oil" is capable of yielding valuable products, particularly lubricants, when properly treated. According to Pogue, a vast volume of potential lubricants is burned annually in the form of fuel oil. The refineries of the country produce only 5.4 per cent of lubricants, while the average lubricating content of the crude oil supply of the country is probably above 25 per cent. In other words, four-fifths of the potential lubricants is used in other ways, much of it burned in the form of fuel oil.[10] Even some of the practical oil men have seen the waste involved in oil-burning, as it has often been carried on. Mark Requa has often spoken of it as inexcusably wasteful, and E. L. Doheny has declared that the amount of oil burned must be reduced.[11] Secretary Lane, of the Department of the Interior, insisted that no oil should be burned until every valuable product was squeezed out of the crude.[12]

The unfortunate cheapness of the product is here again the cause of waste. With crude at from 10 cents to 50 cents a barrel, and gasoline and other products proportionately cheap, it was unprofitable to refine the heavier grades of oil, and they were burned, in competition with coal. Every time that over-production has brought down the price of crude oil, producers have carried on an energetic campaign to stimulate the use of oil for fuel, have spent thousands of dollars in advertising, in an effort to induce the public to consume this irreplaceable natural resource for one of its lowest uses. Thus, in October, 1921, when operators were producing more oil than there was any proper use for, the National Petroleum Marketers' Association was formed to push a "burn oil" campaign; and the next year a number of such "campaigns" were inaugurated. Of course, the oil-burner equipment manufacturers were interested in such a movement, and the National Oil Burner Equipment Association was formed in Chicago in June, 1922, for the purpose of bringing together all the elements in the industry to promote the use of fuel oil for heat and power. As one of the oil journals described the situation in July, 1921: "It is an unaccountable spectacle, is that mad orgy among refiners to sell fuel oil at a price 50 per cent below the dead line of coal competition."[13]

Even the lighter grades of oil have been burned at times, and without refining, millions of barrels of it, and this can only be regarded as criminal waste. As Requa has said: "Burned under boilers, fuel oil is at best wastefully consumed, and at worst is losing in every barrel 20 per cent to 30 per cent of lubricating stocks that we shall some day need, and need badly. Over any long period of time its future use in this manner is without excuse or justification."[14] A vast amount of lubricating stock has been burned under boilers, consumed for a "50-cents-a-barrel" want, when it might in the future have served a "50-dollar want." Our policy has been like that of the man who burned mahogany furniture in his fireplace because there happened to be an oversupply of such furniture in stock.

H. L. Doherty has stated the case against oil-burning in the following forceful words:

That we should permit conditions which allow enormous quantities of this, the most valuable and perhaps the least plentiful of our natural resources, to be wasted is astonishing. Perhaps there is nothing else to do but to rape these various fields, and waste this oil by burning it for purposes for which inferior fuel, to be had in abundance, would be just as valuable, but it is certainly remarkable that the oil industry has never sat down and tried to figure out how this flagrant and lamentable waste can be prevented.[15]

The fuel oil burned under boilers in the generation of steam is wastefully consumed anyhow, because of the small percentage of its power that is rendered available when burned in this way. The average marine boiler is only about 60 per cent efficient, because of the thickness of metals through which heat is transmitted to the water inside the boilers, and because of the many losses on account of air leaks, radiation, soot accumulation, and the like. This means that for every barrel of oil burned under the boilers, under the most favorable circumstances, only about three-fifths of the practical available power in the fuel is secured in the form of steam. About 40 per cent of the available power in the steam is then lost, by radiation, condensation, and in other ways, in transmission of steam to the engine. Another loss of approximately 12 per cent is encountered in overcoming friction of moving parts and operating pumps attached to the main engine, and in converting the up-and-down motion of the pistons to the rotary motion of the shaft. These calculations indicate that under most modern methods for burning oil under marine boilers, the power transmitted to the propeller shaft represents less than 32 per cent of the available power in the oil. Under the most inefficient methods, the percentage is of course much lower.

The use of the Diesel engine would increase the power recovery to 75 per cent, or more than double it. Requa says it would triple it. One pound of oil fired under the boilers of a modern marine steam plant results in approximately .9 shaft horse power, whereas that amount in a modern marine Diesel engine produces about 2.2 shaft horse power. "I seriously question," declared Mark Requa, in 1920, "whether the marine steam engine burning fuel oil is not today, as a matter of fact, an obsolete piece of machinery."[16]

Thus, we have been wasting more than half of the energy latent in the hundreds of millions of barrels of oil that we have burned as fuel, by burning it under boilers instead of in Diesel engines. Furthermore, much of the oil burned has been wastefully consumed, for the additional reason that much of it has been burned in badly constructed boilers, and under inefficient superintendence. Oil has been too cheap to be worth saving.[17]

If oil is to be burned, the principle of priority should doubtless be applied in its use, in order that the maximum satisfaction may be derived from it. It would be possible to draw a scale of fuel uses, from the highest, most important, or most advantageous, to the lowest, least important, or least advantageous. Oil fuel is more important for submarines than for battleships, more important for battleships than for merchant ships, more important, or more advantageous, for merchant ships than for locomotives, and more advantageous for locomotives than for stationary engines or for brick kilns, more advantageous for private homes than for hotels or large institutions. It is true that price is an effective, if not always intelligent, selector of uses; and as the price of fuel oil rises, the less important or advantageous uses will tend to be the first eliminated. It is entirely possible, or even probable, however, that government intervention will be invoked to make the selection of uses more certain and more intelligent, just as in the case of natural gas. As long as the stupid faith in the inexhaustibility of our oil resources is firmly held by the public, no such step is possible, but when oil production is once definitely on the decline, there will be more uneasiness about certain future uses, as, for instance, those concerned with the national defense, and the government will perhaps intervene.

Oil-burning is not the only low use to which crude oil has been put. Large amounts have been used in the oiling of roads; and the asphalt so much used in road-building has valuable lubricants in it. Railroads have even used crude oil to oil their roadbeds, to lay the dust. The Southern Pacific and the Santa Fé railroads used large amounts of oil in this way at least as early as 1904, and a number of eastern roads have used oil thus for a number of years. Petroleum undoubtedly is a valuable material for road-making, and for laying the dust on railways, but these are low uses that have seldom been justified, and will almost certainly be discontinued when the price reaches a reasonable level.

The development of motor transportation during the past two decades has been one of the most rapid and spectacular economic developments in all history, comparable in its magnitude with the development of steamships, railroads, or electric utilities. In 1899, 3,700 passenger and commercial cars were produced in the United States, worth less than $5,000,000 wholesale. Ten years later the production had increased thirty times, to 127,731 cars, valued at $165,000,000; and ten years later this had increased over fifteen times again, to 1,974,016 cars, worth nearly two billion dollars. In 1923, only four years later, the total production of motor vehicles had almost doubled again, with a production for the year of over 4,000,000 motor vehicles, including passenger cars and trucks; and the output was still growing. The Ford Motor Company alone produced 7,500 cars a day in December, 1923. This industry, which amounted to next to nothing in 1899, within two decades became perhaps the greatest of American industries.

There are in the United States at present over 15,000,000 passenger cars and over 2,000,000 commercial cars or trucks—83 per cent of the world's supply. This is an average of one for every eight persons in the United States. In some states the ratio of motor vehicles to population is much higher than this, as the following table shows (for the year 1924).[18]

Number of Persons per Passenger Car in the United States

State	Pop. per Pass. Car	State	Pop. per Pass. Car
California	3.38	Arizona	7.58
Iowa	4.29	Idaho	7.63
Oregon	4.64	Delaware	7.93
Nevada	4.76	Rhode Island . . .	8.01
Nebraska	4.80	Utah	8.02
Kansas	4.85	Maryland	8.04
South Dakota . . .	4.99	Connecticut	8.05
Colorado	5.02	Massachusetts . . .	8.28
Michigan	5.07	New Jersey	8.34
Indiana	5.30	Pennsylvania . . .	8.66
Minnesota	5.37	Montana	8.75
Wyoming . . .	5.45	West Virginia . . .	9.20
District of Columbia .	5.55	New York	9.21
Ohio	5.69	New Mexico . . .	9.33
Wisconsin . . .	5.77	North Carolina . . .	9.78
North Dakota . . .	5.97	Virginia	10.89
Washington . . .	6.04	South Carolina . . .	11.89
Vermont	6.18	Kentucky	11.92
Oklahoma	6.30	Louisiana	12.26
Florida	6.46	Tennessee . . .	13.02
Texas	6.68	Arkansas	14.49
Illinois	6.94	Mississippi . . .	14.66
New Hampshire . .	7.03	Georgia	16.53
Missouri	7.04	Alabama	17.49
Maine	7.19		

The consumption of oil products by automobiles has increased prodigiously. In 1920 approximately 25 per cent by volume, and 49 per cent by value, of the output of the American oil industry, was used by automobiles, and it has increased considerably since. In the decade from 1911 to 1921, the quantity of fuel and lubricants annually consumed by automotive transportation in this country increased from 3,000,000 barrels to approximately 100,000,000 barrels; while during the same period the value of the oil products consumed each year by automotive transportation increased from $9,000,000 to approximately $1,000,-000,000, and it has increased considerably since then. It has been estimated that the automobiles in southern California consume an average per car of between 800 and 1,000 gallons of gasoline annually. The total consumption of gasoline in the United States in 1923 was nearly seven billion gallons.

With the vast increase in automotive transportation has gone a phenomenal development of subsidiary and supplemental industries.

Steel, copper, and aluminum have been demanded in rapidly increasing quantities for the building of automobile bodies, while vast amounts of leather and other fabrics have been needed for upholstering. One of the most spectacular developments has been that of rubber manufacturing. Imports of rubber increased more than 600 per cent in the ten years between 1912 and 1922, from 101,000,000 pounds to 674,000,-000 pounds; and over 50,000,000 rubber tires are now manufactured annually, to keep our automobile wheels running smoothly.

The manufacturers of automobiles are devouring all sorts of raw materials in vast and rapidly increasing quantities. It is estimated that in 1924 they used 80 per cent of the total rubber supply consumed in this country, 52 per cent of the plate glass supply, 7.5 per cent of the copper supply, 46 per cent of the aluminum, 11 per cent of the iron and steel supply, and 69 per cent of the upholstery leather supply.

For the selling and care of our vast fleet of automobiles, a great army of people are employed, in thousands of business establishments. It is estimated that there are 181,000 motor vehicle dealers and salesmen in the United States, 50,000 public garages, 67,000 service stations and repair shops, and 65,000 supply stores. The tourist who drives through the automobile supply district in a city like Denver or Los Angeles can scarcely help wondering whether the people ever buy anything but automobiles and automobile supplies. It is estimated that more than 3,000,000 persons are employed in various branches of the automobile industry.

Automobiles call for good roads, and vast amounts of money have been spent for highway improvement. It is estimated that expenditures in 1923 amounted to over one billion dollars. In 1914, they had amounted to $240,000,000. No other nation ever built roads as fast as the United States has built them since automobiles came into use.

Along with all this has gone an amazing transformation in our economic life. Horses have almost disappeared from our cities; and a considerable part of the business section of every city is devoted to the needs of motorists. "Filling stations" have occupied almost every corner in some parts of our cities.

The benefits arising from the wide use of automobiles are not to be disregarded: the saving of time; the opportunity afforded many people in large cities to live in the outskirts of town, away from congested apartment districts, and yet reach their business places easily; the opportunity to drive out into the country and get a breath of fresh air; the privilege of travel in a relatively inexpensive way, open to people of moderate means; the opportunity afforded country children to attend good consolidated schools, through the use of this rapid means of transportation; the prompt delivery of mail, milk, groceries, and all sorts of commodities; a thousand benefits that we now take as a matter of course, yet would miss greatly if they should be withdrawn. The recreational use of automobiles has become very important. Literally millions of people now leave the warm and humid regions of the United States each summer and go to the mountains, the lakes, or the seashore, for a vacation. A total of 1,422,353 visitors were recorded in the National Parks in 1924; and all of them rode in automobiles—their own or national park cars. More than 315,000 automobiles entered these parks in 1924; and over 10,000,000 motorists visited the various national forests of the country in the same year.

Doubtless we are more efficient, and have a greater per capita income, because of our immense fleet of automobiles, trucks, and tractors. Careful statisticians have estimated that the per capita income of the United States, in goods as well as in money, increased appreciably between 1909 and 1919;[19] and it is not unreasonable to suppose that the automobile, truck, and tractor had something to do with this increase. In 1924 there were in the United States over 17,000,000 motor vehicles, with an average horse power unquestionably above 20 (Ford cars are rated at 22 horse power), a total of probably 300,000,000 potential horse power of energy, far more than the total horse power of all the locomotives on our railroads (if locomotives could be rated in horse power), and more than thirty times as much as the total developed water power in the United States. This amounts to an average of about three horse power per capita, or twelve horse power per family. It is easy to believe that the man with twelve horse power of energy at his disposal is more efficient than he was before this was made available.

Perhaps the most beneficial result of the introduction of automobiles, trucks, and tractors has been to relieve horses of a tremendous burden of cruel work. The task of bus and cab horses, clattering up and down the hard brick pavements of our cities, was a killing task, one that ruined the best of horses in a very few years. In the small country towns, the long rows of horses tied to the hitching posts on Saturdays, many of them tired, underfed, and overworked, have now been displaced by rows of automobiles, with a resulting vast diminution in the hardships borne by horses. Even in the fields, the tractor has taken over some of the hardest work that horses previously had to do. The automobile, truck, and tractor have probably done more to reduce animal hardships and suffering than all the humane societies in the country.

Much of the opposition to automobiles has arisen from somewhat the same causes that have been responsible for opposition to most other improvements in human history, particularly from the conservatism and fear of change which is strong even in changeful America. The introduction of steamships and railroads was at first viewed with considerable distrust by many people. Buggies, especially rubber-tired buggies, bicycles, telephones, moving pictures, pianos, organs in churches, Victrolas, even public education, etc., have at first been looked upon by some people as new-fangled notions, temporary crazes, in some cases even dangerous to the public morals. It is recorded that bathtubs were for years looked upon as mere expressions of extravagance and snobbery and that there were even some efforts to legislate against them. Probably some of the hostility to automobiles, as far as there is such a hostility, is similar to the earlier opposition to some of these other innovations. And, while in the transitional period, when people are getting accustomed to higher speed, some of them seem inclined to lose their heads, in the "long run" they will probably take their automobiles much as a matter of course, in fact, most of them already do that. If automobiles could be made cheap enough so that people of moderate means could really afford to drive them, and if they could be looked upon as a permanent part of our equipment, rather than as a transitional phenomenon, we could hardly fail to welcome their use.

Considering the fact, however, that oil is a strictly limited natural resource, indispensable for certain purposes, as far as our present knowledge goes, we shall have to confess that much of our automobile consumption of oil is unjustifiable from every point of view. Much of it represents extremely unimportant uses. Vast amounts of our irreplaceable reserves of oil are being used—we may well say wasted—by fat-bellied bankers and *bourgeoisié* riding back and forth, to and from their business, when they need nothing so much as a walk; by gay boys and girls in questionable joy rides, when they most need a little more companionship with their parents and contact with their homes; by youthful, and by mature, smart Alecks, who find here an exceptionally flashy and effective way of flaunting their wealth before those not so fortunate as themselves; and by all manner of men and women, who ride about upon our roads and streets in stupid, thoughtless, aimless, pointless diversion, the essential element of which diversion is the entire absence of any serious tax upon the mind, when they might be improving their minds with good books or with anything that might tend to mental enlightenment. Many people are terribly bored whenever they are thrown upon their own resources for entertainment, and for such the automobile has unquestionably satisfied a want, however unfortunate it may appear that part of our precious oil resources should have to be put to so low a use as the entertainment of such minds.

Some will deny that our automobiles, with all the speed and zest they have added to our life, have greatly raised the quality or increased the intelligence of our citizenship; and certainly many Americans who find time to learn all about spark plugs, carburetors, differentials, and magnetos, could scarcely tell whether Lithuania is the name of a physic or a European state, or whether the Ruhr is a new Ford attachment or an industrial region. A fairly plausible argument might be built up to support the thesis that the automobile has in many ways tended to make our lives cheap and shallow and superficial, has fostered a general craze for speed and change, and a general neglect of thrift and economy and some of the modest virtues. The effect of the automobile craze on the American home would make a rather interesting study. One thing, however, seems reasonably certain, without much study. Many people who are complaining of profiteering landlords and of the hardships of apartment life, would have homes of their own if they had not invested the price of their homes in automobiles which no longer function.

The tendency of the automobile to encourage extravagance is certainly one of its worst aspects. The total annual cost of the automobile in the United States, in 1921, was estimated by Dr. Edwin G. Slossen at $7,783,000,000.[20] The items in this total run as follows:

New cars	$1,448,000,000
Depreciation	1,800,000,000
Interest	295,000,000
Fires	450,000,000
Gasoline	823,000,000
Oil	175,000,000
Garage	552,000,000
Repairs and supplies	1,000,000,000
Insurance	185,000,000
Taxes	275,000,000
Drivers' salaries	600,000,000
Road maintenance	180,000,000
	$7,783,000,000

Whatever may be said as to the accuracy of some of these items, the total is probably not far from the truth. Assuming a total national income of perhaps 60 billion dollars, we may conclude that our automobile expenditure represents about one-eighth of our total income.

It may perhaps be argued that consumers know what they want, and may be trusted to buy that which affords them maximum utility. The difficulty is, however, that many people—certainly the vast majority of them—do not have even a vague idea as to what their automobiles cost. Most people know that they have to pay the initial price of their machines, and that they must buy gasoline and oil and tires from time to time, but few of them allow properly for interest, depreciation, insurance, garage rental, and incidental expenses. Many owners of automobiles underestimate their expenses about 50 per cent, and so are in no position to tell whether they are getting their money's worth or not. Not a few are like the lady who kept a strict account of her expenses on her new car, but found that it spoiled her pleasure entirely, so threw away her account book—instead of her car—and had a "perfectly grand time" thereafter. The records of chattel mortgages everywhere show that many people are buying automobiles who have not even enough to pay the first price of a car. A prominent Kansas City banker once reported the failure of a Kansas bank which had over one-third of its loans and discounts in the form of automobile paper. It has often been said that three-fourths of the automobiles sold are sold "on time," and it is reported that there are in New York City alone at least fifteen finance companies, each with a capital of $1,000,000 or more, whose function it is to finance such purchases of automobiles.[21]

Bankers have recognized from the very first that people are likely to buy automobiles beyond their means. In the early days of the motor industry, bankers often refused to grant credit for such purchases, and the purchase of a car did not help any man's standing with his bank. To some extent, this attitude still prevails. Only three or four years ago, the bankers of the United States became so alarmed at the number of automobiles purchased on the installment plan, by those who could not really afford them, that special restrictions were placed upon bank loans for the purchase of automobiles. Even the Federal Reserve Board—a group of fairly sensible men—found that "keeping up with Lizzie" was too much for the American people, and adopted regulations to save them from their own extravagance.

Foreigners visiting the United States are always impressed by the tremendous number of automobiles, and many of them indulge in more or less philosophizing as to the influence of these on American life. Some Europeans pronounce us "motor mad." A rather sane English view runs as follows:

Undoubtedly many people in the United States own cars who would be much better off without them. None the less, individually, there is everything to be said in favor of any man possessing one who can legitimately afford and has proper use for it. It is an immense economizer of time and, what is more important to the worker, of physical strength. It gives him and his family the means of rational recreation in the fresh air and of much increased social contact with their friends. In the aggregate, however, there is no question that the motor car adds greatly to the wastefulness and fever-ishness of American life. The possession of a car inevitably tempts to much extravagant expenditure. The demands of the industry raise the prices of many things—like rubber and plate glass, which are essential to other fundamental industries. Most serious of all, perhaps, is the immense wastage, or non-essential use of gasoline. It has been estimated that this waste runs to some 3,000,000 gallons a day, taking the country as a whole.[22]

As already stated, if we could reasonably look forward to an indefi-nite supply of this cheap power and lubrication, we should rejoice in each car added to our sputtering millions, but here is the crux of the whole question. It is a certainty that our oil supply will gradually approach exhaustion, and that we shall not be able to maintain indefi-nitely our present vast number of automobiles; and it is a practical certainty that we shall some day be unable to provide some of the very essential needs, because we are now satisfying both our essential and non-essential needs so lavishly. We are now providing for the most frivolous and unimportant automotive wants, in the face of a certainty that we shall some day have to pay for it by going without some of the essential services that automobiles afford. As George Otis Smith said not long ago: "The use of gasoline to serve our pleasure can not go on unchecked—the joy ride is not the kind of 'pursuit of happiness' regarded as an 'inalienable right' by our Revolutionary fathers."[23]

Not only have we used tremendous quantities of oil and oil products for unimportant purposes, but we have used them wastefully. Gasoline has been wasted by careless carburetor adjustment, providing too rich a gas mixture, by allowing engines to run while standing, and by the various acts and omissions that are encouraged by the cheapness of the fuel. One writer has estimated that 30 per cent of the total heat in the gasoline is wasted, largely through the exhaust, from improper carburetor adjustment. In the same way, lubricating oil is used prodi-gally wherever it is needed, because it is cheap. No one can be around a garage long without noticing this. But perhaps the greatest waste, or extravagance, arises from the use of automobiles which consume ex-cessive amounts of gasoline and oil. Many automobiles are used which consume three or four times as much gasoline and oil as are necessary for comfortable travelling. On this point *The Lamp* has been quoted as follows:

The United States is the only country in the whole world that insists on high-powered heavy motor cars. It is time that we stopped it. The man who gets from six to eight miles from a gallon of gas, in order to have reserve power for occasional uses that will enable him to push his car along the road at from 50 to 75 miles an hour, is an offender against the public good. For all practical purposes, an automobile that makes 25-30 miles an hour would serve every need. Such a car could be made to average better than 25 miles, and probably closer to 30 miles on every gallon of gasoline.[24]

Perhaps exports of oil should be regarded as waste. Exports have been a result of the constant over-production and extreme cheapness of oil and all its products. Exports have always been extremely heavy, and have grown rapidly in recent years. In 1909 the United States exported over 68,000,000 gallons of gasoline, naphtha, benzine, etc., over 1,000,000,000 gallons of illuminating oil or kerosene, and 161,-000,000 gallons of lubricating oil. In the twelve months ending with June, 1923, exports of gasoline and similar products had increased about 1,000 per cent, to 672,000,000 gallons; kerosene exports had decreased slightly, while exports of lubricating oil had risen to 356,000,-000 gallons. A total of nearly 3,000,000,000 gallons of refined petroleum, and over 552,000,000 gallons of crude oil, were exported during this period. This great total represents a much smaller proportion of our total domestic production than the exports of earlier years did. In 1900, 40 per cent of the total marketable products derived from petroleum was marketed abroad. In general, our exports have included valuable products, particularly some of the finest grade of lubricating oil. A large proportion of the world's machinery is lubricated by American oil.

From the strictly nationalistic point of view, these exports will have to be regarded as mostly waste, for they represent a large deduction from an exhaustible and irreplaceable natural resource, a resource which is absolutely essential and always will be absolutely essential, not only to our national defense but even to our economic life. As Alfred Wallace has said, exports of minerals are in a sense "a permanent and irretrievable deterioration of the country." It represents, let us note also, a deduction for which no adequate compensation has ever been received, for American producers have always sold abroad, as at home, for a price which was far too low. Already students of the oil industry, not only the academic geologists but practical oil men as well, are pointing out the necessity of securing foreign sources of supply to supplement the dwindling American reserves; but when the United States finds it necessary to import oil, from countries other than Mexico or those nearby to the southward, she will have to pay far more than she ever received for the oil she has been forcing upon the world's markets during the past half-century. Admiral Benson pointed out in 1920 how the Shipping Board was unable to secure oil in most parts of the world, even in the ports of great oil-producing countries, except at extremely high prices, and in small quantities. The Shipping Board had to establish its own bunkering stations in all parts of the world and haul oil to these stations from home ports or from Mexico; and all this time American refiners were selling their products in foreign markets for about any price they could get.[25]

It might perhaps be argued that we should not look upon this question from a narrow nationalistic point of view, but should rather treat oil as any other commodity, let it find its way to all markets, and rejoice that American kerosene has lighted the whole world for half a century, that American gasoline is driving automobiles in every country on the globe, and that American lubricating oils are lubricating the wheels of industry wherever wheels turn. It is more than doubtful whether foreign countries that are husbanding their petroleum resources now will treat us in this way when it is our turn to buy in the world's markets. There are many reasons for believing that oil will never be left entirely to unfettered economic sale and purchase.

Since the time seems to be approaching when the United States must become an oil-importing nation, however, it behooves us to be careful not to set any bad examples, and for that reason, probably the best policy would be to try to use our present exports to secure promises of reciprocal treatment when we shall later need to import, or to export only to those countries that will agree to grant the United States the same privilege when she needs it. Whatever success we might have in such negotiations, we could hardly help regretting that we have so lavishly supplied the world's markets for the past half-century.

Doubtless the best method of curtailing exports of oil would be, not to tax or forbid exports, even if such action were constitutional, but to curtail domestic production to such an extent that exports would be unprofitable. This could be done without raising any delicate diplomatic questions.

Most oil producers, and most other men as well, have been entirely unable to see the problems of the industry from the point of view of society at large. They have been unable to see that this industry, like every other, exists, or should exist, for the purpose of meeting certain social wants, and not for the purpose of making profits for landowners and operators. Most people are infected with the business man's outworn theory that business should be carried on in the interests of business men, and have been unable to see that there is anything more important than the private interests of the men engaged in any industry. The various wastes mentioned above have contributed to the private fortunes of the oil operators. When there was too much oil in the market and prices were low, it was profitable to export it, at almost any price, to burn it, or use it for any low and unimportant want that it would serve. It was good for "business" that new uses—and in most cases lower uses—should be developed. Oil was to be exploited in such a way as to make business profitable, and not in such a way as to contribute most to human welfare.

It is not in regard to the oil industry alone that we have adopted the views of business men, but in the oil industry the narrowness and selfishness of that view stand out more glaringly than almost anywhere else. A number of years ago the Governor of Wyoming called the oil men of Wyoming together, and the Wyoming Oil Men's Association was formed, the object being to open up the oil fields and *advertise the oil resources of the state*. At that time oil was being sold in some fields of the United States for 39 cents a barrel—practically being given away; and the Governor and the oil men of Wyoming were really trying to give away the oil in Wyoming, as if it could never be of any use to the people there.[26] In the same way Oklahoma politicians and business men have spent a vast amount of energy in trying to devise ways of inducing people to come into the state and use the state's valuable resources, especially oil and gas, at prices often one-tenth, or even one-fiftieth, of what these resources will some day be worth.

But every state and every community that has had oil or gas has been guilty of the same stupidity. Wasting of irreplaceable resources has almost everywhere been considered "good business," "bringing money into the country," "keeping money in circulation," and "giving employment to labor"; and those who could waste the most resources have not only been the leaders in business, but for their great public service have frequently been rewarded with the highest political and social recognition in the gift of the people.

Some of the words and phrases used in the oil business illustrate very well the failure of those in the industry to think socially or even logically. The word "develop" is an example of this. Oil men always speak of "developing" a field, when they mean exploiting a field, and this usage has become general, in fact, it is found in various places in the preceding pages. Yet, it is a complete misrepresentation of the process that it is intended to describe. We do not speak of "developing" an orange when we suck the juice and throw the rind in the wastebasket. We do not speak of "developing" the coal pile in our basement when we throw it into the furnace. We do not "develop" the forests that we cut down. Yet oil producers are said to "develop" a field when they drain—and largely waste—all the valuable oil deposits of that field. Notable private fortunes are often "developed" in this process, but the oil is merely consumed, that is to say, destroyed.

Oil men often speak of those resources as useless or as wasted which are not now in the immediate process of being "developed." One of the congressional reports, referring to oil deposits in the public lands, bewailed the fact that "for want of adequate legislation too many of our resources are now in a state of nonuse," meaning, of course, that some of the oil was unfortunately being left for future use.[27] A western oil journal pointed out the unfortunate results that might have come from the exclusion of foreign oil companies from American fields: "Had the policy been enforced, the opening of Signal Hill might have been delayed for years." (Signal Hill was opened by the Shell Company, a foreign-owned corporation.) An outsider might well have asked, "What could have been more fortunate than just this?" The price of oil would not have gone so low, less oil and less capital would have been wasted, more oil would have remained for the future when oil will be greatly needed, and even the oil producers would have been far wealthier in the long run, and probably even in the short run. A most happy circumstance indeed, if Signal Hill had not been discovered for twenty years; and Santa Fé Springs and Huntington Beach likewise, and Tonkawa and Burbank and Smackover and Powell, and every field that has been discovered in the past five years.

Before finally disposing of the question of waste of oil, it will be pertinent to define waste of oil as accurately as possible. Not all loss of oil can justly be called waste, of course. For instance, in the capping of some of the wild wells of Louisiana, a great deal of oil was lost; yet, not all of it could properly be said to have been wasted, because the saving of all of it would have called for an expenditure out of all proportion to the value of the oil.[28]

Following this general line of reasoning, some have argued that the loss of oil was not waste, when it would have cost more to save the oil than it was worth in the market. According to this reasoning, very little of the loss of oil which has been described in the preceding pages can properly be called waste, because it would generally have cost more to save the oil than the oil was worth at the time. In other words, it would have cost more to save the oil than to lose it. Mark Requa has expressed this idea clearly:

Much has been said in the past in criticism of the extravagant and wasteful manner in which America has developed and exploited her mineral resources. It was inevitable, I think, that it should have been so. And yet, was it really waste? The very profusion and seeming limitlessness of our resources made for extravagant development and consumption. Why should we conserve, when products were a drug on the market and the available supply apparently without limit? There was apparently neither reason nor necessity. Nor must we forget that the very speed with which we grew was made possible only by ignoring waste and making prodigal use of what we had at hand.[29]

The vast losses of oil may be *explained* in this way, to be sure. When oil has been allowed to run down the creeks, or evaporate in open sumps, it is obviously for the reason that the producer could not afford to save it. Business men can usually (although not always) be relied upon to avoid such losses if they can do so profitably. And when 30-cent oil has been burned as fuel, it was because it was profitable to burn it. No condemnation need be offered the individual operator who failed to save oil or gas at a financial sacrifice, for he could not, do so in a competitive industry. The American people have been extravagant in the use of oil and other resources because oil was cheap, and have been saving of labor and capital because labor and capital were relatively dear. This, as Requa has said, was inevitable in the exploitation of a new continent rich in natural resources, by a sparse and energetic population.

It is thus easy to explain our extravagance in the use of oil, but to justify it is quite another matter. There can be no social justification for the system that permits such extravagance as has marked our use of oil. The individual has failed to conserve oil because it was too cheap to be worth saving; but the system which has permitted such rapid exploitation that oil was not worth saving must be regarded as a gigantic system of wrong. The individual operators of California who allowed millions of barrels of oil to run down the creeks and gullies each year were not really criminals, but the people of California who allowed them to do it, who allowed them to drill for oil when it was not worth saving, cannot easily escape the charge of criminal imbecility.

We may then define waste as *any loss or use which fails to afford the maximum social utility or want satisfaction, for the present and for future generations.* The wants of future generations must be discounted somewhat, because of various uncertainties, but they must be considered. Thus oil which was allowed to flow down the ravine because it was worth only 10 cents a barrel and could not be saved for that much expense, must be regarded as wasted. The hundreds of millions of barrels of cheap oil which have been burned as fuel must be classed as mostly wasted, because their use satisfied a far less important want than they would have satisfied if this oil had been left for the future. The billions of gallons of gasoline and lubricating oil consumed in senseless joy-riding represent largely waste, because they satisfy a "10-cent" or "20-cent" want, where they might have been saved for a "50-cent" want, or, at least in the case of lubricating oil, a "two-dollar" want. From this point of view, we may perhaps estimate that at least four-fifths of the utility in the oil we have thus far consumed has been wasted; and perhaps an estimate of nine-tenths would be nearer the truth.

It is true that the proper conservation and development of human energy is a problem of far greater importance than the conservation of any mineral resource. As Professor Carver says:

The most valuable resource of any country is its fund of human energy, that is, the working power, both mental and physical, of its people. It is safe to say that any capable race of men who will conserve, economize and utilize that fund will be able not only to extract a living but actually to prosper in the midst of poor natural surroundings.[30]

From this it might seem that Americans have been wise in wasting oil, which was plentiful and cheap, to the extent that by doing so they were able to conserve human energy, which was relatively scarce and dear. For the long run, however, this view is entirely incorrect. In the first place, in our exploitation of oil, we have wasted energy almost as lavishly as oil, as will be pointed out in the next chapter; and in the second place, the choice of the American people must be, finally, not between wasting oil and wasting energy, but between wasting them both and conserving them both. Our wasteful use of oil during the past half-century will ultimately have to be paid for by a vast waste of human energy. Having burned millions of barrels of lubricating oil, we, or our children, will some day have to squander great amounts of labor and energy in an effort to produce some sort of substitute—let us say in the form of castor oil. Having consumed billions of gallons of gasoline in senseless motoring, we will some day have to produce some substitute, perhaps alcohol or benzol, at a high cost in labor and energy. Or perhaps we shall be unable to find any substitute at all for some of the oil products, for instance, certain kinds of lubricants, in which case our labor will always be employed at a disadvantage. It is generally if not universally true that those peoples are most energetic and efficient who have the most natural resources with which to work. It is for that reason mainly that the people of the United States, Canada, and some other new countries, exhibit more energy and produce more largely than the people of Europe. If all the countries of the world were listed according to their wealth in natural resources, and according to their labor productivity, most of them would occupy somewhat the same position in both lists; and the position of each country in both lists would correspond rather closely with its standard of what is commonly called civilization.

Rich natural resources, offering a liberal reward to human industry and enterprise, are a great stimulus to human industry, while poor natural resources have almost everywhere meant a lazy and inefficient people. When lubricating oil becomes so scarce and dear that our machinery must be only scantily lubricated, and when gasoline is so scarce and dear that we can hardly afford the most important uses of it, the Americans may be less energetic than they are now. In the long run, waste of oil, or of any other natural resource, means waste of energy.

The idea has been advanced that our extravagance in the use of our resources was not really waste, because it made possible the development of many industries which increased the wealth of the nation. For instance, the editor of the *Engineering and Mining Journal* pointed out that "cheap coal, metals and petroleum are converted into other forms of wealth, and the prosperity of the nation as a whole increases." Another writer advanced a similar view: "While the using up of natural resources is deplorable, still the material gains resulting from their exploitation should not be overlooked, for in a large measure the gradual exhaustion of American petroleum has been offset by the advances of American industries this made possible."[31]

This view is a characteristic business view, and has a superficial semblance of logic, but it cannot be successfully defended. It is true that some industries have been developed on the basis of cheap oil which could not have been developed had oil been expensive. On the Pacific Coast, for instance, particularly in the vicinity of Los Angeles, many industries have been established on the basis of cheap fuel oil, with a resulting large increase in population and aggregate wealth. In the long run, however, probably California would have been wealthier and more prosperous if this oil had been reserved for higher uses, as far as it is adapted to higher uses. Cheap oil is only a passing phenomenon, and can be the basis only of a transient industry. Furthermore, as will be presently seen, our waste of oil has gone hand in hand with a tremendous waste of capital.

For a while, to be sure, our extravagant exploitation of our oil resources may furnish the mechanical energy with which to dig deeper and deeper and so find more and more oil to take the place of that which has been used. That has been the history of the past. It might even be true in the immediate future, but it is not possible that there is an unlimited supply of oil in the earth, no matter how deep or how energetically we may go after it.

A more plausible argument might be built up to prove that there is at least a possibility that the vast floods of cheap oil which have been available for light and power and heat during the past half-century have taught us various uses of oil products which would never have been developed had oil been dearer, and which we may be able to preserve for ourselves from substitutes when the oil supply is exhausted. Perhaps cheap gasoline may have rendered a very great service in making possible the perfecting of gasoline engines and automobiles; and if scientists are later able to find cheap and satisfactory substitutes for gasoline and lubricants, when our oil supplies are exhausted, there will be reason to rejoice that oil was once cheap. The modern automobile would never have been developed with petroleum selling at $20 a barrel.

This line of reasoning, however, proceeds on the theory that satisfactory substitutes will be found, and we are hardly justified in assuming anything which is, to the present time, contrary to fact. We have not yet found any satisfactory substitute for some of the mineral oil lubricants, and until we do, we are not justified in assuming that we shall.

Most of the waste of oil, it must be confessed, was not the subject of much apologizing or philosophizing. We have wasted oil because we were too selfish to consider the interests of future generations. Senator Thomas of Colorado once said that he at first welcomed conservation, but later, when he "saw what it meant," turned to an attitude of hostility. He doubtless "saw that it meant" a measure of self-denial by the present generation, in the interests of future generations, for that is what conservation means, and it is not a popular philosophy. "What do we care if millions of barrels of crude oil are wasted and billions of cubic feet of natural gas go up into the air? After us the deluge!" is the way a western oil journal expressed it. A California operator put it in about the same words: "They say it is a good thing to look out for those who come along next, but I believe in looking out a little for yourself first." Another operator slanted his ideas somewhat differently: "Probably some man a hundred years ago wanted to conserve wax so that his posterity would have candles."[32] Probably most people do little conscious philosophizing, but in most of us the spirit of the ancestral orang-outang doubtless prompts a conscious or unconscious indifference to the needs of posterity.

Without waxing very enthusiastic about those "pioneers" and "venturesome spirits" who wasted the oil resources of Pennsylvania so prodigally, we may recognize for their methods a certain sort of justification which does not apply to the exploitation of later fields. In the early years of the oil industry, almost nothing was known of the nature of oil or of the most efficient and economical means of securing it. The fact that it was an exhaustible resource had to be demonstrated by the gradual exhaustion of individual wells and fields. The knowledge of the nature of occurrence of petroleum, of the best means of drilling, of storing, transporting, and refining petroleum could be learned only by the wasteful method of experiment. In later development, however, much of this was known, and a few far-seeing men were pointing out the necessity for a better system, so that there was no longer the same justification for waste.

With all the lamentable waste of oil, it must be conceded that the oil industry of the United States has been technically a very efficient industry. Most private producers have not wasted outright any great proportion of their oil, for generally they would have found it unprofitable to do so.[33] The really great waste of oil has been in its use for unimportant wants. David T. Day has pointed out that the methods of using it are more satisfactory than the methods of using most other minerals. The oil industry of the United States is unquestionably the most efficient in the world, and in some respects perhaps the most economical. In no other great oil-producing country, perhaps, has a smaller share of the oil been wasted outright, in capping wells, in evaporation, in wasteful methods of drilling, of transportation, or refining. Even burning of oil has been no more conspicuous a waste in the United States than in many other countries, although we have had less excuse for burning oil than countries ill supplied with coal.

The efficiency of American oil operators is the very cause of our greatest waste, because it is the reason, or at least one reason, for the tremendous rapidity with which our oil resources have been exploited, the reason for the cheapness of oil, and therefore the reason why so much of our oil has been devoted to low uses, and why so little is left. Paradoxically, the efficiency of our oil producers in the exploitation of our resources has been the cause of our tremendous waste of oil. The United States would be in a far better position if her oil producers had been less efficient in finding and exploiting her oil deposits, even if, by inefficiency, they had wasted a larger proportion of what they produced.

NOTES

1. *Oildom,* Oct., 1923, 24. It is interesting to note how great is the per capita consumption of oil in the United States, compared with that in other countries. According to statisticians in the Geological Survey, the per capita consumption in the United States in 1923 was about 225 gallons, while that of the United Kingdom was only 31 gallons, France 12 gallons, and Germany less than 3 gallons. The per capita consumption in the United States was more than 27 times as great as that of the rest of the world, as the following table indicates (from *Oil Weekly,* Sept. 19, 1924, 30):

ESTIMATED WORLD CONSUMPTION OF PETROLEUM AND PRODUCTS, 1923

Country	Estimated total oil consumption (millions of gallons)	Production of crude (millions of gallons)	Population (thousands)	Estimated consumption per capita
United Kingdom	1,486	..	47,308	31.4
France	480	21	39,403	12.2
Russia	1,153	1,603	93,388	12.3
Germany	167	15	59,857	2.8
Poland	106	210	27,778	3.8
Italy	175	1	37,528	4.7
Spain	49	..	20,784	2.4
Belgium and Luxemburg . .	78	..	7,743	10.7
Netherlands	183	..	6,841	26.8
Roumania	277	456	17,393	16.0
Canada	715	7	8,769	81.5
Mexico	476	6,278	15,502	30.7
Cuba	275	..	2,889	95.1
Argentina	418	137	9,000	46.4
Brazil	109	..	30,636	3.6
Chile	217	..	3,755	57.7
Venezuela	18	160	2,412	7.5
Other South America . . .	113	290	17,906	6.3
Central America	15	..	5,772	2.6
Dutch East Indies	260	630	50,000	5.2
Japan and Formosa . . .	166	71	60,615	2.7
China	253	..	302,110	0.8
India	471	318	319,075	1.5
Australia	50	..	5,437	9.2
New Zealand	25	..	1,219	20.3
Philippines	83	..	10,351	8.0
Union of South Africa . . .	38	..	6,923	5.5
Egypt	95	44	12,710	7.5
Bunker oil at Mexican ports .	292	..	....	..
Oil handled at Panama Canal .	559	..	....	..
Bunker oil at U. S. ports* . .	1,579	..	....	..
All other countries, other ship consumption, etc.	2,628	1,352	365,116	7.2
Total outside U. S.	13,018	11,592	1,588,000	8.2
United States	25,297	30,870	112,000	225.8
World total	38,315	42,462	1,700,000	23.8

Note: Rumanian consumption figures converted from statistics in the *Moniteur du Pétrole Rumain.* Mexican consumption figures from report by Vice Consul Wm. A. Dunlap, Tampico. World total and U. S. consumption estimated by Mr. W. C. Hill, Petroleum Economist, formerly with U. S. Bureau of Mines. Consumption estimate for Japan from Trade Information Bulletin *Petroleum in Japan* by A. T. Coumbe, Jr. Crude production figures from American Petroleum Institute estimates, converted to gallons.

---

* Laden on vessels engaged in the foreign trade.

2. *Eng. and Min. Jour.*, Jan. 7, 1904, 56. Long before this, in 1882, Thomas Urquehart, Superintendent of Motive Power of the Griazi-Tsaritzin Railway of Russia, had converted 143 Russian locomotives from coal to oil.

3. Andros, *Fuel Oil in Industry*, 159.

4. *The Lamp*, August, 1923.

5. Pogue, *Economics of Petroleum*, 163. See also *Min. and Sci. Press*, Nov. 18, 1911, 642.

6. Pogue, *Economics of Petroleum*, Ch. XXIII, also p. 160; *Sci. Am.*, June 22, 1918, 394.

7. *R. of Rs.*, Oct., 1913, 454.

8. *Hearings on H. R. 24070*, 61 Cong., 2 sess., May 13 and 17, 1910, p. 32.

9. *Oildom*, Nov., 1922, 29.

10. Pogue, *Economics of Petroleum*, 166.

11. *Petroleum Magazine*, July, 1920, 22.

12. *Rept.*, Sec. of Int., 1915, 14.

13. *Oil and Gas News*, July 28, 1921, 15.

14. Am. Petroleum Inst., *Proceedings*, Annual Meeting of 1920, 58.

15. *Natl. Petroleum News*, Sept. 26, 1923, 32 A.

16. *The Lamp*, Dec., 1923, 22; *Natl. Petroleum News*, July 9, 1924, 114; *American Petroleum, Supply and Demand*, by the Committee of Eleven of the Am. Petroleum Institute, 21.

17. U. S. Bu. of Mines, *Bul. 156; Sci. Am. Supp.*, April 11, 1914, 226.

18. *Facts and Figures of the Automobile Industry*, 1925 edition, National Automobile Chamber of Commerce.

19. *Income in the U. S.*, National Bureau of Economic Research, Vol. I, 77, 144.

20. Caldwell and Slossen, *Science Remaking the World*, 30. See this for an interesting discussion of the automobile and its effect on American life.

21. *The Burroughs Clearing House*, April, 1923, 22. In a pioneer community in Idaho, so many people were found to be unable to pay their ordinary bills, because of the purchase of automobiles, that the Chamber of Commerce of the town adopted a rule that the stores should extend credit to no one owning an automobile, unless special exception were made.

22. *Oil and Gas Jour.*, Aug. 13, 1920, 76.

23. Address before the Am. Iron and Steel Institute, N. Y., May 28, 1920.

24. *Oil and Gas Jour.*, Nov. 28, 1919, 80; Sept. 24, 1920, 83; Oct. 22, 1920, 82; Oct. 29, 1920, 80; Am. Petroleum Inst., *Proceedings*, Annual Meeting, 1920, 57; U. S. Bu. of Mines, *Rept.*, 1917, 7; *Eng. and Min. Jour.*, Nov. 1, 1890, 511; June 10, 1911, 1147; *Natl. Petroleum News*, Dec. 13, 1922, 40 B; *Am. Petroleum, Supply and Demand*, Appendix C. For a discussion of the various ways in which automotive engines could be made more efficient and economical in their use of gasoline, see Pogue, *Economics of Petroleum*, 296, 297.

25. Am. Petroleum Inst., *Proceedings*, Meeting of 1920, 43. Dr. David T. Day has often expressed opposition to exports of oil products. (*Oil and Gas Jour.*, June 15, 1916, 24; *Sci. Am. Supp.*, March 28, 1914, 194.)

26. *Mineral Industry*, 1910, 522.

27. *H. Rept. on S. 2812,* 65 Cong., 2 sess., p. 9.

28. Professor Ely has defined "waste," or "wasteful production," as "production which yields a total return to all the factors of production less than that which some other available employment of the same factors would yield at a particular time and place." This definition does not, however, answer our needs at this point. The question as to just what "waste" is has been discussed rather carefully by various writers. (Ely, Hess, Leith, and Carver, *The Foundations of National Prosperity; Sci. Am. Supp.,* Jan. 23, 1909, 63.)

29. Am. Petroleum Inst., *Proceedings,* Annual Meeting, 1920, p. 56.

30. Ely, Hess, Leith, and Carver, *The Foundations of National Prosperity,* 275.

31. *Oil and Gas Jour.,* Feb. 10, 1922, 91.

32. *Inland Oil Index,* Aug. 19, 1922; *Hearings on H. R. 24070,* 61 Cong., 2 sess., May 13 and 17, 1910, p. 63.

33. For an excellent discussion of this point see the *Oil and Gas Jour.,* June 25, 1920, 2.

CHAPTER XVI

WASTE OF CAPITAL AND ENERGY

AS in Pennsylvania, there has everywhere been a vast waste of capital and energy, as well as oil, in the private exploitation of our petroleum resources. This waste of capital and human energy has arisen from the activity of so many men in the field who were ignorant of oil matters; from the secrecy with which the business was carried on; from speculation and fraud; from neglect of other less spectacular industries; from over-production, and the necessity of building immense amounts of storage; from the drilling of too many wells; from the over-development of refineries, service stations, and other capital equipment; from wasteful litigation; and from the unhealthy social conditions surrounding the industry everywhere. Almost all of the evils of private exploitation in Pennsylvania have been duplicated in the other fields as they were exploited.

Any speculative business attracts venturesome men from other callings, and the oil industry has always and everywhere drawn a great many men who knew absolutely nothing about the oil business. If the industry had been left to experienced men—and there have been many such men since the Pennsylvania fields served as a training school—hundreds of millions might have been saved. But men from other callings hurried in to get their share of the "quick money": merchants and farmers and bankers; lawyers and physicians; preachers and professors; widows and orphans and spinsters; the lame and the halt and the blind; the wise and the witless and the half-witted. Dupes by the thousand grasped eagerly the chances held out by smooth-tongued promoters; and the money which should have gone into insurance or homes or bonds went into expensive holes in the ground—or in many instances did not get farther than the pockets of the promoters.

Even where no fraud was involved, amateurs have carried on their oil exploration most wastefully. Knowing little or nothing of the manner of occurrence of oil, or of the geology of the oil regions, they have generally wasted much time and money following down wholly unimportant signs: sometimes taking the scum on iron springs to be evidence of oil; often assuming that all the land for miles about an oil spring was oil producing; assuming, in many instances, that there was no necessity for using geological advice and assistance. Thousands of wells were drilled where competent geological advice would have indicated no chance whatever of finding oil.

In spite of the increasing scientific knowledge available in the oil industry, the "oil smellers" and "doodle bugs" have played some part in almost all of the oil fields. The peach twig has been the favorite instrument in this work, but all sorts of devices have been used. A small tin can filled with some kind of chemical and suspended from a string, the can revolving whenever suspended over an oil pool; a forked instrument with a pointer, one fork to be held in the teeth and the other in the hand, the pointer turning downward when oil was found; a tube with some sort of concoction in the end which was supposed to be drawn downward when over oil deposits; a bottle of crude oil and a piece of twine, the oil to be used as a sort of magnet, etc. One oil finder in Wichita, Kansas, used a bottle of crude oil suspended by a *silk* string, and when he passed over a subterranean accumulation of petroleum, the oil in the bottle became magnetized, and an electric thrill was transmitted to his nervous system. A Long Beach inventor made a mechanical instrument which was supposed to locate oil 28 miles away. Some oil finders used far more complicated instruments, many of them electrical; but some used no instruments at all.

Some have been actual "oil smellers," gifted with so keen a smell as to detect oil through several thousand feet of rock and shale. Some have had "magnetic disturbances" when in the presence of oil deposits. It was said of a certain woman in Wyoming: "Anyone could spread a map out before her, when she would take a pencil and trace it over the map. Whenever the point would reach a locality where there was oil, she would throw a fit."[1] In the Los Angeles fields, it has been claimed that some of the promotion companies even paid 20 per cent commissions to "doodle bug" fakirs who helped them sell stocks, and that the leader of a fashionable occult society in Pasadena was on the sales force of such a company. One of the most prominent mediums in Los Angeles claimed to have made an average income of $6,000 a day for three weeks during the "clean up," or financing of a well.[2]

Blatchley describes one of these wizards who operated extensively in Indiana in 1903-1904. He was sent out by a Chicago firm which advertised stock for sale and promised to guarantee the finding of oil to all who would invest. The instrument which he used was claimed to be electro-magnetic in character, and the wizard said no one but himself could operate it. His theory was that all oil flows in streams, which are continuous from one state to another, and also from one geological formation to another. There were some thirty-five such "oil streams" entering Indiana, mostly from the northeast. With this instrument, he claimed to be able to locate the "oil streams" wherever he went, whether on railway train, on horseback, or afoot. An oil geologist recently told the writer an amusing incident of one of these oil "witchers" who rode along with him for several miles just above the main of a large pipe line which followed the road, all the time declaring that there was no oil "within a mile."[3]

There are many men, some of them intelligent men and even oil operators, who have faith in the efficacy of certain sorts of "doodle bugs." It has even been rumored that some of the large oil companies were experimenting with them, or were actually using them. Some who would not attach much value to the instruments in use at the present time believe it likely that useful contrivances will be invented. J. B. Rathbun, technical editor of one of the leading oil journals, has ventured the opinion that an instrument which would detect oil deposits would be no more wonderful than the X-ray or the radio. He suggests the study of electro-magnetic waves as likely to lead to results. Other writers have suggested the possibility that careful studies of the force of gravity in the vicinity of oil deposits may yield significant discoveries. Certain instruments can be used to advantage in the search for magnetic minerals, such as iron and nickel; and it has seemed to some that instruments would yet be invented which could in similar fashion be used to find oil. It would be a bold man who would set limits to the discoveries that science has in store for us.[4]

Nevertheless, the use of "doodle bugs" in the search for oil has been discredited by scientists and by most oil operators. The United States Geological Survey has consistently warned the public against their use, and the Federal Trade Commission not long ago issued an order forbidding oil companies to advertise that they had found oil deposits by any sort of "divining rod" or other instrument.[5]

The policy of secrecy which has been maintained by operators everywhere has greatly increased the cost of producing oil. Under our system of exploitation, secrecy has been absolutely necessary, but it has involved a tremendous duplication of effort and expense. In some fields, a number of oil companies have prospected, doing several times over the work of mapping out geologic strata, and of collecting well logs. In some instances two or three wells, or even more, have been drilled to secure information that one well would easily have afforded. In many instances, several companies in turn have tested a certain field, where the first test would have been sufficient, had it been accessible to the public. In Wyoming, accurate statistics of oil production were for a long time unavailable, because the oil companies divulged information only to government officials, and under very strict limitations as to how it should be used. Thus, if an oil company failed to strike under certain geological conditions, that fact was not readily obtainable by other companies working under similar conditions, although it might have saved many fortunes in useless drilling.

It may be noted, too, that it was not the oil men themselves who saw most clearly the waste involved in such a system. When agitation arose in California for better logs of wells drilled, and for a law requiring these logs to be filed in some public office, the first argument brought up against these proposals was that they would interfere with the operators' right to secrecy regarding their own business.

Speculation and fraudulent promotion have been rampant in almost every oil field discovered since 1859. F. H. Oliphant, of the United States Geological Survey, estimated that in 1901, 1,578 oil companies were organized, representing over $669,000,000 in stock, and that over one-third of these companies produced "more or less petroleum," while a large majority of the rest had very little value to represent their capital, and many existed only on paper. In Texas alone, according to Oliphant, 619 oil companies were organized in that year, with capital of $283,000,000. This was vastly more than the year's oil production of Texas was worth. It has been stated that 400 oil companies were organized within two months after the Lucas well spouted.[6] There were over 2,000 oil companies organized and operating in California at the close of 1900, and the number was growing rapidly.

H. L. Doherty gave figures for new oil companies from 1915 to 1920, as follows:[7]

	No. companies formed	Capitalization
1915	196	$ 80,533,000
1916	240	419,000,000
1917	512	840,000,000
1918	820	1,430,000,000
1919	1,629	3,786,000,000
1920 (first 10 months)	1,526	2,411,944,000

The National Petroleum News gives figures for later years as follows:[8]

	No. companies formed	Capitalization
1920 (entire year)	1,712	$2,786,814,000
1921	936	1,255,657,000
1922	784	1,639,693,000

Almost every important oil field has had its outburst of speculative activity. Spindletop, Cushing, Burkburnett, Mexia, Signal Hill, Huntington Beach, Santa Fé Springs, and Smackover have perhaps been wildest in this respect, but nearly every field has had its day of oil inebriety.[9]

It has often been stated that Texas was the scene of more dishonest promotion schemes than any other state, although Oklahoma and California would certainly stand "high" in this respect. Blatchley complained a great deal of the operations of Texas promoters in Indiana. One Texas company sold Texas oil tracts in Seattle by means of free moving pictures of the Texas oil industry, disposing of practically worthless plots of land for $150 each. It was reported in December, 1920, that during the preceding six months, 300 Texas oil companies had ceased operations. They had been formed during the boom in North-Texas oil operations, and, having no assets of value, were unable to "operate" after the excitement had died down. Some of them were doubtless not demonstrably fraudulent, but just failed to find oil.[10]

One of the most interesting and extensive promotion schemes ever unearthed was recently brought to light in Texas and Oklahoma. A ring of promoters organized a stock company during the Burkburnett excitement in northern Texas, and when this business "went stale," they divided their properties into a Texas company and an Oklahoma company, and sold stock in these companies for a while. Presently outside brokers were found to be selling the stock of these two companies so cheap that they could make little selling it themselves, so they hit upon a syndicate plan whereby they could get money from new investors and even from their old stockholders. Under the syndicate plan, the parent companies were to receive 50 per cent of the net earnings, in return for operating the properties in the syndicate, while those who invested in the syndicate units were to receive their pro rata share of only one-half of the money which their dollars earned. The lure to the stockholders in the old parent companies was that they not only would receive a direct return from the money invested in syndicate units, but also would participate in the other 50 per cent through the stock which they already held in the old company. The number of units to be issued was not stated, so that investors could not tell how many of these shares were retained by the promoters.

In the Texas office, 150 persons were employed, of whom about 125 devoted their entire time to sending out letters and follow-up literature. This "mail order department" was in charge of an expert who had been a newspaper man and could paint glowing pictures of the opportunities open to purchasers of units. This man worked on a commission basis. In order to get new names for their "sucker list," promoters published at heavy expense an elaborate monthly magazine, with all kinds of clever articles to help in the selling campaign; and when the Mexia field was doing its best they took in hundreds of thousands of dollars each month—as high as $50,000 in one day. With some of this money they purchased a very small tract in the heart of the Mexia field, just enough for one or two wells, and when, one evening after dinner, the first well was brought in at an "estimated" flow of 10,000 barrels a day, the office force was called in and that night 25,000 telegrams were sent out, telling about the great gusher. Of course this brought a flood of money during the next few days; but the well, situated on a small drainage area, was short-lived, soon dwindled to 200 barrels; and investors salvaged something like 16 per cent of their investment.[11]

One financier of Los Angeles estimated that between January 1, 1918, and December 31, 1921, 273 Texas oil promotions were offered in Los Angeles and vicinity, of which, up to 1922, 241 had proved worthless, 12 had paid dividends for a short time, chiefly from the proceeds of stock that was being sold, and 10 were worth a few dollars each in the open market. None were really good investments.

Dr. Frederick Cook, the bogus discoverer of the north pole, has been one of the energetic figures in Texas promotion circles. Cook began operations in Texas in 1919, when he launched the Texas Eagle Oil Company, with $300,000 capital. Soon afterward he floated the Texas Eagle Producing & Refining Company, with $2,500,000 capital. Soon these concerns were consolidated into the Texas Eagle Oil and Refining Company, with $5,000,000 capital. Widely heralded plans of Cook to construct a huge refinery failed to materialize, and in 1921, as one of the journals expressed it, "the last of the Eagles went to the wall." Receivership proceedings were followed by the merging of the remnants with the Revere Oil Company, on the well-known "25 per cent plan." The Revere Oil Company was one of the pioneer Texas "mergers."

In April, 1922, Cook launched the Petroleum Producers' Association, which was modelled after the Revere Oil Company, and brought perhaps larger profits than any other Texas promotion in recent years. Between April, 1922, and the end of the year, Cook is said to have "merged" nearly 300 defunct or badly crippled companies. In each case he applied the "25 per cent plan," which consisted in the exchange of his own stock for the worthless stock of the merged companies, upon purchase for cash of 25 per cent additional stock in Cook's Association.

Although Cook's publicity department reported the earning of large profits, he made extremely meager oil production reports to the office of the Texas Comptroller. Two dividends of 2 per cent each were paid in the fall of 1922. These were six weeks apart, although the literature sent out bore constant reference to "monthly dividends." After paying the second dividend, he announced to the stockholders that in the future all dividends would be paid quarterly, but still on the basis of 2 per cent per month. The first quarterly dividend of 6 per cent was due February 1, 1923, but was not paid. At that time Cook was said to have had nearly $2,250,000 in stock outstanding.[12]

When the government recently brought suit against some 16 Texas promotion companies, it brought to light many interesting facts. The cash receipts of these 16 concerns amounted to over $7,000,000, while the par value of the stock outstanding was over $17,000,000. Some of these companies had taken in over $1,000,000 in their operations. The

Revere Oil Company had stock outstanding with a par value of $8,000,-000. The companies under indictment had "merged" a total of 458 oil companies, in which some 136,000 "investors" had sunk a total of $136,-000,000. It is said that the government indictments of some of these crooks resulted in a great reduction in the amount of stenographic work done in Fort Worth, and a considerable reduction in the amount of mail handled in that city.

In 1917, Casper, Wyoming, was the center of considerable excitement, with half a dozen brokerage houses dealing in Wyoming oil stocks. One reporter stated that "The Casper citizen who hasn't a hatful of certificates of oil stocks under his pillow can't sleep at night." Promoters were said to be making handsome fortunes in their operations. As to the value of the stock floated, the following report is probably fairly judicious: "Much of it is worthless, much more is of little worth, some of it is worth a fair percentage of the par value, a small percentage of it is worth par value, and a fraction of one per cent of it is worth more than par value." At one-hundredth of par value, the stock would have been worth one-half more than all the oil produced in Wyoming during that year. "Only a slight knowledge of conditions in Wyoming," declared a writer in the *Oil and Gas Journal,* "is necessary to beget the suggestion that the average oil company, on the date of its organization, does not have assets equal to one-tenth of the average capitalization of $847,784, or one-twentieth of that amount, or yet much less than one-twentieth."[13]

In the new California fields, perhaps the wildest speculation has prevailed. At Huntington Beach, for instance, it was reported that people almost generated a riot in their anxiety to get into the "oil game," lining up along the sidewalks in front of brokers' offices, awaiting their turn to purchase stock. Even at the present time, a Californian who can talk sensibly about Signal Hill or Santa Fé Springs or Huntington Beach is about as rare as one who can talk judiciously of the California climate; but sanity and oil do not go hand in hand in any part of the country.[14] One writer on the California oil industry estimated that perhaps 1,000,000 people in California are interested in oil. He doubtless exaggerated when he said that it was "exceedingly rare to meet any person in southern California who is not interested in an oil well."

In the fields of southern California hundreds of crooked promoters have been operating during the past few years. Conditions here favored their work: a great number of small lots under diverse ownership, spectacular gushers, and the "boosting," exuberant, speculative western spirit, with many visitors and tourists to be initiated into the mysteries of oildom. Sight-seeing busses of large capacity have been used by some of the promotion companies—known as the "bally-hoo crowd"—to convey people from the inland towns to the oil fields. Many of these companies even sent private busses around to get anyone who wanted to go and look at the oil fields. "Our private car will call and we will be only too glad to show you our property." So common have been the big busses and other paraphernalia of promoters that one modest promoter announced: "We have no bus, lecture or lunch," as if to show the good faith of his own company by such simplicity. The companies here have usually been organized with stocks of $1 par value, but some have tried to "put their opportunities within reach of the poor man" by issuing 10-cent "units" or stock. One company at Huntington Beach, with 3,000,000 shares of 10-cent stock, paid 1,500,000 shares for a lease on three town lots—about one-fifth of an acre—at one-third royalty, gave 20 per cent of the remaining 1,500,000 shares to the stock salesmen, and most of the proceeds of the sale of the rest of the stock for drilling one well. This company was of course never prosperous, but it was less rankly fraudulent than hundreds that were operating. Another company in southern California was found to have put into drilling less than half a cent out of each dollar subscribed. Another capitalized a piece of ground 100 by 170 feet at more than $1,000,000, subdivided it into units and sold out to the public at a total selling expense of 90 per cent. The average selling expense of 104 outlaw companies investigated was 50 per cent.[15]

The "unit" system was the common form of organization, "units" or undivided interests, being issued instead of stock. This form of organization had certain legal advantages, but it proved a snare and a delusion to some, for in September, 1922, the Standard Oil Company and other large purchasers of oil announced that they would take no more oil from certain of the "unit" companies, because they could get no legal title to it.[16]

With the tremendous speculation in oil stocks or "unit" shares went a similar wild speculation in real estate. It was said that more than 27,000 licenses for realty brokers were issued in one year in California, and all kinds of real estate gambles were placed before the investing public. The constant rise in real estate values in some of the oil towns of the Los Angeles Basin furnished the necessary atmosphere for a typical boom. Real estate values rose to almost unbelievable proportions, and some observers claimed that real estate operations yielded greater profits than oil speculation. Thousands of tourists who were too timid to take a "flyer" in oil stocks could not resist the temptation to buy a city lot or so "with the oil rights"; and a great many natives owned enough lots and building sites to provide a change for each month of the year. The rapid growth of the city of Los Angeles has of course made real estate speculation unusually profitable, and when the glamour of oil was added to the pull of the urban increment, it created an irresistible appeal to many people with a little surplus money. Lots were often sold for extraordinary prices because the "oil rights" were included.

In the Butler County field, in Kansas, the "unit" system was worked a great deal. The promoter would purchase perhaps one-tenth or one-eighth interest in a producing lease for $10,000, and then capitalize this fraction into thousands of "units," each purporting to be an undivided interest in the fraction. In one instance, a $10,000 fraction was made the basis for 100,000 units which were sold for $20 each. In another instance, the fraction was divided into 5,000,000 units at 10 cents each. Complaints to the "Blue Sky Board" showed that some owners of units would not receive as much as 1 per cent on their investment until the property, yet to be "developed," yielded $3,000,000 worth of oil. Widows' estates were sometimes invested in just such securities as these.[17]

Some of the Texas promotion artists used the "trustee" form of organization in their operations. They filed a "declaration of trust," under which they became "trustees," and as trustees they might do almost anything imaginable, and keep out of the penitentiary. A trustee might issue $1,000,000 of stock, put $1 in the treasury and the rest in his pocket. He could sell to his company the assets, including the "sucker list," of any defunct company he might have been connected with. He might transfer funds from one company to another. He might pay for practically all of the stock with an acre of land 100 miles from an oil field. The stock was his, and when he sold it, the money was his. The trustee usually took care of his own interests out of the sale of stock, and was not at all dependent on the finding of oil. He usually held some leases somewhere, which he could turn over to the company at a high figure. He often donated a little to the company for the sake of appearances, as long as the "suckers" were at the nibbling stage. When they swallowed the hook, the company for which he was trustee was usually taken off his list of charities.

The "merger" scheme of Dr. Cook and others has been described. A somewhat similar plan engineered by a company in Fort Worth, aimed at trapping the holders, not only of worthless oil stock, but of all kinds of stock. This company operated extensively in Iowa, among the farmers who had bought worthless packing house stock, offering to "merge" or take over the packing house stock, with a $25 contribution from the investor. *Wallace's Farmer* exposed this game very soon.[18] The merger type of organization had many advantages, as a promotion scheme. It started with an excellent list of "suckers," of people who had already shown their weakness for quick money. To be sure, some of them had learned their lesson, but many of them were so anxious to salvage something out of their previous ventures that they would bite at almost anything. The promoter's story of the necessity of large combines and "mergers" in modern industry, and particularly in the oil game, made a rather effective appeal.

A rather ingenious scheme was worked out by a Fort Worth lawyer. He wrote what purported to be a personal letter to his brother in New York, advising him confidentially of an opportunity for a clean-up, and urging him to send $100 or so, and then sent typewritten copies of the letter to several score of his fellow barristers throughout the country. He figured that the men receiving this letter would think the mailing of the letter an honest mistake and would rush a check to Fort Worth to get some of the easy money. Inspectors in the Post Office nipped this little game at its inception.

In May, 1922, 300 oil companies were reported to be in trouble with the United States postal authorities, for using the mail to defraud.

Many of the promotion companies were organized with stocks of very low par value—one dollar, ten cents, or even one cent. One Texas company issued 35,000,000 shares of a par value of one cent each. This was done to give even the poorest people a "chance."

Many promoters sold stock on the partial payment plan, and the initial payments accepted were sometimes very small. In California shares were often sold on payments as small as $5, or even $1 or $2. This "partial payment plan," as one company called it, enabled the purchaser to "control five times the amount of stock" that he could control if he purchased outright.

In a California case, the promoters issued 5,000,000 shares of stock, declaring that 3,000,000 shares had been placed in escrow in Los Angeles until the stockholders should be paid what they had invested, but at the trial evidence showed that this provision had not been observed. In another California case, the promoter represented himself as "President of the College of Petroleum Engineering and Geology," as well as an official in the "International Oil and Land Syndicate," with lands in Montana, Texas, and Mexico, as well as in California; and, with this substantial backing, sold some stock without using the form approved by the Commissioner of Corporations, as a result of which he presently landed in the Los Angeles jail, because unable to furnish bail in the sum of $10,000. Another California "Doctor" got into trouble with the federal government for sending out prospectuses claiming that he had 5,000 acres of oil land, when, as one reporter tersely put it, "he had no land and it wasn't oil."

Stock promotion companies seldom failed to point enthusiastically to the success of other companies. They often made tables showing what an investment of $100 in certain companies had brought in returns to investors. Often some sort of "Trust Company" was referred to as connected with the company, to give it dignity. In some instances banks and bank officials actually worked in cahoots with the crooks. One of the companies investigated was putting out its advertising over the signature of a man who bore one of the most honored names in America. This man was represented as being in complete charge of its affairs. When the inspector of the Federal Trade Commission came to interview him, he found an old man, far gone in dotage and poverty. "These are mighty good men," he told the inspector, "they know that an old man's misfortunes are not always of his own making, and so they pay me $30 a month just for coming down every day and signing my name to a few papers."

Promotion companies liked to emphasize in their literature that they had production and a large acreage. With reference to production, the promoter frequently bought a few acres in proved territory at an exorbitant price, simply so that he might have some proved territory. Then he secured thousands of acres in leases outside, or wildcat territory, at a few cents an acre. The few acres in proved territory gave quality, the remainder afforded quantity, for his appeal to the investor. One of the favorite methods of promotion concerns was to claim an entire lease or the entire production of a lease whereas they often owned but a fourth, eighth, or sixteenth of it.

Telegrams were used by promoters a great deal, because telegrams would sometimes get results where letters would be ignored. It is reported that one company spent as high as $16,000 in a day for telegrams, but got $100,000 in returns. These telegrams usually hinted at an immense well which was either just about to be brought in, or was already brought in, but was being kept secret until a few "near friends" could "get in on the ground floor."

One Kansas company acquired an old refinery in Kansas, poorly located with regard to producing territory and market, which had never been able to pay expenses, paid 12 per cent dividends a time or two, and so sold the stock to the public at less than par value. This company sent a salesman to deliver the dividend check personally to the owner, and thus was able to induce many stockholders to buy more stock, this time at par. Dividends were often declared, before earned, to influence purchasers to buy more stock. Sometimes dividends were guaranteed— as high as 500 per cent; and when government officials interfered with this scheme, the guaranty was changed to state merely that investors should have dividends before the promoters received any. Some of these guarantors were later placed in jail.

Promotion costs have always been very high—often as high as 70 per cent of the money received from investors, and even as high as 99½ per cent, in at least one case. A group of Texas promoters dug a well 20 feet deep and sold the "property" to the public for $500,000. A Sour Lake "developer" bought a 40-acre lease for $40,000, and dividing it into 144 lots, sold them for a total of $1,750,000. The lease never produced half as much as it sold for. In another case, an oil company made a contract turning over to a salesman for 40 cents a share oil stock which he in turn sold to the public for $2.50 a share. Promotion expenses in this company were 84 per cent. An investigator in one oil field reported that it was "common to find that a leasehold worth a few thousand dollars has been capitalized for a million." One oil company with 60 acres of land was capitalized at $61,000,000. In a certain Oklahoma company, it was ascertained that it cost the stockholders $555 for every dollar's worth of oil produced. In the Los Angeles Basin a rather common method of the stock selling promoter was to get a lease on oil land at a certain royalty—perhaps 15 per cent and a bonus. Having done this, he would turn it over to the company on a higher royalty—say 40 per cent, and thus pocket a portion of the receipts in case oil was found. In several Kansas cases two or three men got some oil leases, organized a company, elected directors, and then got the directors to buy the leases at from two to ten times the price originally paid for them, the "swag" being divided among all the parties to the transaction.

The *Oil and Gas Journal* has summarized some of the findings of the Federal Trade Commission on the misrepresentations of oil promoters as follows: alleging that syndicates and companies were incorporated in a certain state when they were not incorporated anywhere; alleging a production of 2,000 barrels a day from a single well that the (Federal Trade) Commission's experts found never had produced in excess of five barrels; alleging the payment of 24 per cent dividends by a company which had never paid one cent of dividend; alleging a large number of producing wells when there had never been a single producer; falsely asserting all sorts of things about the operations of the companies and about plans for development; alleging that the companies had hundreds of salesmen and hundreds of brokers, giving the approximate number of such salesmen and brokers, when they had never had a single one, the sale of stock being only by the men cited to appear before the Commission; claiming royalty interests in specified counties, when there appeared to be a single royalty interest, and that worthless; alleging that tracts owned were surrounded by producing wells when they were not near production; alleging enormous estimated values of leaseholds which in fact were found to have no value whatever; alleging all sorts of equitable distribution of stock to assure fair dealing, when none of these statements were true.[19]

When stock in oil companies could not be sold easily, some promoters have sold city lots with the oil rights attached, "real homesites where you get all the oil," perhaps giving away a lot or two by lottery to stimulate interest. In this way a great deal of worthless land has been disposed of at handsome prices. In Texas when the Associated Advertising Clubs got onto the trail of the promotion stock salesmen, many of them dropped their stock selling business and began to peddle leases. These leases were often hundreds of miles from production and of course practically always worthless.[20]

Many of the prospectuses and advertisements are interesting studies in psychology, and reflect little credit upon American business intelligence: "Here at last"; "Your ground floor opportunity"; "Your big chance"; "The cleanest, squarest proposition in oil you ever looked at"; "Two hundred feet from a producer and your chance to make your clean-up in oil"; "Straight ahead to dividends, . . . and no side tracks on the way"; "Only selling 2,000 shares, and they are going fast"; "Oil, the Master Fortune Builder"; "Fear, procrastination or a bunch of flimsy alibis never get you anywhere"; "Daily expecting this well to hit the deep oil sands and start pouring oil profits into the pockets of the fortunate lot owners"; "We can't miss the oil"; "Prepare to get your share of the happiness that oil brings"; "As soon as our well is fully drilled in, and we know exactly how large it is, I expect to get out an announcement shooting up the price of stock or taking it off the market altogether"; "My personal advice to you is to wade in and take all you are able to handle"; "Come with me to the golden lanes of wealth where the oil of old Mother Earth pours forth its riches"; "Mike O'Roke the man who opened the Sunrise pool is ready to open another bigger one"; "Jim McCleery, big-hearted, far-seeing Jim, Jim the man, Jim who loves the clang of the tool dresser far more than the ring of a dollar"; "Smackover, the Marvellous: Prolific Production and Gigantic Dividends (cash) are Keeping Step to the Tune of 'Onward Ever' in the Smackover Sectors of Ouachita and Union Counties, Ark." "In comparison with Smackover all previous boom days are as the miniature stream oozing from the mountain side to the spouting voluminous geyser in Yellowstone Park." "In nine months this Garden of Gurgitating Gushers, whose production rolls in like billows from a sea of gold and with such roaring volcanic sounds, that investors' dollars ring like pebbles upon the golden shore, there has been paid out in cash dividends to better than 15,000 investors throughout the length and breadth of the United States and in remote parts of European countries better than $3,000,000."

The man who guaranteed 500 per cent dividends backed his guarantee with the following literary "gusher": "Meanwhile my giant drill is nearing the monster gusher depths. If you could be here when I bring in this great potential gusher, thundering and rumbling to high heaven, and bellowing to hades, with a wealth of liquid gold which staggers the very powers of human comprehension, and if you could know the gigantic, staggering, immense, stupendous dividends which are just ahead, which should make our little 500% guarantee look like a very small disbursement, then you would say if you *knew how much of a cinch it is to win, and win big, and win quickly*, you would say, 'I will invest not hundreds with you, Shallcross, but thousands.' "

It might perhaps be suggested that speculation has harmed none but the speculators, and that there is no valid reason for denying to all Americans the great democratic privilege of making as big fools of themselves as they wish. It is not true, however, that speculation has harmed only the speculators, for it has often interfered with legitimate oil development. As R. C. Moore, of the Kansas Geological Survey, has said, speaking of oil development in Kansas:

Lease speculation has become a rather lucrative business at the present time, especially in the newly opened districts. The speculator closely watches the prospecting and development, and on news of an oil strike rushes to the locality and leases all available property with as little expense as possible. When the operators who really wish to drill come to the field later, they are forced to pay the speculator's price. Examples of this speculation may be seen in every newly opened oil district, and have been common in the Kansas and Oklahoma fields.[21]

Other men have written of the "swarms of lease brokers," "old and young, male and female," in the various fields. In the Depew, Oklahoma, field there were said to have been a great many of these "waiters," men who came to Depew several years before the strike and bought leases and property and then waited until someone else should do the work necessary to make their property valuable. Of course this sort of business was not always profitable, and the "waiters" who made fortunes were much more written about than those who "waited" in vain.[22] Perhaps the strongest argument against this sort of activity is that it is socially unproductive.

The swindling of innocent dupes has sometimes been thought a mere private matter between the swindler and his victim; but it is unfortunately not so simple as that. Not a few men in positions of trust have been led into the mazes of oil speculation, and have landed eventually in Canada, or in the penitentiary, after having lost the painfully accumulated savings of many innocent people. In some instances bankers and trustees have staked the lifetime savings of others upon what seemed a promising venture in oil, and have guessed wrong.[23] Some pathetic stories of this kind might be told. Then there is another public aspect of the matter. The loss of hundreds of millions of dollars in unwise and fraudulent oil ventures has meant much more than the loss of this money. It has meant a great loss of confidence in all kinds of securities, not only legitimate oil stocks and bonds, but stocks and bonds of all kinds. Some of the large investment houses have often complained of the flow of money into worthless oil stock, not only because it meant so much of a subtraction from the funds available for productive uses, but because it made many people suspicious of all kinds of securities.[24]

"The curse of all this crooked exploitation is that it is so senseless, so useless," as one journal has well said, ". . . There are abundant honest opportunities for every investor without lying and gypping. The amount of energy and promotion ability expended to hook and rob the unsuspecting public would be a tremendous community asset if expended along community lines."[25]

As to the total amount lost in speculation, and in fraudulent oil companies, no estimate can be more than a wild guess. Felix Renick, secretary of the New York Petroleum Exchange, estimated that worthless oil stock cost investors in the United States nearly $500,000,000 in 1919. He pointed out that only 15 per cent of the 30,000 oil companies in the United States were listed in the New York Stock Exchange.[26] Another writer has estimated the loss at $250,000,000 a year.[27] The Postmaster General estimated in May, 1923, that oil stock promoters in Texas alone had taken more than $100,000,000 from the gullible public in the preceding five years. Others have placed this figure considerably higher.[28] These are all mere guesses, and nothing better can be offered. It is certain, however, that the losses have been very great, and it is probable that most people would underestimate them. Many people have lost more or less money in oil stocks—not only stenographers, doctors, teachers, farmers, and professors, but business men as well. The number of losers, and the amount of the losses, is likely to be underestimated because those who lose do not usually talk much about it. It is reasonably certain that the total losses sustained in this way in the past would have to be measured in hundreds of millions of dollars, and possibly they would run into billions.

Some efforts have been made to curb the activities of swindling promoters. More than twenty years ago, Blatchley began an energetic campaign against them in Indiana. His various reports bristle with denunciations of these promoters. "Better it would be for the person who has money to invest," he wrote in 1903, "to buy grass seed and scatter it abroad for the sparrows than to invest in the stock of these or other similar companies. . . . The oil business is a big enough gamble within itself—that is, the risks of losing in the legitimate companies are great enough—without taking a thousand-to-one shot in the stock certificates of those get-rich-quick concerns whose members do not know a walking-beam from a derrick."[29] In other states a number of public-spirited individuals, newspapers, and other private agencies have taken up the war against promoters. About the same time that Blatchley wrote, the Post Office Department started a campaign against fraudulent promoters, and was claimed to have accomplished a great deal.[30]

Within the past two or three years, oil stock salesmen have found business increasingly hazardous in many parts of the United States, because of the action of various state and federal authorities, aided by a few newspapers, advertising clubs, public spirited individuals, and especially by some of the officials of the big oil companies, who realized the injury done by these wolves to the legitimate oil industry. The National Vigilance Committee of the Associated Advertising Clubs took a very strong initiative in this matter, and the American Petroleum Institute appointed Henry L. Doherty as representative to work with the Vigilance Committee. This combination coöperated with the federal Post Office authorities and with the Department of Justice, and accomplished a great deal in cleaning up some of the worst nests of crooks. State Corporation Commissioner Daugherty, of California, announced a similar campaign against a large number of promoters in California. In this he was assisted by the federal authorities, and he accomplished something in spite of great obstacles.

Within recent years many of the states have passed laws providing some sort of state agency to protect the people from fraudulent stock salesmen. So-called "blue sky" laws have been passed in many of the states, and have even been enforced in some states. This departure from the good old time-honored tradition of "laissez faire" has no doubt saved a great deal to the investing and speculating public. The Secretary of State of Illinois, where fraudulent promoters have generally been treated more severely than in many other states, recently stated that the blue sky law of that state had saved its people more than $40,000,000 a year during the three years of its operation. The Kansas Charter Board has done a great deal to suppress some of the worst iniquities of stock salesmen.

In California, in 1911, a campaign was inaugurated against mining and oil stock promoters, by the State Mineralogist, Lewis Aubury, in coöperation with the federal government, and much good work was done. Later California passed a blue sky law, but when the state authorities began to clamp down too hard on the oil promoters, the promoters conceived the idea of organizing under the "unit" system, issuing "units" or "interests" in the business, instead of shares of stock. In this way they were for a while able to evade the law requiring a permit from the State Corporation Commissioner, but the Supreme Court of California finally decided that the sale of "units" required a permit, the same as the sale of any other kind of securities. The enforcement of the California blue sky law has not been greatly to her credit, but something has been done. Enforcement was so lax that the city of San Diego found it necessary to pass an ordinance of her own, requiring unit operators to pay a fee of $50 a day for the privilege of soliciting on the streets or from house to house, or of operating free bus excursions to any alleged oil property for the purpose of selling stock or units therein. A similar ordinance was considered in Long Beach, but, while the city officials admitted that the state laws were ineffective, and that the city was getting a bad name because of the extensive operations of gangs of oil stock pirates, no action was taken.

It is perhaps easy to see why San Diego should be more prompt in such a matter. San Diego has no oil, and her citizens would naturally not enjoy seeing San Diego money taken up to Long Beach for investment in Signal Hill lots. In Long Beach, it was at least a case of "money spent at home"—something very important to the average economist on the streets; and if the promoters caught a few tourists in their nets, that was somewhat analogous to a favorable balance of trade with "money coming into the country," and "circulating around therein." Long Beach finally banished the free busses from her streets.

A great many cities outside of California tried to suppress oil stock frauds within their limits. In May, 1918, the city officials of Denver put on a campaign in which a great many frauds were brought to light and punished. Even booming, bustling Tulsa finally saw that frauds did not help business, and organized a Tulsa "Better Business Bureau," one of the functions of which was to help drive out oil stock promoters. Many other cities have taken similar action.

The federal government, through several of its branches, has tried to suppress fraudulent oil stock promotions. As early as 1904 the Post Office Department waged a fairly effective war on fraudulent mining and oil stock promoters, and caught a number of them in California and elsewhere. This work the Post Office Department has kept up pretty well ever since. Late in the year 1922 the department started a "drive" at Fort Worth, Texas, with 129 fraud cases in that city. In the entire Austin division there were 446 cases under investigation at that time, most of which related to oil stock sales. During the Great War the Capital Issues Committee put a ban on the sale of certain kinds of oil stocks, and for a time this ban exercised a salutary influence, but at the close of the war the committee was abolished and oil speculation resumed its normal course. In 1922 the Federal Trade Commission began the investigation of a great many oil companies, under its power of preventing unfair practices. In May, 1922, 20 oil companies were under investigation in Texas, mainly around Houston and Fort Worth. In April, 1923, federal warrants were issued for some 25 promoters in Fort Worth, and 14 of them were immediately arrested, including Dr. Frederick Cook, and a number of others engaged in similar business. A number of these, including Dr. Cook, were promptly convicted, with the result that Fort Worth lost one of her important industries.

In its efforts to convict swindlers, the government has often been handicapped by the difficulty of securing convictions in the vicinities where oil swindlers live. The operations of these men bring money into the communities where they live, and some of it is, of course, "put into circulation" there. One government official estimated that the activities of some 300 promotion companies under investigation in July, 1922, had netted a total of $75,000,000 for the men involved, 95 per cent of whom lived in Texas, while almost all of the money was secured from people outside of the state. In other words, these gentlemen were putting $75,000,000 of Kansas, Missouri, and Iowa money "into circulation" in Texas; and when the Department of Justice brought suit against them, the local juries promptly acquitted their "benefactors." When the federal government started its campaign against the Fort Worth operators, it was hindered by local sentiment, and was fought effectively by the men themselves, who were well organized, with strong political influence, and with abundant funds to hire the best legal talent and to put up a strong fight against the government.

A similar situation has existed outside of Texas, and in some instances the government has had to bring suit outside the communities where the crooks lived, in order to have any chance for conviction. For instance, in a case where the swindler lived in Iowa, and the fraudulent literature was circulated in North Dakota, suit was brought in North Dakota, to avoid the sentiment of the Iowa community.

Not all losses were due to fraud. One reason why most speculators in oil stock lost their money was that many of the small companies which sold stock to the public drilled in "wildcat" territory. (Wildcat territory is territory not yet known to contain oil, and not near enough to proved territory to be recognized as probably oil-bearing.) Wildcat drilling is usually expensive because it is away from oil well supplies of all kinds, and crews often have to wait for tools. Within recent years large oil companies are more prominent in such operations than they were previously. Benedum and Trees of Pittsburg and the Texas Pacific Coal and Oil Company have been primarily wildcatters, and many of the large companies have done considerable wildcatting. The Empire Gas and Fuel Company discovered the Eldorado field in Kansas; the Shell Company discovered the Long Beach field, and the Humble Oil and Refining Company discovered the Powell field. As oil becomes more scarce, the larger companies will perhaps seek oil more energetically than they have in the past.[31]

NOTES

1. *Wyoming Oil Index,* Sept. 16, 1922.

2. *Cal. Oil World,* April 26, 1923, 1.

3. Indiana Dept. of Geology and Natural Resources, *Rept.,* 1903, 95.

4. *Wyoming Oil News,* July 9, 1921; Aug. 20, 1921; *American Petroleum, Supply and Demand,* by the Committee of Eleven of the Am. Petroleum Inst., 13.

5. *Cal. Oil World,* Aug. 3, 1922; *Nat. Gas,* Nov., 1921, 9, 12; *Oil and Gas Jour.,* March 15, 1917, 28; Nov. 8, 1917, 36; June 25, 1920, 3; Sept. 17, 1920, 3; Sept. 9, 1921, 75; July 20, 1922, 16; *Eng. and Min. Jour.,* Sept. 18, 1920, 555; *Petroleum Age,* Oct. 1, 1924, 25. For an extended discussion of the history of the use of the divining rod in finding minerals, water, etc., see U. S. Geol. Survey, *Water Supply Paper 416.*

6. *Mineral Resources,* 1900, 583; 1901, 528.

7. There is an apparent discrepancy between these figures and those of Oliphant, for it is difficult to believe that more oil companies were formed in 1901 than in 1915, yet 1901 was a year of tremendous excitement in Texas and in California, while 1915 was a year of great uncertainty, and the following years, 1916, 1917, and 1918, were war years, when the government was watching all sorts of promotion enterprises very carefully. (Am. Petroleum Inst., *Proceedings,* Annual Meeting of 1920, 51.)

8. *Natl. Petroleum News,* Feb. 15, 1922, 109.

9. Indiana Dept. of Geology and Natural Resources, *Rept.,* 1903, 96-98.

10. *World's Work,* July, 1919, 307; *Oil and Gas Jour.,* Dec. 3, 1920, 3.

11. *Natl. Petroleum News,* Jan. 10, 1923, 41.

12. *Natl. Petroleum News,* April 4, 1923, 32.

13. *Oil and Gas Jour.,* May 3, 1917, 28; Sept. 27, 1917, 32.

14. *Oil and Gas Jour.,* June 10, 1921, 38.

15. Albert Atwood, in the *Saturday Evening Post,* July 14, 1923, 101.

16. *Cal. Oil Fields,* Nov., 1921; *Oil and Gas Jour.,* May 13, 1921, 34; July 15, 1921, 32; *Cal. Oil World,* May 26, 1921, Sept. 14, 1922; *Natl. Petroleum News,* May 9, 1923, 35.

17. *Oil and Gas Jour.,* April 23, 1920, 3.

18. *Wallace's Farmer,* April 13, 1923, 3.

19. *Oil and Gas Jour.,* May 18, 1922, 62.

20. *Natl. Petroleum News,* Dec. 6, 1922, 30, 90.

21. Kans. Geol. Survey, *Bul. 6,* Pt. I, p. 62.

22. *Oil and Gas Jour.,* July 25, 1919, 70; *Petroleum Age,* Nov. 15, 1922, 20.

23. *Oil and Gas Jour.,* Feb. 4, 1921, 72. For a rather pathetic letter of a washerwoman regarding her "investment," see the *Oil and Gas Jour.,* June

22, 1921, 2. An investigator in southern California pictures this very clearly in his description of the tent lectures at Santa Fé Springs:

We were both disgusted with the noise and din of the place, with the heat and flies, with the cheap hectic atmosphere of it all, with the pity of such ignorance and sordid greed. I looked across at another table where a pathetic looking, poorly dressed young woman, with a face half gone, evidently burned off in a cookstove explosion, was handing a liberty bond over to the salesman. The husband, dressed in the roughest working clothes, sat by with an expression as intelligent as that of a faithful ox. In the woman's eyes was a pitiful look of mingled hope and fear, much as a deer or a white baby might have shown as an Apache warrior raised his ax, uncertain perhaps whether to strike. I wanted to get out; the air was choking me. (Albert Atwood, in the *Saturday Evening Post,* July 14, 1915, 92.)

24. *Oil and Gas Jour.,* May 17, 1917, 2.
25. *Saturday Evening Post,* July 14, 1923, 105.
26. *Oil and Gas Jour.,* March 19, 1920, 3.
27. *Financial World,* Jan. 27, 1923, Pt. I, 109.
28. *Oildom,* May, 1923, 64.
29. Indiana Dept. of Geology and Natural Resources, *Rept.,* 1903, 96-98.
30. *Pacific Oil Reporter,* May 28, 1904, 3.
31. *Wall St. Jour.,* Sept. 18, 1920, 1.

CHAPTER XVII

WASTE OF CAPITAL AND ENERGY

(Continued)

THE great, impelling incentive for speculation, as for wildcatting, and so for all development, has been the chance—small though it usually was—of making very large fortunes. The stories of some of the fortunes made in the oil "game" read like fables, in fact many of them are to some extent fables. We read for instance of the West Virginia driller who opened the Pine Island field, in Louisiana, and made several million dollars; of Harry Sinclair, who was a drug clerk not many years ago, and is now president of a $200,000,000 oil company; of Colonel Humphreys, who was bankrupt at least once, yet made a fortune estimated as high as $50,000,000 in the Mexia field; of E. L. Doheny, who was a poor miner, and later became one of the wealthiest men in the United States, through his perseverance and good luck; of the Reed family which was reported to have accumulated Salt Creek holdings worth many millions; of the Pittsburg operator, J. S. Trees, who was said to have cleared $6,000,000 in two years in Louisiana oil lands; of the banker who made several millions in a Butler County lease, in Kansas; of the Eldorado "slicker" who bought a farm in the Eldorado field from an ignorant farmer for $15,000 and sold it the next day for $400,000; of the two Bakersfield men who bought a section of land near Bakersfield for $2.50 an acre, and sold their interests in a few years for $5,000,000, after having received $2,000,000 in dividends; of the three men who paid $25 for a lease in the Healdton field and later sold it for $250,000, receiving $10,000 for each dollar invested (this story is claimed to be strictly true);[1] of the two geologists who studied structures in some of the untried fields of Oklahoma, much to the amusement of the people, and later sold out for $500,000; of the men who contributed $100 each to drill a well on the Fowler ranch, in the Burkburnett field, and afterward received $12,000 for each $100 invested; of the cattle king who owned the 600,000-acre ranch on which the Electra field was opened, and built a 20 story building in Fort Worth with only a part of his income; of the two high school teachers who had ranches at Santa Fé Springs worth $1,000,000 each, after oil was discovered; of Alphonzo E. Bell, whose 150 acres in the center of the Santa Fé Springs field netted him $100,000 a month for a while; of the farmers and barbers and street car conductors and people from all walks of life who had happened to buy leases or stock or land, and later found themselves the recipients of princely incomes. Unfortunately, there are a hundred times as many stories of money lost, as of money made, but they are not for the most part available for publication. Even if they were, they would make rather dispiriting reading. Most speculators had to satisfy themselves with "enjoying happy anticipations."[2]

All industries but the oil industry have been neglected wherever oil has been discovered. The lure of quick fortunes has everywhere attracted men and capital from other industries. The steady and uninteresting operations of farming and merchandising, and even the professions of law and medicine and teaching, have suffered in competition with the spectacular operations of the oil producer; and men and capital much needed in other industries have poured into the oil fields, where to some extent they were not only not needed, but actually a nuisance. If every farmer, merchant, lawyer, and professor had remained at his own job, and left the oil industry to the men who understood it, the ordinary business of the country would have been done in better fashion, and the oil industry would have had a healthier development. Perhaps the oil would not have been exhausted so rapidly, but it would have been handled more sanely, and far less capital and energy would have been wasted in the securing of it.

It is not only through speculation in leases that human energy has been wasted in the oil fields. Many capable and energetic people have been turned from productive labor by the "windfalls," the unearned fortunes that abound in the oil fields. These fortunes accrue to various classes—to fortunate speculators in leases or stocks or royalties, and to landowners. If we assume that the fortunate speculators have earned their fortunes by risking their funds in a hazardous lottery, it leaves the landowners as the chief recipients of these windfalls or unearned incomes. Under our land policy, hundreds of millions of dollars, perhaps we may say billions of dollars, have been paid to owners of oil lands, who have performed absolutely no service for their gains, and have, in some instances, done no useful service any time thereafter, being able to live in idleness the rest of their lives, on the income from their lands.

The total unearned income accruing to owners of oil lands probably amounts to over $100,000,000 a year at the present time. The total oil production of the United States in 1923 amounted to about 735,000,000 barrels, and, of this, probably more than 100,000,000 barrels should be regarded as royalty oil. The royalty is usually one-eighth. It is of course difficult to figure average royalties, but much of the immense California production pays 25 per cent, and some of it pays large bonuses in addition. In some other fields, royalties run as low as 8 per cent, but the higher royalties prevail in the more productive fields, and considerable bonuses and rents have been paid in some instances where no oil was found. Even after subtracting the production on federal and state lands, 100,000,000 barrels seems a modest estimate of the total royalties accruing to landowners. It would be very difficult to calculate an average price, in fact it would be impossible to do it accurately; but if we assume that an estimate of $1 per barrel would not be too high, this royalty oil would probably be worth more than $100,000,000. In addition there must be included a large amount for bonuses and rents. Altogether, it is certain that much more than $100,-000,000 was turned over to landowners in 1923 in the form of entirely unearned "windfalls" or "finds," the effect of such windfalls being merely to distract the minds of the general public from ordinary productive labor, and, in some cases, to rob society of the services of the recipients of these unearned gains. Such gains, from the economic point of view, represent worse than absolute waste.

The total amount wasted in the United States in this way since 1859 would certainly be a staggering figure, if it could be accurately computed. The total value of the oil produced in the United States is estimated by the Geological Survey at nearly 10 billion dollars. Royalties alone on this production could not well amount to less than one billion dollars, and they have probably been somewhat more than that. In addition to this, there have been bonuses and rentals, perhaps greater in amount than the royalties, although figures on this point are difficult to obtain. A few years ago one of the staff members of the *Oil and Gas Journal* estimated that at least 200,000,000 acres of land were under lease by oil companies. In Oklahoma, about 25 per cent of the area of the state is under lease; and C. C. Osborn, economist for the Marland Oil Company, has computed that if the same proportion of the land in the 17 oil-producing states is under lease, a total of over 200,000,000 acres of leases is indicated. Osborn estimates the average bonus at $8 per acre, and the average annual rental at $.75 per acre. A questionnaire sent out by the Marland Oil Company to representative oil companies in 1923 disclosed an average bonus of $13 per acre and an average rental of $1.37 per acre. If the first mentioned figures be used for bonuses and rentals, it would make a total of $1,600,000,000 for bonuses on the land then under lease, and total annual rentals of $150,-000,000. Some bonuses would of course be paid more than once, as leases expire from time to time.

Following these figures, Osborn, and his chief, E. W. Marland, have estimated that "citizen landowners" in the United States have received net incomes of $4,100,000,000. Marland's opinion, and that of his economist, C. C. Osborn, on any question relating to oil, should carry great weight, and perhaps this estimate is not too high. It is true, at any rate, that the total unearned income accruing to landowners in the United States, because of oil and gas production, has been a fabulous sum. If Marland's estimate is not too high, the amount thus wasted, if invested in conservative bonds, would yield an annual income of about $200,000,000 per year, or very nearly enough to pay the total annual cost of the universities, colleges, and professional schools of the United States.[3]

Ordinary economic reasoning would perhaps lead to the conclusion that the owners of oil land would in the long run absorb a very large share of all the profits of the oil industry, that competition among operators would raise the price of desirable leases so high that there would be small if any net profits for the operators. In some fields, there is no doubt that landowners really have absorbed all net profits, and in some fields have even got more than that, have received good incomes while operators worked at a loss. Blatchley reported in 1896 that it was the farmers who had gained most from the discovery of oil in Indiana, since "they had everything to gain and nothing to lose in leasing their property." Blatchley estimated in 1904 that the landowners of Indiana had received about $6,000,000 for their royalties; and he was very doubtful whether the oil business of Indiana had yielded a net profit for any other class.[4]

Marland has expressed similar views regarding the oil industry as a whole. He has estimated that during the past history of the industry in the United States, operators have sold crude oil at a loss of $4,900,-000,000, while, as stated above, landowners received royalties of $4,-100,000,000.[5]

Thousands of stories might be told of the dazzling fortunes that dropped into the laps of landowners in the oil fields. Some such stories have been recounted above. There is also the story of the truck farmer in the Orange, Texas, field, whose farm netted him $2,000 a day when oil was struck; of the poor farmer in the Cushing field who rated as a millionaire after that great field was opened; of the farmer in the Augusta, Kansas, field whose rocky quarter was presently earning him $1,500 a day, and of other men in the same field who received bonuses running into hundreds of thousands of dollars, in addition to their royalties; of the man who bought a five-acre tract in Augusta, called the church-yard property, and sold it two years later for $100,000; of the woman who refused $1,800,000 for her royalty interest in a farm in the Towanda field, which was netting her $3,500 a day; of the man who bought a section of land in a California field for $15 per acre, and later sold to a British syndicate for $1,500,000; of the negro woman whose land in the Homer field made her a multimillionaire. But these are the mere ordinary everyday stories of all the oil fields. The true stories of all the fortunes accruing to landowners in the oil fields would make as interesting a series of stories as any novelist ever wrote, and it would have to be a book much larger than the present volume.

Some of the most conspicuous illustrations of unearned land incomes have been found among the Indians in Oklahoma. Some Indians have received as much as $50,000 a year from their oil royalties, and one Indian tribe is said to have by far the highest per capita wealth in the world. One of the most noted of the Indian millionaires, Jackson Barnett, a Creek "incompetent," and the owner of an allotment in the Cushing field, was said at one time to be receiving royalties of $50,000 a month. He was wholly unfitted to care for his fortune, so that the government had to appoint a guardian for him. When he was married, there was much talk of fraud and "undue influence," and even some talk of contesting the validity of his marriage, but with the aid of a lawyer, who received a $50,000 fee, his wife was able to establish some kind of claim to his "affections." Unlike a great many of the rich Indians, he was apparently not ruined by his flood of wealth, for he preserved his simple manner of living. Indian Commissioner Charles H. Burke cited an Osage family which received an income of $60,000 a year, and at the end of the year was $20,000 in debt. Commissioner Burke has often expressed great concern regarding the demoralizing influence of large incomes on the Indians, and has even favored limiting the amount of money that any of them should receive.

The demoralizing influence of suddenly acquired wealth has certainly not been so great among the white people as among the Indians, but it is doubtful whether most people of any race receive any benefit, in the long run, from their unearned gains. It is said that many of these "newly rich" landowners save their money, perhaps pay off mortgages or buy more land and remain at work as before.[6] Others are differently affected, and the showy and ostentatious expenditures of some of these have done much to pauperize and cheapen the life of some American communities.

According to one observer, a large proportion of the newly rich oil magnates in southern California bought big automobiles with the first money they received. A Long Beach salesman for one of the most expensive makes of cars declared that he had several times been offered the assignment of the first dividend check as a payment on a car. In some instances the cars were bought before questions of title were even settled sufficiently to make the lease a perfectly safe business venture. One clerk in an office was said to have bought two expensive cars and to have promised $10,000 toward the building of a new church before he received his first dividend check. Many landowners went to Europe with their first oil money. It is said that some oil companies complained that the lessors of oil land sometimes left for Europe so soon after receiving their first check that it was difficult to carry on necessary negotiations with them regarding their business.

Scarcely a town in the oil regions or even anywhere near the oil regions but has its oil rich, with "incomes of $65 a minute," who set a standard of extravagant expenditures that makes the standard of ordinary business incomes look tawdry and cheap. These large oil incomes differ from most other large incomes in several ways. They usually accrue more suddenly, and therefore beget a sort of intoxication that is not quite so likely to result from the slow development of ordinary business success. When a man earns a fortune in the course of ordinary business, he usually gets somewhat accustomed to it as it grows, and usually knows how to handle it; but when it falls suddenly into his lap, he is frequently confused or intoxicated, and very often loses or spends it about as fast as he gets it. Also, the oil-land owners have not earned their incomes, and so tend to be more hilarious over their luck, and more hilarious in their manner of spending. One such millionaire is reported to have bought fifteen automobiles out of his first year's income.

In almost any oil town in Oklahoma or Kansas, visitors will see new and splendid residences erected by oil men, owners of oil land and successful operators, houses extravagant in size and in appointments, and dwarfing all other residences in town. One of the essential appurtenances of such houses is often a large garage, or perhaps the house may rather be said to "pertain to" the garage, for the big automobile is the center around which the lives of many such revolve. Or perhaps it is not one automobile, but a dozen or so, and they are of a size which makes all the Dodges and Overlands and Buicks look pitifully cheap and insufficient. And some of these rich folks buy furs and silks and gowns and glad raiment which make the poor merchants and doctors and lawyers and their families feel ashamed to venture outside their humble doors; and they hire servants and chauffeurs and maids and lackeys and flunkies in such numbers that ordinary folk hardly dare wash their own faces.

As far as these oil rich have been successful operators, we must allow them to spend their money in any way that suits their fancies. The economic theory is that only thus can we induce them to render their valuable services. Furthermore, it is often stated by oil men that few oil operators retire to enjoy their earnings, that they almost always stay in the game until age or bankruptcy drives them into retirement. "Once an oil man, always an oil man," is said to be the rule. Since the landowners, as landowners, have rendered no valuable services, there is no economic justification for a system which permits them the use of any of their wealth.

In some of our universities the evil effects of these oil fortunes have become more and more evident. The presidents of several universities in oil-producing states have been compelled to prohibit students from keeping automobiles, in an effort to preserve some semblance of democracy.

Not only has oil made a few millionaires. It has made a vast number of people of moderate circumstances poor, or at least poorer than they were, and has thus increased social inequalities by working at both ends of the scale. For every man who has got rich through oil, a hundred or so have got somewhat poorer. So, not only do the modest incomes of the ordinary *bourgeoisié* seem smaller because of the extravagant expenditures of the rich, but they actually are smaller, by the amount of the losses incurred by ordinary investors in oil stock.

Over-production of oil has been not only wasteful of oil, but expensive in money. In December, 1922, W. S. Farish, of the Humble Oil and Refining Company, estimated that it cost approximately $200,-000,000 a year to carry the gross stocks of crude and refined oil at that time in storage, for interest, taxes, insurance, evaporation, and other charges.[7] The cost of carrying the flush production of the three great southern California fields alone runs into many millions of dollars. It was estimated in September, 1923, that $40,000,000 had been spent in this way; and much of the storage will probably never serve any useful purpose again, because it is unlikely that so much storage will ever again be needed in these fields. Too much was erected anyhow, for in February, 1924, after production had been declining for several months, some 25,000,000 barrels of storage was reported never to have been filled. In their tremendous campaign of construction, the big oil companies had overestimated the amount of storage needed, and as a result millions of dollars had been wasted.

Too many wells were drilled in most of the important fields, just as in Pennsylvania, and in some regions overdrilling was far worse than in Pennsylvania. In a few places, it is estimated that there were ten or even twenty times as many wells as were needed to drain the oil sands. In the Bremen field, in Ohio, wells were not only drilled on adjacent town lots, but in at least one case, two were drilled on the same lot. Sometimes the derricks were so close together that workmen had difficulty making room for the tools. In one Oklahoma field, half a million dollars were spent on unnecessary wells.[8]

Perhaps the worst case of overdrilling, however, was at Spindletop, where over 1,200 wells were drilled on an area of about 170 acres, besides many dry wells outside of this producing area. At Spindletop, drilling was done on tracts of one-sixteenth of an acre, and in a few cases even as little as one-sixty-fourth of an acre—about 25 square feet. At some points there were four or five wells to the acre, with wells

Warren M. Sargent, Photographer

CROWDED DERRICKS AT LONG BEACH

almost touching each other. Ex-Governor Hogg, and others in the Hogg-Swayne Syndicate, were among the extensive operators in small lots in the Spindletop field. Within 18 months pumping machinery had to be installed at almost every boring, and 750 wells had been abandoned. Within four years, more than 1,000 of the wells were abandoned, and the other 95—all of them "pumpers"—were producing less than 6,000 barrels per day. This was all investors had to show for an investment of $10,000,000, of which at least $1,000,000 had been expended in useless borings, $1,750,000 in producing wells, and the rest in pipe lines, reservoirs, purchase of land, and the like. Much of this capital was practically useless when the oil production declined, and had either to be abandoned or moved at great expense to other fields. All the oil in Spindletop could have been secured at a small fraction of the expense incurred, had it been in the hands of one well-managed company.[9] In the Sour Lake field, overdrilling also caused the rapid exhaustion of the oil, but here there were not so many wells, because one large company owned considerable territory, and could drill conservatively.

In the Burkburnett Townsite pool, 885 wells were drilled in a proved area of 1,265 acres—an average of 1.4 acres per well. It is probable that a normal development program for the field would have called for about five acres per well. In other words, 250 wells, or one-third of the number actually drilled, should have been sufficient to drain the pool. The daily production per well, at the end of less than two years, was less than 5 barrels. In the Northwest Extension pool, at the end of June, 1920, 1,251 wells had been drilled in a proved area of 2,440 acres—an average of 1.8 acres per well.[10]

In all three of the new California fields far too many wells were drilled. The Signal Hill field, discovered in June, 1921, was so riddled with holes in a little more than a year that in one area of 205 acres of land there were 93 wells producing and 40 drilling—an average of one well to 1.54 acres. In one ten-acre tract in this field, 23 wells were drilled—more than two per acre. In another ten-acre tract, 20 wells were drilled, and in one five-acre tract, 12 wells were drilled. In some individual leases the congestion was still worse, and it was not unusual to find as many as three wells on a half acre of land. There were instances of drilling on strips of land that were originally intended for street purposes, 30 feet wide, or half the width of an ordinary street.[11]

In the Santa Fé Springs field, a similar situation developed, with an average of one well to 1.74 acres of land in one 80-acre tract in the center of the field. The Standard Oil Company reported in August, 1923: "In a locality where, theoretically at least, a single well, if given time, could drain the oil from five acres of sub-surface, there are places where there are on an average five wells to the acre."[12] The Standard

Oil Company has considered it good practice in California, where conditions permit, to drill one well to about eight or ten acres. This is estimated to be enough to drain about 40 to 50 per cent of the total recoverable oil during the first year, and the remainder during a period of 15 to 20 years.

The Standard Oil Company made some rather interesting computations on the question of profits in some of these closely drilled fields. In the Santa Fé Springs field, for instance, it was estimated, from government charts and surveys, that there should be not over 174,000 barrels of recoverable oil per acre, three-fourths of which would be the share of the operator, assuming 25 per cent royalty, or a total of 130,500 barrels. The cost of drilling wells in this field averaged $120,000 each, which did not include bonuses paid for the land, camps, roads, and other improvements, nor the cost of producing and handling the oil. With oil at prevailing prices, it is easy to conclude that operators' profits were extremely moderate in a field as closely drilled as this. The Standard Oil Company has for years generally avoided such fields, because of the unprofitable nature of such operations.[13]

The great southern California fields, with their deep sands, of course, needed more wells for a given area than fields with sands one-tenth as thick, so it is rather difficult to say what would constitute overdrilling there. Not all oil companies agree with the Standard Oil Company as to the number of acres that should be allowed per well.

Figures for other California fields indicate that many of them were subject to the same evil. The following table, taken from the *Oil Weekly* of February 10, 1923, proves that too many wells were drilled in many of the California fields:

Oil field	Proved acreages, March 1, 1922	Number of wells drilled within proved areas	Average acreage per well basis average spacing of wells
Fresno County:			
Coalinga	14,365	1,723	6.1
Kern County:			
Belridge	2,195	262	5.4
Devils Den	30	26	.5
Elk Hills	9,080	180	8.0
·Kern River	7,325	2,525	2.4
Lost Hills	2,280	390	4.2
McKittrick	1,655	421	2.0
Midway	40,291	2,659	6.2
Sunset	6,010	779	3.4

Oil field	Proved acreages, March 1, 1922	Number of wells drilled within proved areas	Average acreage per well basis average spacing of wells
Los Angeles County:			
Beverly Hills	110	20	5.0
Long Beach	508	58	†2.3
Montebello	1,160	177	4.5
Newhall	380	137	2.7
Salt Lake	934	317	2.5
Santa Fé Springs ...	716	49	*3.5
Whittier	564	252	2.2
Orange County:			
Brea Olinda	1,963	571	1.7
Coyote Hills	2,160	286	5.2
Huntington Beach ..	2,151	235	‡3.2
Richfield	1,292	175	4.4
Santa Barbara County:			
Casmalia	1,920	112	4.8
Cat Canyon	1,630	79	6.7
Lompoc	1,193	37	19.4
Santa Maria	4,600	287	10.5
Summerland	120	387	.3+
Santa Clara County:			
Sargent	80	8	4.4
San Luis Obispo County:			
Arroyo Grande	600	34	3.3
Ventura County:			
Bardsdale	560	174	1.4
Conejo	80	98	.5
Ojai	420	89	2.8
Piru	498	145	2.3
Santa Paula	570	140	2.0
Simi	606	69	5.8
Sespe	451	84	3.4
South Mountain	630	39	3.7
Ventura	265	44	4.2
Totals	109,392	13,068	4.2§

† Partly town lots—some offset locations only 50 feet apart.
* Partly town lots—some locations as close as three wells per acre.
‡ About one-third town lots—many locations less than 100 feet apart.
§ Average acreage for California.

If we take the Standard Oil estimate of 8 to 10 acres as an ideal, it will be seen that there are at least twice too many wells in these fields, but the oversupply is less than in Signal Hill or the other fields discussed above. In other words, too many wells have been drilled in most of the fields, but Spindletop, Burkburnett, Signal Hill, Huntington Beach, and Santa Fé Springs represent the worst, and not the average.[14]

In *The Lamp* for August, 1923, is found a computation regarding the cost of exploiting the Long Beach field. According to this computation, wells already drilled and drilling (in August) in Signal Hill cost not less than $50,000,000, to say nothing of additional wells by the score that were later drilled. The cost of tanks, pipe lines, loading facilities, pumping stations, and other equipment needed in getting the oil out was not less than another $50,000,000 or a total cost of at least $100,000,000, to August 1, 1923; with the probability that this cost would increase to somewhere between $100,000,000 and $200,000,000 before the drilling epidemic ended. A fair proportion of this expense could be regarded as nothing whatever but waste.[15]

In October, 1923, there were 630 wells in a proved area of 1,200 acres in the Long Beach field. The field could probably have been drained by half that number of wells, so that approximately 315 wells could be regarded as superfluous—wells which cost perhaps $100,000 each, or a total of probably more than $30,000,000 wasted, not including very nearly an equal amount wasted in unnecessary pipe lines, storage, gasoline plants, and other equipment. As *The Lamp* expressed it:

In too rapid exploitation where work that should be spread over a period of five years is crowded into one year, the money must be forthcoming, not only to carry on the drilling but to purchase the oil when it is produced and carry it in storage until there is a market for it. This is a very heavy drain.

It is probable that the total amount of capital wasted in the southern California oil fields, including excessive wells, refineries, pipe lines, storage and other plant and equipment, and including capital wasted in unwise speculation, would be almost enough, invested at a reasonable rate of interest, to pay all the costs of higher education for the entire state for the indefinite future. Almost infinite good might have been accomplished with what was ignorantly wasted, to the absolute detriment of the people.

Such overdrilling is sometimes wasteful of oil, as well as capital. Speed in getting wells into the pay sand is of the utmost importance, because the first well in any locality gets most of the flush production, and in this race of drillers there is often a great deal of careless drilling. In September, 1923, one authority estimated that there were 390 operators in the three southern California fields, a large proportion of whom had had no previous experience in California oil field practice; and under such circumstances there is almost sure to be more or less hurried and careless drilling.

In other cases, however, it appears that close drilling, if not done carelessly, does not reduce the amount of oil recovery, but rather increases it. It has been estimated by competent authorities in the United States Bureau of Mines that the total production of at least some of the California fields will probably be greater than it would have been with fewer wells. The waste in the storage and transportation of the oil would, however, have been less, and the waste in the use of oil would have been far less, had oil never become so cheap.[16]

The situation prevailing in town-lot drilling has been well described by Max W. Ball:

Ignorance there may be, carelessness there undoubtedly is, but back of ignorance, of carelessness, of reckless, headlong methods, is the real cause— the fact that the average holding is so small that speed is the owner's sole protection. Let him be careful if he can; let him be economical if he can find a way; but careful or careless, reckless or conservative, he must be speedy if he would survive. The small holding is his master.[17]

Another writer, referring specifically to conditions in the Mexia field, drew a similar picture:

In such a field caution must be thrown to the winds. If you want to get in on this game, you have to think fast and act quickly. There is no time to seek advice or call a meeting. Acreage that is offered today at $300 is $600 tomorrow and gone at $1000 two days later. The surveyor has only time to drive his first stakes before the teamsters are dumping their material and the erectors are calling for men. If it could be arranged that one association controlled all the acreage with full authority to act for the best interests of a multitude of the owners, all of the oil that lies under the ground at Mexia could be taken out as needed over a period of years, without any waste and at but a fraction of the expense of the work as it is being done. But it is idle to think of costs and economics when the man who owns the adjoining lease may be bringing in a 10,000 barrel well while you are adding up totals and checking estimates.[18]

Much of the cost of our insanely rapid exploitation of each new field is, of course, borne in the long run by the public. The oil producers bear some of the unnecessary cost, but not all, for the recurrent periods of flood and famine in the oil industry simply increase the cost of production, and so are in large measure paid for by the consumers.

The "offset" wells, so often spoken of in the oil world, represent a waste of about half the amount invested. It is customary for the holder of a tract of oil land to begin by drilling a well at the edge of his tract, or, better still, in the corner. In this way he can drain the oil from his neighbor's holding. His neighbor, to protect himself, of course drills what is called an "offset" well, just across the line. In fact, an unwritten law among operators in most fields requires the lessee to drill offset wells where necessary. It has been legally determined that a landowner can bring suit to compel a lessee to drill offset wells or surrender his lease.[19] Leases have often been drawn in such a way as to put too much pressure on the lessee to drill regardless of market conditions. This has been one of the reasons for the complete demoralization of the industry in times of over-production.[20] In many oil fields, the wells commonly appear in pairs, at the lines of adjoining tracts, the interior of each tract being reserved for later exploitation. In one instance, in the Hewitt field in Oklahoma, four wells were drilled into the Hewitt sand within a week, and all of them found edge water. Any one of these wells would have given the desired information. Sometimes it is possible for adjoining owners to make satisfactory agreements to avoid the necessity of boring offset wells, but most fields show a great deal of such duplication of expense.[21]

One of the worst examples of offset drilling is shown by developments in the Powell field in the latter half of 1923. At the time the field was discovered, the market was already flooded with oil, and the operators tried to get some sort of agreement which would obviate the necessity of an expensive drilling campaign. They were unsuccessful, however, and soon a drilling campaign was on which raised the production of the field above 300,000 barrels daily for a while. The Humble Oil Company had a number of scattered tracts and drilled offset wells by the dozen.[22]

It would be unfair to picture such horrible examples as Burkburnett, Powell, and Long Beach as typical of the oil industry in general. Unfortunately such examples have been far too common, but it is well to bear in mind that many fields, such as Tonkawa, Burbank, and Cromwell, and perhaps even more particularly Salt Creek, have been exploited in fairly orderly fashion, with a reasonable number of wells and reasonable expenditure of capital.

These offset wells were an unusually important problem in the case of some of the railroad grants, particularly the Southern Pacific grant in southern California. The grant of alternate sections, checkerboard fashion, made a tract which was very hard to manage. Owners of the alternate sections, which were oil bearing, hastened to drill as soon as oil was discovered, and the Southern Pacific soon found itself forced to drill against other producing wells, although the company announced that it would not drill until work had been started upon adjoining property. It seemed unfortunate that some equitable arrangement could not

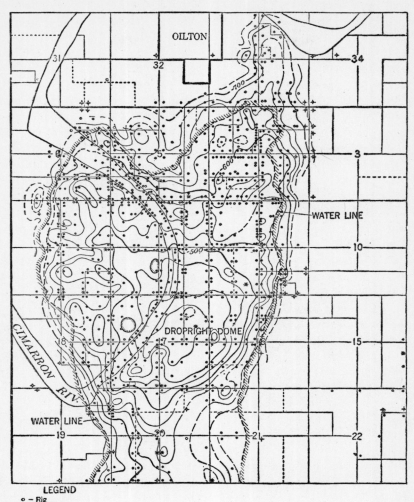

LEGEND

o – Rig
● – Drilling well
✦ – Dry hole
◉ – Oil well
✴ – Gas well
✦– Abandoned oil well
✴ᴛ–Abandoned gas well

MAP OF A PORTION OF THE CUSHING OIL POOL, OKLAHOMA,
SHOWING THE SUBDIVISION OF THE AREA INTO
SMALL PROPERTIES, WITH NUMEROUS
OFFSET WELLS

Reprinted by permission from *Economics of Petroleum*, by Joseph E. Pogue,
published by John Wiley & Sons, Inc.

be made that would permit the railroad company to keep its oil for future requirements.[23]

The law pertaining to oil deposits has been an absurd, almost idiotic, conception anyhow. It is difficult to understand how it could have been tolerated by an intelligent people. There have been instances where oil companies have drained the oil belonging to adjacent owners, refusing to buy their land or transport their oil. In an interesting Pennsylvania case, one man used a gas pump to suck the oil into his own well, and increased the flow, but decreased the flow of a neighbor's well. The neighbor sued, but the Court of Common Pleas of Pennsylvania held that the first man was entirely within his rights. In an Oklahoma case, an independent operator held a very good oil lease which he desired to develop. The only pipe line in this field was owned by a Standard Oil subsidiary, which held leases on all the neighboring land. The independent operator could not afford to build a pipe line of his own, and the corporation refused to transport his oil, and refused to buy his land. The Standard leases were lower in the dip, and so they were able to drain his oil entirely.

Under the law as it has stood, the problem of each oil operator is a twofold problem of defending his own property from drainage by neighbors, and of draining his neighbors' properties of as much oil as possible. As one California operator expressed it: "We all try to get the oil out of each other's land." This problem is frankly stated by practical men in the field. One treatise reads as follows:

Petroleum mining differs from other mining in that there is no legal liability for the operator who depletes a neighboring property, while in mineral mining no operator can obtain ore from neighboring property without being responsible for the mineral so taken.

Close drilling and rapid exhaustion of an oil field always meant a heavy depreciation charge against the equipment used. Much of this equipment, including casings, is durable enough to last many years, but if a field is exhausted quickly, there is nothing to do but pull up the casings, if they are worth pulling, and move them, with other equipment, to another field. This involves a considerable expense to those in the industry.

Thus our oil has been secured at an immense cost in unnecessary wells, a cost that would have to be measured in hundreds of millions, or even, for the entire history of the oil industry, perhaps in billions. A total of over 600,000 wells have been drilled in the United States, and the average cost per well would not have to be very high to run into billions of dollars. It has been estimated that the wells drilled in the year 1922 cost $529,000,000. It is not improbable that because of our system of small holdings and offset wells, the cost of extracting the oil has been twice as high as would have been necessary under an intelligent system. A business man who expended twice as much as was necessary in his business, or, in other words, who wasted half of all that he spent in the prosecution of his business, would be rated very poor or insane, but our oil producing industry represents just about that grade of efficiency.

There were almost always far too many refineries, also. Hundreds of small refineries were built in the oil regions, by men who had little means and no experience in the business, and who therefore fell into difficulties with the first serious depression that swept over the industry. H. G. James has estimated that there are over 500 refineries in the United States, 400 of which might be called independent, and that nearly half of these 400 refineries were not operating in March, 1924. Most of these small plants were not only financially unprofitable but, as suggested in another connection, were wasteful of the oil resource. Many of them were mere skimming plants or skimming and lubricating plants. They were almost entirely dependent on current production of oil, because they had very small storage facilities, and of course this made their situation particularly precarious. The loss of capital represented by excess refinery capacity, and by the high failure rate among refineries has been a debit item in the oil industry which has probably amounted to hundreds of millions of dollars.

A great many students of conditions in the oil industry have insisted that there are far too many service stations. The attorneys-general of some 20 states who were investigating the oil industry in the fall of 1923 included among their recommendations a recommendation that special attention should be given to the prevention of unnecessary and wasteful duplication of service stations; and many practical oil men have directed attention to the same question. The excessive number of service stations, and the extravagant service rendered, suggest a possibility that there are artificial forces at work keeping the price of gasoline above a competitive level, and thus making the retailing of gasoline too profitable.

David W. Moffitt, Vice-President of Cosden and Company, recently criticized the excessive investment in retail service in rather witty fashion:

This department has developed from the delivery into a tin can at the corner grocery to delivery into the gasoline tank of a limousine, from a service station with liveried servants to test your air, examine your radiator, fill your crank case, hand the baby a stick of candy and the lady in the party a fragrant rose or carnation. A colored orchestra at the right and a free picture show on your left while you wait is just a slightly delayed improvement now being prepared. . . . Complaints are beginning to be heard in some quarters that there are not sufficient corners in the cities as now laid out to furnish locations for service stations and a movement is expected soon to open streets where there are now alleys, in order that service station locations may keep pace with the needs of promoters in this line. We are rapidly approaching that stage in the marketing end of the petroleum industry where 50 per cent of the people will be consumers and the other 50 per cent vendors in charge of service stations.[24]

The logic of general reasoning would lead easily to the conclusion that if there are too many service stations, it must be because the margin allowed for the retailing function is too wide. The business is not attractive, like the grocery business, by reason of the small investment required, for most service stations are rather expensive affairs. Of course the larger oil companies feel constrained to maintain a considerable number of stations, in order that they may have an outlet for their refined products, but they could not afford all that we have in most cities without an outrageously wide margin of profit. The business of retailing gasoline looks very much like competition gone wild, or, perhaps better, like competition badly hobbled, for there is generally little competition in price. Some service stations make immense profits, in spite of the great multiplicity of such establishments.

There has been an enormous waste of capital and social energy in litigation and in contests over titles to oil lands. The oil fields have not been the only places in America with a surplus of shyster lawyers and of cooked-up litigation, but they have had far more than their share of these American products. The immense values involved in the oil fields make litigation a very profitable business for the lawyers. There is a story about a lease in the Butler County field, in Kansas, reported to be worth $2,000,000 which, after two years of operations,—largely legal operations,—had not paid a dollar to anyone but the lawyers. Lillie Taylor was said to have had to give one-half of her estate in Louisiana to the lawyers who were to prove it was hers.

Oklahoma, with her tangled Indian titles, was long the rendezvous of enterprising title and lease lawyers, and the scene of a great deal of litigation. One writer described this situation as follows:

Taking advantage of unsettled conditions, contingent-fee lawyers filed all kinds of suits in all of the courts of the state, and in the federal courts. An ingenious mind, not necessarily learned in the law, could find a technical excuse for attacking almost any title in the state. This is particularly true of the titles to valuable oil properties which at the time were thought to be worthless, or at least of little value. As long as these lands, most of them allotments in the rough, hilly country, remained undeveloped, or the surrounding country remained undeveloped, there was no question raised as to the title or as to legality of the allotment. But immediately upon the discovery of oil and gas, a cloud of claimants arose with all manner of contests, armed with all manner of processes.[25]

Leases in the western part of Oklahoma were said to command better prices than those in the eastern part, because titles were not so tangled. "In the Old Indian Territory," as one correspondent reported, "the towns swarm with lawyers who are willing, for any sort of a contingent fee, to bring suits attacking on any sort of flimsy grounds. . . . Not until legislation straightens out the tangle in eastern Oklahoma, will the oil country be without shyster, grafting lawyers and bogus claimants, cluttering up the dockets of the courts, interfering with the development of the greatest natural resources of Oklahoma, and generally making nuisances of themselves."[26]

The situation in Texas has been somewhat the same, much trouble arising from the titles which date back to the days of Mexican domination, or to French or Spanish grants. Increased land values, shyster lawyers, faulty transfers, missing heirs, prosecuted claims, blackmail fees,—it is the same story all over again.[27]

So great has been the waste of capital and human energy in the exploitation of our oil resources, that it is not certain that these resources —perhaps the richest, and certainly the richest accessible resources in the world—have even afforded the exploiters a net profit for their operations. In fact, many observers believe that oil production, like gold mining, has been carried on at a loss. In ordinary competitive business operations, cost of production is supposed to govern or approximate the normal value of products; but in the production of oil there are peculiar conditions which prevent this law from applying. On the one hand, the monopolistic position of the Standard Oil Company has in many fields tended to raise values far above cost of production and to make possible enormous profits; and on the other hand, the speculative glamour of the business has attracted so many into it that losses have been enormous.

Whether there have been net losses or net profits in the oil industry is a much mooted question. Blatchley thought that in the Indiana fields there were at least no considerable profits.[28] Professor Knight, writing in 1901, was of the very definite opinion that the Wyoming fields had been operated at a loss up to that date, although it is doubtful if he would hold the same opinion now. One writer estimated in 1899 that the average cost of producing a barrel of petroleum in the United States was 65 cents. This was probably not far from the average price received at that time, but like most computations, it failed to include many indirect costs, and many indirect gains from the use of petroleum.[29] Another writer estimated in 1910 that the output of the California fields yielded an income of only 7½ per cent on the amount of capital invested.[30] In 1904, however, Professor Erasmus Haworth, State Geologist of Kansas, made the following computation regarding the Kansas wells:

Our 1116 wells have averaged a cost of about $1500 each or a little more than $2,500,000. To this should be added the cost of all dry holes, which would average less than $1000 each, as all casings are drawn from such wells for future use. I am confident that thus far not more than 10 per cent of all wells drilled have been failures. This would increase the cost to about $3,000,000. Should every drill stop today, the wells already made would more than pay this entire bill in one year, if the price of oil remains as it is now.[31]

E. W. Marland expressed his opinion in October, 1923, that the producing branch of the oil business had been operating for 50 years at a total loss of approximately $2,500,000,000; that in the years 1922 and 1923 producers had lost approximately $500,000,000 each year; that the oil produced in the Burbank field, up to October, 1923, had netted a total loss of $70,000,000, in spite of the fact that it was one of the most conservatively drilled and operated fields in the history of the oil industry in Oklahoma. A few months later, apparently when in a despondent mood, Marland raised his estimate of the total net losses in the industry to more than 4½ billion dollars. In his later calculations he estimated that approximately 12 billion dollars had been placed in the legitimate channels of oil field development and operation in the United States, while only 7½ billions have been returned from the sale of the crude oil, leaving a deficiency of 4½ billion dollars—without including the money put into fraudulent promotion companies.[32]

The Texas division of the Mid-Continent Oil and Gas Association recently compiled some figures to show that the cost of labor, bonus and rentals to landowners, and special taxes on the Texas oil production, amounted to $1.37 a barrel, not including the cost of material, machinery, insurance, and regular state and federal taxes. A cost of $1.37 a barrel was not far from the average price of Texas oil during the period under consideration.

Unfortunately, some of these figures have to be taken with a few barrels of salt. Many of the oil producers are anxious to impress the public with the smallness of profits in the oil industry; and the figures of the Mid-Continent Oil and Gas Association were compiled expressly for presentation to the state legislature, as an argument against increased taxation. Furthermore, these figures did not include as part of cost the losses from speculation, fraud, neglect of other industries, etc.[33]

Mr. H. L. Doherty, of the Cities Service Company, has expressed his belief that the oil industry was being conducted at a loss, for the same reason that the gold mining was supposed to have been carried on at a loss. Doherty pointed out that people would buy tickets in a lottery where the chances of loss were greater than the chances of gain, and, by this psychological fact, justified his belief that people would invest in an unprofitable industry.[34] A considerable number of practical oil men have indicated their belief that the oil industry was carried on at a loss. The American Petroleum Institute submitted figures before the Senate Committee on Manufactures in 1922 to show that the average net earnings of all the oil companies in the United States for which published balance sheets were available, amounted in 1921 to only 4.4 per cent on the total "net worth" of the companies.[35]

It is easy to point out the vast profits of some of the great oil companies. H. L. Doherty made computations in 1920 of the fiscal operations of 108 companies organized prior to 1912, and of 142 companies formed since that time, showing that at the close of 1911, the 108 companies operating at that time showed a capital investment of $717,-098,563, while at the end of 1919 these companies, with the 142 companies formed since that time, had a capital investment of over $2,500,-000,000. Thus, during these eight years the total capital investment of the entire group was increased by $1,784,000,000, to some extent from earnings, while in the same time the companies paid cash dividends of $906,000,000.[36]

A writer in the *Oil and Gas Journal* has presented some very different figures, purporting to show that the securities of 73 leading oil companies, 34 Standard companies, and 39 independents, were worth over $5,000,000,000. Much of this, particularly in the case of the Standard companies, represented the reinvestment of earnings.[37] Several competent authorities have estimated the present value of all the oil investments in the United States at over $8,000,000,000.

Against the billions in dividends disbursed, and surpluses accumulated by the successful oil companies, however, must be placed the vast amounts lost by investors in fraudulent and unsuccessful companies, and a great many other items. As pointed out elsewhere,[38] there is no way of estimating even approximately all the losses incident to the conduct of the oil industry, but it is entirely possible, if not probable, that they are greater than the total gains.

Just as in Pennsylvania, in oil fields farther west social conditions have usually been very bad. The speculative nature of the business, and the character of the people who are attracted to such a business; the temporary nature of the industry, arising from the rapidity with which many fields are drained, and the consequent lack of confidence in future values, and unwillingness to invest in substantial improvements; the extremely rapid growth of population, with the resulting difficulty in working out efficient systems of police and government; the transient nature of the population; the great wealth found in such communities and its unequal distribution; all these and other factors tended to make the oil town anything but a healthy social unit. Many forms of wickedness have flourished: cheating, gambling, stealing, drinking, bootlegging, prostitution, robbery, and murder.

Beaumont was often called the "Hell Hole" of Texas during its golden era; Burkburnett was the scene of many murders—four in one night; Tulsa once claimed the distinction of having more automobiles stolen than any other city in the United States, and insurance companies threatened to withdraw from the city. In some Kansas and Oklahoma fields, it was said in 1916 that many thugs discarded ordinary "stick up" methods, as taking too much time and trouble, and just hit their victims over the head without any preliminaries, and took their money. One oil journal suggested the wearing of steel helmets by those who worked in the oil fields. The district attorney at Wichita Falls reported that there was a strong and effective organization among the criminals in the oil fields, and claimed that it was almost impossible to suppress the prevailing lawlessness. Conditions in Mexia once became so bad that the state militia was called upon, and actually took charge of the town for some time. In El Dorado, Arkansas, conditions were almost as bad, and in nearby Smackover, the lawless element of the population became so obnoxious that after a particularly busy week of robberies and murders, a citizens' vigilance committee was organized to clean up the town and vicinity. This committee flogged several undesirables, tarred and feathered others, burned one gambling resort, and shot the owner of another resort, in what newspapers described as a "battle" between the "saints and sinners." Not only social conditions, but, according to an investigation by the Bureau of Mines, sanitary conditions as well have generally been bad; in fact there are few respects in which the typical oil town would be given a high rating.[39]

NOTES

1. *Oil and Gas Jour.*, March 15, 1917, 3.
2. For some of the stories of fortunes lost in Pennsylvania, see *Oil and Gas Jour.*, Dec. 31, 1920, 64.
3. *Natl. Petroleum News*, April 30, 1924, 61; *World's Work*, July, 1919, 307.
4. Indiana Dept. of Geology and Natural Resources, *Rept.*, 1896, 95; *Mineral Industry*, 1904, 346; *Eng. and Min. Jour.*, Jan. 26, 1905, 202.
5. *Natl. Petroleum News*, April 30, 1924, 60.
6. *Oil and Gas Jour.*, July 18, 1919, 53; Sept. 24, 1920, 2.
7. *Natl. Petroleum News*, Dec. 13, 1922, 33, 35.
8. Ohio Geol. Survey, Fourth Series, *Bul. 12*, p. 11.
9. *Mineral Resources*, 1904, 712; *Oil and Gas Jour.*, April 12, 1917, 3.

10. *Pac. Oil Reporter,* Feb. 6, 1904, 4; *Cal. Oil Fields,* Jan., 1921; *Eng. and Min. Jour.,* Dec. 25, 1920, 1221; *Oil and Gas Jour.,* April 22, 1921, 76.

11. *Oil Weekly,* Oct. 20, 1923, 57; *Oildom,* Aug., 1923, 14.

12. *The Lamp,* Aug., 1923.

13. *Standard Oil Bulletins,* Sept. and Oct., 1922; *Natl. Petroleum News,* Jan. 10, 1923, 60; Cal. State Oil and Gas Supervisor, *Third Ann. Rept.,* p. 10. See also U. S. Bu. of Mines, *Bul. 194,* 18, 25; *The Lamp,* Aug., 1923, 14; *Oil Weekly,* Oct. 20, 1923, 57.

14. *Cal. Oil Fields,* Jan., 1921, 7; Aug., 1921, 5-9.

15. *The Lamp,* Aug., 1923, 14.

16. *Standard Oil Bulletin,* Dec., 1923, 1, 2.

17. Pogue, *Economics of Petroleum,* 32.

18. *The Lamp,* Dec., 1921, 18.

19. Ill. State Geol. Survey, *Bul. 22,* 151.

20. Johnson and Huntley, *Oil and Gas Production,* p. 103.

21. Swigart and Schwarzenbek, *Petroleum Engineering in the Hewitt Oil Field;* U. S. Bu. of Mines, in coöperation with the State of Oklahoma, p. 28; Ok. Geol. Survey, *Bul. 2* (1911), 220. (Good map to show offset wells in Glenn pool.)

22. *Oil Weekly,* Aug. 25, 1923, 33.

23. *Mineral Industry,* 1911, 560, 561.

24. *Oil and Gas Jour.,* Oct. 4, 1923, 72.

25. *Oil and Gas Jour.,* Sept. 16, 1915, 24.

26. *Oil and Gas Jour.,* Oct. 7, 1915, 2.

27. *Oil and Gas Jour.,* Jan. 2, 1920, 18. See also *Oil and Gas Jour.,* July 6, 1916, 2. For an account of tangled titles in Wyoming, see *Bulletin 670* of the U. S. Geol. Survey, p. 5-9; and for an account of litigation in the Smackover field, see *Oil and Gas Jour.,* June 3, 1921, 66.

28. Indiana Dept. of Geology and Natural Resources, *Rept.,* 1896, 95-.

29. *Mineral Industry,* 1899, 452.

30. *Min. and Sci. Press,* June 11, 1910, 859.

31. *Pac. Oil Reporter,* April 16, 1904, 3.

32. *Natl. Petroleum News,* Oct. 17, 1923, 25.

33. *Petroleum Age,* March 15, 1923, 25.

34. Am. Petroleum Inst., *Proceedings,* Ann. Meeting of 1920, p. 46.

35. *Oildom,* Sept., 1922, 60.

36. *Wall St. Jour.,* Nov. 20, 1920, 5.

37. *Oil and Gas Jour.,* June 15, 1922, 82.

38. See above, pp. 43, 44.

39. *Good Words,* May, 1903, 327, 334; *Oil and Gas Jour.,* Nov. 16, 1916, 2; Oct. 17, 1919, 66; Jan. 9, 1920, 75; Jan. 28, 1921, 60; June 3, 1921, 15; Sept. 28, 1922, 21; K. C. *Times,* Jan. 2, 1922; Oct. 27, 1922; Nov. 29, 1922; Nov. 30, 1922.

CHAPTER XVIII

MONOPOLISTIC CONDITIONS IN THE OIL INDUSTRY: THE STANDARD OIL COMPANY

ONOPOLISTIC conditions have developed in most later fields, just as in Pennsylvania; and in many fields the dominating power has been the Standard Oil Company, or one of its subsidiaries or allies. As stated in an earlier chapter,[1] the Standard Oil Company, the "trust" as later known, was organized in 1879 and 1882. Ten years later the trust agreement was declared illegal by the Ohio courts, and finally in 1899 the company was reorganized as a holding company, under the laws of New Jersey, with a capital of $10,000,000, which was soon increased to $110,000,000. In 1904, the Standard Oil Company produced, in plants either directly controlled by it or affiliated with it, over 86 per cent of the country's output of kerosene—the most important oil product at that time,—and probably one-third of the remainder was produced by companies to some extent under Standard control. The 9 or 10 per cent produced by independents was scattered among about 75 small concerns, whose combined output was altogether less than that of either the Bayonne or the Philadelphia works of the Standard Oil Company.

From 1899 to 1907 it (The Standard Oil Company), with subsidiaries, produced more than one-tenth of all the crude oil in the country, transported over four-fifths of the oil from the Pennsylvania and Indiana fields, manufactured more than three-fourths of the crude oil refined in the United States, owned and operated more than half of the tank cars used to distribute its products, marketed more than four-fifths of all the illuminating oil sold in the United States, exported more than four-fifths of all the illuminating oil sent forth from the United States, sold more than four-fifths of the naphtha sold in the United States, and sold more than nine-tenths of all the lubricating oil sold to the railroad companies in the United States.[2]

The troubles of the Kansas oil producers in 1904 and 1905 led to a House Resolution in February, 1905, requesting the Secretary of Commerce and Labor to investigate the causes of the low price of oil in Kansas, and to investigate the general situation in the oil industry. As a result of this investigation, the *Report of the Commissioner of Corporations on the Transportation of Petroleum* was published in 1906, disclosing the railroad rebates and advantages of various kinds enjoyed by the Standard Oil Company. Not long afterward, the Bureau of Corporations issued another report, on the *Position of the Standard Oil Company in the Petroleum Industry*, and in August, 1907, a report on *Prices and Profits in the Petroleum Industry*. The publication of these reports led to stricter regulation of the railroads, and to increasing hostility toward the Standard Oil Company. Within the next three or four years, that company was almost constantly in court on charges preferred by the federal government or by various state governments; and in these various trials further information was secured regarding its position and business practices.

For a number of years the Standard Oil Company held this dominating position in the oil industry, but it was the constant object of attack in Congress and in the courts, and in May, 1911, the Supreme Court of the United States ordered its dissolution. The court rendered its decision of little immediate effect, however, by permitting the parent company—the Standard Oil Company of New Jersey—to distribute the shares of subsidiary companies ratably among its shareholders. By doing this, the Standard Oil group were enabled to preserve a very effective community of interest among the various companies, since the same group of men now owned much of the stock of each of them. According to a report of the Federal Trade Commission in 1915, there was, at least in the sale of gasoline, practically no competition among the various Standard Oil units. They divided the country up into districts, within which there was no competition between Standard companies.[3]

The stock remained for some years in the same hands that had held it before the decree. In 1915, four years after the dissolution, a certain group of interests were found to have 55 per cent of the stock of the Atlantic Refining Company, the Standard of Indiana, the Prairie Oil and Gas Company, and the Prairie Pipe Line Company; 54 per cent of the stock of the Standard of New Jersey; 53 per cent of the stock of the Standard of Ohio, the Standard of Kentucky, and the Standard of New York; and 52 per cent of the stock of the Continental Oil Company. Thus this group of interests, which represented 29 individuals, controlled the important Standard units. Only a few years ago, Walter Teagle, President of the Standard Oil Company of New Jersey, admitted that there was practically no competition among the various Standard units.[4]

After the so-called dissolution decree in 1911, the Standard companies declined somewhat in importance, relatively to the independents, yet in 1919 they controlled 23½ per cent of the total crude production of the country, operated about 68 per cent of the pipe lines, refined almost 44 per cent of the total quantity of crude refined, and produced 60 per cent of the gasoline, while the Standard marketing companies were the dominating factors in both the domestic and export trade of the entire country.[5]

The absolute wealth and financial power of the Standard companies increased prodigiously during these years; and it is probable that what is called the Standard Oil interests represent now by far the greatest financial power in the United States, or perhaps in the world. The Standard Oil Company of New Jersey alone, the parent company, now has assets of nearly a billion dollars, and several of the other larger Standard companies have assets nearly half as large. The total assets of these companies amount to several billion dollars.

The Standard companies are a vast power, not only in every department of the oil business, but in other kinds of business as well. They control oil lands, wells, storage tanks, refineries, pipe lines, steamship lines, retail filling stations—everything necessary to produce oil and bring it to the final consumer; and Standard Oil influence and control extend to many other industries. For years the Standard Oil group have been heavily interested in railroad companies, particularly the Union Pacific, the Chicago, Milwaukee and St. Paul, and the New York, New Haven and Hartford Railroad, and in gas and electric lighting companies. The Consolidated Gas Company of New York City, for example, was once, and perhaps still is, a Standard Oil affair. The same group of men control a large number of shares of the United States Steel Corporation, the Amalgamated Copper Company, and the Corn Products Refining Company; and in the field of banking they have been very influential. The National City Bank of New York—the largest bank in the western hemisphere—is often spoken of as a Standard Oil bank.[6]

In most of the oil fields, the Standard companies represent the dominant interests. In the Mid-Continent field, the Standard Oil has long been the dominant purchasing interest through the Prairie Oil and Gas Company. Professor Haworth reported in 1906 that other smaller concerns had little influence on the market, which was practically controlled by the Prairie Oil and Gas Company. He reported that there was an idea abroad that if any producer once sold oil to an independent concern, the Prairie Oil and Gas would not take his supply or any part of it, and that this belief prevented some operators from selling oil to any other refining or pipe-line companies. In 1908, the Prairie Oil and Gas was buying about two-thirds of the Mid-Continent production—33,000,000 barrels, of a total of 50,000,000 barrels. Previous to 1907, the Prairie Oil and Gas operated the only pipe line in Oklahoma, but during 1907 the Gulf Oil and Pipe Line Company and the Texas Company each completed a pipe line from the Glenn field to the coast of the Gulf of Mexico, and built a large number of large storage tanks.

The predominance of the Prairie Oil and Gas in the Mid-Continent field has decreased considerably in recent years.[7]

Standard interests dominate the oil industry of Wyoming more completely than that of any other oil region of the United States. The isolation of the Wyoming fields, and the great expense involved in drilling and in constructing pipe lines and refineries, has thrown the industry into the hands of large companies, with abundant capital. The Midwest Refining interests, now controlled by the Standard Oil Company of Indiana, controlled between 65 and 69 per cent of the oil production of the state during the period from 1917 to 1919, when the Federal Trade Commission made its investigation.[8] The Ohio Oil Company, a Standard company, had the largest owned production in the state, and, in addition, controlled considerable production through working agreements. These two Standard interests controlled altogether between 93 and 97 per cent of the total production of the state.[9] Two Standard companies, the Midwest Refining Company and the Illinois Pipe Line Company, operate practically all the pipe-line mileage directly, and transport nearly all the crude petroleum marketed in Wyoming. In 1921, the Midwest Refining Company owned and controlled under lease about 46 per cent of the trunk pipe-line mileage of Wyoming, and the Illinois Pipe Line Company operated 53 per cent, while the two together controlled practically 99 per cent of the total mileage.[10] At the same time, the Midwest Refining Company, and its subsidiary, the Utah Oil Refining Company, owned and operated 85 per cent of the refining capacity operating on Wyoming crude petroleum. The Midwest and other Standard refineries owned and operated 90 per cent of the refining capacity dependent on Wyoming crude oil. Since that time, the Texas Company has built a refinery at Casper, and the White Eagle Oil and Refining Company now has a refinery there, so that the proportion of oil refined by Standard companies may now be lower than it was in 1921.

There is reason to believe that the Standard has used its strong position in Wyoming to make very large profits in the refining of oil. The general policy of the company has been to maintain a low price for crude, while selling gasoline at a highly remunerative price in the mountain states.

In California, Standard Oil interests have been very strong for many years, although by no means so nearly dominant as in Wyoming or in the Mid-Continent fields. Standard interests entered early into the transportation, marketing, and refining of California oil, and built pipe lines into every important producing district, with refineries at Los Angeles, Point Richmond, El Segundo, and San Francisco. In recent years, it has entered extensively into the production of crude oil.[11]

The report of the Federal Trade Commission credited the Standard Oil Company with the ownership, in March, 1920, of only about 11 per cent of the proved oil lands of the state; but this one company produced over 25 per cent and refined over 47 per cent of the total production of the state. It owned 913 miles of trunk pipe lines, of a total of 2,439 miles in the state. These figures have, of course, changed greatly with the exploitation of the three great fields near Los Angeles, but the relative importance of the Standard Oil Company has probably not decreased greatly.

In California, as elsewhere, Standard power has grown rapidly during years of over-production and instability. In 1900, the opening of the Kern River field increased the output so greatly that prices declined to 20 cents a barrel, and even less. This was just the situation that the Standard has everywhere been able to take advantage of. During these years the company fought hard to squeeze out the small operators in the Los Angeles field.[12] The dissolution of the Standard in 1911 effected no change in the strong position of the company, and was even regarded as unfortunate for the other oil companies. Like many of the Standard Oil companies, the Standard Oil Company of California enjoyed very great prosperity after the dissolution. Its capital of $25,000,000 in 1911 was increased to $50,000,000 in 1912, to $100,000,000 two years later, and to $250,000,000 in 1922—an increase of 1,000 per cent in a little over a decade.

It is claimed that the Standard Oil Company fixes the price of crude oil in California, and the price of oil products as well. The price announced by this company is considered the market price, and generally other marketers name the same price. Of course, the Standard Oil Company cannot name a price far out of line with existing conditions of supply and demand, but it usually takes the initiative in making price changes.[13]

In the Texas coastal region the Standard Oil Company was never as strong as in some other regions, especially for a few years after the Texas anti-trust campaign of 1907 and 1908, and the dissolution of several Standard companies; yet several of the largest companies operating in this region are Standard controlled, and some of the larger independents generally work in harmony with Standard interests. In the Louisiana fields, Standard interests overshadow everything else.[14] Unfortunately, no government agency has carefully investigated these regions, as they have those of Wyoming and California, and detailed information is not available.

The old Standard Oil Company controlled markets abroad, in many quarters of the globe. As early as 1894, the Standard was negotiating with Russian petroleum exporters for a division of the world's markets, negotiations being on the basis of 70 per cent for the Standard and 30 per cent for the Russian producers. As early as this, the Standard had a strong position in the German market, through the German-American Petroleum Company; in the English market, through the Anglo-American Oil Company; in the Dutch and Belgian markets, through the American Petroleum Company; in the Scandinavian countries, through the Danish Petroleum Company; and in the Italian market, through the Italian-American Petroleum Company. In fact, the Standard Oil Company was for many years almost a world monopoly.[15]

Several of the present Standard companies have large producing and refining interests abroad. The Standard Oil Company of New Jersey, for instance, owns producing companies in Peru, Roumania, Mexico, Venezuela, Colombia, Sumatra, and Java, with a total daily production of about 47,000 barrels; and owns through subsidiary companies, a total of 80,000 barrels of refining capacity, in Canada, Peru, Argentina, Italy, Cuba, Spain, France, Colombia, Mexico, Roumania, Sumatra, and Java. This company has at times carried on negotiations for Persian concessions and for concessions elsewhere in Asia.[16] The Standard Oil Company is reported to have secured the De Mares concession of 2,000,000 acres of oil lands in Colombia, for a consideration of $40,000,000.[17]

The Standard Oil Company of California has interests, not only in various parts of the United States, but in the Philippine Islands, and in Colombia and Venezuela.

For the conduct of their world-wide trade in oil products, the Standard companies operate a large fleet of tankers. The Standard Oil Company of New Jersey alone owns and operates under the American flag 37 tank vessels, with a total dead-weight of 444,847 tons. This company and its subsidiaries operate 10 per cent of the world's tanker tonnage.[18]

The story of Midas becomes a common and uninteresting fable compared with the record of Standard Oil profits. Formed in 1870 with $1,000,000 capital stock, the company has grown, largely out of reinvested profits, into a large number of great allied corporations with combined assets of several billion dollars, while paying dividends, which aggregated over two thousand times as much as the original capital. What is known as the Standard group of oil companies paid out dividends of over $129,000,000 in 1922, over $138,000,000 in 1923, and $150,000,000 in 1924, exclusive of interest on bonds. This would represent 4 per cent dividends (Standard stocks usually sell on a 3 or 4 per cent basis), on a principal of $3,750,000,000. It is safe to say that the market value of all the securities, including bonds, of what are known as the Standard Oil group, is not far from $4,000,000,000. This figure must be subjected to some discount, however, for two reasons. In the first place, not all of this is reinvested earnings, for a small part of it has been subscribed by stockholders from time to time. In the second place, some of the companies known as Standard companies are only partly owned by Standard interests. On the other hand, it must be noted that Standard Oil interests own stock in many other companies of various kinds, including some of the independent oil companies. For instance, Standard interests own about 25 per cent of the outstanding stock of the Pacific Oil Company of California, and doubtless substantial amounts of the stock of various other independent oil companies. According to one estimate, the oil industry in the United States represents an investment of about $8,000,000,000, 40 per cent of which is found in the Standard group.

While the Standard Oil companies were developing assets worth billions of dollars, they paid dividends aggregating over two billion dollars. During the years 1882-1906, the trust reported net earnings of $838,783,783, and paid out dividends amounting to $548,436,446; and during the period of something over a decade, from the time of the dissolution in 1911 to January, 1924, the Standard group of oil companies paid a total in cash dividends of $1,619,956,009. Thus the Standard Oil Company, and the various Standard companies into which it was dissolved, have paid a total of well over $2,000,000,000 in cash dividends since the formation of the company fifty years ago.[19]

Stories of the growth and earnings of some of the individual Standard companies also read like fables. The Standard Oil Company of New Jersey, the parent company, was organized in 1882 with a capital of $3,000,000. In 1922 this capital had been increased to $625,000,000 (authorized), about four-fifths of the increase representing reinvestment of profits. The Standard Oil Company of California grew from a $1,000,000 corporation in 1879 to a $250,000,000 company in 1922, $152,000,000 of the increase representing profits plowed under. The Standard Oil Company of New York began business in 1882 with an authorized capital stock of $5,000,000. Its present outstanding capital stock is $225,000,000, of which $210,000,000 represents earnings distributed by means of stock dividends. A share of stock in this company, worth $400 at the time of dissolution, would now represent a value of $2,760.[20] The Standard Oil Company of Indiana has been one of the most spectacular money makers in the history of business. This company started in 1889 with a capital stock of $500,000. Three years later this was doubled; in 1912 it was increased to $30,000,000, and in 1917 to $100,000,000; and today it stands at $220,000,000. Of this increase, $1,736,357 was issued for cash, $34,354,462 for property, and $184,090,819 represented stock dividends. This represents an increase of 50,000 per cent, mainly through the investment of earnings and their capitalization through stock dividends. The approximate book value of the stock at the time of dissolution in 1911 was $2,520. That share of stock, with its increments from stock dividends, would today amount to 600 shares, worth about $60 a share, or $36,000, while cash dividends received from 1912 to 1922 would have amounted to $8,220— a total of $44,220 realized from one share of stock, in a little over ten years. During this time, the total net earnings of the Indiana company amounted to $239,024,534.[21]

The following record of dividends paid by the Standard Oil Company of Indiana is copied from Moody's *Analysis* (1923):

Prior to disintegration of Standard Oil Company of New Jersey, dividends of 850% in 1903; 450% in 1906 and 110% in 1911 were reported to have been paid. Since disintegration, dividends have been paid as follows: 1912, 13% and 2900% in stock, May 15, 1912; 1913, 32%; 1914, 25% 1915 and 1916, 12% each; 1917, 1918, and 1919, 12% regular and 12% extra each year; 1920, 12% regular and 16% extra in cash and 150% in stock on December 17; $4 in 1921 on shares of $25 par value; 1922, $4 and 100% in stock on December 27; March, 1923, 62½ cents.

The record of the Standard Oil Company of California is somewhat less dazzling. This company was organized in 1906 as a consolidation of two concerns, the Standard Oil Company of Iowa and the Pacific Coast Oil Company. The latter company had been organized in California in 1879 with a capital of $1,000,000 and had passed under control of the Standard Oil Company of New Jersey in 1900. The authorized and issued capital stock of the Standard of California in 1906 was $17,000,000. Thirteen years later, in 1919, the authorized capital stock of the company had increased to $100,000,000, $49,686,655 of the increase representing stock dividends; and in December, 1922, the capital stock was increased to $250,000,000.[22]

These vast profits have not all accrued to the few wealthy men who composed the original Standard Oil group, for within recent years these stocks have been rather widely distributed. For instance, on October 31, 1922, the Standard Oil Company of New Jersey reported 11,013 holders of common stock and nearly 40,000 holders of preferred stock. Only six of the holders of common stock owned 1 per cent or more of the common stock of the company, and these six owned 28.4 per cent. At the time of the dissolution in 1911, this company had had only 6,078 stockholders, twelve of whom owned a total of over 50 per cent of the stock. One popular writer has stated that there are more than 300,000 stockholders in the various Standard companies that once comprised the Standard Oil trust.[23] Thus there has been a considerable diffusion of the stock during the last decade or more.[24] Several of the

Standard companies have tried to facilitate this distribution, by splitting the high-priced shares into smaller units, so that people of smaller means could afford to buy, and by selling stock at reduced prices to employees. The former policy was probably calculated to enlist public sympathy, or at least reduce public hostility, and the latter was designed, of course, to secure the effective interest of the employees.[25]

It has been reported that the senior John D. Rockefeller is no longer a holder of large amounts of Standard Oil stock, that he no longer owns as much as 1 per cent of the stock in any of the Standard companies. At the time of the dissolution he owned 24.8 per cent of the stock of the Standard Oil Company of New Jersey, and so of course got a similar share of each of the various units, but since then, according to a Senate investigation, he has disposed of almost all of his holdings. John D. Rockefeller, Jr., was found to own $410,000,000 worth of stock (market value on June 30, 1922) in the various Standard companies. He owned 23.4 per cent of the stock of the Standard Oil Company of California, 24.9 per cent of the stock of the Standard of New York, 11.4 per cent of the Standard of New Jersey, and 11 per cent of the Prairie Oil and Gas Company.[26]

The Standard Oil Company does not hold the same dominating position now that it has held in the past; in fact, its relative importance, at least in the refining business, has been gradually declining for some years. In 1906, the Standard Oil Company produced 85 per cent of the gasoline produced in the United States. In 1915, this proportion had declined to 65 per cent, and in 1919 to about half, or less than half.[27]

It even appears recently that there is some measure of competition among the various Standard companies, that the Standard Oil combination is gradually breaking up. This process of disintegration has been going on for some years, in fact, since the very day of the dissolution decree. Immediately after that decree, Standard Oil business was carried on much as before, the various companies working in complete harmony, under orders from "26 Broadway"—the headquarters of the old trust. For a moment or an hour all subsidiary companies had identical lists of stockholders, and coöperation was easy. But a change set in immediately. Some stockholders in the various companies began to sell; others bought stocks. Individuals died, and their estates were distributed. Some stocks were given to charitable foundations; and the general public took an increasing interest in Standard Oil securities. These stocks, which had once been very closely held, presently represented a wider and wider ownership.

For some years this shift in the ownership of Standard stocks was not sufficient to disturb the harmony existing among the various companies. The Prairie Oil and Gas Company produced, bought, and stored crude oil; the Prairie Pipe Line Company transported oil; the Standard Oil companies of New Jersey and Indiana did mostly refining; each unit had its own function to perform, and when there was disagreement among the units, it was settled at "26 Broadway." In the summer of 1923, however, conditions arose which strained friendly relations, and revealed the fact that all was not well in the Standard camp. When the oil production of California began to upset the oil industry, the eastern Standard refiners promptly threw up their connections with the Mid-Continent producing units, and bought oil from California, and even, through subsidiaries in the Mid-Continent field, ordered reductions in crude prices, to bring them to a basis that would compete with California oil. This was a serious matter to the Mid-Continent units, particularly the Prairie Oil and Gas Company, which had a vast amount of crude oil in storage and did not want prices lowered. The Standard of Indiana was financially tied up with the Sinclair Company, which had a large amount of production and storage. The efforts of the Prairie Oil and Gas Company to keep crude prices up, and the attempt of the Standard of Indiana to hold gasoline prices up, have been discussed elsewhere.[28]

About this time, competition appeared between the Prairie Oil and Gas and some of the other Standard units, in the Mid-Continent oil fields. For years the Prairie Company had been exclusively a producing, purchasing, and storage company, with no refining interests, while refining activities were turned over to the Standard of Indiana, the Standard of Kansas, and, farther east, the Standard of New Jersey. This division of functions was maintained for a while, but gradually the Standards of New Jersey and Indiana invaded the production business in the Mid-Continent, through subsidiaries. The Standard of New Jersey secured several important producing companies, and entered actively into the production of crude oil. In 1920 the Standard of Indiana got control of the Midwest Refining Company in Wyoming—an important producing, as well as refining company—and bought a half interest in the Sinclair Crude Oil Purchasing Company and the Sinclair Pipe Line Company. In this way, the Indiana Company got control of a considerable production. Thus the movement of the Standard companies in the direction of self-sufficing integration left the Prairie Oil and Gas Company high and dry, with a heavy production and huge storage and no assured market for its oil. To meet this situation, in December the Prairie Company took over the Producers' and Refiners' Corporation, a refining and producing company, thus definitely entering the refining business. About this time, the executive offices of the Prairie company were transferred entirely to Independence, Kansas. Some of the offices had been at 26 Broadway.

These are not the only instances of competition between Standard units. Other instances could be found in Louisiana, and elsewhere. The great Standard companies are losing their close interdependence, and are becoming individual integrated companies, somewhat like the larger independents.[29]

It is to be noted however, that while the Standard Oil companies have been declining in relative importance in the refining business, they have entered more extensively into the production of crude oil. Several of the Standard companies that formerly produced no oil at all are now heavily interested in producing properties. The parent company, the Standard Oil Company of New Jersey, entered the business of oil production only recently. The general policy of the company had been to let other companies do the producing, since that was generally unprofitable, and to stick to the refining and transportation end of the business. In 1917, however, when Walter Teagle became president, the company embarked upon a policy of developing its own production, and it now controls a very large amount. It is stated that the production controlled by the company quadrupled between 1917 and 1920.[30]

The reasons for the tremendous profits and growth of the Standard Oil Company in later fields have been somewhat the same as in Pennsylvania. Unfair practices have been part of the Standard method of doing business everywhere, until recent years. Railroad rebates, suppression of competition through monopoly of pipe lines, unfair discrimination in sales, bribery and corruption of rival employees, of railroad employees, and perhaps even of legislators, have been common Standard practices. The records show that there is almost no unfair business practice of which it has not been guilty. The Standard Oil Company had a very efficient service department, which kept records of purchases and sales of gasoline everywhere. Through bribery or otherwise, it was able to find out whenever any customer bought oil products from its competitors; and whenever competing products appeared in any market, the Standard stopped at nothing to weed out the competitor. If railroad rebates and discriminations could not do the business,—sometimes the railroads gave these rebates by buying lubricating oil at excessive prices—the monopoly went directly to the customer, and tried by bullying or by bribery to prevent him from using the rival product. In a Tennessee case, the Standard was found to have bribed a customer with a gift of a certain amount of oil to cancel a contract with a rival refiner. Later the price was, of course, raised. In some instances, where local hostility to the Standard Oil Company was too strong, the trust sometimes worked through a bogus independent company, which sold oil products very cheaply in any locality where competition appeared. When the independent competitor had been driven out, this bogus company disappeared. Through its extensive use of all sorts of unfair and underhanded business practices, the Standard Oil Company was for many years a corrupting influence on business.

Concerning Standard influence in politics, little can be stated positively. The record of Standard practices in business leaves no room for doubt that the monopoly would have bribed legislators whenever and wherever it was safe and profitable to do so. The record of many legislators a few decades ago makes it easy to believe that legislative corruption would not have been a difficult matter. The action of some legislatures on questions that involved the Standard Oil Company is difficult to understand on any theory other than the theory that Standard money was freely employed, perhaps indirectly, to help the legislators get the right point of view. In many cases, Standard interests were lined up with other interests, such as the railroads, the packers, the steel, woolen, and other big interests. It is established that the Standard Oil Company was in "sympathetic accord" with the Mark Hanna political machine, and this fact is not without significance.

While unfair practices have been the most important reason for the growth and power of the Standard Oil Company during the first few decades of its existence, the moral turpitude involved in these practices has usually been stressed far too much. From the point of view of the business ethics prevailing in 1926, the early business policies of the Standard Oil Company seem reprehensible enough indeed; but so do the policies of most other big business concerns of the seventies and eighties, or even later, or, for that matter, of small business of the same period. Many of the railroads were in the hands of large-scale thieves much of the time; and "stealing a railroad" was not looked upon as a particularly reprehensible offense. Men have been elected to Congress as a recognition of just about that type of public service. "The public be damned" was not the slogan of the elder Vanderbilt alone; it extended into business of all kinds. The railroads were partners in guilt with the Standard Oil Company much of the time anyhow, and their business policies were not generally on a high plane. In some ways they were on a lower plane, because, while the Standard Oil Company was crushing competitors and (perhaps!) fleecing the public, the railroad magnates were doing all this, and in addition were robbing their smaller stockholders.

Other big industrial combinations of the time were guilty of many of the offenses that darken the record of the Standard Oil Company. The big packers had their pools and agreements, and received rebates from the railroads, just as the Standard Oil Company did; and they were quite as inconsiderate of the public interests and of public opinion. Great line elevator systems, in some instances controlled by the packers or by railroads, robbed grain farmers of about everything above a mere existence; and line creamery companies served the dairymen of some regions in about the same way. The records of the American Sugar Refining Company, the United Shoe Machinery Company, the National Cash Register Company, the anthracite monopoly, some of the lumbermen's associations, and of a great many other industrial concerns, are not much better than that of the Standard Oil Company. It is further to be noted that there were few other industries in which monopoly had as many advantages as in the oil industry; or, in other words, few industries in which there were so many good reasons for crushing out competition. Finally, there were few if any other monopolies that preserved the same efficiency that the Standard Oil Company has shown, after eliminating competition. In many industrial combinations, monopoly soon bred inefficiency, but not so in the Standard Oil Company.

Unfair practices, discrimination, and oppression of the public were not confined to big business either, in the good old days. The methods followed by many small retail stores were on a par with those of the big monopolies. In the retailing of lumber, for instance, a great many handsome fortunes have been made, by unconscionable overcharging of consumers, where competition did not exist. It is very doubtful indeed whether the average consumer had more reason to complain of the Standard Oil Company than of some of his local retailers, either on grounds of service and of prices or of ethical standards in business. In short, the practices of the Standard Oil Company were common to almost all business of the time. If the great trust was more unscrupulous than other business in some respects, in some other respects it set standards above those generally prevailing.

One important reason for the vast Standard profits is found in the possession of valuable patents. The Burton process for "cracking," or breaking up the heavier components of crude oil into lighter compounds, particularly gasoline, has been controlled by the Standard Oil Company of Indiana, and has contributed immensely to its profits.

In the early days of the Standard Oil Company, it was often stated that Rockefeller and the Standard Oil group were the keenest group of men in business. For many years they bore that reputation, and no doubt deserved it. They seem also to have been among the most "ambitious," resolute, grasping, pertinacious, and perhaps we may say ruthless, groups of men in business life. Many anecdotes are told to show the clear vision and shrewd judgment of the elder Rockefeller in formulating business policies, and his resolute and inflexible determination in following them out. There is some evidence that the younger Standard Oil group lack the almost infallible business judgment of the older group, just as there is abundant evidence that their conception of their social responsibilities is more than a generation in advance of that of the Rockefeller-Rogers-Archbold combination.

The wisdom of the Standard Oil men has been shown in their policy of keeping large cash reserves. Rockefeller saw the importance of this from the very first, and the Standard Oil Company has always followed this policy. It seldom or never needed to call upon banks for financial assistance, no matter how serious business conditions were.

The strong cash position of the Standard company was an important factor in its success. The oil business has always been about as unstable and precarious as any business could well be, and in every period of depression many oil companies, refining as well as producing companies, were always on the verge, or over the verge, of bankruptcy. In such a situation the Standard Oil Company, with its great supply of ready cash, was able to profit in two ways. In the first place, it was able to take over the most valuable of these weak companies, often for a small fraction of the real value of their assets; and, in the second place, it could store vast quantities of cheap oil, to sell it later in the form of refined products, when the market was better. There was no depression so severe that the Standard Oil Company did not have money, and at such times, because of its excellent credit, it often even successfully floated bonds, always at low interest rates, to increase its funds for the purchase of cheap oil; and thus out of each depression emerged relatively stronger than before. Probably this factor in the success of the Standard Oil Company has not generally received the attention it deserved.

The immense profits that the Standard Oil Company made in its first decade or two in the eastern fields, of course gave the company a tremendous advantage in the exploitation of later fields to the westward. Most of the operators in every new field were poor in purse and lacking in experience, and in periods of stress the Standard found them "easy picking."

The wisdom of the Standard Oil men is shown in other ways than in the policy of keeping large cash reserves. Perhaps no other large business in the United States has ever been conducted with the same general business efficiency that has characterized Standard Oil management. The very best independent oil companies have possibly been as efficient as the Standard, but most independents certainly have not. Standard methods in every respect have usually been as good as the best. Standard refineries have not only operated at low cost, but have generally excelled in the production of high-value products, and have been commendably economical in their utilization of crude oil. In the retail distribution of gasoline, the Standard Oil Company of California affords the highest type of retail service that the writer has ever seen in the marketing of any commodity.

The wisdom of the Standard Oil management was shown in the scope of business covered. The Standard early left the producing end of the business—the most speculative and the least profitable—to independent enterprise, but ruthlessly crushed out almost all competition in refining and transportation. The Standard Oil Company was a leader in "direct-to-the-consumer" merchandising, with its tank wagons and, more recently, its service stations.

The oil industry is in many respects a natural monopoly. In other words, the production, transportation, refining, and perhaps the marketing of the oil in any given region, are more cheaply and efficiently done by a single monopolistic unit than by a number of competing units. In few fields of enterprise are the wastes of competition greater than in the oil industry.

In the production of oil, a monopoly has a tremendous advantage over a number of competing units. Where a given field is entirely under the control of a single company, it can drill a conservative number of wells, and it need drill no more than is necessary to drain the sands; it can chart the field and drill in the best locations, without any consideration of party lines; it can stop drilling whenever market conditions make such a course profitable; it can guard as effectively as it wishes against water infiltration and similar injury; it can operate the wells at much less expense than a number of small companies could. That is the reason why most fields tend to gravitate into a few hands, after the first spurt of speculative drilling. Some of the worst evils that have characterized the exploitation of our oil fields have been the result of our reliance upon competition in the production of oil; and these evils—over-production, instability of prices, and the like—have furnished the conditions which made the development of the Standard Oil monopoly inevitable, and even desirable. The Standard did not control production directly, it is true, but through its pipe lines and refineries it exercised a large measure of control in most fields; and in recent years it has controlled some fields directly. It is doubtless unfortunate that the Standard Oil Company did not control production as completely as it controlled transportation and refining.

The transportation of oil is almost a complete natural monopoly, and should have been recognized as such from the start. One pipe line is usually enough to carry the oil from an oil field, and duplication of lines generally represents wasted capital. In this respect, the transportation of oil is similar to railroad transportation, or transportation of most other kinds; and our reliance on competition here represents the same mistaken policy that was long followed in dealing with the railroads.

The Standard Oil Company saw from the first the importance of controlling the transportation of oil. It secured control of the first lines built in Pennsylvania, by typical Standard methods, and in most later fields it has tried to prevent the construction of independent pipe lines, or has tried to secure control of those that were built, by somewhat similar methods. In 1904, the monopoly controlled 88.7 per cent of the pipe-line business of the Pennsylvania field, and 93.5 per cent of that of the Lima-Indiana field, while there were no competing lines in Illinois, and up to 1906 almost none in the Mid-Continent field. In 1906 the government finally recognized the fact that pipe lines are essentially monopolies, by declaring in the Hepburn Act that they should be common carriers. About the same time, some of the states adopted similar laws. For several years, the Standard Oil Company evaded these laws, and the constitutionality of the federal law was not finally decided until 1914. Thus, after a period of about fifty years, we have finally recognized that the transportation of petroleum is a natural monopoly.[31]

The refining of petroleum is not so certainly a monopoly, but in some respects it presents monopoly characteristics. The integrated companies, with crude production, with pipe-line transportation, with refineries, and with marketing facilities, have a great advantage, although not always a decisive advantage, over companies which operate only refineries. This is shown clearly by the present drift in the direction of integration. Since the production of the oil in a given field should be controlled by one, or at least by a very few large producers, and, in fact, tends to be so controlled, and since pipe lines are an almost absolute monopoly, it is easy to see why the refining business should have been monopolized to some extent. It is difficult to imagine monopolized transportation with competitive refining; and, of course, monopolized marketing naturally went with monopoly conditions elsewhere in the industry. It is true that in the past, monopoly conditions throughout the industry have been founded largely upon monopoly in transportation, but monopoly in the production of oil, within any given field, should also have been recognized as desirable, if not inevitable.

The Standard Oil Company, in short, acquired its monopolistic position partly because the oil industry was really a natural monopoly. Even if there had been no John D. Rockefeller, and no Standard Oil crowd, it is almost certain that a large measure of monopoly or unified control, would have developed in the industry, because there were so many conditions favorable to its development. The Standard Oil monopoly represented to some extent the handiwork of selfish and unscrupulous men, but to some extent it represented a natural economic evolution.

There have been numerous revolts by independent producers, and by state and federal governments, against the power of the Standard Oil Company. Some of the efforts of Pennsylvania producers to throw off the yoke have already been discussed.[32]

In California, the independent oil companies for years carried on a bitter struggle with the Standard Oil Company. The oil situation in California has always been complicated by the fact that some of the most important producers of oil were the railroad companies, the Southern Pacific, and the Atchison, Topeka, and Santa Fé, which were also the most important consumers of oil. The Standard Oil Company was in close working harmony with these railroads; in fact, Standard Oil capitalists owned $47,000,000 of the stock of the Santa Fé. Because of the lack of pipe lines, and because of the viscosity of most of the California oil, much of the oil was moved by railroad, in the early years of development, and consequently the rebates offered by the railroads to the Standard Oil Company, and to themselves as shippers, were a very serious matter for the independent producers. The Commissioner of Corporations reported in 1906 that these rebates had amounted to about $200,000 a year, of which about $100,000 went to the Standard.[33]

For some years the independent oil producers of the Kern and Coalinga fields were absolutely at the mercy of the Standard Oil Company, and of the Associated Oil Company, which owned the only pipe lines leading from these fields. The independent producers either had to sell to these companies, at prices named by them, or not sell at all, and some producers actually shut down their wells altogether. Such conditions led to the formation of the Independent Oil Producers' Agency, and the Coalinga Oil Producers' Agency, the latter of which was presently absorbed by the former. This organization arranged for the building of a pipe line from the San Joaquin Valley to the coast in 1909, and made a contract with the Union Oil Company, by which the latter was made the exclusive sales agent for the fuel oil produced by the Agency. This agreement with the Union Oil Company enabled the independent producers to present a stronger front in their struggle with the Standard, because the Union Oil Company was a powerful organization. In California there have been attempts to secure the building of a state refinery, but these attempts have failed in the legislature.[34]

The story of the struggle of the state of Kansas against the domination of the Standard Oil Company is rather dramatic. The Standard Oil Company had appeared in Kansas in 1895, when it bought out

Guffey and Galey, eastern wildcatters, who had found a small amount of oil in southeastern Kansas. The Forest Oil Company, the Standard company here, built a small refinery at Neodesha, and connected it with the nearby wells, and was soon making good profits. For several years there was no noticeable increase in production, but in 1901 the production began to climb, and for several years rose very rapidly, the production coming from the discoveries of small independents, since the Standard had little success in drilling.

These producers immediately faced a serious problem in disposing of their oil, and one refinery was begun at Humboldt in 1903. The Standard Oil Company, however, needed oil at this time, since the eastern fields were beginning to show signs of depletion, and it began to build extensive pipe-line connections, storage tanks, and a large refinery at Kansas City. When it became known that the Standard was spending some $15,000,000 in Kansas, the state went oil crazy. Scores of oil companies were formed, and soon production was rising to new levels. In this situation, a slump in oil prices was inevitable.

The position of oil producers would have been bad enough at best, but to make it still worse, the Standard Oil Company put into operation some of its unfair practices, which had worked so well in Pennsylvania. In order to compel producers to sell to the Standard, that company had, in 1902, induced the Santa Fé Railroad to raise the freight rates on oil, through the subterfuge of a change in its classification. This increase in rates was enough to ruin some of the independent operators, but some survived it, and in 1904 another increase in oil freights was ordered. This increase brought rates to a prohibitive level. It was ordered just at the time that the Standard finished its pipe line to Kansas City; and it was freely charged that it was the result of a pooling of oil freight between the Santa Fé and the Standard pipe line, the intent being to force all oil shipments to the pipe line. The railroad was supposed to have received a certain payment—perhaps ten cents per barrel—for all oil going through the pipe lines. The Standard Oil interests had some stock holdings in the Santa Fé Railroad, even had one of its most prominent men, H. H. Rogers, on the board of directors of the Santa Fé a little later, while John D. Rockefeller and William Rockefeller were directors of the Missouri, Kansas, and Texas. The Standard pipe line was largely laid on the Santa Fé roadbed; so it is easy to believe that there was collusion between the Standard and the railroads.

All this looked like the old days in Pennsylvania. Nearly all producers were forced to sell to the Standard, for the pipe lines were not yet common carriers. To further strengthen its position, and to prevent absolutely all sales of oil to independent refineries or other purchasers, the Standard Oil Company refused to take any oil whatever from producers who sold any part of their oil to other purchasers, except on payment of penalty for such sales. One company was fined twenty cents a barrel for having sold to another purchaser.

Yet two independent refineries, at Humboldt and at Muskogee, Oklahoma, managed to get some oil, and built up a sort of market for their products. The next step of the Standard was to try to force these refiners out by cutting the price of kerosene in the markets they had developed. Prices were cut very low in Emporia, for instance, where the Humboldt refinery had a fair trade, the Standard recouping itself by selling at remunerative prices elsewhere. With serious over-production of oil, with low crude prices, and with railroad rebates and other unfair practices of the Standard Oil Company to fight, operators had enough to make them thoroughly disgruntled, and in 1904 they began to try to get together. They formed an association, but that did not help prices; in fact, prices continued to decline. The Standard Oil Company at first lowered prices by a blanket cut. Next it classified oils according to gravity, and cut the prices of low-gravity oil drastically. Not long afterward all gravities were cut again and again. Of course, the entire blame for the situation was placed on the Standard Oil Company rather than on the over-production of oil; and various schemes were advanced for fighting the trust. One extravagant proposition was launched to organize a company with a capitalization of $50,000,000, modeled after the Pure Oil Company of the eastern fields, which should handle production, transportation, manufacturing, and marketing, and "run the Standard out of the state."

The excitement spread from the oil fields all over Kansas. Not only the oil producers, and the one or two refiners still solvent, but thousands of stockholders in the numerous small oil companies, were seething with wrath and indignation. It was estimated by the Topeka *Capital* that in the 400 oil companies in Kansas at this time there were probably 20,000 stockholders, while there were perhaps 5,000 individual operators and 6,000 landowners who had given oil leases. Thus there were, according to this estimate, 31,000 people in the state who were in some way interested in the oil business. Consumers believed that, with the drastic declines in crude prices, the prices of kerosene and gasoline should decline more than they did; and the Standard policy of cutting prices drastically wherever competition appeared, while holding prices up elsewhere, was of course known to every farmer in the state.

The result of all this discontent was that the political campaign of 1904 hinged largely on the fight against the Standard Oil Company. One candidate for governor, Ed Hoch, based his campaign partly on the scheme of a state oil refinery, to be built as a branch of the state penitentiary, employing convict labor, as the state twine plant had done for some years. The proposal was popular, and Hoch was elected. The movement for a state refinery soon grew into something much broader. The producers called not only for a state refinery, but also for a law making all pipe lines common carriers, for a law forbidding the sale of products at a lower price in one portion of the state than in another, and for just and reasonable transportation rates by railroad and pipe line. The refinery bill met strong opposition, on the grounds of feasibility and constitutionality, but letters poured into the legislature by the hundred, urging support of the bill.

The Standard Oil Company unwittingly helped to secure the passage of the refinery bill by the tactics it used to fight it. It sent lobbyists to Topeka, and a rumor soon arose that they were using money, that they had received two bags of gold by express. The papers took up the matter, and the Speaker of the House of Representatives promptly appointed a "smelling committee" or "boodle committee," with orders to investigate not only the Standard Oil lobby, but railroad lobbies and lobbies in general. Whether there was anything to the story of the bags of gold or not, there was said to be a considerable exodus of hangerson; and, of course, the hostility to the Standard increased greatly.

The Standard Oil Company cut the price of refined oil from two to four cents all over the state, but this did not seem to conciliate anybody. On the contrary, it was regarded as a proof that the prices had been too high. Then presently the Standard stopped work on all pipe lines and storage tanks throughout the field, calling in the crews and ordering them home. Presently the oil towns were filled with crowds of unemployed workmen. Soon afterward the company ordered a boycott of Kansas oil, that is, refused to take it except in cases of contract. About this time the company declared a 15 per cent quarterly dividend. All this was too much for the legislators to withstand. The boycott order particularly angered the people; and the refinery bill passed by a large majority, appropriating $400,000 for the construction of a state oil refinery. At the same time the other oil bills demanded by the public were also passed.

The legislators left Topeka, much pleased with their work, and the people of the state were for a while happy in the belief that they had brought the greatest of all monopolies to its knees. Their joy was premature, however, for the operators were still without a market for their oil, and construction work was at a standstill. To make the situation still less cheering, the state refinery law was presently declared unconstitutional.[35] It is true that under the laws passed by the legislature, requiring that pipe lines be made common carriers, that rates be reasonable, and that there should be no unfair discrimination in the sale of oil products, independent operators had now a fair competitive chance; but this was of no avail in the sale of oil or oil products outside of the state, because the state laws could have no effect in interstate shipments. Thus the Kansas producers had fair competition insured them only in the sale of products to the rather limited market within the state. The situation is concretely shown by some of the rates in force after the passage of these laws. The independent refiners asserted that while they could ship oil from Chanute, Kansas, to Webber, Kansas, a distance of 254 miles, for $10\frac{1}{2}$ cents per 100 pounds, to ship it seven miles farther on the same road, to Superior, Nebraska, cost 30 cents— nearly three times as much. The Kansas producers appealed to the federal government for measures to supplement the state laws, but effective federal coöperation was slow in coming. One of the first results of the Kansas fight was, however, a careful investigation of the Standard Oil Company by the United States Bureau of Corporations, and an investigation of freight rates on oil by the Interstate Commerce Commission. The section of the Hepburn Law making pipe lines common carriers was a direct result of the Kansas situation.

In a later lawsuit against the Standard Oil Company—the Prairie Oil and Gas Company—the state of Kansas was successful, the Prairie company agreeing not to engage in the refining business, and promising to "behave" well, on condition of being permitted to remain in the state.[36]

During the years from 1907 to 1910, the Standard Oil trust was almost constantly in the courts. In 1906 and 1907, the federal government brought suit against the Standard Oil Company of Indiana on criminal charges of accepting unlawful rebates from certain railroads. The Standard of Indiana was the nominal defendant, but it was recognized by the court that the Standard of New Jersey was the real controlling power, and the latter company assisted in the defense. In January, 1907, the trust was found guilty on 1,462 counts; and Judge Kenesaw Mountain Landis imposed his famous fine of $29,240,000— $20,000 for each offense. This heavy fine created a great sensation, although it was a mere bagatelle compared with the vast profits that the Standard had made by its unfair practices.

The Standard Oil Company appealed from the Landis decision and penalty, and secured a reversal on various grounds, particularly on the ground that Judge Landis had assessed the penalty improperly, since each separate car shipped did not constitute a separate offense. The case was retried a number of times, but somehow, although there is no doubt that the monopoly was guilty of the offenses charged, it succeeded in escaping punishment.[37]

The Indiana company was in trouble elsewhere at the same time. In Tennessee, it was charged by the federal government with receiving rebates, but was acquitted.[38] About this time the parent company was also in the Tennessee courts, charged by the state with violation of the state anti-trust law. In this case a dealer in Gallatin, Tennessee, who had ordered some oil from an independent refiner, was interviewed by a Standard agent, and was persuaded, in consideration of a gift of 100 gallons of oil, to cancel his order. The Standard Oil Company and the agent of the company were both convicted in the circuit court, but the decision was later reversed, in February, 1907, as to the company itself, the agent being held for the offense. In this case appeared some of the familiar Standard tactics: the operations of the secret service, in discovering promptly any order given to an independent company; discrimination among different localities, as competition appeared in some of them; and ruthless suppression of competition, in this case by underhanded and unfair methods. The evidence in the case indicated that, because of lack of competition in the town of Gallatin, the company sold an inferior grade of oil, at excessive prices.[39]

Not long after this, another Standard company—the Standard Oil Company of Kentucky—was convicted of violating the state anti-trust law of Tennessee, and was ousted from the state. The company appealed to the United States Supreme Court, but that court affirmed the conviction.[40]

In March, 1907, the federal government secured an injunction against the Waters-Pierce Oil Company and the Galena Signal Oil Company— the Standard of Indiana was also mentioned in the case—enjoining them from carrying out agreements dividing territory among themselves, contrary to the anti-trust law. Three years later, the federal government brought suit against the Standard of New York and the Vacuum Oil Company, on a number of charges of receiving rebates, and secured a decision and a fine of $55,000.[41]

In 1907, the state of Texas prosecuted and convicted the Waters-Pierce Oil Company, a subsidiary of the Standard Oil Company, for violation of the state anti-trust law, fined that corporation $1,623,900, and placed its affairs in the hands of a receiver appointed by the state. In the same year, the state sued the Security Oil Company, the Union Tank Line, and the Navarro Refining Company, corporations subsidiary to the Standard Oil, on the same charge, and issued an injunction forbidding the removal of any of their property from the jurisdiction of the court, thus preventing the shipment of oil outside the state, and tying up all the tank cars used in interstate shipment. The result of this drastic order was to compel the closing of several refineries until the injunction was modified. This suit resulted in a cancellation of the charters of the corporations and a perpetual injunction from doing business in Texas. The Union Tank Line Company was fined $75,000, and its cars in custody were ordered sold.

The Waters-Pierce Oil Company appealed from the decision of the state court, but the Supreme Court of the United States confirmed the state court on all points. The fine was duly paid to the state treasurer, and the property sold. The Standard Oil Company withdrew from Texas, following this drastic treatment, but a new corporation—the Magnolia Petroleum Company—was formed to buy the property of the ousted corporations, and in the new corporation two Standard men —Archbold and Folger—were reported to have important stock interests. The state was suspicious of the dismembered fragments of the Standard subsidiaries, and countenanced the promotion of the Magnolia Company only upon the unique agreement that the Archbold and Folger stock should have no voting power or right to control the affairs of the company, and that the state should have the authority to participate in the selection of the men who were to manage its affairs. The new company was not a corporation, but was what is called in Texas law a joint stock association, and was controlled by trustees, in the selection of whom the state and the minority stockholders exercised full power. In recent years some of the rules and restrictions regarding these companies have been modified.[42]

Even very recently the state of Texas has tried to maintain her reputation as a "trust buster." Early in September, 1923, the state filed suit against four of the large oil companies: the Texas Company, the Gulf Refining Company, the Producers' Refining Company, and the Humble Oil and Refining Company—the latter a Standard subsidiary. The allegation in each case was violation of the state anti-trust laws, the particular complaint against the Humble Company being that it was owned by the Standard Oil Company of New Jersey. Much more than half of the Humble stock was admitted to be owned by the Standard, but the state was unable to prove that the Standard had tried to exercise any control over the operations of its subsidiary, and its suit therefore failed.

Missouri has likewise had her troubles with the oil monopoly. In April, 1905, the Attorney-General of the state filed an information against the Republic Oil Company, and the Waters-Pierce Oil Company, alleging a combination in the oil business. While the case was under consideration, the anti-trust law was amended in 1907, so as to provide that if any corporation should be found guilty of a violation of the provisions of the act, its charter or license should be forfeited, and the court might also forfeit any or all of its property to the state, or cancel its right to do business, or the court might assess a fine. After the passage of this amendment, the commissioner made his report, finding against the defendants on the law and the fact. Less than a month afterward, on June 22, 1907, the Republic Oil Company filed a notice of its withdrawal from the state; but the court ousted both defendants, and fined each of them $50,000. In view of the capital of the Republic Company and the profits that had been made during the period of combination, some members of the court expressed the opinion that the fine should be $1,000,000. The Waters-Pierce Company paid its fine and complied with conditions, and was allowed to continue to do business in the state.[43] Kentucky sued the Standard Oil Company of Kentucky in 1913, for discrimination in the sale of oil, but failed to convict.[44]

Thus there is abundant evidence as to the early practices of the Standard Oil Company. There is good reason to believe, furthermore, that, even in recent years, the Standard companies have been strong enough to exert a measure of monopoly control in most fields. The question as to what extent they have used or abused their power in recent years is not so easy to answer; although it may safely be stated that, during the past decade or more, their record has been one that demands little apology or defense. Investigations, indictments, and prosecutions by the dozen, there have been; in fact, these companies are being investigated a fair share of the time. Almost no session of Congress has been regarded as complete without a few proposals to investigate the Standard Oil Company, either directly or by some commission. Yet very little evidence of improper practices has been unearthed in recent years, in spite of the fact that some of the investigating bodies were pledged and "hell-bent" to find something of the kind. The reports published have generally pointed to the great resources and earnings of the Standard companies, and the lack of competition among them, but have brought out no evidence of illegal collusion or discrimination.

In 1915, the extraordinary fluctuations in the price of gasoline led to an investigation by the Federal Trade Commission, and to the publication, two years later, of a *Report on the Price of Gasoline*, in which the Commission announced the following findings: that the Standard companies had divided the United States up into districts, marked roughly by state lines, and that there was practically no competition among these companies; that the Standard companies had the power to exercise considerable control over the price of crude oil and oil products, and that they generally set the prices which governed the market, although "more or less competition" was offered by the independents, who sometimes even cut below the Standard price; that the Standard of Indiana maintained very low prices in its own territory, while selling to the other Standard companies at such a price as to recoup itself, to some extent at least, for the losses; and that this company later raised prices to an artificially high level. The Commission admitted, on the other hand, that the natural conditions of supply and demand had been partly responsible for the conditions which existed, and that the decline in Cushing production had much to do with the rise in prices in the latter months of 1915. The Commission further conceded that it had "found no direct evidence of collusion among the several Standard companies in violation of the dissolution decree."

In the same year, charges of discrimination in the Healdton oil field led to an investigation by the Bureau of Corporations, with the result of completely exonerating one Standard company—the Magnolia Pipe Line Company. The report of the Federal Trade Commission on the *Pipe Line Transportation of Petroleum*, issued in February, 1916, likewise divulged little that was new except that pipe-line companies were unreasonable, not only in their charges, but also in some of their shipping requirements.

The following year, in response to a Senate resolution, Secretary Lane of the Department of the Interior made a report, in which he stated, among other things, that the selling price of gasoline was "fairly responsive to the supply and demand for gasoline." In other words, he found no evidences of monopoly control.[45]

In 1919, an investigation of the oil industry of California was ordered, under a Senate resolution introduced by Senator Poindexter,[46] and two years later the report of the Federal Trade Commission on the *Pacific Coast Petroleum Industry* was published. This exhaustive report contained much valuable information regarding the situation in California, regarding some of the major companies operating there, and their costs, investments, and earnings; but no evidence of improper practice was divulged.

The report of the Federal Trade Commission on the *Petroleum Industry of Wyoming* was issued early in 1921, and this report, like the report on the Pacific Coast industry, contained much information as to the control of the Standard Oil interests, but no evidence that this control had been used improperly.

The La Follette report, made by Senator La Follette, from the Senate Committee on Manufactures, in March, 1923, was strongly hostile to the Standard Oil interests, but it brought out little that had not been known before. It had long been known that the Standard units had not generally competed with each other, that they had made extraordinary profits, that they set prices and the other companies almost always followed. Doubtless it was not generally known how many large salaries were paid to Standard officials. The Standard Oil Company of New Jersey, for instance, had six men who drew salaries of $100,000 a year or more. President W. C. Teagle, and A. C. Bedford, chairman of the board of directors, drew $125,000 each; four others drew $100,000; one drew $85,000, and five others drew $50,000 each. In the Standard of Indiana, two men, R. W. Stewart, chairman of the board of directors, and President W. M. Burton, drew $100,000 each. In the Standard of

California, the highest salary paid was only $75,000, and there were only four salaries above $50,000. In the Standard of New York, President Folger drew $100,000 a year, and three others drew $50,000 or more. The La Follette committee seemed to believe that these high salaries showed the extravagance of the Standard companies. As the committee expressed it: "The extravagant salaries paid by most of the companies to their officers and directors are typical, and a little indication of the lavish and wasteful manner in which their business is conducted when the public pays everything and is consulted about nothing." Whether these salaries are higher than business efficiency justifies is a question that few men are qualified to answer.[47]

The La Follette committee maintained that the price changes in the Mid-Continent field in 1920, 1921, and 1922 showed conclusively that "arbitrary manipulation" and "not the law of supply and demand" was at work. The committee reported that there had been price agreements between the Standard of Indiana and certain other companies, basing this statement on the testimony of T. S. Black, of the Western Petroleum Company. It practically asserted that it was the policy of the Prairie Oil and Gas Company—the Standard purchasing company in the Mid-Continent field—to buy large amounts of crude when prices were low, advance the price, sell its storage, reduce the price, and fill again. It indicated its belief that the Prairie Oil and Gas Company was strong enough to change prices in this way, and that it found it profitable to do so. One prediction of the committee which brought much ridicule from oil men, was the prediction of $1 gasoline "before long," "if a few great oil companies are permitted to manipulate prices for the next few years as they have been doing since January, 1920." Gasoline began a spectacular decline to extremely low prices very soon after the report was published.

The La Follette report has been subjected to very bitter criticism by oil men everywhere. Standard officials claimed that one of the witnesses who had testified as to the manner in which the Standard of Indiana dictated the price of gasoline in its territory, had testified falsely, and they asked for a rehearing, in order that they might prove this. Standard Oil officials denied that they had in recent years exercised any arbitrary control over prices, or that they had even had the power to do so. They claimed that the oil industry is highly competitive, and that the price of crude oil and of oil products is fixed entirely by the law of supply and demand. Officials of the Standard Oil Company of Indiana flatly denied the accusation that they had ever entered into any agreement to fix prices.[48]

An exhaustive discussion of the question of the Standard Oil monopoly is quite beyond the scope of this book, yet a few conclusions as to the practices of Standard companies in recent years seem justified by the evidence at hand. In the first place, there is little doubt that the Standard companies, at least in some sections of the country, have the power to crush out competition and to control prices to some extent. They have seldom competed among themselves, and in many regions they are so much stronger than their largest competitors that they have a voice of great authority. In the Wyoming fields, for instance, where Standard interests control almost all—93 to 97 per cent—of the gasoline output, it can hardly be doubted that the small independent interests live by Standard sufferance; and it is likewise certain that Standard interests can to some extent control the output and the price of oil. As a matter of fact, they have done so. During the recent depression in oil prices, the curtailment movement in Wyoming was conceded to be effective, and this was certainly due to unified control—Standard control. In no other important field are Standard interests as strong as in Wyoming, but in the marketing of gasoline in what is known as Standard of Indiana territory, comprising the states of North and South Dakota, Minnesota, Iowa, Kansas, Missouri, Wisconsin, Michigan, Illinois, and Indiana, the Standard Oil Company of Indiana sells half the total amount of gasoline consumed. Most authorities would concede that a single company with control of 50 per cent of the supply is to some extent in a monopoly position.

There is also plenty of evidence to show that the Standard companies have tried to fix prices of crude oil and of gasoline which were not in accordance with the law of supply and demand. As recently as the summer of 1923, the Prairie Oil and Gas Company tried to do this. In the fall of 1922 and early spring of 1923, this company bought and stored over 20,000,000 barrels of crude oil, in anticipation of an approaching shortage and higher prices. As pointed out in an earlier chapter,[49] instead of an oil famine, the industry encountered one of the worst periods of over-production in its history, and in March and April, 1923, prices began to tumble, and continued to decline throughout the summer. It is common knowledge that the Prairie Oil and Gas Company, which buys a large proportion of the crude oil produced in the Mid-Continent, did not lower its posted price as rapidly as the conditions of supply and demand would have warranted. Apparently its management tried to keep prices up until it should be able to unload some of its heavy stocks of crude oil, although there may have been other reasons for its policy. In some ways its policy is hard to understand. For weeks this great purchasing organization tried to keep prices up, even bought large amounts of crude oil at prices higher than its competitors were paying, and higher than it would have had to pay to get all the oil it wanted. In other words, instead of taking immediate advantage of the over-production to force prices down to the lowest possible figure, the company held prices up for a considerable time, at a heavy cost in its purchases, trying in the meantime to keep production down by a system of proration. It was "price-fixing" of a peculiar kind, and of a kind that the government will not need to be much concerned about.[50]

About the same time that the Prairie Oil and Gas Company was trying to keep crude prices up, the Standard of Indiana, and some of the other big refiners of the Mid-Continent region, were trying to keep the price of gasoline up to a "remunerative level." This was, of course, denied, but there is much reason to believe that the Standard company set gasoline prices that the conditions of supply and demand would not have justified, and then got the other refiners to adhere to this price, either by some sort of agreement or by less democratic "instructions." The testimony of Thomas S. Black before the La Follette commission may or may not be worthy of credence, but the record of the Chicago meeting of the National Petroleum Marketers' Association, in July, 1923, certainly raises a grave suspicion as to just what was going on behind the curtains. There is reason to believe that the Indiana company was making a handsome margin of profit before Governor Mc-Master of South Dakota forced the 6.6-cent cut in gasoline prices.

No one who has talked much with independent oil men needs to be told that only a man of much courage or of little sense will try to "buck the Standard Oil Company," or, in other words, to set a price different from that which the Standard has set. Some independents have done that in the purchase of crude oil, and even in the sale of gasoline, but it is generally regarded as dangerous policy.[51]

No definite proof has ever been produced of agreements among the oil companies in fixing gasoline prices, but there are some reasons for suspecting such agreements, and for believing that the margin of profit in marketing oil products is excessively high. One reason for suspecting this is the tremendous increase in the number of service stations in recent years, and the extravagant service rendered.

The "revelations" made at the Chicago meeting led promptly to an epidemic of investigations of the oil industry. Not only the Attorney-General of the United States, but the attorneys-general of a number of states promised speedy investigation. At the annual meeting of the National Association of Attorneys-General in August, there was much discussion of the question of gasoline price-fixing, and a committee was appointed to conduct an investigation.[52] In October, the attorneys-general of more than twenty states met in Chicago, and after a three-days' conference made public their program, which included the elimination of trade agreements and other unfair practices, the conservation of the nation's oil resources, and legislation making intrastate pipe lines common carriers—interstate pipe lines had already been made so by federal legislation. They also pointed out the enormously wasteful duplication of service stations and the unreasonable expense and extravagance in the marketing of gasoline. They appointed an executive committee to study the question further and to confer with United States Attorney-General Daugherty. In March, 1924, the executive committee made a report calling upon President Coolidge, and the governors of the various states, for a more rigid regulation of the oil industry. The committee urged that the Federal Trade Commission be empowered to investigate, and that Congress immediately appropriate an adequate sum of money for the expenses of the work.[53]

The report of the attorneys-general did not divulge any definite information as to illegal agreements or unfair practices, but the Attorney-General of Missouri, after working on the question for several months, admitted that he had found no evidence of illegal or unfair practices. He declared that, as far as his investigation had revealed, the Standard of Indiana dominated the Missouri market, not by any unfair practices, but by sheer size and efficiency.

The general belief on the part of the public that oil companies were charging exorbitant prices for gasoline and were making extravagant profits, led to much heated discussion and even to state, municipal, and coöperative purchase of gasoline. The legislature of Iowa framed a bill to permit cities, as well as the state, to enter into the purchase and sale of oil and coal, whenever an unlawful combination, pool, or trust was found to exist. Many other states, and Canada, ordered investigations or took some action in the direction of curbing the "gasoline trust." In the city of Omaha, a number of "Muny" (municipal) stations were established, and for a while sold gasoline several cents below the regular service station price. The sale of gasoline in South Dakota has been continued by Governor McMaster, and the state has erected a number of 20,000-gallon tanks for temporary storage. A correspondent for one of the oil journals makes the statement that these tanks are being built with money that was appropriated for the various state institutions.[54]

There is no doubt that the Standard companies have often used their posted price as a means of reducing or stimulating crude production. When the Standard purchasing companies raise crude prices, it is generally understood by those in the industry that the Standard wishes more production, and vice versa. Regardless of what may be in the minds of Standard officials, their price quotations, of course, have the effect indicated, just as prices would, no matter how determined.

While there is considerable evidence that Standard companies have been able at times to exert some influence on prices, there is no doubt that prices in general have always been determined fundamentally by the conditions of supply and demand. Even if the Standard companies had produced 100 per cent of the supply of crude oil, they could have influenced prices only by regulating the amount put on the market; and they have never in any region had direct control over all the production; in fact, Standard companies have in most regions produced a rather small share of the crude output, and have controlled the production only in so far as they have offered the principal refining market for crude. Wildcatting and producing companies have been, and still are, largely independent companies, in spite of the fact that the Standard companies have invested heavily in production in recent years. The supply of crude oil, and of oil products, is largely beyond Standard control, and of course the demand is beyond any possibility of control. Therefore, prices are in general beyond control.[55]

It is hardly believable that the Standard companies, at least in recent years, have ever purposely caused wide fluctuations in the price of crude oil, in order to ruin independent producers, as the La Follette report has charged. Independent operators who have tried to do business on insufficient capital, and have gone down in the first depression that struck the industry, have not infrequently charged Standard interests with deliberately ruining them. In every period of over-production, hard-pressed operators, instead of seeking their own salvation by a curtailment of drilling operations and a reduction in the output of oil, have drilled wells with unabated energy, turned thousands of barrels of unwelcome oil into the pipe lines, and cursed the Standard Oil Company for the resulting low prices; yet the Standard Oil Company has frequently taken the lead in efforts to secure a reduction in the crude output.

During the exploitation of the Cushing field, for instance, when oil was being produced in such quantities that the price went down to almost nothing, and operators throughout the Mid-Continent fields were failing by the dozen, many of these operators charged their distress to Standard manipulation, when as a matter of fact, it was due to their own wild and senseless drilling orgy, which the Standard—the Prairie Oil and Gas Company—was doing its best to stop, and when the Prairie was the only important company willing to buy large amounts of the production at any price.[56]

The La Follette committee charged the Standard with responsibility for the violent fluctuations in the price of crude oil in the Mid-Continent field during 1920, 1921, and 1922, and cited the testimony of certain independent operators to substantiate the charge; but the testimony is not convincing, since it is pretty well established that the Prairie Oil and Gas Company itself lost heavily during some of these years. The Prairie officials seem to have erred in judgment, in basing their policy on the theory of an impending oil shortage, but that they could have been interested in increasing fluctuations in the market, when they themselves lost heavily by those fluctuations, is hardly to be believed.

On the whole, it is probable that the influence of the Standard Oil companies in recent years has been beneficial to the oil industry and to the public. Probably the oil business has been carried on more economically and more efficiently than it would have been without Standard influence, or without some similar influence; and it is even probable that oil products will be cheaper to the consuming public, in the long run, than they would have been without the influence of the Standard Oil Company. No less an authority than George Otis Smith, former chief of the Geological Survey, has expressed his belief that prices of oil products are lower today than they would be, had the Standard Oil Company not existed. The oil industry, as already stated, is in some respects a natural monopoly. It should have been recognized as such in the beginning. We must regret that the vast monopoly profits which have gone to Standard stockholders could not have been retained, at least in part, by the public; yet it is clear that no scheme for accomplishing this could ever have been put into operation in the early years of the industry. The Standard Oil Company, as John D. Archbold once said, was a natural evolution, and in many ways it has played its part well. As will be explained in a later chapter, the great danger to the people of the United States is not that of monopoly in the oil industry. It is fairly certain that there has not been enough monopoly control.

In its policy of conserving oil and oil products, the Standard Oil Company has rendered a service to the people of the United States which will probably be better appreciated when the end of our oil reserves is in sight. The efficiency of Standard Oil refining methods has often been noted. M. L. Requa, a few years ago, declared that in saving the valuable products of oil, the Standard was "better conserving the oils of the United States than any other concern."[57]

A concrete illustration of the benefits of monopoly control in the production of oil may be found in the history of the exploitation of Salt Creek during the past few years. Had the Salt Creek field been covered with a large number of small holdings in the hands of small companies, it would have been gutted long ago, but Standard control has prolonged the life of the field, has prevented disastrous over-production, with its accompanying waste of oil, and has meant a great saving in the cost of wells and other equipment.[58]

NOTES

1. See above, pp. 47, 48.

2. U. S. vs. Standard Oil Co. of New Jersey, *173 Fed Rep.*, 183.

3. U. S. vs. Standard Oil Co. of New Jersey, *173 Fed. Rep.*, 177; 221 U. S., 1-; *S. Rept., 1263,* 67 Cong., 4 sess., p. 6.

4. Federal Trade Comm., *Rept. on the Price of Gasoline,* 1915, 145; *S. Rept. 1263,* 67 Cong., 4 sess.

5. Federal Trade Comm., *Rept. on the Pacific Coast Petroleum Industry,* I, 21, 27; *Rept. on the Price of Gasoline,* 5, 15.

6. *Wall St. Jour.,* Sept. 14, 1920, 2; Nov. 11, 1921, 3; Jones, *The Trust Problem in the U. S.,* 90.

7. *Mineral Industry,* 1906, 625; 1908, 682; Ok. Geol. Survey, *Bul. 1,* 22; *Bul. 2,* 179; *Bul. 22,* 89.

8. During the year 1920 the Standard Oil Company, which had maintained close business relations with the Midwest company since its formation, purchased 205,053 shares of its stock, and since then it has bought practically all of the rest of the outstanding stock.

9. Federal Trade Comm., *Rept. on the Petroleum Industry of Wyoming,* p. 8.

10. Since the Sinclair interests have laid the pipe line from Salt Creek to Kansas City, these figures must be revised.

11. *Wall St. Jour.,* Nov. 10, 1921, 10; Federal Trade Comm., *Rept. on the Pacific Coast Petroleum Industry,* 1921, Parts I and II.

12. *Eng. and Min. Jour.,* Oct. 25, 1902, 546; May 25, 1905, 1013; *Min. and Sci. Press,* Jan. 6, 1912, 22.

13. *Mineral Industry,* 1911, 561; 1912, 632; 1914, 556.

14. *Mineral Industry,* 1910, 520.

15. *Mineral Industry,* 1893, 530.

16. *The Lamp,* Feb., 1923, 8; *Oil and Gas Jour.,* Oct. 5, 1922, 90.

17. *Wall St. Jour.,* Jan. 20, 1920, 10.

18. *The Lamp,* Feb., 1923, 8.

19. Pforzenheimer and Company, *Weekly Summary,* Aug. 25, 1923; *Natl. Petroleum News,* Jan. 9, 1924, 92; Jones, *The Trust Problem in the U. S.,* 89; Federal Trade Comm., *Rept. on the Price of Gasoline,* 1915, 108; *Rept. on the Pacific Coast Petroleum Industry,* Pt. I, 27, 89, 95.

20. *S. Rept. 1263,* 67 Cong., 4 sess.

21. *S. Rept. 1263,* 67 Cong., 4 sess.; *Oil and Gas Jour.,* Jan. 27, 1922, 54; *Oildom,* April, 1923, 43.

22. Federal Trade Comm., *Rept. on the Pacific Coast Petroleum Industry,* Pt. I, 89.

23. Isaac Marcosson, in *Sat. Evening Post,* May 3, 1924, 12.

24. *The Lamp,* Feb., 1923, 9.

25. *Natl. Petroleum News,* May 16, 1923, 23.

26. *Barrons,* April 9, 1923, 8; *Oildom,* Aug., 1923, 34.

27. Federal Trade Comm., *Rept. on the Price of Gasoline in 1915,* 143, 144; *Rept. on the Advance in Price of Petroleum Products,* 51; *Rept. on the Pacific Coast Petroleum Industry,* Pt. I, 206. The latter reports credit the

Standard refineries with the consumption of 43.8 per cent of the crude oil consumed in refineries, but this might differ from the percentage of gasoline produced, because the Standard refineries are generally more efficient and have better methods of extracting gasoline than have the independent refineries.

28. See above, pp. 116-118.

29. *Petroleum Age,* Nov. 15, 1923, 11; Feb. 15, 1924, 24; April 1, 1924, 102; *Sat. Evening Post,* May 3, 1924, 12; *Natl. Petroleum News,* Feb. 6, 1924, 35; *The New Republic,* Nov. 14, 1923, 300.

30. *Wall St. Jour.,* Nov. 11, 1921, 3.

31. *204 Fed. Rep.,* 798; *234 U. S.,* 548.

32. See above, Ch. V.

33. U. S. Comr. of Corporations, *Rept. on the Transportation of Petroleum,* 1906, pp. xxiv, 16, 394.

34. *Eng. and Min. Jour.,* Feb. 16, 1905, 330; March 2, 1905, 426; Jan. 7, 1911, 89; Feb. 18, 1911, 386; *Mineral Industry,* 1910, 515; Cal. State Mining Bu., *Bul. 69,* p. 481.

35. *Kansas Reports,* 71, 811.

36. On the Kansas episode, see U. S. Comr. of Corporations, *Rept. on the Transportation of Petroleum,* pp. xxiv, 24, 370; *Kansas Historical Collections,* Vol. IX, 94-100; *The Oil Trust and the Government,* Francis Walker; *Pol. Sci. Quar.,* March, 1908, 18; *Mineral Resources,* 1905, 854; *Kansas and the Standard Oil Co.,* Ida Tarbell; *McClure's Magazine,* Vol. 25, p. 469, 608; *Outlook,* Vol. 80, pp. 19, 427; Vol. 81, p. 55; Vol. 82, pp. 283, 820; *World's Work,* May, 1905, 6155; *R. of Rs.,* April, 1905, 471; *Arena,* May, 1905, 500.

37. *Fed Rep.* 148, 719; 155, 305; 164, 376; 170, 988.

38. *Fed. Rep.* 154, 728; 183, 223.

39. Standard Oil Co. vs. State, *100 S. W.,* 705.

40. Standard Oil Co. of Ky. vs. Tenn., *120 Tenn.,* 86; *217 U. S.,* 413.

41. *Fed. Rep.* 152, 290; 158, 536; 179, 614; 192, 438.

42. *107 Tex.,* 1; *S. W. Rep.* 103, 836; 105, 851; 106, 326, 918; *212 U. S.* 86, 112; *222 Fed. Rep.,* 69; *Mineral Industry,* 1907, 758; 1908, 687; 1909, 570; *Wall St. Jour.,* Sept. 28, 1920, 1.

43. *218 Mo.,* 1; *224 U. S.,* 270.

44. *61 So.,* 981; *65 So.,* 468.

45. *S. Doc. 310,* 64 Cong., 1 sess.

46. *S. Res. 138,* 66 Cong., 1 sess.

47. *S. Rept. 1263,* 67 Cong., 4 sess.

48. *Natl. Petroleum News,* Oct. 10, 1923, 95; *Oildom,* Sept., 1922, 13.

49. See above, Ch. XII.

50. *Oil Weekly,* Sept. 1, 1923, 16, 17; *Natl. Petroleum News,* Aug. 29, 1923, 18, 69; *Petroleum Age,* Sept. 1, 1923, 26; *Report of the Federal Trade Commission to the President on Gasoline Prices in 1924.*

51. *Oil Weekly,* Sept. 1, 1923, 16; *Natl. Petroleum News,* March 14, 1923, 17; Aug. 15, 1923, 18, 19, 20; Aug. 22, 1923, 17, 21; *Oildom,* March, 1923, 46, 47; Federal Trade Comm., *Rept. on the Price of Gasoline,* 157; *Rept. on the Pacific Coast Petroleum Industry,* Pt. II, 4; *S. Rept. 1263,* 67

Cong., 4 sess., p. 44, 45; *Report of the Federal Trade Commission to the President on Gasoline Prices in 1924.*

52. *Natl. Petroleum News,* Aug. 29, 1923, 17.

53. *Oildom,* March 19, 1924, 1.

54. *Natl. Petroleum News,* May 14, 1924, 18; July 9, 1924, 29; Aug. 13, 1924, 25.

55. *Cal. Oil World,* Aug. 3, 1922, p. 1; *Oil Weekly,* Jan. 20, 1923, 72; *Natl. Petroleum News,* May 24, 1922, 20; *The Lamp,* June, 1924, 4.

56. Federal Trade Comm., *Rept. on the Price of Gasoline,* 49, 50.

57. Hearings before the Public Lands Committee of the Senate, on S. 45, 65 Cong., 1 sess., Pt. 2, p. 82; *S. Doc. 363,* 64 Cong., 1 sess., 16.

58. On the Standard Oil Monopoly, see: Tarbell, *History of the Standard Oil Company; Jour. Pol. Econ.,* Oct., 1907, 449; April, 1916, 404; Ryan, *Distributive Justice,* 76; Moody, *The Masters of Capital,* Ch. IV; Lloyd, *Wealth against Commonwealth;* Montague, *The Rise and Progress of the Standard Oil Company;* Jones, *The Trust Problem in the United States,* Ch. V (has a brief bibliography). Also see cases above cited.

CHAPTER XIX

MONOPOLISTIC CONDITIONS IN THE OIL INDUSTRY (Continued): LARGE SCALE ORGANIZATION IN GENERAL

THE Standard Oil Company is not the only example of the tendency toward large scale operation in the oil industry. In all fields the tendency has been toward absorption of the smaller companies by the larger units. In Pennsylvania this movement was pretty well completed many years ago.

In the ownership of oil lands and in the production of crude oil, there is almost everywhere a tendency toward concentration in few hands. The Texas Gulf Coast fields are mainly controlled by three companies: the Gulf Production Company, a subsidiary of the Gulf Oil Corporation; the Texas Company; and the Humble Oil and Refining Company. There are a number of other companies, but their proportion of the production is small. In the Saratoga district of Texas the operations were from the first carried on, for the most part, by a few large companies. The Sinclair Oil Company practically owns the Damon Mound field. In the Beaumont field, the small operators soon gave way to those of larger capital. The Jennings field in Louisiana has from the first been in the hands of strong companies—one reason why the oil was more intelligently handled than at Spindletop. In the Haynesville field, the Ohio Oil Company, a Standard company, produced in January, 1923, about half the total production.[1]

The ownership of California oil lands has been rather thoroughly investigated, and considerable information is available. In 1914 the largest oil-land owner, the Kern Trading and Oil Company (Southern Pacific interests) controlled 23 per cent of the proved oil lands, and three other large concerns owned 8 per cent each.[2] In 1918 the ownership of proved oil lands was found distributed as follows:[3]

Company	Acres of proved oil land in the state
Associated Oil Company	7,347
Doheny companies	4,286
General Petroleum Corporation	2,584
Honolulu Consolidated Oil Company	2,701
Atchison, Topeka, and Santa Fé Railway Company	3,097
Shell Company	2,442
Southern Pacific Company	18,267
Standard Oil Company	8,187
Union Oil Company	8,198
Other companies	30,171

In 1918, 65 per cent of the proved oil lands were held by only nine companies, and they were gradually enlarging their holdings by purchase from the smaller companies.

A more recent report of the Federal Trade Commission, issued in April, 1921, indicates that the holdings of the large interests had increased between 1918 and 1920. In March of the latter year, seven large interests controlled 68.1 per cent of the total proved area of the state. The ownership of these lands is shown in the following table:[4]

Acreage of Oil Lands and Acreage of Proven Oil Lands Held by the Seven Large Interests, and the Acreage of Proven Lands Held by Other Producers

Interests	Total oil lands, Dec. 31, 1919 Acres	Proven oil lands, Mar. 1, 1920 Acres	Per cent
Union Oil Co. of California	286,417	9,184	10.0
Associated Oil Co.-Southern Pacific interests*	154,112	28,055	30.6
Standard Oil Co. (California)	95,413	10,017	10.9
Doheny companies	30,332	4,761	5.2
General Petroleum Corporation	20,443	3,559	3.9
Santa Fé Railway Co.	15,970	4,043	4.4
Shell Co. of California	9,474	2,893	3.1
Total seven large interests	612,161	62,512	68.1
Other producers	......	29,280	31.9
Total	......	91,792	100.0

* Southern Pacific lands and control of Associated Oil Co. taken over by the Pacific Oil Co., Dec. 3, 1920.

The ownership of oil lands in the Mid-Continent and some of the other fields has not been investigated carefully, as in California, but a somewhat similar situation prevails. Perhaps the large companies do not control relatively quite as much of the proved lands in most other fields as they do in California, because railroad land grants, which have been the basis of some of the large holdings in California, have not been so important elsewhere. Also, in the Mid-Continent fields, Indian lands have been the source of much of the production, and the Government controls the ownership of the leases, so that large holdings have not developed to the same extent as in California. In Illinois, on the other hand, a very large proportion of all the oil leases and production are said to be controlled by one company—the Ohio Oil Company, a Standard company.[5]

The production of oil, as well as the ownership of the lands, has passed largely into the hands of large companies. E. W. Marland, of the Marland Oil Company, declared, in October, 1923, that the smaller independent companies were producing only 25 per cent of the oil produced in the United States, whereas seven years before they had produced 85 per cent of it. The large, integrated companies were rapidly taking over the oil-producing business.[6] In Wyoming, as already pointed out, one group of interests—Standard interests—control between 93 and 97 per cent of the total production. In California, eleven companies, with production of over 1,000,000 barrels each, produced, according to the Federal Trade Commission, 76.3 per cent of the reported production in 1914, and twelve companies produced 80.4 per cent in 1919.[7] Thus, the proportion produced by the larger interests has apparently been increasing.

A few large companies produce most of the crude oil produced in California, and their share of the total production is apparently increasing, as the following table shows:[8]

Production by Seven Large Interests, in Barrels of 42 gallons, 1914 and 1919

Interests	1919 Barrels	1919 Per cent	1914 Barrels	1914 Per cent
Standard Oil Co. (California) ..	25,484,984	25.2	13,550,809	13.1
Southern Pacific-Associated Oil Co.	18,444,055	18.2	17,319,452	16.7
Union Oil Co. of California	8,705,447	8.6	6,329,978	6.1
Shell Co. of California	6,702,971	6.6	4,322,833*	4.2
Doheny companies	5,836,799	5.8	4,980,659	4.8
Santa Fé Railway companies ...	3,754,299	3.7	3,170,517	3.0
General Petroleum Corporation .	3,387,261	3.3	....	..
Total	72,315,816	71.4	49,674,248	47.9
Other companies reporting to Commission	6,987,712†	6.9	16,208,127	15.7
Companies not reporting to Commission	21,918,256	21.7	37,741,320	36.4
Total production in California	101,221,784	100.0	103,623,695	100.0

* Production of Shell interests taken over by Shell Co. of California in 1915.
† First 6 months of 1919.

Even in the great fields of the Los Angeles Basin, where there were hundreds of small companies operating, with many town-lot leases, the production has been mainly in the hands of a few major companies, as the following table indicates:[9]

Daily Production in March, 1923

Company	Santa Fé Springs	Huntington Beach	Long Beach	Other Fields	Total bbls.
Standard Oil	22,500	26,000	.....	17,640	66,140
General Petroleum Co. ...	30,800	.....	1,200	3,500	35,500
Associated et al.	2,875	32,540	.....	4,000	39,340
Shell Co.	12,900	11,442	26,800*	....	51,140
California Petroleum ...	18,800	10,900	15,200	5,500	50,400
Union Oil	15,500	8,370	750	10,800	38,020
Miscellaneous	59,100	32,000	86,500	3,500	181,100
Totals	162,475	121,252	120,450	44,940†	461,640

* Shell has estimated 18,000 barrels a day shut in temporarily.
† About 22,000 barrels a day shut in.

For some of the Texas fields, information as to production in January, 1923, is available in the *Oil Weekly*.[10] In the Stephens County field, seven companies produced considerably over half; in the Electra field, five companies produced considerably over half; in the Mexia field, one company, the Humphreys Oil Company, produced nearly half, and the ten largest companies produced nearly all; in the Humble field, and in the West Columbia field, the Texas Company produced considerably over half; in Goose Creek, two companies produced about 90 per cent; in the Orange field, five companies produced over two-thirds; in the Hull field, one company produced nearly half, and six companies produced practically all. In the Powell field, later in the year, the Humble Oil Company produced about one-third, and a half-dozen companies produced over one-half of the total production. In several of the other fields, however, the production was far more scattered; in Eastland, Ranger, Desdemona, Pioneer, Burkburnett, and Young County. M. M. Doan, Chairman of the Mid-Continent Oil Operators' Association, once stated that in the Gulf Coast country 80 per cent of the development work was being done by three large companies.[11]

The ten leading producing companies of Texas, in the fourth quarter of 1924, reported a production of over 16,000,000 barrels, of a total of 31,000,000 barrels for the state:[12]

Company	Production, Fourth Quarter 1924
Humble Oil & Refining Company	3,428,657
Gulf Production Company	2,538,216
The Texas Company	2,237,132
United North and South Oil Company	2,188,246
Pure Oil Company	1,856,850
Magnolia Petroleum Company	1,418,382
The Sun Company	794,944
Republic Production Company	786,237
Atlantic Oil Producing Company	704,883
Simms Oil Company	606,530
Total, 10 companies	16,560,077
Total, all companies	31,194,183

During the same period, the 13 leading companies in Oklahoma produced over 23,000,000 barrels, of a total of 46,000,000 barrels produced by all companies in the state, or about half of the total:[13]

Company	Production, Fourth Quarter 1924
Comar Oil Company	5,018,220
Gypsy Oil Company	2,884,592
Phillips Petroleum Company	2,558,694
Carter Oil Company	2,170,066
Prairie Oil & Gas Company	1,785,623
Magnolia Petroleum Company	1,658,326
Cosden Oil & Gas Company	1,500,019
Sinclair Oil & Gas Company	1,329,603
Roxana Petroleum Corporation	....
Amerada Petroleum Corporation	1,231,415
Wentz Oil Corporation	1,185,749
Skelly Oil Company	1,084,025
The Texas Company	1,066,194
Total, 13 companies	23,472,526
Total, all companies	46,845,989

For the United States as a whole, the Federal Trade Commission has made no careful investigation of this question, but some information is available for the year 1919. The Trade Commission received reports covering a total production of 239,038,000 barrels of crude petroleum, or 63.3 per cent of the total production of the United States, and of this reported production, 224,190,000 barrels, or 59.4 per cent of the total production of the United States, was produced by 20 different interests, comprising altogether 32 companies. Nine of these companies were Standard companies, and they produced 21.3 per cent of the total production of the United States, while 23 large independents produced 38.1 per cent of the total.[14]

In the refining of petroleum, the preponderance of the large units is greater than in the production of crude oil. In California, five companies—the Standard Oil Company, the Union Oil Company, the Associated Oil Company, the Shell Company, and the General Petroleum Corporation—owned 94 per cent of the gross investment in refineries in 1919, operated 89 per cent of the rated daily capacity, and consumed 91 per cent of the crude petroleum refined. In Wyoming, in the years 1918 and 1919, one single company—the Midwest Refining Company, with its subsidiary, the Utah Oil Refining Company—purchased and refined approximately 94 per cent of all the Wyoming crude oil refined, except that refined in Canada. The refining of Illinois crude was reported in 1913 to be practically all done by Standard interests.

The Mid-Continent and Texas refining industry has not been investigated carefully, but a few large companies dominate the refining business largely there as elsewhere. Probably the relative importance of the few largest companies is not so great as in California, Wyoming, or Illinois.

For the United States as a whole, the transporting and refining business is more completely in the hands of large companies than is the producing end of the industry. Whereas Standard companies controlled in 1919 about 23½ per cent of the total crude production, they operated 68 per cent of the pipe lines and refined about 44 per cent of the total crude oil refined. A group of large independent refiners refined over 41 per cent. Altogether 39 companies refined 85 per cent of the crude oil refined in the United States, and some of these refiners represented identical interests.[15]

About 1915 there developed a very strong tendency in the direction of consolidation of the producing and refining interests and their joint expansion under centralized control. The years following witnessed a number of consolidations of individual producing, transporting, and refining interests, into such powerful organizations as the Empire Gas and

Fuel Company, the Empire Pipe Line Company, the Empire Refining Company, Cosden and Company, the Sinclair Oil and Refining Company, the Oklahoma Producing and Refining Company, and the Pan-American Petroleum and Transport Company. During these same years came the aggressive expansion of such existing corporations as the Tide Water Oil Company, the Midwest Refining Company, and the Magnolia Petroleum Company, to include important pieces of property not formerly owned by them. This movement has been fortunate for all concerned, since it tended to give more stability to the industry. It is even fortunate for the individual producer, who benefits more than anyone else from increased stability, and who has more markets open to him with the development of additional strong companies.[16] Perhaps the greatest oil merger in many years has been the recent purchase by the Standard Oil Company of Indiana of the foreign properties of the Pan-American Petroleum and Transport Company. This purchase makes the Standard of Indiana one of the largest oil companies in the world, perhaps second only to the Standard of New Jersey.

In explaining the increasing importance of large companies in the oil industry, it is unnecessary to dwell long on the wickedness of the Standard Oil Company, or of the large independents. This movement toward concentration in large units is one which has already gone far in most of the industries which are concerned with the exploitation of natural resources, as it has indeed in industries of many kinds. It was long ago known that the forest lands were drifting into large holdings, and the same is true, to a lesser extent, of coal lands, and of various kinds of mineral lands.

In the exploitation of oil resources, large units have many advantages. First of all, large companies have advantages in the production of oil. There are many advantages in drilling. The large company, with large lease acreage, can survey more cheaply, can do its own drilling, or if it contracts its drilling, can secure better rates than smaller companies. It can avoid the expense of many offset wells that smaller holdings would make necessary. Owning large contiguous leases, instead of a few scattered leases, the large company can "feel out" from producing wells, location by location, unhampered by property lines. It will less frequently be forced by lease provisions to drill before needed information is available. The large company usually keeps more careful logs of geological strata, and of course can use its own logs without difficulty. Among many small producing companies there are always some that keep poor logs or keep them secret, or in some cases there are even some who falsify their records.

Large companies, in the second place, have advantages in the operation of their wells and their business generally. They can economize labor. They can employ more highly specialized men and give wider application to their activities. They can use one pumper to attend several neighboring powers. They can save in teaming costs by maintaining well-distributed and well-stocked warehouses. They can utilize power better by connecting up the greatest number of wells that each power is capable of pumping. They can afford more continuous utilization of plant and equipment, such as, for instance, pulling machines. They can install gasoline extraction plants, and thus save a valuable by-product, which is often wasted by smaller companies. They can conserve pressure and use water-flushing, where they control an entire pool, and can reduce the danger from premature flooding by water from improper casing or plugging. They can try out new methods and carry on experiments beyond the purse of small companies; they can buy supplies at lower prices by buying in large lots. In the sale of products they, of course, enjoy a great advantage in advertising.[17]

Not all the advantages, to be sure, are with the large companies. The large companies have the disadvantages that go with large-scale organization: the evils of absentee ownership, the lack of personal interest in the employee management, "red tape" and bureaucratic inefficiency, somewhat similar to that of government undertakings. The small companies, at their best, have the advantage of more direct and vital interest on the part of the management, better opportunity to watch expenses and grasp favorable opportunities. As Professor Marshall says, referring to small units in general:

The small employer has advantages of his own. The master's eye is everywhere; there is no shirking by his foreman or workmen, no divided responsibility, no sending half-understood messages backwards and forwards from one department to another. He saves much of the book-keeping, and nearly all of the cumbrous system of checks that are necessary in the business of a large firm.[18]

The management of a small company, has, of course, greater flexibility, and can sometimes strike out a bold course which would be impossible to the clumsier and more conservative large company. Thomas O'Donnell has claimed that the big oil companies cannot produce as cheaply as the small ones. At any rate, those small companies which have efficient management certainly have some advantages over the very largest companies.[19]

In general, however, it is clear that the advantages are mainly with the larger companies. This seems to be proved by the fact that the larger companies generally make larger profits. According to the Federal

Trade Commission, five large oil companies, with investment of over $100,000,000 each, earned in 1919 an average of 24.6 per cent, while 31 companies with an average investment of between $1,000,000 and $5,000,000 each, earned only 15.4 per cent. Those companies with less than $1,000,000 investment each, earned an average of 17.3 per cent, slightly more than the group of companies next larger, but much less than the very large companies.[20]

Classifying the oil companies of the United States on the basis of the percentage of profit earned in 1919, we see again the advantage of large units in the production of oil:[21]

Group	Percentage profit	Average gross production Barrels
1	Loss	117,022
2	Less than 10	497,703
3	10 to 20	631,105
4	20 to 30	1,409,244
5	Over 30	1,655,408

Earnings in the producing business in California point likewise to the efficiency of large units:[22]

Year	Group 1a Per cent	Group 2b Per cent	Group 3c Per cent	Group 4d Per cent
1914	8.9	10.6	Loss .8	Loss 4.0
1915	5.1	9.5	Loss .8	Loss 4.2
1916	15.1	12.0	3.	Loss 4.3
1917	22.4	19.2	5.5	Loss 1.7
1918	27.5	28.1	11.4	Loss .4
1919*	29.7	27.3	12.3	.7

a Producing 1,000,000 barrels or over.
b Producing 250,000 to 1,000,000 barrels.
c Producing 50,000 to 250,000 barrels.
d Producing under 50,000 barrels.
* First six months on an annual basis.

In the production of crude oil in California, the cost of production for companies producing over 1,000,000 barrels per year was about one-third as high as for companies producing less than 50,000 barrels, as the following table shows:[23]

Cost of Production per Barrel

Year	Group 1a	Group 2b	Group 3c	Group 4d
1914	$.25	$.29	$.49	$ .72
1915	.28	.27	.43	.98
1916	.30	.33	.53	.96
1917	.34	.38	.62	1.09
1918	.38	.47	.69	1.19
1919*	.43	.50	.75	1.21

a Producing over 1,000,000 barrels per year.
b Producing 250,000 to 1,000,000 barrels per year.
c Producing 50,000 to 250,000 barrels per year.
d Producing under 50,000 barrels per year.
* First six months.

Figures for California showing the net investment of various companies, in proportion to their production, likewise show the economy of large units. The following table illustrates this:[24]

Net Investment per Barrel, Based on Annual Production, by Groups

Year	Group 1a	Group 2b	Group 3c	Group 4d
1914	$2.43	$1.97	$3.71	$ 7.10
1915	2.69	1.90	3.20	8.99
1916	2.61	2.19	3.64	8.58
1917	2.77	2.16	4.10	10.97
1918	2.77	2.28	4.26	9.25
1919*	2.79	2.55	3.80	10.52

a Producing 1,000,000 barrels or over.
b Producing 250,000 to 1,000,000 barrels.
c Producing 50,000 to 250,000 barrels.
d Producing less than 50,000 barrels.
* First half of 1919, on annual basis.

Similar figures for the Mid-Continent field tend to show that the very largest companies make less profits than the moderately large companies, but that the small companies are at a very great disadvantage:[25]

Group	Percentage of profit	Number of companies	Average per co. Barrels
1	Loss	6	49,026
2	Less than 10	14	571,955
3	10 to 20	9	1,034,617
4	20 to 30	5	2,413,396
5	Over 30	8	1,440,097

In the refining of petroleum, as in the production of oil, large companies are more efficient and more profitable than small companies, although the very highest profits were made in 1919 by a group of moderately large companies. The following table, for the entire United States, illustrates this:[26]

Group	Percentage of profit	Number of companies	Crude consumed Average per company Barrels
1	Loss	28	323,562
2	Less than 10	39	1,259,225
3	10 to 20	26	2,739,271
4	20 to 30	17	6,210,279
5	30 and over	28	2,441,290

Since larger companies make larger profits in the production and in the refining of petroleum, there is a very strong tendency for both production and refining to gravitate into the control of such companies. In competition between large and small companies, the advantage lies definitely with the large units, and therefore there is a constant tendency for the small companies to be absorbed or driven from the field by the large companies.

There is nothing unfortunate about this, for it is socially desirable, as well as economically profitable, to have oil produced, transported, and refined by large companies. Large companies do the work, not only more cheaply but, as already pointed out, with less waste of oil and of capital.

One writer has described the operations of the small oil companies in the following caustic terms:

The little companies, especially the multitudinous stock promotions, have a hard time generally, on account of their bad management, and the latter is due often as much to lack of experience as to lack of honesty. A professional oil man hates to have these new fellows as neighbors; they don't know how to drill their wells; they set their casings improperly and let salt water into the oil sand; they don't know what to do with the oil when they get it; they drill many more wells than necessary, forcing the experienced neighbors sometimes to do likewise.

These inexperienced companies will build a refinery anywhere, regardless of crude oil supply or railroad advantages; demoralize the crude oil market by paying high premiums and the refined products market by dumping their products whenever in need of working capital, which is often. They will spend huge sums for inventions and "processes," will pay ridiculously high prices for oil properties in new districts, forcing the professional men to do

likewise, or in many cases to stay out, and otherwise make themselves unmitigated nuisances to the legitimate industry. The result is these inexperienced companies play out very rapidly—to the accompaniment of angry
howling from their "stung" stockholders.[27]

In the oil industry, as in industries generally, there has been a strong
drift, not only toward larger companies, but toward integrated companies, that is, companies controlling crude production, transportation, refining, and marketing—every process between the well and the
final consumer. The reason for this drift has been the fact that integrated companies have many advantages over companies which engage
in only one or two processes of the industry. During periods of over-
production, producing companies have often been embarrassed by lack
of a market for their crude oil, while in periods of relative scarcity the
refiners have become nervous for fear of a failure of their crude supply.
Often the companies without marketing facilities have found it difficult
to dispose of their product. During the over-production of 1923, Cosden
and Company, one of the largest and most efficient refining companies
in the Mid-Continent field, found it very difficult to dispose of its
products, because it had no service station outlets. The large integrated
companies had too much of their own products to sell, and were unwilling to buy from independent refiners. In the all-too-numerous periods
of over-production of crude oil, producing companies have always found
much difficulty in disposing of their oil; in fact, producing companies
without any refineries have always been regarded as occupying a very
precarious position. The large integrated companies, on the other hand,
can always depend on their own crude material, and can market their
products advantageously. They are in a relatively secure position, and
can effect substantial economies in the transformation of crude oil into
consumable products. As pointed out elsewhere, the Standard companies
have shown the drift toward integration as clearly as the big independents.[28]

NOTES

1. *Oil Weekly,* Jan. 20, 1923, 120.
2. Cal. State Mining Bu., *Bul. 69,* p. 15.
3. Cal. State Oil and Gas Supervisor, *Third Ann. Rept.,* p. 21; *Eng. and Min. Jour.,* Jan. 18, 1919, 146.
4. Federal Trade Com., *Rept. on the Pacific Coast Petroleum Industry,* Pt. I, 45.
5. Ill. State Geol. Survey, *Bul. 22,* p. 146.
6. *Natl. Petroleum News,* Oct. 10, 1923, p. 32N.

7.

PRODUCTION OF CRUDE PETROLEUM IN CALIFORNIA BY GROUPS OF COMPANIES
REPORTING, IN BARRELS OF 42 GALLONS, 1914–JUNE 30, 1919

Year	Group	Range of production	Number of companies	Quantity Barrels	Average per company Barrels	Proportion by groups Per ct.	Proportion of state total Per ct.
1914	1	Over 1,000,000	11	50,265,379	4,569,580	76.3	48.5
	2	250,000 to 1,000,000	18	9,207,863	511,548	14.0	8.9
	3	50,000 to 250,000	53	5,290,753	99,826	8.0	5.1
	4	Under 50,000	43	1,118,379	26,009	1.7	1.1
		Total	125	65,882,374	527,059	100.0	63.6
1915	1	Over 1,000,000	8	42,550,368	5,318,796	71.8	47.5
	2	250,000 to 1,000,000	20	10,572,668	528,633	17.9	11.8
	3	50,000 to 250,000	42	4,454,160	106,051	7.5	5.0
	4	Under 50,000	62	1,653,930	26,676	2.8	1.8
		Total	132	59,231,126	448,721	100.0	66.1
1916	1	Over 1,000,000	10	50,834,439	5,083,444	76.3	55.3
	2	250,000 to 1,000,000	19	8,879,891	467,363	13.3	9.7
	3	50,000 to 250,000	51	5,296,988	103,863	8.0	5.8
	4	Under 50,000	64	1,585,126	24,768	2.4	1.7
		Total	144	66,596,444	462,475	100.0	72.5
1917	1	Over 1,000,000	10	58,823,426	5,882,343	77.5	60.5
	2	250,000 to 1,000,000	19	9,255,178	487,115	12.2	9.5
	3	50,000 to 250,000	57	6,088,825	106,821	8.0	6.3
	4	Under 50,000	78	1,705,704	21,868	2.3	1.7
		Total	164	75,873,133	462,641	100.0	78.0
1918	1	Over 1,000,000	10	63,202,366	6,320,237	77.4	62.2
	2	250,000 to 1,000,000	22	10,541,776	479,172	12.9	10.4
	3	50,000 to 250,000	53	6,043,233	114,023	7.4	5.9
	4	Under 50,000	84	1,896,566	22,578	2.3	1.9
		Total	169	81,683,941	483,337	100.0	80.4
1919*	1	Over 1,000,000	12	33,968,353	2,830,696	80.4	67.1
	2	250,000 to 1,000,000	22	4,911,694	223,259	11.6	9.7
	3	50,000 to 250,000	45	2,556,486	56,811	6.0	5.1
	4	Under 50,000	72	825,211	11,461	2.0	1.6
		Total	151	42,261,744	279,879	100.0	83.5

* First 6 months.

(From Federal Trade Comm., *Rept. on the Pacific Coast Petroleum Industry*, p. 62.)

8. *Ibid.,* p. 64. Other figures apparently differ considerably from those of the Trade Commission. The California State Mining Bureau reported production control for the first half of 1913 as follows: (*Bul. 69,* p. 30.)

CONTROL OF CALIFORNIA PRODUCTION
FIRST HALF OF 1913

Company	Barrels	Per cent
Standard	12,672,122	28.64
Union-Agency	9,640,444	21.78
Associated	9,169,453	20.71
Kern Trading and Oil Co.	4,554,093	10.29
General Petroleum	1,771,050	4.
Santa Fé	2,376,541	5.38
Amalgamated	2,228,747	5.03
Misc.	1,857,803	4.17
	44,270,253	100.00

The difference between the figures of the Federal Trade Commission and those of the California Mining Bureau is explained by the Trade Commission as due to the fact that the Trade Commission includes only the *production* of crude petroleum by each of the large companies, while the figures of the California State Mining Bureau refer to the amount of production *controlled* by these companies. Also, the figures of the Trade Commission are one year later in time; and, finally, the interests listed are not identical in the two tables.

Figures from the California State Mining Bureau for a later year do not differ so widely from the figures of the Trade Commission: (Cal. State Oil and Gas Supervisor, *Third Ann. Rept.,* p. 19.)

CALIFORNIA OIL PRODUCTION, JUNE, 1918

Company	Oil (barrels) daily	Per cent
Associated Oil Company*	25,583	9.1
Doheny companies†	20,959	7.3
General Petroleum Corporation	12,082	4.3
Honolulu Consolidated Oil Company	3,632	1.3
Atchison, Topeka and Santa Fé Railway Company (Fuel Department)‡	11,163	4.0
Shell Company of California	19,233	6.8
Southern Pacific Railway (Fuel Oil Department)	24,096	8.5
Standard Oil Company	63,964	22.6
Union Oil Company of California	22,931	8.1
Other companies	79,029	28.0
Totals	282,672	100.0

* Includes Amalgamated Oil Company.

† Includes American Oilfields Company, American Petroleum Company, California Star Oil Company, Doheny-Pacific Petroleum Company, Midland Oil Company, Midland Oilfields Company, Niles Lease Company, Pan-American Petroleum Company, Pan-American Investment Corporation, Petroleum Midway Company, Ltd., Red Star Petroleum Company.

‡ Includes Chanslor-Canfield Midway Oil Company, Petroleum Development Company.

The figures of production in barrels are not comparable with those of the Trade Commission, but the percentage of the total production for each of these companies approximates roughly the percentage indicated by the Trade Commission. The difference of one year between the two sets of figures would of course account for some difference.

9. *Oil Weekly,* March 31, 1923, 54.

10. Jan. 20, 1923, 109.

11. *Oildom,* Oct., 1922, 27.

12. *Oil Weekly,* March 13, 1925, 35.

13. *Oil Weekly,* Feb. 20, 1925, Supplement, p. 2.

14. Federal Trade Comm., *Rept. on the Pacific Coast Petroleum Industry,* Pt. I, 27.

15. *Ibid.,* Pt. II, 21.

16. *Mineral Resources,* 1915, 562; 1916, 681.

17. Bacon and Hamor, *American Petroleum Industry,* I, 434.

18. Marshall, *Principles of Economics,* 284.

19. *Natl. Petroleum News,* Jan. 23, 1924, 35.

20. Federal Trade Comm., *Rept. on the Pacific Coast Petroleum Industry,* Pt. I, 29.

21. *Ibid.,* p. 212.

22. *Ibid.,* p. 146.

23. *Ibid.,* p. 133.

24. *Ibid.,* p. 142.

25. *Ibid.,* p. 213.

26. *Ibid.,* p. 217.

27. *Wall St. Jour.,* Sept. 4, 1920, 1, 11.

28. Bacon and Hamor, *American Petroleum Industry,* I, 435.

CHAPTER XX

DEVELOPMENT OF CONSERVATION SENTIMENT

THUS, the results of private ownership and exploitation of oil were almost everywhere the same: instability in the industry, over-production, wide fluctuations in prices, with prices always far too low; curtailment campaigns carried on in a generally vain effort to secure stability and reasonable prices; waste of oil by the millions of barrels; waste of capital by hundreds of millions of dollars; waste of human energy; speculation, and fraud, and extravagance, and social inequality; and finally, the development of monopoly conditions as the only means of escape from the intolerable conditions of private competition. The question arises: how early were these unfortunate results of our oil policy recognized?

Conservation views regarding oil were rather late to appear, largely because oil was not an important resource until after the Civil War. Donaldson's *Public Domain*, published in the early eighties, does not even mention it. The conservation of coal was an important topic in England early in the nineteenth century, and in Pennsylvania the approaching exhaustion of anthracite was discussed before the Civil War; but oil received relatively little attention until 1900 and after.

There were, nevertheless, some expressions of concern for the future oil supply almost from the first. In October, 1861, a writer in the *Derrick* reported: "Fears are entertained that the supply (of oil) will soon be exhausted if something is not done to prevent the waste." Von Millern, in his *All About Petroleum*, published in 1864, raised the question, "Will the supply hold out?" but he seemed to think it would, basing his opinion on the long time that the wells of Baku had been producing. Wright, in his *Oil Regions of Pennsylvania*, written in 1865, mentioned the gross waste of oil and gas, and pointed out that all wells give out finally; but he was not seriously concerned about the future supply. About this time, the question of the duration of the supply of oil was discussed a great deal in Pennsylvania, many people finding cause for concern in the gradually declining flow of all wells and the increasing number of wells required to maintain production. *Derrick and Drill* devoted considerable space to this question in 1865.[1]

Even the Commissioner of the Land Office seemed interested in the question of oil deposits, for in 1865 he directed the register and receiver of Humboldt, California, to withhold from disposition all land valuable for petroleum deposits.[2] Charles Harris, in his *History of the Venango Oil Regions*, published at Titusville in 1866, seemed more concerned about the waste of capital than about the waste of oil. He thought that "as a business operation" the production of petroleum had "generally been carried on wastefully," and commented on various "unthrifty methods" of exploiting the oil fields.

In 1869 R. W. Raymond, Commissioner of Mining Statistics, in his report on the *Mineral Resources of the States and Territories West of the Rocky Mountains*,[3] pointed out that mining is different from agriculture, in that it is taking part of an irreplaceable resource. Some of his remarks, although intended to apply to the precious metals, rather than to oil or coal, sound a note of broad and generous wisdom, in their bearing upon the conservation of all national resources:

In view of these peculiar relations of mining, it is evident that governments are in a certain sense trustees of the wealth stored in the mineral deposits of their realms—trustees for succeeding generations of their own citizens and for the world at large.

Unfortunately, Raymond did not follow his own remarks to a logical conclusion, but urged the sale of mineral resources with the surface of the soil.

As early as 1871 the *Weekly Report* of Petrolia, Ontario, noticed the decline in the production of the Canadian oil field, and wondered what could be done "to supply the increasing demand." The Geological Survey of Pennsylvania showed a desire to assist the oil industry as early as 1874. In that year Henry Wrigley made his *Special Report on the Petroleum of Pennsylvania*, and in this report he expressed great concern about the waste of oil:

Without considering the question of blame, or possibility of a remedy, it seems to be a fact that merits serious attention, that we have reaped this fine harvest of mineral wealth in a most reckless and wasteful manner.[4]

Wrigley seemed to see that the state had a duty in the matter, although his suggestion was rather guarded:

Whether any protection to our general interests in this matter as a state is possible, is a question that present abundance has caused to be put out of sight.

In 1883 the Pennsylvania State Geologist, Peter Lesley, sounded a distinctly warning note:

The next generation will gather from our oil history, with angry astonishment, a lesson of warning in political economy, only useless because coming too late. It is certain that petroleum is not now being produced in the Devonian rocks, by distillation or otherwise. What has been stored up can be got out. When the reservoirs are exhausted, there will be an end of it. The discovery of a few more pools of two or three million barrels each can make little difference in the general result.

About this time John F. Carll, in charge of oil investigations of the Pennsylvania Geological Survey, wrote:

There are not at present any reasonable grounds for anticipating the discovery of new fields which will add enough to the declining products of the old, to enable the output to keep pace with the shipments or consumption.[5]

An excerpt from the census of 1880 indicates that there was some concern about the future of the oil supply at that time.

While it is not probable that the deposits of petroleum within the crust of the earth are being practically increased at the present time, there is reason to believe that the supply is ample for an indefinite period. When prophecy, indulged even by the most sagacious producers of long experience, proves so futile, I think I am warranted in expressing the opinion that, as regards the future supply of petroleum, the drill alone gives valid testimony. Yet this fact is worthy of the most serious consideration: the production of petroleum, as at the present conducted, is *wasteful in the extreme*. No thoughtful person can escape the conviction that future generations will want what this present generation is destroying to no purpose. "After us the deluge" is written all over the oil region in the destruction of forests, and in the waste of the oil itself.[6]

The Pennsylvania Geological Survey evinced considerable interest in the conservation of oil throughout the early eighties. Dr. Charles Ashburner expressed in 1885 the opinion that the boundaries of the oil regions were well established, and that there was no reasonable expectation that any new or extensive field would be found—a rather unfortunate prediction in the light of developments of the next few years.[7] In 1886 John F. Carll, of the Pennsylvania Survey, wrote at length of the impending exhaustion of oil:

The great Pennsylvania oil fields, which have supplied the world for years, are becoming exhausted, and cannot respond to the heavy drafts made upon them many years longer, unless reinforced by new deposits from deeper horizons.[8]

In the reports of Professor Edward Orton, of the Ohio Geological Survey, are found some of the earliest expressions of concern over the wasteful use of our oil and gas resources, and some of the most advanced views on the question of the proper policy to follow in the use of such resources. Orton became Assistant State Geologist of Ohio in 1869, and State Geologist in 1882; and his enunciations on conservation were widely quoted by those who later became interested in this question.

At the annual meeting of the American Institute of Mining Engineers in February, 1882, Henry Wrigley read a paper in which he estimated the amount of oil remaining in Pennsylvania and New York at 96,000,-000 barrels. In closing, he voiced the following warning:

We have a certain fixed quantity of oil placed to our credit in the Bank of Nature, on which we have been drawing checks for twenty years; in 1875 we were checking out six millions, and in 1882 we are checking out twenty-five millions annually. Some day the check will come back indorsed "No deposit"; and we are approaching that day very fast.[9]

His estimate now seems a wild guess, and in the proceedings of the same association three years later, Charles Ashburner, State Geologist of Pennsylvania, criticized his estimates and his conclusions in the following judicious words:

That the product has passed its meridian there is no question, but what the total aggregate for the future will be, it would be folly to estimate. That the product per well will be less, and the cost of producing one barrel much more than in the past, experience would seem to prove. These, with other collateral facts connected with the production, manufacture, and consumption of the product will make the exhaustion of the field a gradual one; and it is probable that long before every barrel of oil shall have been taken out of the oil-sands, the cost of production per barrel will be so great that the oil-men's occupation in Pennsylvania and New York will be gone.[10]

In 1883 S. H. Stowell, of the United States Geological Survey, seemed somewhat apprehensive concerning the future oil supply, and foresaw that the price must ultimately rise high enough to cut down unimportant uses. Mr. Stowell recognized the possibility that new and prolific fields might be discovered, but thought that possibility rather remote.[11] In 1884, the California State Mineralogist, Henry G. Hanks, evinced an interest in "economizing our petroleum resources," which we were "exhausting at a rapid and in some instances reckless rate."[12]

A United States Consular report of 1888 refers to the numerous prophecies of the exhaustion of the Russian petroleum.[13] S. S. Gorby, of the Indiana Department of Geology and Natural History, in the same year, ventured the prophecy, "The stock of petroleum will be entirely dissipated after a time."[14] In 1889 and 1890, the *Engineering and Mining Journal* and the *American Geologist* began to show concern over the declining production of Pennsylvania. An editorial in the *American Geologist* in February, 1890, ran as follows:

The probability that there will be any permanent and considerable increase in the supply is slight. In this condition of the petroleum industry, it is plain that very shortly we shall reach an oil famine, when the use of the oil in certain cases will be a luxury instead of a necessity. This will be regulated by a curtailment of the demand through increased cost, tending to equalize the supply and demand.

In 1893 Joseph D. Weeks, of the United States Geological Survey, made one of the characteristic predictions of the times:

While it would be rash to say that the limits of the oil fields of the United States are well defined, the writer is of the opinion that the oil producing localities of the future will be those at present recognized, or their extensions.

Weeks sensed the rashness of these predictions, however, for he added:

However, there have been so many surprises in petroleum that these statements must be regarded as only setting forth the indications as to producing localities at the present time.

Weeks evinced considerable concern about the future oil supply during his term in the Geological Survey.[15]

Mr. S. F. Van Oss, in his *American Railroads and Investments*, published in 1893, raised the question, "Whether the supply (of oil) will last for many years to come," and decided it was "impossible to say." A writer on fuel oil, Clayton O. Billow, pointed out in 1894 that oil should be used for very high fuel uses, where coal could not serve. This distinction between "high" and "low" uses of oil was destined to become more important in later discussions. A writer in *Chambers' Journal* the following year foresaw an impending "permanent increase in price," and spoke of this as something which had considerably "agitated the public mind." The Pennsylvania fields were beginning their long decline at this time, and if there had been no fields farther west, there would have been much reason for concern.[16] During the next decade there was a great deal of discussion of the failing Pennsylvania supply. It was even said that Rockefeller and other Standard Oil officials were alarmed at the situation, and that quotations on Standard Oil stock fell 200 points because of the difficulty of finding adequate production.[17]

When W. S. Blatchley became State Geologist of Indiana in 1895, he immediately began a most energetic campaign for the education of public opinion in the matter of conservation. On every possible occasion he preached against the waste of oil and gas and all natural fuels; and some of the best statements ever made of the principles that should govern the use of our resources are to be found in his annual reports. In the first page of his first report he pointed out that the fuels of Indiana—coal, gas, and petroleum—were valuable only for the stored energy they contained; that man cannot invent any new forms of energy, nor produce any energy, but can only devise machines for transmuting or changing forms of energy already existing; that the supply of these fuels is limited and can never be increased; and that their duration depends wholly upon the rate at which they are used.

In his report for 1896 Blatchley wrote as follows:

These fuels are the most valuable resources of our state today. We are drawing upon them with a lavish hand. They come to us without great labor, as comes oftentimes the accumulated riches of a toiling and thrifty parent to a spendthrift son, and, as with the latter, "come easy, go easy," seems to be our motto. Once again would I repeat that no coal, no natural gas, no oil is being formed beneath the surface of our state today. Our present supply of each of these fuels will never increase, but ever diminish. Each constitutes a great reservoir or deposit of reserve energy upon which the people of the present generation are daily drawing without adding thereto. Like a bank account under the same conditions, it is only a question of time until it will become exhausted. . . . As Professor Orton has well said: "Every producer of petroleum knows that a field of petroleum begins to die the moment it begins to live." . . . The time will come, and that before many years, when the stored reservoirs of these great resources will have been drained, and only the dregs be left as a reminder of the plenty that has been.[18]

These warnings Blatchley repeated insistently as long as he was in office, and he succeeded in getting a state law passed, prohibiting the waste of gas.

In the nineties the exhaustion of the oil supply was a common subject of discussion in California; but the rapidly increasing production of 1901 and 1902 stopped this discussion for the time. In January, 1904, C. T. Deane wrote very cheerfully of the outlook, "The permanence of the supply (of California oil) need not trouble this generation."[19] Some writers, however, saw that California oil men were drilling too many wells; and in other parts of the country, there were many signs of an increasing interest in conservation.

In 1907 and the years following, interest in the general subject of conservation developed very rapidly, under the influence of Gifford Pinchot and Roosevelt. In a message to Congress in February, 1907, President Roosevelt gave particular attention to the question of conservation:

This government should not now repeat the mistakes of the past. Let us not do what the next generation cannot undo. We have a right to the proper use of both the forests and the fuel during our lifetime, but we should not dispose of the birthright of our children.[20]

Roosevelt's Conference of Governors, and his various conservation conferences and congresses were typical spectacular moves of that ardent friend of conservation; and he helped to stimulate interest in the conservation of oil, although oil was hardly mentioned at the Conference of Governors. The *Engineering and Mining Journal* and the *Mining and Scientific Press* both came out shortly after the conference with strong editorials on conservation. The *Mining and Scientific Press* quoted George Otis Smith's picturesque dictum:

Future historians will date the end of barbarism from the time when generations begin to feel that they rightfully have no more than a life estate in this sphere, with no right to squander the inheritance of their kind,

adding to it:

This is the new gospel of the practical world, which restates the golden rule and insists with the old militant spirit of every prophet who has had a true message for the world, that it must be observed.[21]

This journal for several years showed a very strong interest in conservation, and in it appeared some of the ablest editorials that have been written on the subject. One editorial, in the issue of October 16, 1909, pointed out very clearly the various factors at work in the fight for conservation: the deep-seated American distrust of monopoly, the growing public interest in the question of appropriating unearned increments, and, on the other hand, the "morbid interest" which every city and town in the West has "in its own growth in wealth and population."[22]

M. T. Bogert, writing for the *School of Mines Quarterly* in April, 1909, discussed in detail the waste of oil and the possibility of finding substitutes when the supply should be exhausted. At a joint meeting in March, 1909, of the American Society of Civil Engineers, the American Society of Mechanical Engineers, and the American Institute of Electrical Engineers, several addresses were devoted to the conservation of resources. Little of the discussion pertained directly to oil, however, and not all the sentiments expressed were favorable to the policy of conservation.[23]

During these years there were many well-thought-out expressions of the duty of conserving our resources. Dr. I. C. White, one of the greatest oil geologists of America, and a pioneer in the conservation movement, wrote, in February, 1909:

Just as sure as the sun shines and the sum of two plus two is four, unless this insane riot of destruction and waste of our fuel resources which has characterized the past century shall be speedily ended, our industrial power and supremacy will, after a meteor-like existence, revert before the close of the present century to those nations that conserve and prize at their proper value their priceless treasures of carbon.[24]

John Hays Hammond, one of the most eminent engineers in the United States, in a speech at the Conference of Governors in 1908, complained of the general indifference to the conservation of resources:

It has unfortunately become the popular custom to speak of the natural resources of our country as illimitable, and consequently to regard the discussion of the conservation of these resources as academic, or at best as scientific speculation. All efforts heretofore of a few enthusiastic theorists—as they were considered—have availed naught to disturb this imperturbable complacency and optimism.[25]

The question of conserving oil, as already stated, was not discussed much in the first years of the conservation movement. Conservation was first of all applied to forests, under the wise and energetic leadership of Gifford Pinchot. Next came coal. The Ballinger-Pinchot controversy hinged mainly on the conservation of coal. Shortly after this, however, oil came into the spotlight, and has held a prominent place on the stage ever since. With the interest in oil, or perhaps even preceding it in some regions, came an interest in the conservation of natural gas.

Within the past few years the conservation views expressed have been too numerous to be noted here. Geologists have performed an immense public service, in pointing out the geological factors involved. These geologists have not always been correct in their estimates, and in the predictions they based upon them; in fact, some of them have made rather wild guesses, in discussing the duration of the oil supply; but geologists have been correct in pointing out that the supply of oil is limited, and that it will ultimately be exhausted. Whether this exhaustion comes in twenty or fifty years is a relatively unimportant matter, from a broad social point of view, and many of the distinguished geologists of the country—I. C. White, George O. Smith, M. L. Requa, and David T. Day—have done notable service in stirring up public opinion on this question. Following them have come the newspapers and periodicals, government officials and politicians, and a few economists.[26]

A gradually developing appreciation of the value of our oil resources, and of the necessity of at least avoiding unnecessary waste, is shown by the history of state legislation. Laws against waste of oil were enacted first in New York and Pennsylvania, where, as early as 1879 and 1881, legislation was enacted regulating the method of plugging abandoned oil wells, to prevent water from reaching the oil-bearing sands. The Pennsylvania laws were extended to gas wells in 1885. The Pennsylvania statute of 1883, "To prevent waste by the production of petroleum from lands in controversy, etc.," probably does not indicate any particular interest in saving oil, but was merely an effort to prevent financial loss to owners of land. Perhaps the same may be said of some later statutes, as for instance the law of 1907 regarding forest fires on oil lands. The New York law was amended and expanded a number of times in the next few years after its passage.[27]

Under the influence of Blatchley and other far-seeing geologists in her state service, Indiana was rather prompt in her adoption of legislation to conserve oil and gas, particularly gas. In 1889, an act was passed, establishing a Department of Geology and Natural Resources, one of the functions of which was to "preserve" the natural resources of the state; and in 1893 a law was passed, forbidding the waste of oil or gas from open wells. Through the Department of Geology and Natural Resources, this law was really enforced to some extent. Ten years later, this act was elaborated and strengthened, and the State Natural Gas Supervisor was empowered to enforce it. The Indiana laws were concerned mainly with gas, since gas was the more important product in this state, but some of them applied also to oil wells. A rather curious law of 1889 and 1903 seemed to aim at suppressing oil and gas stock frauds, by limiting the amount of capital stock that any company might issue.

West Virginia passed a law as early as 1891, regulating the drilling and operation of producing wells, and the plugging of abandoned wells. This law was revised and expanded in 1897, to cover a great many detailed rules against the waste of oil and gas. Kentucky passed a similar law in 1892, but the mildness of some of the provisions of these laws almost elicits a smile. Under the provisions of the West Virginia law of 1891, and the Kentucky law of 1892, owners of wells were given three months to make some utilization of escaping gas, and in case a well was operated as an oil well, the gas might be wasted indefinitely. In 1906, Kentucky passed a law to protect the oil sands, and, in 1917 and 1918, passed a law imposing a tax of 1 per cent on the gross oil production of the state.

A Kansas law of 1891 required wells to be properly cased, to prevent water from reaching the oil sands, and a law passed ten years later required owners of new wells to confine them within two days after striking. These laws were not very carefully observed, and in 1913 a law was passed, authorizing the board of county commissioners of any county to appoint a supervisor to "personally direct and supervise the plugging of oil and gas wells."

When oil was first discovered in Texas, it was exploited so wastefully that the state legislature soon afterward declared that the situation constituted an "emergency," and, by an almost unanimous vote, hastily passed a law regulating in various ways the exploitation of oil and gas. Six years later, the same "emergency" seemed to prevail, since there was no effective agency for enforcing the law, and the law of 1899 was amended to allow the District Court of each county in the state to enforce the law and to appoint officers to assist. Later the Railroad Commission was given the duty of enforcing the laws against waste. One of the interesting features of the regulations issued by this Commission is the requirement that no well shall be drilled less than 300 feet from another well, or less than 150 feet from the lease boundary. In the Kosse field, where there were some city-lot holdings, the owners were compelled by this law to pool their interests in drilling wells.

In California there was some agitation very early for a law to prevent the oil measures from water infiltration—the greatest source of oil waste in California—but it was opposed by most of the oil men, and nothing was done until 1903, when a law was passed, providing various rules for the casing and plugging of wells. In 1909, another act was passed, requiring owners to keep logs of their wells,[28] these logs to be kept on file for the inspection of the oil commissioner, and requiring the board of supervisors of any county to appoint a commissioner to enforce the provisions of the act if three or more oil companies should petition for such an appointment. California was one of the leading states in her requirements of records and publicity regarding oil well logs and figures. In 1915, the legislature created a department in the State Mining Bureau for the protection of the oil and gas resources of the state, and provided for the appointment of a State Oil and Gas Supervisor, at a salary of $4,500 (later raised to $6,000). The supervisor was given authority to supervise the drilling, operation, maintenance, and abandonment of oil and gas wells in California, and to prevent, as far as possible, waste of crude petroleum and gas. This law of 1915 was one of the most advanced statutes yet passed by any state, and two years later it was amplified by another statute, giving the supervisor even greater authority.

Under the California law, the expense of the support of the Petroleum Department of the State Mining Bureau and of the necessary repairs of oil wells is levied on the industry. In 1920, the cost amounted to something less than $150,000 a year, one-tenth of which was levied on the oil land of the state, and nine-tenths of which was levied on the oil and gas produced.[29]

Conditions in the California oil fields have been much better, as a result of these laws. The Oil and Gas Supervisor has exercised great influence in oil development, an influence which has been highly beneficial to the operators generally, although many of the operators have at times done about all they could to obstruct the supervisor in his work. A detailed record of every important phase of drilling and operating results is kept, the geological conditions are carefully studied, and the oil and gas producers have the benefit of recommendations and suggestions of an engineering staff whose state-wide observations enable them to give the best scientific and technical information regarding drilling, operating, and maintaining oil and gas wells.

Immediately upon the discovery of oil in Illinois in 1905, the legislature of that state passed a law regulating the drilling, operation, and abandonment of wells, and, in the same year, Arkansas, still many years from production, passed a similar law. About the same time, Oklahoma, rapidly developing into a great oil-producing state, passed a law regulating the drilling and plugging of oil and gas wells. After several years of experience with this law, the state found it necessary to provide some effective means of enforcement, and an inspector was provided for in 1909. The following year an attempt was made to secure more publicity regarding Oklahoma geological strata, by requiring that all logs of abandoned wells should be open to public inspection, and a law passed in 1917 provided for the compiling and publication of a record of abandoned oil and gas wells.

In 1915, as a result of the great waste of oil and gas in the exploitation of the Cushing field, a law was passed "declaring an emergency," and providing that "the production of crude oil or petroleum in the state of Oklahoma, in such manner and under such conditions as to constitute waste," should be prohibited. "Such conditions as to constitute waste" were declared to include not only outright waste, but also the production of oil "at a time when there is not a market demand therefor at the well at a price equivalent to the actual value of such crude oil or petroleum," whatever that may mean. The Corporation Commission of the state was given authority to make such investigations and establish such regulations as might be necessary to prevent waste of all kinds. A curious provision of this law stipulated that "whenever the full production from any common source of supply of crude oil or petroleum" could "only be obtained under conditions constituting waste," any producer should be permitted to take "only such proportion of all crude oil and petroleum . . . that may be produced therefrom, without waste, as the production of the well or wells of any such person, firm, or corporation, bears to the total production of such common source of supply." This provision was evidently directed at the too energetic and wasteful exploitation of oil fields, and practically authorized the Corporation Commission to enforce a proration of runs, when the production should be so great as to involve waste of oil. Unfortunately, this law has not been utilized as much as it should have been, for Oklahoma has needed it much of the time since 1915.

Few states have moved faster than Louisiana in efforts to conserve oil and gas resources. The gross waste of gas in some of the early "wild wells" led to the passage of a law in 1906, authorizing the Governor to send the State Board of Engineers to take charge of operations where gas was being wasted, and, two years later, the legislature created a Commission for the Conservation of Natural Resources, to guard against waste of oil and gas. About this time the State Geological Survey, in coöperation with the United States Geological Survey, began a careful investigation into the oil and gas situation in Louisiana, publishing a report two years later.[30] In 1910, the legislature levied a tax on oil and gas production, as well as on the production of other natural resources, to create a "conservation fund." This "severance tax," on the business of "severing" minerals and timber from the soil was at first fixed at two-fifths of a cent per barrel, for oil, and one-fifth of a cent per thousand feet, for gas, and was later raised considerably. (In 1916 the oil tax was placed at three-fourths of a cent per barrel.) In recent years, it has yielded a handsome revenue—enough to improve greatly the state school system. In 1914 the state legislature passed an act, authorizing the Police Juries of the various parishes of the state to levy a similar tax, not exceeding that levied by the state. Thus Louisiana has proceeded on the theory that the exploitation of natural resources is a kind of enterprise which should be discouraged, at least if a revenue can be raised by doing so. Louisiana also passed laws in 1906 and 1910 regulating the drilling, operation, and plugging of oil wells, and in 1918 and 1920 added further careful regulations.

Ohio was slow to give any attention to the waste of oil and gas, but passed a law in 1889, providing rules against the waste of gas, and in 1893 regulated the methods of drilling oil wells. The slowness with which the public interests get attention is well indicated by the fact that Ohio legislated as early as 1885 to provide for the recording of oil leases—to prevent loss to individuals, and did not do anything to protect the public interest in oil until several years later.

Wyoming saw the need of legislation as early as 1905, when a law was passed, imposing a penalty on any owner of a well who should allow oil or gas to escape unused for a period of over thirty days. This law was amended in 1913; and in 1919 a stricter law for the conservation of gas was passed. Montana, as might be expected, was late in passing any conservation laws; however, the first law regulating the drilling and plugging of oil and gas wells was passed in 1917, before there was any important production. In 1921 an act was passed, requiring owners of abandoned wells to file affidavits of their location. Colorado, although one of the early oil producers, was very slow in legislating for the protection of her oil deposits, passing her first law in 1915.

Some states without any important oil production have passed laws in anticipation of development. Utah and New Mexico have laws regarding the casing and plugging of oil wells; and even Nebraska, entirely devoid of oil, as far as present information goes, passed a law in 1917, providing that the State Conservation and Welfare Commission should preserve the logs of all deep wells drilled in the state. Nevada offered a bounty in 1901 to anyone who should discover oil or gas in the state; but Washington, about the same time, provided for a system of leases of state lands, and Oregon was so illiberal as to demand a license tax of 3 per cent of the gross earnings of oil companies doing business in the state.

Thus most of the oil states have shown some interest in the conservation of oil and gas. Some of the laws have merely prohibited waste. Some have fixed a time within which new wells must be controlled, and the time permitted has varied from two days to three months. Many of the laws have prescribed certain regulations regarding the casings of wells in drilling, to shut out water from the oil sands; and many of them have prescribed with some minuteness the manner in which abandoned wells should be plugged, with the idea of guarding the oil sands. A number of the laws have recognized that the rights of adjoining owners are involved, and have permitted them to enter and plug a well properly, if the owner refuses to do it. Some states, as for instance Louisiana, Oklahoma, and California, have provided an efficient agency to enforce the laws against waste, but many have failed to do this.

An increasing number of states have seen the logic of a leasing policy for lands belonging to the state. As early as 1883 Texas passed an act reserving to the state all the minerals in her public lands (when Texas was admitted to the Union, she retained her own public lands), and in her University and Asylum lands; and laws passed in 1907 and 1913 provided that in future sales of state lands, all minerals should be reserved to the state, and should be extracted under a system of leases. These laws have been expanded and elaborated since. It has often been said that the Texas leasing laws are so drastic as to be rather uninviting to oil prospectors. There is nothing unfortunate about this, for the state will be wealthier in the long run for every oil field that remains for the present undiscovered. Texas has several million acres of state lands, some of which are probably oil-bearing.[31] In 1901, Washington provided that the State Commissioner of Public Lands might execute leases for the extraction of petroleum on state lands.

In 1909 Oklahoma passed a law, providing that state lands should be subject to lease, and in 1917 adopted a complete leasing law. In 1913, the legislature passed a resolution, submitting to a vote of the people the question of leasing river beds belonging to the state, but the Governor disapproved of this. In 1917, a law was passed providing for the leasing, for oil and gas exploitation, of "any of the school or other lands owned by the state"; and in 1919, the leasing of river beds for similar exploitation was provided for. Some very important oil territory happened to be situated along the Red River, and, since Texas also had a law claiming river beds as state land and providing for leasing by the state, the contest over the bed of the Red River became a state affair. One quarrel between Oklahoma and Texas became so bitter that armed men appeared on the scene, and the affair took on the aspect of civil war before the Supreme Court of the United States finally decided it. Oklahoma owns some valuable oil lands which she received from the government as school lands, when she entered the Union, and the revenues from the leasing of such lands are a considerable item in the state budget.

Wyoming has some state oil lands, including the valuable Section 36 in the Salt Creek field; and the state royalties from leasing of such lands have amounted to a total of several million dollars. Section 36 was leased for some years to the Midwest Refining Company at a royalty of 33⅓ per cent, the company also contracting to buy the royalty oil at a minimum price of $1.50 a barrel. In April, 1923, this section was reported re-leased to the Midwest, at an extraordinary royalty of 65 per cent of the oil produced. In the competition, the Midwest submitted two other bids, one of them offering a cash bonus of $5,000,000 and a royalty of 12½ per cent. With her receipts from state oil lands and from federal lands within the state, Wyoming enjoys a very handsome state income; and some of the natives look forward to a future "taxless" state. Montana likewise leases some of her state lands, although her royalties are not yet of any importance.

California unwisely disposed of most of her valuable state lands, but she now has a leasing law for mineral lands belonging to the state; and there has been some agitation for the leasing of small tracts of land which the state happened to own in the Southern California fields.

Michigan never had any important oil or gas production, but in 1913 the legislature of that state passed a law reserving all minerals, including oil and gas, in future sales of state lands. In 1912 Louisiana passed a law authorizing the Governor to lease mineral lands belonging to the state; and in 1915 a complete leasing law was passed, providing for the leasing of state lands, including lake and river beds and "other bottoms." In 1915, Ohio also provided for leasing her state lands, and in 1919, Kansas passed a similar law. Not long afterward, Utah authorized the State Board of Land Commissioners to sell or lease state lands. New Mexico likewise recently provided for the classification and leasing of her state lands. Oregon has considerable areas of timber lands which may some day contribute materially to the support of her educational system.

It may be worth noting that at least two states have provided for leases of other resources than oil. Georgia, as early as 1884, passed a law permitting any discoverer of phosphate in navigable streams of Georgia, or in the public land on the margin of streams, to mine the phosphate on payment of a fee or royalty of $1 a ton to the state. Florida provided in 1911 that where state lands were sold, there should be a reservation to the state of a three-fourths undivided interest in all phosphate, minerals, and metals in the land, with the privilege of mining the same. Minnesota has for many years reserved to the state the title to all iron, coal, copper, gold, and other valuable minerals, as well as water powers, on all state lands sold or otherwise disposed of.

Not only states but cities have "oil domains." Long Beach, California, has leased some oil lands belonging to the city at a 40 per cent royalty, and is enjoying a princely revenue, so great indeed that the city officials have been much at a loss as to how to spend it. Huntington Beach has been receiving a handsome income from royalties on school lands, the royalty in at least one case amounting to 50 per cent of the production. Redondo also has some land, but has wisely declined to lease it until the era of 60 cent and 80 cent oil is past. The city of Los Angeles has some land which would perhaps yield oil, if leased.

The gross production taxes levied by some states on all the crude oil produced is doubtless to some extent indicative of an interest in the conservation of this resource. To some extent, such taxes have been levied to secure revenues, but the idea of conservation has generally been present also. These will be considered in a later chapter.

The United States Geological Survey and the Bureau of Mines have for many years done a great deal of conservation work. These government bureaus have furnished much of the leadership in the movement for conservation of oil. Such men as George Otis Smith, David T. Day, David White, and many others have been responsible for much of the public interest in the conservation of oil and gas.

The United States Bureau of Mines has given a great deal of attention to the investigation of waste of oil and gas and of methods of preventing such waste. The Petroleum Division of the Bureau of Mines was created in 1915, and in the years following it devoted much attention to the problems of elimination of waste in production, underground wastes, increasing recovery from the oil sands, and other problems. In December, 1917, a petroleum experiment station was established at Bartlesville, Oklahoma, under the direction of the Bureau of Mines, to study various questions relating to oil utilization. Congress appropriated $250,000 for salaries, equipment, and maintenance, the legislature of Oklahoma appropriated $25,000, and the Bartlesville Chamber of Commerce provided $50,000 for the erection of buildings.[32]

The Bureau of Mines has done much of its work in coöperation with various state and private interests. Such work has been carried on in nearly every oil-producing state. In 1919, for instance, two government experts were sent to the Wyoming fields to investigate conditions there, the Midwest Refining Company and the Ohio Oil Company setting aside $60,000 for the expenses of the work. Coöperative work of this kind is generally more effective than work done entirely at the expense of the government, because it enlists better support from those in the industry.

Finally, it should be said that many private operators have evinced considerable interest in the conservation of oil, the Standard companies taking the lead in this. The American Petroleum Institute, the national association for the oil industry, has worked for better conditions in the industry.

NOTES

1. Second Edition, p. 234-.
2. U. S. Geol. Survey, *Bul. 623,* p. 22.
3. p. 175.
4. p. J. 8.
5. Pa., Second Geol. Survey, 1880 to 1883, *Geol. Rept. on Warren County and Neighboring Oil Regions,* p. XIV, IIII.
6. *Tenth Census,* Vol. 10, 267.
7. *Mineral Resources,* 1892, 610.
8. Pa. Geol. Survey, *Ann. Rept.,* 1886, Pt. II, 627. During the eighties the average production of the Pennsylvania wells was decreasing very rapidly. From an average of 7.8 barrels in 1875, the production declined to 3.15 barrels in 1884, and to 2.80 barrels in 1886. This decline attracted considerable attention. (*Mineral Resources,* 1886, 449.)
9. *Transactions,* Am. Inst. of Mining Engineers, Vol. X, 1881-1882, 354, 360.
10. *Transactions,* Am. Inst. of Mining Engineers, Vol. XIV, 1885-1886, 427.
11. *Mineral Resources,* 1883, 210.
12. Cal. State Mineralogist, *Fourth Ann. Rept.,* 1884, 280.
13. *Mineral Resources,* 1887, 459.
14. Indiana, Dept. of Geology and Natural Resources, *Rept.,* 1888, 201.
15. *Mineral Resources,* 1893, 462; 1895-1896, 653.
16. *Chambers' Journal,* Oct. 26, 1895, 678-.
17. *Pac. Oil Reporter,* Nov. 14, 1903, 7; Nov. 21, 1903, 3; Nov. 28, 1903, 7; Jan. 16, 1904, 4; *Cassier,* Dec., 1902, 292.
18. Indiana, Dept. of Geology and Natural Resources, *Rept.,* 1895, 13; 1896, 7, 11, 94.
19. *Eng. and Min. Jour.,* Jan. 7, 1904, 56.
20. *Cong. Rec.,* Feb. 13, 1907, 2806.
21. *Min. and Sci. Press,* Sept. 5, 1908, 305; *Eng. and Min. Jour.,* May 23, 1908, 1061.
22. *Min. and Sci. Press,* Oct. 16, 1909, 510, 511; Jan. 1, 1910, 3.
23. *Eng. and Min. Jour.,* Apr. 3, 1909, 716.
24. *Eng. and Min. Jour.,* Feb. 20, 1909, 421.
25. *Proceedings,* Conf. of Governors, p. 52.
26. *Mineral Resources,* 1915, 559.
27. All citations to state laws have been omitted here, in order to conserve space. Since the dates are usually given, however, it will be easily possible to refer to the proper session laws.
28. A log is a record of the geological strata encountered in drilling.
29. *Cal. Oil Fields,* Aug., 1920, 11.
30. U. S. Geol. Survey, *Bul. 429.*

31. It is not true that Texas has passed no title to mineral deposits since 1883, for during a number of years previous to 1913, it appears that public lands which had not been classified as mineral were sold with no reservation of the minerals they might contain.

32. *Oil Weekly,* Oct. 17, 1924, 29.

CHAPTER XXI

THE PUBLIC OIL LANDS: MANNER OF DISPOSITION: THE PLACER LAW

THE history of oil development on private lands has been traced; and it is not a story of which thoughtful Americans will be very proud. It will now be pertinent to trace the history of oil development on the public lands.

In the land policy of the United States, the general tradition has been that of private ownership. Under the common law of England, mines of gold and silver were the exclusive property of the Crown, and did not pass under a grant of land by the King. Thus the colonies inherited a certain tradition in favor of public ownership of mineral deposits, and for a while some of the colonies adhered to this policy, as to gold and silver. Later, however, they all abandoned it. In the territory ceded by the individual states to the United States, however, and disposed of under the Land Ordinance of 1785, "one-third of all gold, silver, lead and copper mines," were reserved to the federal government, "to be sold or otherwise disposed of," as Congress should direct; and salt springs, saline lands, lead, copper, and other mineral lands were reserved in certain states or territories. During the next half-century or more, however, practically all such mineral lands were sold, and all of the minerals passed into private ownership. Thus oil exploitation in most of the states, as described in preceding chapters, was carried out on privately owned lands, the owners of which held the title to all the deposits beneath the surface.[1]

If all of the United States had been fit for cultivation, probably all of the land, even in the western states, would have been taken up by settlers, under laws which gave the surface owner title to all the mineral deposits; and the federal government would never have owned any oil lands. It was, in general, only lands not known to be of any particular value which were left to the government. Those which were thought to be worth taking were taken by private interests, by bona fide settlement, by grant, by purchase, or, in not a few instances, by fraud and theft. Some of the ways in which valuable oil lands on the public domain passed into the hands of private owners must be considered in detail.

Railroad land grants have been a very important factor in determining the present ownership of oil lands, just as they were in determining the ownership of timber lands; and these grants have been the basis of some large oil-land holdings, just as they were the basis for some of the largest timber-land holdings. In California, in particular, railroad grants were responsible for some of the largest holdings of oil lands— those of the Southern Pacific, Santa Fé, and, to a lesser extent, the Standard Oil and other companies. The Federal Trade Commission estimated that in March, 1920, the Southern Pacific interests, or- ganized as the Pacific Oil Company, with the Associated Oil Company, held over 28,000 acres of proved oil land in California, and the Santa Fé Railway owned over 4,000 acres. These two railway interests held 35 per cent of the proved oil lands of California, and some of the hold- ings of other companies could be traced to railroad land grants. In no other region have railroad grants been so important, but the Union Pacific has owned some oil lands in Wyoming, and the Northern Pacific has some land in Montana which may prove to be valuable for oil.[2]

The government tried for years to regain title to some of the South- ern Pacific lands in California. The Act of July 27, 1866, under which the Southern Pacific Railroad had received patents to the land, ex- pressly excepted and reserved mineral lands, except those containing coal and iron, from the operation of the grant. Notwithstanding this, the railroad filed its selected list for agricultural lands and received patent. When it presently appeared that the lands contained valuable oil deposits, Attorney-General Wickersham brought suit to regain 6,000 acres of land in the San Joaquin Valley, alleging that the railroad officials and agents knew that the lands were mineral in 1904, when the company selected them. This was known as the Elk Hills suit. Six other suits were later filed, involving about 160,000 acres in the San Joaquin Valley, which had been patented in the early nineties. These six suits were afterward consolidated and heard as one case.

The first bill of complaint in the Elk Hills case was filed in December, 1912. This case involved lands said to be worth $18,000,000, and it was stubbornly fought by the Southern Pacific, and came up for hearing a number of times. Finally, in June, 1915, Judge Bledsoe, of the federal district court, rendered a decision favorable to the government, holding the lands to be mineral in character, and holding that the patents had been procured through fraud, since the railroad officials knew that the lands were valuable for oil when patents were applied for. The company appealed to the circuit court of appeals on February 1, 1916, and on May 6, 1918, this court reversed the decision of the district court. On August 3, 1918, Attorney-General Gregory docketed an appeal to the Supreme Court of the United States, where a decision was rendered on November 17, 1918, reversing the circuit court of appeals, and re- turning the lands to the government.

The consolidated suit, involving about 160,000 acres of land, said to be worth over $500,000,000—probably a gross exaggeration—was submitted to the district court in April, 1919, and on August 28, 1919, a decision was handed down by the district judge, upholding the patents to the Southern Pacific. This decision was based upon evidence that the railroad company had offered to sell, and had even actually sold some of the lands at a few dollars per acre, and could therefore not be assumed to have known that they were oil-bearing. Some of the land in the city of Los Angeles which was later found to be oil-bearing was acquired from the Southern Pacific. Attorney-General Palmer therefore announced that it would be useless, or perhaps even "frivolous," to appeal.[3]

Whether Palmer was justified in refusing to appeal this case is a question which might well be probed a little. He has been severely criticized for his failure to make more effort to save these lands for the government. Senator Thomas Walsh indicated his very definite belief that the facts in this latter suit were the same as in the suit which the government finally won, and that the government probably would have regained at least three sections of the land involved in the suit, if Palmer had appealed to the Supreme Court. Whatever may be said as to Palmer's record in this matter, it is unlikely that it is any worse than his record on many other questions. The Southern Pacific Railroad Company still owns most of its original large holdings—the largest acreage of proved lands held by any oil company in California.

The position of the Southern Pacific had been somewhat fortified by a decision of the United States Supreme Court, rendered several years before. A man by the name of Burke had contested the title of some of the Southern Pacific lands alleging that he had rights in them by virtue of early placer claims, but the Supreme Court held that the patent which had been given the railroad could not be attacked by any but the government, and dismissed his suit.[4]

It has been charged that there was fraud at various points in the Southern Pacific case. One charge made was that the California Miners' Association had tried to get the lands examined by the government and the mineral lands segregated, but that the influence of the Southern Pacific in Congress was strong enough to prevent such action.[5] Whether there is any truth to this—and it is easy to believe—it is certain that the railroad company tried to select mineral lands, in violation of the terms of its grants. Several cases involving the question of their right to do this were carried up to the Secretary of the Interior for decision.[6] It has also been charged that, for some reason, government officials were only half-hearted in the prosecution of the railroad company. It was charged that there was fraud in the classification of some of the railroad lands, fraud running into various divisions of the Land Office; but of course the government could not well offer as a ground for cancellation of the land patents the fraud of its own agents.

As to the justice of the final decision in the Southern Pacific case, the editor of the *Mining and Scientific Press* expressed a well-reasoned and judicious conclusion:

> In view of the fact that the intention of Congress was not to give mineral lands, and that the officials of the company in accepting them evidently believed that only agricultural lands were received, there would seem to be an equitable reason for bringing the matter before the courts and restoring to the United States the title to the mineral in the ground if possible. The railroad would retain all that it bargained for, no more and no less. It is not the part of good government to unsettle titles unnecessarily, but if the title conveyed be in fact, as it seems, a restricted one, there is no injustice in enforcing its terms.[7]

Some of the independent oil producers of California claimed, in the years 1903 and 1904, that the railroads of California, particularly the Southern Pacific and the Santa Fé, with the Standard Oil Company, were trying to gain complete control of the important oil lands of that state, and that they had even purposely tried to depress oil prices in order that they might buy more advantageously. The United States Commissioner of Corporations indicated his belief that there might be some justice in this charge.[8]

The Southern Pacific Railroad is not the only railroad that has secured oil lands in its grant. The Union Pacific grant running through southern Wyoming covered considerable areas of land later found to be oil-bearing. The railroad company has claimed the title to the minerals in its grant, at least against mineral locators who have tried to prospect on these lands.[9]

The Atchison, Topeka, and Santa Fé Railroad Company owns extensive areas of oil lands in California. The Federal Trade Commission, in March, 1920, credited this railroad with 4,043 acres of proved lands, or 4.4 per cent of the proved lands of the state. In the year 1919, this railroad company produced 3,754,299 barrels of oil in California, or 3.7 per cent of the total for the state. The Santa Fé also has oil lands in Oklahoma.

The Missouri, Kansas, and Texas Railway Company tried to lease its right of way near Cleveland, Oklahoma, to an oil operator, on the theory that its title in the right of way amounted to a fee; but the Secretary of the Interior held that the railroad's title in its right of way was only an easement, and that it could not be used except for railway purposes.[10]

Grants to the states, like grants to the railroads, practically always reserved mineral lands, yet some valuable oil lands passed to the states, and thence into private hands, through such grants. When the federal government gave school lands to the various states, it declared that such lands, or the "indemnity" lands (which were sometimes selected where the sections originally granted were not available) must be non-mineral; and the discovery of minerals on such lands before patent had passed rendered them unavailable for selection. In some instances, however, mineral discovery was not made until the state had received patent, and in such cases the title of the state was held to be good. Perhaps the most important case of this kind was Section 36 in Naval Reserve No. 1, in California. This school section was patented to the State of California, and later found its way into the hands of the Standard Oil Company. In 1919, the Standard Oil Company found oil on the section, and it presently proved to be very rich in oil, but Secretary of the Interior Fall decided it was impossible to question the title to the land, since it was not known to be oil-bearing at the time the state received patent.

The State of Wyoming also has some valuable oil lands which it received before they were known to be of any particular value, including one section in the Salt Creek field which yields the state a handsome royalty. Oklahoma has a large amount of school lands which have been proved as oil-bearing, and several western states own lands which may prove to contain oil.[11]

Thus some of the oil lands on the public domain passed into private hands through the operation of various public land laws, and through railroad grants and state grants. In order that title might pass under such laws or grants, it was necessary that the oil lands should not be known as such. Some of the oil lands were known as oil lands, however, and they were disposed of under different laws, or were finally reserved by the government.

The first official report of the occurrence of oil on the public domain seems to have been that of January, 1865, in a letter from the register and receiver at Humboldt, California, to the Commissioner of the Land Office. The Commissioner in reply directed that tracts of land valuable for petroleum deposits be withheld from disposition unless further specific instructions were issued. How long this suspension was in force is not evident from available data.

For many years after oil had been noticed on the public lands, there was no definite, consistent policy in the Department of the Interior as to how it should be disposed of. The Placer Act of 1870[12] had provided for the mining of "claims usually called 'placers,' including all forms of deposit excepting veins of quartz, or other rock in place." This law was doubtless meant merely to include placer deposits, as contrasted with lode deposits, but its language, covering "all forms of deposit," was of course broad enough to take in practically everything; and the Land Department very soon applied the new law to various minerals, including oil. On July 15, 1873, Commissioner Drummond, of the Land Office, issued a circular in which he stated that "lands valuable on account of borax, carbonated soda, nitrate of soda, sulphur, alum and asphalt, as well as 'all valuable mineral deposits' " might be patented under the Placer Act. On January 30, 1875, Commissioner Burdett ruled that oil might be patented under the law; and the first patent to an oil claim, according to one report, was issued on March 22, 1880, in the Los Angeles district, California. Two years later, Commissioner McFarland again declared the rule that petroleum lands might be entered under the Placer Law.[13]

These decisions were, however, all announced by the Land Office, and were, therefore, subject to review by the Secretary of the Interior. In December, 1883, Secretary Teller referred to the question as to whether oil should be included as a mineral under the Placer Law as "an undetermined question." During these years, in spite of his uncertainty, he allowed oil lands to go to patent under this law; and in December, 1885, Secretary Lamar ruled that oil lands might be patented thus. This decision stood for over a decade, but in August, 1896, it was reversed by Secretary Hoke Smith, who, after a careful consideration of previous decisions, ruled that land containing petroleum did not fall within the contemplation of the mineral laws, and could not be located and entered under the Placer Law.[14] This decision was afterward reversed by the Secretary of the Interior, but at the time it fell like a bomb among some of the oil prospectors of the West, and several of the western men in Congress promptly introduced bills to permit the entry of oil lands under the Placer Law. Senator Clark and Representative Mondell, of Wyoming, and Senator White of California, agitated the matter earnestly, and secured such a law without difficulty, and almost without debate. It was clear that Congress did not recognize the evils of the Placer Law in its application to oil lands.[15]

The Placer Law provided that any individual might file a location on 20 acres of mineral land, or an association of eight or more persons might file on eight claims aggregating 160 acres, and by expending a certain amount each year in "assessment work" and by finally making a discovery, might secure title to the land on payment of $2.50 per acre. The law was not adapted to the exploitation of oil and gas for several reasons: in the first place, it gave prospectors no definite rights until discovery; in the second place, it required the performance of assessment work regardless of the need for oil; in the third place, it provided for the disposition of tracts too small for efficient operations, and so made it necessary for the oil operators to use dummy entrymen to get large enough tracts.

The Placer Law gave no rights to prospectors until discovery was made. In placer gold mining, this was a reasonable provision, but in drilling for oil, it was absurd. As Thomas O'Donnell once said:

The placer miner looking for gold, could go along with a shovel and turn over a little gravel, and get a color of gold, and he had then made the necessary discovery. But our petroleum in California is in many instances 4,000 feet under the earth.[16]

The drilling of a well, perhaps 4,000 feet deep, required several months' time and the expenditure of a large sum of money—sometimes as much as $50,000. Yet, during their discovery work, operators had to drill on land to which they had no title or right whatever, as against the government, and to which they had no rights under the law, even against strangers, until the courts conferred an uncertain measure of protection. Thus operators under the Placer Law were always subject to the danger that, after they had expended thousands of dollars in drilling, someone else might make discovery first and so take the claim. In a few cases two operators drilled on the same claim, the first to discovery getting the title. In many cases, oil prospectors found their land taken from them by claimants who asserted discovery of asphaltum, or gypsum, or some other mineral mentioned in the Placer Law. A western oil journal gives an interesting story of the acquisition of some oil lands in Wyoming by use of the coal land laws.

In parts of California, the iniquities of the Placer Law were joined to other unfortunate provisions of the land laws to create a situation closely bordering on chaos. In the spring of 1899, the discovery of petroleum in the Kern River field, adjacent to Bakersfield, led to tremendous excitement and to a great rush for the lands—lands which had previously been about worthless because of the aridity of the climate. In this region, alternate sections of the land belonged to the Southern Pacific Railroad, but title to the government sections was sought in two ways: first, through the Placer Law, and, second, through the use of various kinds of "scrip" and land rights.

There were several kinds of scrip and land rights available: "old soldiers'" scrip, or military bounty warrants, originally issued for military service, to the amount of over 61,000,000 acres; railroad indemnity rights, accruing to land-grant railroads for land within the limits of their grants found to be in the possession of settlers; and Forest Lieu rights secured under the Forest Lieu Act of 1897. Much of the "old soldiers'" scrip was assignable, so could be used anywhere; and the land-grant railroads had millions of acres of rights which they tried to apply to valuable lands of all kinds. The Forest Lieu Act had provided that the owners of land included within the boundaries of forest reserves might release their land and select elsewhere "vacant and unoccupied" land anywhere on the public domain. Entrymen under the Placer Law had no title whatever until they made actual discovery, while the "scrippers" had only to show that the land was "vacant and unoccupied." Many of the "scrippers" were not the original holders of their scrip or rights, but were speculators who had bought from the original holders, and were using it to levy tribute on the mineral entrymen. Since most of the land in California was almost worthless for agriculture, the scrippers could make no use of the scrip themselves, under any pretense of selecting agricultural lands, but could sell it to oil operators, or compel them to pay for the release of their rights. These scrippers infested many of the oil fields of the West for years, in fact, they frequented all kinds of lands, ready to seize valuable tracts wherever possible.[17]

Among the important scrippers in the Kern River field were William H. Crocker, a prominent capitalist of San Francisco; William Singer, Jr., of the Southern Pacific Law Department, and a number of lawyers—for a situation like this always makes business for lawyers. One agent of the Crocker interests, while trying to file on lands near Bakersfield, was set upon by a hundred men, but was released uninjured. Certain commissioners sent to the fields to report on the situation were reported to have been bribed by the mineral locators to report the discovery of oil. In one district of California an association of claim owners employed a thousand guards to keep the scrippers off. In the San Joaquin Valley, there was much trouble with "claim jumpers," some of them armed, and the Governor was appealed to for aid in restoring order. In many instances, oil companies had to take up "gypsum claims" in order to hold their land until they could test it. In some instances placer mining locators, and scrippers—or "jumpers," as they were sometimes called—drilled on the same location, and the one who struck oil first got title to the land.

The difficulty of securing peaceable possession of claims sometimes led to physical threats and violence; and even where no violence arose, efforts of two different prospectors or drillers to find oil on the same claim involved a wasteful duplication of labor and expense. Furthermore, the necessity of making the discovery as soon as possible under such circumstances, led to hurried and reckless drilling, without proper regard for the protection of the oil sands.[18]

Against agricultural claimants, placer prospectors were finally given a measure of protection by court decisions. In the case of the Olive Land and Development Company vs. Olmstead,[19] the court held that, previous to discovery, a placer entryman had no rights, either against the United States or against other entrymen; but four months later, in November, 1900, in the case of Cosmos Exploration Company vs. Gray Eagle Oil Company,[20] which was affirmed by the Supreme Court of the United States,[21] the court declared that land which was being explored under the Placer Law was not "vacant and open to settlement" by entrymen using Forest Lieu scrip. This decision would be somewhat more significant if oil had not been discovered before the agricultural selection had been approved; but a number of later cases followed the same rule.[22] Much of the oil land of the western states was worthless for ordinary agricultural production, but that did not always discourage entrymen. Only the most serious obstacles are effective in discouraging men who are seeking title to oil lands.

It may be noted also, that while, under the decision in the Gray Eagle case, the prospector was protected against adverse agricultural claimants, there was nothing of record in the Land Office to show that the land which he occupied was not "vacant and open to settlement," and such agricultural filings were likely to be made at any time, subjecting him to the expense of maintaining contest proceedings before the Land Office or in the courts.

The extent to which homestead entries, enlarged homestead entries, and desert land entries, were used in an effort to secure oil lands may be judged by the number of homestead entries in many parts of Wyoming, where there is not a possibility of successful agriculture. On a day's travel through the state one may see a hundred little shacks on abandoned claims, many of them evidently never designed for habitation. Some of these claims were filed by hopeful men who thought they could make a living by dry farming; and some were filed by entrymen who sought land for grazing purposes; but some of them were filed by men who hoped thus to get title to land which might later prove to be valuable for oil. Many western "pioneers" were naturally hopeful.[23]

In order to protect oil prospectors better, the Commissioner of the Land Office issued a number of withdrawal orders in the years 1900 and following, withdrawing from agricultural entry considerable areas of land thought to be valuable for oil. These withdrawals did not stand long, however, for they included some agricultural land, and agricultural entrymen called insistently for their restoration. Furthermore, there was soon a reaction against them in the Department of the Interior. The agents sent out to investigate the fields reported as non-oil land nearly every tract upon which there were no derricks; and many of the withdrawals were presently restored to agricultural entry.[24]

During the very time when these latter withdrawals were being restored to agricultural entry, geologists from the United States Geological Survey, investigating other oil fields in California, were recommending further withdrawals. David T. Day, Ralph Arnold, and A. C. Veatch were in California at this time investigating oil resources for the forthcoming report of the Conservation Commission. When they saw the situation existing on the public oil lands, they urged that some of them be withdrawn from agricultural entry; and in 1907, certain lands were withdrawn, including some that had just been restored to agricultural entry, and some that had not previously been withdrawn. All these lands were withdrawn to protect the oil operators and not to conserve the oil. Additional withdrawals in California were recommended by the Geological Survey and were approved by the Secretary of the Interior in 1908 and 1909; a petition from a number of people in Oregon resulted in the withdrawal of 74,000 acres in Oregon; and 6,599 acres in northwestern Louisiana were withdrawn in December, 1908, on recommendation of the Geological Survey, to prevent waste of natural gas.[25] In the Louisiana withdrawal appeared for the first time the purpose of conserving the mineral resources. The other withdrawals were made with the idea of protecting oil prospectors from agricultural entrymen.

The fight with the scrippers got into Congress, as a great many oil troubles did. Representative Dick, of Ohio, introduced a bill in Congress to permit the defeated scrippers to go before the Court of Appeals of the District of Columbia for a rehearing, but the California Petroleum Miners' Association held a meeting in San Francisco shortly afterward to fight the measure, and it received no further consideration.[26]

Against other mineral claimants who attempted by forcible, fraudulent, surreptitious, or clandestine means, to violate the possession of an oil prospector, the courts finally afforded protection; but they at the same time declared that, prior to discovery, the possession of a mineral claimant was not good against that of an adverse claimant who entered in good faith.[27] A great deal of oil land was taken up under various mineral land laws. One large area was held for some years under building stone placer claims, although much of the area was covered by dune sand, and no building stone of particular value was known in the region. In some of the fields, especially in California, there were deposits of gypsum of fair quality upon the surface of valuable oil lands, and a great many gypsum claims were patented, when it was common knowledge that oil was the valuable resource sought. In this way, land worth thousands of dollars per acre was acquired from the government for a consideration of $2.50 per acre, and without any development at all.

A second fault of the Placer Law was that it did not permit the locator to hold his land under any circumstances without doing the actual work of development. The law required that discovery must be made, and at least $500 (later $100) expended on each claim; and this requirement demanded an expenditure in what was sometimes worse than useless drilling. For instance, after the opening of the Lake View gusher, nothing was so little needed as more oil, but placer locators were required (at least nominally) to perform their development work just the same. In 1885, Secretary Lamar liberalized this provision somewhat, by ruling that, in case of application for patent to several locations, the expenditure of $500 need not be made on each location, if enough were expended on one of the locations to average $500 for all of the locations involved.

The Supreme Court of California, in September, 1903, likewise held that where eight men formed an association and entered 160 acres of land, one discovery for the entire 160 acres was sufficient.[28] This ruling was clearly contrary to the letter of the law, and it was not consistently followed by the Department of the Interior. In 1896, efforts were made in Congress to change the Placer Law so that it should require discovery only on each 160 acres, but without success.[29]

In 1903, however, Representative Sutherland of Utah secured a law known as the "Five Claims Act," providing that the annual assessment work on oil claims might be done upon any one of a group of not more than five claims "lying contiguous and owned by the same person or corporation." This law passed Congress, it may be pointed out, without a single word or suggestion indicating any concern for the public interests involved in oil land legislation.[30]

Requirements as to assessment development work were generally not strictly enforced. In some of the fields, particularly in the Rocky Mountain States, almost any kind of pretended development work was for some years sufficient to hold a claim for an indefinite period. Here many subterfuges were adopted by those who wished to hold possession of valuable petroleum lands without the expenditure of the large sums necessary to make discovery. Rigs capable of drilling only a few hundred feet were erected for the purpose of doing assessment work on lands beneath which the oil sands were known to lie at depths of thousands of feet. Tunnels were dug in hillsides to tap oil springs or seeps in the attempt to obtain, without much expense, enough oil to constitute a discovery. In at least one western state, large areas were claimed for years under assessment work in the form of road building. Some of the roads thus built facilitated transportation to the oil fields, but some of them had only the merit of entailing the expenditure of the required amount of money. In 1901, W. C. Knight described conditions in the Arago field, in Wyoming, as follows: "The development has not amounted to anything up to date. The numerous shafts and tunnels answered for assessment work only, and did not in the slightest degree improve the property." Knight found a similar situation in the Dutton field.

The Placer Law, as applied to exploitation of oil, was seriously defective in providing for the disposition of tracts altogether too small for an efficient industry. The law permitted individuals to file on tracts no larger than 20 acres; and eight individuals might file on a claim of 160 acres. Even a tract of 160 acres was entirely too small for efficient exploitation, and prospectors had no preferential rights in the adjoining claims. In case they struck oil, they were almost sure to find claimants on adjoining territory, unless they had wisely and illegally, and successfully, initiated some sort of claim by means of dummy entrymen. Thus the law provided for holdings entirely too small for an efficient or economical system of oil exploitation. Unless prospecting companies, in direct violation of the law, were able to secure fair-sized, solid blocks of land, the discovery of oil could result in nothing else than a typical American drilling spree.

The limitation of claims to 20 acres, or 160 acres for an association, was a typical example of congressional edicts contrary to economic law; and it was constantly evaded. Oil operators wanted and needed larger tracts, and they secured them by various means, sometimes in evasion of the law. Secretary Teller held, in 1883, that patent could issue to only 160 acres for an individual or corporation, but, two years later, Secretary Lamar ruled that an application for patent might embrace more than one placer location, and, in a case brought before him, specifically authorized the issue, to one individual, of patents to four claims of 160 acres each. This was practically an authorization of patent for as many locations as anyone happened to want to buy. As one writer expressed it, "If a prospector is so inclined and has no competition, he may keep staking out claims until he or the mineral lands is exhausted." Representative Smith of California once confessed that "a man with a reasonable supply of relatives" could get about all the land he wanted.[31]

It was common practice for individuals to file locations, and then transfer to oil companies, which were in this way able to secure large holdings. Thomas O'Donnell once described the way in which, in some of the little towns in the California oil fields, on midnight of January 1, many saloon keepers and hangers-on would go out and "locate the whole country," and then ask a bonus from the operator, or try to organize a small company to drill on the lands.[32] Following the discovery of oil in Kern County, California, in 1899, the entire countryside for miles around was plastered with locations and relocations. One group of speculators, employing 25 or 30 locators, located approximately 50 square miles. Everything unpatented was taken in.[33]

. One report from Kern County showed that in that county alone, within a year, 103 persons located 8,248 placer mining claims—an average of over 80 claims, of 20 acres each, or 1,600 acres per person, covering a total of 164,960 acres. In this list were five persons whose claims numbered between 200 and 250; seven with from 150 to 200; 13 with between 100 and 150; 41 with between 50 and 100; and 37 with 30 to 50 claims. Three members of one family located 220 claims, or an average of 73 each; six members of another family located 565 claims, an average of 94 each; seven members of another family located 885 claims, an average of 126 each; and four members of another located 292 claims, or an average of 73 each.[34] Thomas O'Donnell stated before a committee of Congress in 1914 that over 90 per cent of all placer locators in California were "dummies."

A writer in the *Pacific Oil Reporter* in March, 1904, described the operation of the Placer Law as follows:

In not one case in a hundred is the work done with any view of discovering oil and is a means where conscienceless land hogs get large areas with no idea of developing. We know of one locality where an oil discovery was made in 1900, and all the government land for 50 miles around was filed on and is held in this manner; and in some cases 40,000 acres have been filed upon by one party through the use of "power of attorney." A party has only to get seven disinterested parties (often relatives from the East) to sign a paper giving him power of attorney and he gets busy and files on every quarter of a section he can get hold of, using the seven names and his own, giving each party the 20-acre placer claim allowed by law. Then he deeds the whole to himself and there you are. It is no unusual thing to hear of some syndicate taking up a hundred thousand acres of land this way.[35]

In 1889, 90 placer locations, covering or controlling 145,000 acres of oil lands were reported to have been made in Wyoming by a group of men from Omaha. In 1901, when oil was reported near Moorcroft, in eastern Wyoming, 60,000 acres were promptly claimed in the vicinity, mostly under the Placer Law. At this time the Geological Survey reported that nearly all the available petroleum lands in Wyoming, and probably a large amount of land that never would be found to contain oil, had been located.

In the past two or three years there have been many reminders of old fraudulent placer claims in Wyoming, particularly in the Salt Creek field. This field was suspected of being valuable for oil long ago, and during the eighties many claims were filed, although relatively few passed to patent. When Salt Creek rose to prominence as an oil producer, many of these claims were unearthed, and, with the aid of enterprising lawyers, were made the basis of lawsuits. In December, 1922, a Denver attorney, John T. Bottom, filed suit for $2,500,000 against H. N. Isenberg, who had bought some of the old Iba placer claims. Several other suits involving Salt Creek lands were likewise based on the old Iba claims, but most of them did not fare well in court. Bottom was, however, successful in a suit against the Midwest Refining Company, prosecuted in behalf of F. G. Bonfils, one of the owners of the *Denver Post*. A few months before this, Henry Miller filed suit against the Midwest and several other companies for oil lands in Salt Creek, and for oil produced therefrom, basing his suit upon placer locations made 35 years before. This suit was said to involve equities worth $43,-000,000, although this was probably an exaggeration. About this time, Patrick Sullivan, a prominent Wyoming politician and a personal friend of President Harding, filed suit in equity against the Mammoth Oil

Company and the Pioneer Oil and Refining Company, for 960 acres of land on Teapot Dome, to which he claimed title under the so-called "Roundup" oil placer locations made in 1907. Sullivan employed as his attorney John T. Bottom of Denver, who had become more or less of a specialist in Wyoming placer locations.[36]

These placer claims in the Salt Creek and Teapot Dome fields were almost uniformly fraudulent. One investigator reported on claims in Salt Creek and on Teapot Dome as follows:

As to assessment holes . . . only a very few witnesses could state that they had seen such assessment holes therein, and in that they were very indefinite as to location. It was the practice in Salt Creek to dig assessment holes four by six by ten feet deep. Although these holes tended in no way to develop the claims or to discover oil, they were considered in the early days to be the equivalent of $100 worth of work. . . . A hole of these dimensions could be put down in one or two days by an ordinarily vigorous laborer.

Much of this work was done after the land was withdrawn.[37] In many cases the Placer Law merely deprived the government of the lands, without conferring any benefit upon the locators, who exhausted their small means drilling, and then abandoned the land or sold out to some large company, often for less than they had spent.

For several years the oil operators seemed able to get about all the land they wanted, by a generous use of dummy entrymen, but suddenly, in July, 1909, a decision of the Secretary of the Interior threw some of the placer law locators into consternation. In the famous Yard case, Assistant Secretary Pierce held that placer locations transferred before discovery had been made were invalid, and could not be perfected by the transferee upon subsequent discovery. This meant that many oil companies, mainly in California and Wyoming, could not secure title to their land, even if they discovered oil. It made it almost impossible for some of these companies to get outside capital for oil development.[38]

The Yard case itself throws much light upon the public land situation in some parts of the West. The decision concerned 85 placer locations in the Plumas National Forest, in the Susanville land district in California, but these 85 claims were only a small portion of the several hundred placer locations—800 locations, according to one report— made by H. H. Yard in this region. The evidence showed that the entrymen were transferring their claims—many of which were valuable only for their timber—to H. H. Yard, and to the North California Mining Company, which he had organized to assist him.

Yard was trying to secure title to these locations in batches of 85 at a time. He was constructing telephone lines, wagon roads, trails, ditches, dams, and reservoirs, and there is some evidence to show that, while he was making surveys and doing some assessment work, his real design was to prepare a way for a new railroad—the Western Pacific—and that he was acting for the Gould interests. In his operations he was assisted by the California Miners' Association, particularly by the Secretary of that body, Edward Benjamin. Yard realized that it would be well to secure the sanction of some similar body in order to cloak his operations with some appearance of dignity; so he made Benjamin vice-president of the North California Mining Company which he had previously organized. Fortunately, the State Mineralogist of California, Lewis Aubury, became apprised of this situation, and he published some of the facts, and the case finally came up to the Secretary of the Interior, who ruled as indicated above. The Yard case itself involved lands more valuable for timber than for oil, as a matter of fact oil is not mentioned in the case; but the decision affected hundreds of oil claims, in other parts of California, and in Wyoming.[39]

The oil operators promptly took their troubles to Congress, sent a committee to Washington to protest against the Yard decision and to ask Congress for relief. Representative Smith of California introduced a bill authorizing the issue of patents to assignees of locators under the Placer Law, regardless of the Yard decision.[40] Smith claimed for the bill the hearty approval of the Department of the Interior, but Mann, of Illinois, was from the first suspicious of this "pig in a poke." James, of Kentucky, likewise opposed the bill, pointing out how rich corporations, when they found a law in the way of getting things, always came to Congress for special legislation. In spite of the opposition of these men, however, the rules of the House were suspended and the bill passed without even a call for the yeas and nays. In the Senate it passed without opposition and President Taft signed it in March, 1911.[41]

The Assignment Act allowed claims initiated prior to March 2, 1911, to pass to patent in spite of the Yard decision if "in all other respects valid and regular." It was exclusively retroactive; and so the Department of the Interior and the courts continued to require discovery on placer locations before a valid transfer could be made. In the Bakersfield Fuel and Oil Company case, the Secretary of the Interior ruled that the Yard decision still held, that, until discovery, entrymen acquired no interest in the public domain and therefore had nothing to convey.[42] This view was not consistently taken by the Department of the Interior, however, and many claims passed to patent through the location of dummy entrymen. In some cases the circumstances of the entry were not known to the department, but in many cases it was impossible that they should not have been known. It was impossible, however, for the Land Office to make careful investigation into all of the cases coming before it, no matter what its official ruling may have been.

Most of the land claimed or taken under the Placer Law was in California and Wyoming, but vast areas were filed upon in other western states. Secretary of the Interior Albert Fall estimated that in New Mexico alone over 5,000,000 acres of land had been located under this law. New Mexico "placer claims" were being sold in Tulsa in July, 1919, for 50 cents per acre. A number of years ago, an oil spring near Raven Park, Colorado, caused a stampede of oil seekers, and in the neighborhood of 45,000 acres or nearly 70 square miles of the public domain were taken up under the Placer Law. Upon failure to find oil, the locators were unable to get patents, but for years afterward the whole field was covered by unpatented locations, still divided up among the companies that had taken them.[43]

There were always a great many men in the West who defended the Placer Law in its application to oil lands. Representative Mondell never missed an occasion to comment on the excellent results that had arisen from its operation, and many other oil operators defended it. Thomas O'Donnell testified as to its excellence before the Public Lands Committee of the House of Representatives in 1910. The chief virtue in the operation of the law, according to these men, was that it compelled rapid exploitation—they called it "development." Since the operator had no protection from other operators until he found oil, he naturally drilled with great rapidity. The fact that such drilling was likely to be reckless and wasteful was not an important matter to "practical" men. Mondell spoke of the "pliability and freedom" of the Placer Law, in that it was easily adjusted to almost any kind of conditions, and lent itself to the development of new fields. In the hands of some operators it assuredly possessed "pliability" and as much "freedom" as was in accord with the public interest.

In general, the defenders of the Placer Law were those who had a selfish interest in the maintenance of the conditions existing on the public domain. Some were merely promoters who saw that it would be easier to float stock if they had a prospect for a definite title to certain lands, and who saw that if the government abandoned the Placer Law, it would probably adopt the leasehold system and would probably impose strict requirements to protect the public interests. Then there were numerous lawyers, in and out of Congress, who wished to let well enough alone, because they realized the general unsettlement that might result from the passage of any new law, or from any new departure in the government's land policy. Lawyers are usually conservative about any change in statutes, especially where they have interests which might be jeopardized by such change. Among all business classes, there is usually a deep distrust of government intervention of all kinds; and it was generally believed that when the Placer Law was repealed, it would be superseded by a system of government leasing.[44]

NOTES

1. U. S. Bu. of Mines, *Bul. 94,* Pt. II, p. 1080; Treat, *The National Land System,* p. 36-38; Knight, *Land Grants for Education in the Northwest Territory,* p. 90.

2. Federal Trade Comm., *Rept. on the Pacific Coast Oil Industry,* Pt. I, p. 45; *Wall St. Jour.,* Feb. 3, 1921, 1.

3. *Fed. Rep.,* 225, 197; 249, 785; 260, 511; *251 U. S.,* 1; U. S. Atty.-Gen., *Rept.,* 1912, 40; 1915, 37; 1916, 46; 1918, 92; Federal Trade Comm., *Rept. on the Pacific Coast Petroleum Industry,* Pt. I, 77-80, 221. See also the *Hearings* on the leasing of Teapot Dome, pp. 1533, 1581, 1587, 2419-2434; and *S. Rept. 794,* 68 Cong., 1 sess., pp. 27, 28.

4. Burke vs. So. Pac. R.R. Co., *234 U. S.,* 669.

5. *Eng. and Min. Jour.,* Sept. 25, 1909, 621; *Collier's,* Jan. 8, 1910, 18.

6. *Land Decisions,* 25, 351; 29, 269; 41, 264.

7. *Min. and Sci. Press,* Oct. 8, 1910, 461.

8. U. S. Comr. of Corporations, *Rept. on the Transportation of Petroleum,* 1906, 498.

9. *Oil and Gas Jour.,* Jan. 16, 1920, 46.

10. *Land Decisions,* 33, 470; 34, 504; 157 Southwestern, 737.

11. *Land Decisions,* 29, 181; 39, 491. For a history of the Wyoming section, known as 36-40-79, in the Salt Creek field, see *Hearings before the Senate Committee on Public Lands,* on S. Res. 282, Dec., 1923, p. 1599.

12. *Stat.,* 16, 217.

13. Much of the information here is taken from an article by Max W. Ball, in *Transactions of the Am. Inst. of Mining Engineers,* Vol. 48, 451. On the question as to whether oil and gas should be regarded as minerals, see Morrison's *Mining Rights,* 666, *Annotated Cases,* 1913 B, 1214, and *Am. and Eng. Annotated Cases,* Vol. 20, 937.

14. *Land Decisions,* I, 560; II, 707; IV, 60, 284; 16, 117; 23, 222; 24, 183; 25, 351.

15. S. 3471, S. 3475, S. 3551, H. R. 9606, 54 Cong., 2 sess.; Stat., 29, 526. Mondell afterward claimed credit for the law which was passed, and was always a staunch defender of the oil Placer Law.

16. *Transactions,* Am. Inst. Mining Engineers, Vol. 48, 463.

17. Hearings before the House Committee on Public Lands, on H. R. 24070, May 13 and 17, 1910, p. 55.

18. *Rept. Sec. of Int.,* 1915, 13.

19. *103 Fed.,* 568.

20. *104 Fed.,* 20; *112 Fed.,* 4.

21. *190 U. S.,* 310.

22. Miller vs. Chrisman, *140 Cal.,* 440; *197 U. S.,* 313, 321; McLemore vs. Express Oil Co., *158 Cal.,* 559, 562; Weed vs. Snook, *144 Cal.,* 439; N. Eng. etc. Oil Co. vs. Congdon, *152 Cal.,* 211; Smith vs. Union Oil Co., *135 Pac.,* 966; *39 Sup. Ct.,* Rept. 308.

23. Hearings before the House Committee on Public Lands, on H. R. 24070, May 13 and 17, 1910, p. 126, 128; *Oil and Gas Jour.,* Feb. 27, 1920, 77; May 7, 1920, 44.

24. U. S. Geol. Survey, *Bul. 623,* Professional Paper 161, 68.

25. The Louisiana withdrawal was a blanket withdrawal covering 1,500,-000 acres, but it included only 6,599 acres of vacant land.

26. H. R. 11233, 57 Cong., 1 sess.; *Eng. and Min. Jour.,* Aug. 30, 1902, 270.

27. Garthe vs. Hart, *73 Cal.,* 541; Miller vs. Chrisman, *140 Cal.,* 440; Belk vs. Meagher, *104 U. S.,* 279; Atherton vs. Fowler, *96 U. S.,* 513; Nevada Sierra Oil Co. vs. Home Oil Co., *98 Fed.,* 673; *Lindley on Mines,* pp. 378, 379.

28. *Eng. and Min. Jour.,* Dec. 28, 1907, 1214; Miller vs. Chrisman, *140 Cal.,* 440. See also *104 U. S.,* 636; *109 U. S.,* 440; *111 U. S.,* 350.

29. *Land Decisions,* IV, 221; H. R. 7625, 54 Cong., 1 sess.

30. S. 2763, S. 3326, H. R. 9045, 56 Cong., 1 sess.; H. R. 12941, H. R. 15198, 57 Cong., 1 sess.; Stat., 32, 825. On the judicial construction of this act, see Smith vs. Union Oil Co., *166 Cal.,* 217; *39 Sup. Ct.,* Rept. 308.

31. Under the general placer laws, there had been no limit to the number of placer claims that any person might locate. (*Lindley on Mines,* 450; Riverside Sand and Cement Mfg. Co. vs. Hardwick, *120 Pac.,* 323.)

32. *Eng. and Min. Jour.,* July 20, 1889, 57; Jan. 18, 1890, 94. Hearings before the House Committee on Public Lands on H. R. 24070, 61 Cong., 2 sess., May 13 and 17, 1910, p. 10.

33. *249 Fed.,* 789.

34. Puter and Stevens, *Looters of the Public Domain,* 416; Gird vs. Cal. Oil Co., *60 Fed.,* 531. For an interesting account of some of these operations, see *Technical World,* June, 1912, 372.

35. *Pac. Oil Reporter,* March 5, 1904, 3. For a similar story, see *Cong. Rec.,* Sept. 3, 1919, 4752.

36. *Wyo. Oil Index,* Dec. 16, 1922; *Petroleum Age,* Nov. 1, 1922, 27; *Natl. Petroleum News,* Dec. 6, 1922, 18.

37. Hearings before the Senate Committee on Public Lands on S. Res. 282, p. 461.

38. *Land Decisions,* 38, 59; *Min. and Sci. Press,* Jan. 7, 1911, 74; Feb. 18, 1911, 277.

39. For an interesting account of the facts in the Yard case, see Puter and Stevens, *Looters of the Public Domain,* Ch. 27.

40. H. R. 32344, 61 Cong., 3 sess.

41. Stat., 36, 1015.

42. *Land Decisions,* 39, 460; *Cong. Rec.,* Jan. 17, 1917, 1547.

43. U. S. Geol. Survey, *Bul. 350,* 39.

44. *Min. and Sci. Press,* Aug. 6, 1910, 167.

CHAPTER XXII

THE PUBLIC OIL LANDS: WITHDRAWALS

IT has been characteristic of public land legislation that the interests of exploiters of natural resources have always received sympathetic attention before the public interests have received any consideration. That was true in legislation dealing with the public forest lands, and it was true of oil-land legislation. The first change in the Placer Law was not in the direction of better protection of the public interests, but in the direction of greater liberality for locators. The Five Claims Act of 1903 was one such step.

When Roosevelt became president, however, the interests of the general public, and of the future, received a proper measure of attention. Roosevelt's conservation movement—perhaps the most important of his "policies"—did not at first include oil among the important resources to be conserved. Perhaps the real genesis of the conservation movement is to be found in the appointment of the Public Lands Commission in 1903, "to report upon the condition, operation, and effect of the present land laws"; and to recommend needed changes. The commission, composed of W. A. Richards, F. H. Newell, and Gifford Pinchot sat in session at Washington for several weeks, hearing testimony regarding the public lands, and then Pinchot and Newell went west to confer with western representatives of various interests. In March, 1904, a partial report was finished, and a year later the report was completed. It contained recommendations for a great many changes in the public land laws, although it is significant of the lack of public interest in the oil-land question that oil was not mentioned in the report.[1]

In a message to Congress in May, 1906, on the question of Standard Oil misbehavior, Roosevelt declared: "The time has come when no oil or coal lands held by the government, either upon the public domain or in territory owned by the Indian tribes, should be alienated. The fee to such lands should be kept in the United States Government . . . and the lands should be leased only on such terms and for such periods as will enable the Government to keep entire control thereof." In his annual message of December, 1906, and in later special messages, he urged insistently the conservation of the mineral fuels, and called for a leasing law.[2]

Following President Roosevelt's recommendations, a number of bills were introduced into Congress providing for the reservation and leasing of coal and oil lands. In December, 1906, and January, 1907, bills providing for the reservation of such lands were introduced by Senator

Hansbrough of North Dakota, by Representative Volstead and Senator Nelson of Minnesota, by Representative Lacey of Iowa, and by Senator La Follette of Wisconsin. These men wished to reserve the title to all coal, petroleum, and natural gas deposits to the government, and lease them, while disposing of the surface of the land under the various public land laws. None of the bills introduced was even reported from the committee.[3]

To Senator La Follette, of Wisconsin, belongs the credit for some of the first efforts in the direction of an intelligent administration of the natural resources of the country—including oil. During these years he was always at work, appearing at almost every congress with his proposal for a reservation policy. In June, 1906, he offered a concurrent resolution authorizing the President to withdraw from entry and sale all public lands known to be underlaid with coal, lignite, or oil, until Congress could provide properly for their disposal; but nothing came of it.[4]

The public coal lands received far more public attention than oil lands during the years 1905 and 1906; and in June, 1906, President Roosevelt directed the Secretary of the Interior to withdraw all valuable coal lands from entry. Altogether about 64,000,000 acres of land, supposed to contain coal, were withdrawn during the year. It was a characteristic Roosevelt gesture, spectacular, dramatic, and really shrewd and effective. The public lands were passing rapidly into private hands, and the fate of reservation bills that had been presented in Congress indicated clearly that Congress would probably do nothing but talk, and in some of these bills Congress did not even take enough interest to talk. Roosevelt knew that he must act if anything was to be accomplished, and it was his theory of administration that the executive should do whatever the public interest required, unless specifically prohibited from doing so. His action concerned mainly coal lands, but it was a precedent for later oil-land withdrawals, and as such is important here. Furthermore, the oil lands received some consideration in the debates Roosevelt's withdrawal order stirred up in Congress. This order, like Taft's order withdrawing the oil lands three years later, raised the question as to the President's authority to make such withdrawals without authorization from Congress. The congressional debates on the subject illustrate some of the western views on the public land question.

Representative Mondell, of Wyoming, asserted that Roosevelt's withdrawals were "unauthorized and a dangerous precedent." He made a careful citation of the laws and precedents to show that the president's order was beyond his authority. "The plain if not express purpose of these withdrawals is, in effect, to repeal or alter existing land legislation," he declared. "If an executive officer has the power to repeal or alter a land statute, he has the power to alter or repeal any statute."

Mondell admitted that Secretaries of the Interior had sometimes withdrawn land where there was imminent danger of fraud, but asserted that these withdrawals were made to prevent acquiring land in violation of the law, whereas Roosevelt's withdrawals were to prevent acquiring land *"in accordance with the law."*

"And so, Mr. Chairman," he announced, in the course of a long and able speech, "we have the edifying spectacle presented to us of all the land laws being suspended and the rights of all homestead and other entrymen set aside over an area of 700,000 square miles. . . . Thus we have passed, so far as land laws are concerned, from a government under law and by statute to a government under which laws are annulled, suspended or modified at the whim and pleasure of executive officials. I desire to suggest the thought that the very radical departure in policy involved in government ownership and operation of coal, oil, and gas mines and wells is rather too large a one to be undertaken off hand without serious consideration. It is the most tremendous step thus far suggested in the direction of that state socialism, paternalism, and centralization which just now seem to be so popular in certain quarters. . . ."[5] Mondell thought the people of the West were "becoming alarmed at the growing menace of the usurpation of power by executive officers; over the growth of that system that now seems to be so strongly urged by certain government officials, under which our public land system is to be overturned, sales and dispositions of the public lands discouraged, and a system of permanent federal landlordism established, with an army of federal spies, informers, inspectors and tax gatherers to administer it."

Mondell did more than talk about the withdrawal orders. He later introduced a resolution calling upon the Secretary of the Interior to show by what authority he had withdrawn these lands; but later withdrew this resolution, with the explanation that he thought the new administration (that of President Taft) would be more judicious. "It is to be hoped that with the incoming of a new administration of the Department, we shall see an end of prejudice, muck-raking, and hysteria in the administration of the public land laws," he announced, "and we shall have substituted therefor a sane and sensible policy under which the laws will be enforced without malice or sensationalism, and with a view of assisting, rather than retarding the development of the West."[6] Mondell criticized particularly the La Follette reservation bill.

In the California legislature there was some talk of trying to dispute the right of the federal government to reserve these lands, and bills were introduced to tax federal lands and subject them to state control.[7]

In the first year of President Taft's administration, several bills appeared in Congress providing for the disposition of the oil lands; but nothing was accomplished. Some of the California representatives were anxious to do something for the oil men; Lindbergh, of Minnesota, wished to reserve minerals for the states; and La Follette brought in his usual reservation bill, but only one of these bills got any attention. Representative Smith, of California, urged a bill to give prospectors on public oil lands a preference right for a certain time, until they could make discovery; and this was debated at some length. The line of cleavage in these debates seemed to be largely partisan, for on a motion to suspend the rules and pass the bill the vote was almost a straight party vote. The real merits of this important question seemed to be subordinate to politics.[8]

In the next session of Congress, La Follette again appeared with his reservation and leasing bill, but again it received no attention.[9] In the next Congress, he offered it again, and again failed to secure any consideration. Bills introduced by other western men likewise were not reported.[10] Reservation of the oil lands was destined to come in a somewhat different way. Congressional debates seemed to accomplish nothing, but a more effective means of protecting the public oil lands was on the way. Roosevelt's coal land withdrawals, and the Land Office withdrawals of oil lands from agricultural entry have been discussed.[11] As already stated, these latter withdrawals had been made for the purpose of protecting oil operators rather than for the purpose of conserving the oil resources. Except for the Louisiana withdrawal, they had practically all been aimed at the "scrippers" and "oil homesteaders" who had been such a nuisance on the oil lands of the West. They had prohibited agricultural entry or selection, but, except the Oregon withdrawal, permitted mineral locations, so that exploitation of oil was hastened rather than retarded. This at first seemed excellent for the producer, but when production began to outrun demand, and prices declined to almost nothing, it became evident that the withdrawals were beneficial to no one, and geologists and others urged withdrawal from every kind of entry.

Congressman Mondell of Wyoming once stated that the agitation for withdrawals was in some instances started by people who had oil lands located in the vicinity, and who thought it would help in advertising their stock and in getting capital for development, to say in their appeals and in their advertisements that the government had designated it as oil territory by withdrawing it from agricultural entry. Mondell even hinted that such a motive might have been behind the petition for a withdrawal in Oregon; and it was rumored in 1910 that the Standard Oil Company was in favor of the California withdrawals, although this was denied.[12] Day, Arnold, and Veatch, in November, 1908, pointed out the necessity for conserving our supply of oil, stating that this was one of the most important questions before the public—as they expressed it, "in a class by itself."

In February, 1908, the Director of the Geological Survey, George Otis Smith, wrote a letter to the Secretary of the Interior, calling attention to a *Consular and Trade Report* of February 15, on the subject of petroleum fuel for the navy, and asking that a future supply of this fuel be assured for the navy by the reservation of remaining oil lands in the West. No action followed this recommendation, and in September of the next year, Smith again renewed his recommendation, and the new Secretary of the Interior, Richard Ballinger, in turn wrote President Taft, pointing out the necessity of conserving our oil supply, and urging the reservation of the remaining oil lands of California until more adequate legislation could be enacted. In response to this, President Taft, in September, 1909, ordered the withdrawal of a large amount of oil lands, or, to put it more exactly, ordered the conversion of the existing withdrawals in California and Wyoming into withdrawals from all forms of disposition. Some thirteen more withdrawals were made by the Department of the Interior between October 1, 1909, and June 30, 1910. The withdrawal order of September 27, 1909, covered a total gross area of 2,871,000 acres in California, and 170,000 acres in Wyoming. It covered a very large amount of oil land in the West, including almost all of the Salt Creek field and at least 25 per cent of the productive territory on the west side of the San Joaquin Valley, in California.

This order came without warning and caught a large number of operators in every conceivable stage of development, from those who had made only "paper" locations to those who had drilled almost to the oil sands. Of the 169 oil companies in the Coalinga field, it was estimated that about 75, claiming at least 11,000 acres of land, were unable to get patent, although they had collectively spent nearly $2,500,000 in drilling, and had brought in a number of excellent wells. The great Lake View gusher was on land which could not be patented. In the state of California, approximately 100,000 acres were covered by nearly a thousand pending entries, largely oil entries. Figures compiled by the California State Mining Bureau showed that there were 19,745 acres of proved oil land in California, the title to which was affected by the withdrawals. Assistant Secretary of the Interior, Franklin Pierce, and George Otis Smith of the Geological Survey made a special trip out to California in 1910 to investigate the troubles of these oil men.[13]

The order of withdrawal was very generally disregarded by operators, however. Thomas O'Donnell testified before the House Committee on Public Lands that 95 per cent of the work on the reserved lands proceeded as if the order had not been issued. As he expressed it, "They had the bear by the tail, and they could not let go."[14] In 1919, there were 653 wells on this withdrawn land, and the total production from it had amounted to over 76,000,000 barrels. It was estimated that a total of $18,000,000 had at that time been spent in "developing" the lands, much of it having been spent after the withdrawal order.[15] Many operators and lawyers argued that the withdrawals were not valid, and they were able to find much support for their contention, since two of the lower courts held them invalid.

President Taft asked Congress to set all doubts at rest by giving the President the authority to make such withdrawals, and in April, 1910, Representative Pickett, of Iowa, introduced a bill for this purpose.[16] The Pickett Bill confirmed the existing withdrawals, and authorized the President to withdraw lands in the United States and in Alaska "for public uses, or for examination and classification." In another section it authorized him to withdraw lands "for other purposes," "whether classified or not," and "submit to Congress recommendations as to legislation respecting the land so withdrawn." These provisions seemed wise, but there was a proviso that the lands, when classified, should be "restored to appropriation and disposition under the laws applicable thereto." This apparently did not require executive action to secure the restoration of withdrawn lands, for a later section of the bill mentioned three ways in which lands might be restored: by the President, by Congress, "or hereinbefore provided." The intent of this provision apparently was to secure the automatic reversion of withdrawn lands to the general public domain, so they could be entered under the various misfit public land laws. It is fairly clear that some of the advocates of unrestricted exploitation of the public lands had put a "joker" into the bill, but who was responsible the writer is entirely unable to guess. Mondell, of Wyoming, may have had something to do with it—at any rate, he defended it in Congress.

The Pickett Bill was the point around which the various forces interested in public oil lands assembled for debate and discussion; and the debates in Congress brought out very clearly some of the diverse views held regarding the public land policy. When the bill came up in the House, Mondell, Chairman of the Committee on Public Lands, immediately launched an attack on the withdrawals, questioning the right of the President to make them, although he had the bill in charge and did not oppose its passage. Some of the southern Democrats opposed the bill, probably because it was an administration bill. Martin, of Colorado, objected because withdrawal often closed the land to entry for many years before classification could be made. Ferris, of Oklahoma, declared that the bill was fathered by opponents of conservation, objecting particularly to a committee amendment which saved the "legal rights of any settler or entryman initiated prior to the withdrawal," which he declared would allow many fraudulent claims to go to patent. Mondell stated that the bill was desired by President Taft, but Ferris declared that a number of committee amendments were bad and not what the President wanted.

Smith, of California, objected to this sort of a "federal tax," and to the "federal guardian" who was "to supervise and tell you how to run your business, which the American people have been running for themselves since this was a nation." Smith could not see why anyone should "foist any such foreign, new-fangled system upon the people of the West," when "the East, the South and the Middle West went forth and possessed themselves of all that was good and glorious that lay before them." Taylor, of Colorado, objected to the bill for several reasons: because the existing land laws were generally satisfactory; because the bill gave the President too much power; because it was not fair to the people of the West; because there was nothing in the bill requiring speedy examination, classification, and elimination of lands; and because it was a reversal of "the policy of a hundred years." He objected to establishing a "feudal landlord, reigning over unwilling tenants by the agencies of irresponsible bureaus," objected to this "confiscation," "legalized grand and petty larceny," and "bureaucracy gone mad." Like most western men, he objected to anything that would interfere with the "development of the West."

Support of the bill came from various quarters. Mondell, a sworn opponent of conservation, seemed a strange supporter, but his attitude may have been due to party relations with President Taft, who had asked for the bill. Perhaps Mondell may have thought, as later developments proved, that the President already had the power to make these withdrawals, and that the bill was really a restriction, rather than an enlargement of his powers; but he did not express this opinion at any time. He supported the bill only very mildly, anyhow. It was to be expected that such a bill as this should have the support of James Mann, one of the ablest friends conservation has ever had in Congress. Dawson, of Iowa, argued strongly for the bill, asserting that some very questionable entries had been made on California oil lands, and that some of the entrymen were responsible for the opposition to the withdrawals, and to their validation by Congress.

There was considerable disagreement among the congressmen and some hair-splitting legal discussion as to the exact meaning of some of the provisions of the bill, and of some of the committee amendments. On the amendment providing that "such withdrawals shall not affect the legal rights of settlers or entrymen initiated prior to withdrawal," long discussion failed to establish exactly what was meant by "legal rights" or by "initiated," or just what "legal rights" or "valid rights" an entryman had "initiated" or "acquired" when he initiated a placer claim and began drilling. There was some ambiguity as to what was meant by "such withdrawals," since a number of kinds of withdrawals had been made. Mann wondered exactly what was the effect of the proviso "that the lands when classified shall be thereafter restored to appropriation and disposition under the laws applicable thereto"; and no one seemed able to explain it clearly. The suggestion was made that the proviso be changed so that the land should be restored "by the order of the President," but there was some disagreement as to whether this was necessary, as well as to whether it was desirable.

The bill, as it passed the House, was an excellent conservation measure, confirming the withdrawals and the President's right to make withdrawals, but denying some of the privileges the western forces had wanted. The amendment validating the rights of settlers had been voted down, along with some other objectionable amendments. This was the kind of action that might have been expected in the House of Representatives, for the House was at the time generally favorable to the conservation policy. On public land questions, the Senate almost always took a different position. The Senate has generally been more friendly to the interests that desire to exploit the public domain; and in the Senate the western states, with the typical western zeal for "development," have had relatively greater power. The Senate was always hostile to conservation; and the substitute bill that Senator Smoot reported, in place of the House bill, was more favorable to the western view in several respects. In the first place, it opened withdrawn lands to exploration for minerals other than coal, gas, and phosphates. In the second place, it saved the "rights" of those who were on the land at the time of withdrawal and were "in diligent prosecution of work leading to the discovery of oil or gas." Several provisions of the Senate bill were designed to protect the rights of various kinds of land claimants. In the third place, there was a proviso that no more forest reserves should be created in the West—a mere bit of western spleen, since the creation of forest reserves in the important western timber states had been forbidden three years before. Some of the relief provisions of the bill represented the influence of a powerful lobby of western oil men who were hanging around the Capitol, and who had appeared before the Senate committee.

Various arguments were urged by the friends of the substitute bill. Some approved on the theory that the bill would confirm a power that the President had anyhow. Others, Clark, of Wyoming, for instance, approved on the theory that the bill conferred a power which the President did *not* have. Borah, of Idaho, approved because he thought the bill would restrict rather than enlarge the power of the President; although he illogically denied that the President had acted within his powers in withdrawing the lands in the West. Borah was correct, it later appeared, in assuming that the bill would restrict the power of the President. Some of the senators voted for the bill because some sort of trade had been arranged whereby, if it passed, some $20,000,000 or $30,000,000 was to be turned over to the reclamation fund. Perhaps it was the support thus secured which caused the bill to pass the Senate.

Opposition to the bill was based on a variety of grounds. Some senators who were friendly to the conservation movement distrusted the measure, thinking, as Borah did, that it would restrict rather than enlarge the power of the President. Others opposed because of a suspicion that the bill, as the Senate committee had reported it, would validate some fraudulent entries in California. Dolliver did not oppose, but was suspicious of the clause permitting mining on the lands withdrawn, fearing that the miners would get control of water power sites. "Slippery Tom" Carter, of Montana, opposed on the general grounds that private ownership would secure greater "enterprise" and initiative than any system of "landlordism." Bailey, of Texas, wanted public lands given to the individual states, because the states could manage them better than the federal government. He asserted that an amendment giving the lands to the states would probably pass the Senate, and Smoot agreed, but thought it would not pass the House.

Senator Heyburn opposed on all possible grounds. He opposed conservation at the expense of the West, for the benefit of the East. He opposed such "socialistic" legislation; opposed saving too much for posterity, and asserted his belief that Congress had not the power to authorize the President to suspend the operation of the land laws. Some of his comments on saving for posterity seem worth quoting:

Are you going to make life a downy bed of ease for the children of the future, with no problems for them to solve, with no tests of their strength and ingenuity to put them on their mettle? Are you going to soften and degenerate all our descendants in the days that are to come by saving up for them all the good things, while their fathers and their mothers hunger in vain for what is so bountifully about us? My doctrine upon that subject, Mr. President, is that future generations, like the past and the present, must and should grapple with the problems of their day; should develop the manhood, the courage, the womanhood, and the strength which will make them worthy descendants of the founders of this government and of those who have conquered the desert and built up these great states and cities.

Most of the opposition to the land withdrawals, and some of the opposition to this bill, must be considered in connection with the situation regarding the public timber lands. Under the Forest Reserve Act of 1891, over a hundred million acres of timber lands had been withdrawn and placed in forest reserves. In the process of reserving these lands, some hardships had been imposed upon settlers in the West, and of course serious impediments had been placed in the way of stealing or otherwise appropriating the reserved lands. The reservation of timber lands had been pushed so vigorously that in 1907 Congress deprived the President of his right to set aside forest reserves. Just about this time President Roosevelt launched his movement for withdrawing other kinds of lands, and this brought the whole question up. It was not a fight of the administration against the oil men alone, or even against oil and coal interests, but against oil, coal, lumber, grazing, and mining interests; and these latter interests were often joined by certain western railroads, and by "capital" in general. The entire West was always pervaded with the idea of rapid exploitation, and when the principle of reservation was applied in such vigorous and thoroughgoing fashion it was not surprising that many classes and interests should unite in opposition. It is necessary to bear in mind that the oil question was not a separate question, but merely one aspect, and perhaps not the most important aspect, of the great problem of dealing with natural resources.

Almost every senator had some sort of amendment to propose. The committee reporting the bill had changed it very materially, but many further changes were suggested. Clark wanted an amendment providing that withdrawn lands should revert back to the public domain, unless Congress by affirmative action should confirm the withdrawals. Nelson pointed out that this would practically destroy the bill, because Congress was always slow to act; and the amendment was rejected. Heyburn offered a substitute bill providing that the President should have only the power to investigate lands and then make recommendations to Congress, which was likewise rejected. Carter offered an amendment advancing $30,000,000 to the reclamation fund, and this delicate morsel of "pork" was accepted, 57 to 3. Dixon, of Montana, wanted $50,000 for a conservation commission, but this was not so acceptable, in fact, failed by a vote of 37 to 19, after Heyburn had tried to amend it to allow only men living west of the 100th meridian to serve on it. Carter offered an amendment permitting metal mining on withdrawn lands, which failed. Guggenheim, of Colorado, was anxious to save the rights of any who had taken up and improved water power sites; Gore wanted the Philippines included; Smoot wanted to include Alaska; Burton, of Ohio, and Beveridge, of Indiana, wanted a little better conservation provision. Beveridge stated that his amendment was prepared by Mr. Wells, legal adviser to the National Conservation Association, and that it followed the ideas of former President Roosevelt, but it failed. A hot debate developed between Beveridge on the one hand, and Clark and Borah on the other; and Beveridge sustained his right to be regarded one of the ablest friends of conservation in either house of Congress.

There was little general interest in the discussion, however. When Clark's amendment was up, providing for the automatic restoration of withdrawn lands, only 19 votes could be secured, on both sides of the question. Heyburn and Clark complained of this indifference. Most of the discussion that did not relate to the general subject of conservation bore on coal, water power, and forests, rather than on oil, but it has been treated here because of its importance in the general conservation movement.

There is reason to believe that, in the final disposition of the bill, oil interests played a more important part than the debates would suggest. Early in May, before the Senate had begun serious consideration of the Pickett Bill, Flint, of California, presented resolutions of the Chamber of Commerce of Coalinga, California,—one of the important oil towns, —against the bill, particularly praying that the bill be not made retro-active, but that it should save the rights of those who located on the with-drawn land previous to the date of its withdrawal, September 27, 1909. Three days later Perkins, of California, presented a number of telegrams from oil producers in California, praying that the Pickett Bill be not passed until a delegation of oil men could reach Washington and be heard. Senator Nelson later stated that the bill really enlarged the rights of the oil and gas claimants, because under the laws in force at the time there must be an exploration and discovery to give any rights, whereas, under the bill, if prospectors were "in diligent prosecu-tion of work leading to discovery," they might claim, even though there had been no real discovery. Nelson pointed out the provision that the act should not be "construed as a recognition, abridgement or enlarge-ment of any asserted rights or claims initiated upon any oil or gas bearing lands after any withdrawal of lands made prior to the passage of this act," and stated that this had been secured in committee by the oil lobby from California.

The bill passed the Senate with a number of unfortunate amend-ments; and when it reappeared in the House, some of its friends veered around to a position of hostility, while some of the anti-conservationists changed to a favorable attitude. Taylor of Colorado asserted that the Senate amendments were about the same as he had once offered in the House, and declared his approval. Englebright and Smith, of California, likewise spoke of the Senate amendments as favorable to the West. There was some Democratic opposition, and Madison, of Kansas,—a Progres-sive Republican,—spoke of Roosevelt's ideas, and declared they were not in the bill; but it passed 101 to 71. It was clearly regarded as a victory for the oil men and anti-conservationists, and later court decisions proved that it was, but it was signed by President Taft and became law.[17]

Later developments proved that the main provision of the Pickett Act was not the provision confirming the President's withdrawals, since the courts later confirmed them anyhow. The most important section in the act provided: "That the rights of any person who at the date of any order of withdrawal heretofore or hereafter made, is a bona fide occupant or claimant of oil- or gas-bearing lands and who, at such date, is in diligent prosecution of work leading to the discovery of oil or gas, shall not be affected or impaired by such order so long as such occupant or claimant shall continue in diligent prosecution of said work." The California oil men, who appeared before the public lands committees of both houses, pointed out that the withdrawal order, if found valid, would cut off the rights of every prospector who had not made a discovery, although he was in actual bona fide possession of the land diligently drilling for oil. It had long been established that claimants under the Placer Law secured no right or title until they made discovery; but the California operators urged the committees to adopt the views of the California Supreme Court, which recognized the rights of claimants in diligent prosecution of exploration work, as against adverse claimants. They asked that the claimant be given the same rights against the government that they already had against adverse claimants.[18] This was granted by the above relief provision.

Under the new law, the President at once, on July 2, 1910, confirmed the previous orders of withdrawal, but this did not settle all the questions relating to the oil lands, for many operators had initiated placer locations after the withdrawals, and they immediately set up the claim that locations made thus, after the withdrawal orders but before the confirmation order, were valid. The government was forced to bring test suits against several of these; and, six years after the withdrawal order of 1909, after two federal district courts had held the order invalid, the Supreme Court of the United States, three judges dissenting, rendered its decision in the Midwest case, holding the original order valid and effective.[19] The decision was based, not on the President's power as Commander in Chief of the Army and Navy, to preserve a naval fuel supply, nor on any power expressly delegated by Congress. In the language of the court, ". . . long continued practice, the acquiescence of Congress, as well as the decisions of the courts, all show that the President had the power to make the order." During the years when this question was being litigated, thousands of barrels of oil were drawn from the government land.[20]

The withdrawn lands constituted a very complicated and difficult problem for the Land Office during these years. In 1915, the Commissioner of the Land Office reported:

In California alone there are over a million and a half acres within the withdrawn area, of which about a million acres are patented lands. More than one hundred thousand acres are embraced in pending filings, selections, or entries, and there are over four hundred thousand acres for which no filings have been presented to the local land office, but which are doubtless covered by mineral locations recorded in the various mining districts.[21]

Within the six years following the passage of the withdrawal act of 1910, about 35 new withdrawals were made, covering all the public lands in which there was thought to be reasonable prospect of finding valuable deposits of oil and gas. During the same time, all lands which field examination showed to be probably non-oil were restored. The confirmatory orders of 1910 covered approximately 4,697,600 acres; new withdrawals during the next six years covered approximately 3,483,-300 acres; and nearly 2,593,900 acres were restored, leaving a total of about 5,587,000 acres outstanding on January 15, 1916. This was distributed as follows:[22]

Alaska	Unknown
Arizona	230,400
California	1,507,547
Colorado	87,474
Louisiana	414,720
Montana	641,622
North Dakota	84,894
Utah	1,952,326
Wyoming	668,094
	5,587,077

Since 1916, there have been further withdrawals, including two oil shale reserves in Colorado and Utah, one helium reserve, on the Woodside structure in Utah, and one naval reserve in Alaska. During the same time, there have been about as many restorations as withdrawals, but the restorations have been smaller in area, and the total area withdrawn has increased nearly 1,000,000 acres since 1916, exclusive of Alaska. The new Alaska reserve covers 30,000 square miles, or nearly 20,000,000 acres.

The government finally brought suit against some of the men operating on withdrawn lands, and against the pipe-line companies that had bought oil of them. In some of the important oil fields of California, half of the oil lands—the alternate sections—were owned by the Southern Pacific Railroad, some by the Standard Oil Company, and some by the Santa Fé Railroad; and the pipe lines were largely owned by the Southern Pacific and by the Standard Oil Company. These purchasing companies refused to transport the oil produced by operators whose titles were challenged, while they could of course transport their own production, and were drawing the oil out of the ground, leaving the other operators to wait for a court decision on the validity of their entries, or for Congress to grant them relief. In Congress, Representative Church of California and Senators Pittman of Nevada and Walsh of Montana took up their case and finally got a relief act—the Pittman Bill—through Congress. This "Operating Agreements Act" afforded merely temporary relief, for it authorized the Secretary of the Interior to make working arrangements whereby the oil claimants could go on with the production of oil until some permanent leasing law could be secured.

Under the Operating Agreements Act, the Secretary of the Interior entered into over 40 contracts with operators, but none of these contracts covered operations in the naval reserves—some of the withdrawn land was set aside specifically for the navy—because, by agreement between the Secretary of the Navy, the Secretary of the Interior, and the Attorney-General, suits were being brought against claimants in the naval reserves. Under practically all of the agreements made by the Secretary of the Interior with California producers, one-eighth of the proceeds from the crude petroleum produced was put in escrow to await decision as to the ownership of the land. Where there were adverse claimants, however, as there were in many cases in Wyoming, the entire proceeds, above operating expenses, were set aside.[23]

After the Midwest decision had finally settled the question as to the validity of the withdrawals, several suits were brought, most of them in California, to have receivers appointed and to protect the property until proceedings in the Interior Department should determine the ownership of the lands as between the claimants and the government. Only five suits had been brought prior to February 23, 1915, when the Midwest decision was rendered. During the spring and summer of 1915, 25 such suits were filed. Previous to February 28, 1918, 55 suits had been instituted in California, and on June 30, 1920, there were 40 California cases still pending in the district court and two in the higher federal courts.[24]

NOTES

1. Ise, *U. S. Forest Policy*, 149-154; *Chautauquan*, June, 1909, p. 21; S. Doc. 189, 58 Cong., 3 sess.

2. *Cong. Rec.*, May 4, 1906, 6358; Dec. 17, 1906, 450; Feb. 1, 1907, 12077; Feb. 13, 1907, 2806; S. Doc. 141, 59 Cong., 2 sess.

3. S. 7241, S. 7327, S. 7498, S. 8013, S. 8136, H. R. 23207, H. R. 23552, H. R. 23553, H. R. 23554, H. R. 24368, 59 Cong., 2 sess.

4. As early as 1892, Representative Stockdale, of Mississippi, had introduced a bill into Congress to withdraw all public mineral lands and lease them on a royalty basis, but he was apparently interested in getting more money to build levees along the Mississippi, rather than in a more intelligent administration of the public lands. (H. R. 5146, 52 Cong., 1 sess.)

5. *Cong. Rec.*, Feb. 8, 1907, 2617.

6. *Cong. Rec.*, Feb. 12, 1907, 2801; Feb. 21, 3591; March 4, 4666-.

7. *Eng. and Min. Jour.*, Feb. 25, 1911, 434.

8. S. 4168, S. 5890, H. R. 6255, H. R. 20391, H. R. 17827, 60 Cong., 1 sess.

9. S. 7369, 60 Cong., 2 sess.

10. S. 597, S. 438, S. 2623, H. R. 9771, H. R. 9964, 61 Cong., 1 sess.

11. See above, pp. 299, 300, 310.

12. Hearings before the House Committee on Public Lands, on H. R. 24070, May 13 and 17, 1910, pp. 56, 57, 60, 104.

13. *Min. and Sci. Press,* Nov. 26, 1910, 696. Hearings before the House Committee on Public Lands, on H. R. 24070, May 13 and 17, 1910, p. 10.

14. *Ibid.,* p. 38.

15. Fed. Trade Comm., *Rept. on the Pacific Coast Petroleum Industry,* Pt. I, 83.

16. H. R. 24070, 61 Cong., 2 sess.

17. *Ibid.; Cong. Rec.,* May 3, 1910, 5698; May 6, 5870, 5871; Stat., 36, 847. For legal interpretations of this law see *Annotated Cases,* 1918, C 1009.

18. Miller vs. Chrisman, *140 Cal.,* 440, *197 U. S.,* 313; U. S. vs. Stockton Midway Oil Co., *240 Fed.,* 1006, 1009.

19. *236 U. S.,* 459. See also U. S. vs. McCutchen, *217 Fed.,* 650, *234 Fed.,* 702, *238 Fed.,* 575, 579; U. S. vs. Midway Northern Oil Co., *216 Fed.,* 802, *232 Fed.,* 619, 627; U. S. vs. Grass Creek Oil and Gas Co., *236 Fed.,* 481.

20. For a discussion of these withdrawals, see U. S. Geol. Survey, *Bul. 623.*

21. Rept. L. Off., 1915, 7.

22. These figures do not represent public land actually affected by the withdrawals. Most of the orders were drawn in "blanket" form; that is, they covered areas believed to be oil-bearing without regard to ownership, and so included many tracts to which private owners either held title or some sort of valid claim.

23. *Oil and Gas Jour.,* July 11, 1919, 73; S. 5673, S. 5434, H. R. 15661, 63 Cong., 2 sess.; Stat., 38, 708.

24. Federal Trade Comm., *Rept. on the Pacific Coast Petroleum Industry,* Pt. I, 76. See also the annual reports of the Commissioner of the Land Office from 1914 to 1921.

CHAPTER XXIII

THE PUBLIC OIL LANDS: DISCUSSIONS IN CONGRESS

DURING the decade between 1910 and 1920—between the date of the Pickett Act and the passage of the Federal Leasing Act—one of the great questions before Congress almost all of the time was the question as to the disposition of the public oil lands. The withdrawal of these lands had thrown the burden of overcoming congressional inertia upon those who wished to exploit the lands, and they were of course industrious and persistent in their efforts. There were two classes of oil interests desirous of further legislation. There were operators and would-be operators who wanted to exploit the public oil lands, and asked for a law under which they might do this. Then there were the other operators who had already gone upon the reserved lands, before or after reservation, some of whom had found oil, but under such circumstances that they could secure no right to the land. The persistence with which the various "relief" sections in oil-land bills were pressed indicates that the latter class of oil interests played perhaps a more important part in legislation than did those who merely wanted a law for exploiting the general public domain. The list of companies with some sort of an "interest" in the reserved lands included some of the largest oil companies in California and Wyoming.

Some of the lands were soon set aside as naval reserves, for the future use of the navy. In 1912, two such reserves were set aside: No. 1, known as the Elk Hills Reserve, in the Elk Hills, California, covering 38,969 acres (gross, that is, including patented lands); and the Buena Vista Hills Reserve, in the same vicinity, covering 29,341 acres (gross). In 1915 another reserve, the Teapot Dome Reserve, was set aside in Wyoming, adjoining the Salt Creek field, north of Casper. There were considerable areas of patented lands in these reserves, particularly in No. 2, the Buena Vista Hills Reserve; and there were always a large number of claimants who asserted rights of one kind or another. The history of these reserves will be treated more fully in a separate chapter, but they must be considered in this chapter also, because they were probably more important than the ordinary public oil lands. For years it was impossible to secure any general law for the disposition of the public oil lands, because it was impossible to secure any agreement in Congress as to what should be done for the relief of the claimants on these lands.

At the very time, in 1910, when some men in Congress were working for the law to authorize the withdrawals, others were trying to get the withdrawn lands restored to entry; and still others were trying to secure laws for the general disposition of the public oil lands. Some wanted a more careful classification of the public lands, and wished the withdrawals to hold only until the classification could be completed. Gore, of Oklahoma, and Tillman, of South Carolina, were particularly interested in the question of oil reserves for the navy (the naval reserves had not yet been set aside), but Gore's bill was reported adversely by the Committee on Naval Affairs. Newlands, of Nevada, wanted a national commission appointed for the conservation of natural resources; and of course Heyburn, of Idaho,—always the arch enemy of conservation,—and Borah, of the same state, wanted the withdrawn lands restored.

Following the wishes of President Taft, Senator Nelson of Minnesota introduced nine bills, covering various items of legislation needed in the administration of the public lands. One of these bills authorized the Secretary of the Interior to make temporary withdrawals of public land pending examination and classification. This bill did not relate specifically to oil, but seemed to include it by implication. It was debated a little, but was finally recommitted. Nelson's bill for the administration of oil, gas, and phosphate lands received no attention.[1]

In the next session of Congress, Senator Smoot tried to secure a bill authorizing the Secretary of the Interior to grant permits to explore the public lands, including the withdrawn lands, for oil and gas. The bill provided for the granting of exclusive permits to explore tracts as large as 1,280 acres, for a fee of 5 cents per acre, the real intent of the bill being to give the prospector reasonable assurance that he might enjoy the fruits of discovery, if he should make one. The "scripper" troubles in California illustrated clearly enough the unfortunate possibilities of the Placer Act, in its application to oil; but Heyburn opposed Smoot's bill on the ground that it would create a "monopoly"; and Senator Root, of New York, thought the bill was "full of lawsuits." Heyburn stated that he would like to see the government offer a premium to the discoverer of oil or gas, just as the miners used to give the discoverer of gold two claims on the gulch where he discovered it. So little interest was taken in the bill that no quorum could for some time be secured to vote on it, but it later passed the Senate without comment. That was as far as the bill ever got, for it disappeared in the House Committee on Public Lands. A similar bill introduced by Flint, of California, received no consideration.[2]

Thus the three sessions of the Sixty-first Congress passed with no great change in the situation of the oil lands; and that situation was not satisfactory from any point of view. In the first place, the Placer Law was still the law of oil development, on lands not withdrawn, and the faults of that law have been sufficiently discussed. In the second place, there was no law providing for the prospecting or exploitation of withdrawn lands, and this was a constant source of irritation to the people of the West, for these withdrawn lands were not open to either agricultural entry or oil entry. During the next few years oil-land activity in Congress took the form of efforts on the part of the western men to open up the reserves to some form of development. Taft's withdrawal orders had put the burden upon them of securing some kind of agreement among the warring factions in Congress; and later developments showed that this was no easy task. In the second Congress of Taft's administration, there was little interest in the matter. Gronna, of North Dakota, brought in a bill for the disposal of oil and phosphate lands, and one of the Colorado representatives appeared with a bill to cede the public lands to the states, but it got no attention.[3]

During these years some of the western men lost hope of securing any law for the restoration of the withdrawn lands, or for the exploitation of the minerals, and turned their attention to the problem of securing the right to make surface entries, with a reservation of the minerals to the government. There had been a number of "separation acts," applying to various kinds of land. The first of these acts was that of March 3, 1909,[4] which provided that persons who had entered or selected under the non-mineral laws any lands subsequently classified, claimed, or reported as valuable for coal, might elect to receive patent to their lands by reserving all rights in the coal deposits to the United States. This act granted relief only for entries or selections antedating withdrawal. Another act, passed in June, 1910,[5] went a step further with regard to coal lands, and provided that homestead entries, desert land entries, and Carey Act selections, might be made on lands withdrawn or classified as coal, with a reservation of the coal to the government. An act of April 30, 1912, extended this act to include state selections and isolated tracts.[6]

Early in the year 1912, Senator Smoot tried to get through Congress a bill authorizing surface entries on oil, gas, and phosphate lands, and finally got it through Congress, although in badly mutilated form. In the Senate there was considerable suspicion of this bill, but it passed without serious objection. The House committee reporting the bill, however, limited its application to Utah—"decided to try it first on the dog," as Mann expressed it. Congressmen from other states objected to its application to them. Mann was suspicious of the bill and insisted that coal lands should be included, in fact he expressed a decided opinion that all the minerals, and possibly even the metals, should be reserved when title was granted to the surface, but he did not succeed in securing any change in the bill, and it passed with little difficulty.[7]

Not long afterward, Senator Warren brought in a bill providing for the agricultural entry of oil lands in Wyoming, and got it through the Senate, but it never came up in the House. It appears that some of the western men were despairing of ever securing satisfactory legislation dealing with the withdrawals, and were willing to accept even a surface right of entry, as better than nothing.

In the first session of President Wilson's administration, a strong effort was made by several of the western men to secure for Idaho the right to make state selections on oil and phosphate lands. Several congressmen insisted that legislation of this sort should be uniform for all of the states, and Lenroot's objection sent one of the bills back to the calendar once, but it finally passed both houses and became law. Finally, by the act of July 17, 1914, provision was made for all forms of non-mineral entry on lands withdrawn for their minerals.[8]

These separation acts provided well enough for agricultural entrymen, in certain states, but there was still no way by which the reserved oil lands could be exploited; and during practically all of President Wilson's two administrations, the oil men were trying to secure relief from what one of them called the "blighting influence and effect of that selfish, short-sighted and ruinous public land policy." During this time there was always an over-production of oil from privately owned lands, and there was never any need or justification for opening any public oil lands, but Congress was constantly busy with plans for opening them to exploitation.

In the first session of Congress after President Wilson took office, a number of proposals were advanced for the disposition of the public oil lands, not a few of the western men favoring cession of these lands to the states.[9] The "states rightsers" apparently had considerable strength at this time, for in 1916 the Commissioner of the Land Office, commenting on the bills appearing in Congress to cede the public lands to the states, seemed to think them important enough to justify a rather extended argument on the necessity of retaining them in federal ownership.[10]

In the second session of the Sixty-third Congress, the oil lands were the subject of a prolonged debate, centering on the Ferris Bill, which attempted a comprehensive revision of the law applying to coal, oil, gas, phosphate, potassium, and sodium. This bill was an administration bill, drawn up under the direction of Secretary of the Interior, Franklin Lane, and was introduced in April, 1914, by Ferris, of Oklahoma, as chairman of the House Committee on Public Lands, in place of a number of bills referred to the committee. It was debated at great length.[11]

The oil and gas section of the Ferris Bill authorized the Secretary of the Interior to grant the exclusive right to prospect for oil and gas on not more than 640 acres of land, where located within ten miles of any producing well, or not more than 2,560 acres where not within ten miles of such a well. The permittee was required to begin work within four months and drill at least 500 feet, and within two years drill wells aggregating 2,000 feet in depth. In Alaska, permits were to be given for four years, with more liberal provisions generally. There were restrictions as to the shape of these tracts, the length not to exceed two and one-half times the width. In case oil or gas was discovered, the permittee was entitled to a patent to one-fourth of the land embraced in the prospecting permit, or, if he preferred, he might pay a royalty on the oil or gas produced during the remainder of the term of his permit, the royalty to be fixed by the Secretary of the Interior. After the expiration of his license, the land was subject to leasing by competitive bidding, the leases to run for twenty years, with preferential rights for ten years longer.

Thus the Ferris Bill provided for an unfortunate mixture of sale and leasing. It has been stated that the bill was drafted largely by lawyers in the Department of the Interior, who knew little of oil field conditions, and who fashioned the bill somewhat on the plan of the old Placer Law. The bill, as thus drawn, had the defects of both the fee-simple and the leasing systems. Under its provisions, tracts might be so chosen that no point of the remaining government lands would be more than 2,150 feet from the possible location of a well on private land, and the government lands would be drained by private wells. On the other hand, no operator was allowed an acreage, either freehold or leasehold, sufficient to justify adequate exploration, in fields remote from pipe-line and refining facilities. This mixture of sale and leasing appeared in many of the bills presented in Congress during the next few years.[12] Mondell criticized this feature of the bill, but Ferris explained that the offer of a patent to one-fourth of the land would stimulate prospecting more than any offer of a lease, while the reservation of the other three-fourths to the government would insure government participation in the benefits of the oil discovery. Mondell offered an amendment giving the finder of the oil one-half, instead of one-fourth, of the land covered by his license, but this failed.

Mondell offered an amendment granting licenses covering 2,560 acres, and, on discovery, leases to 640 acres, leases to run thirty years, with possible extensions for twenty years longer; but this amendment failed. Mondell had placed the royalty at 10 per cent. He later offered an amendment giving the discoverer of oil not only the patent to one-fourth of the land covered by his license, but a preference right to lease the other three-fourths, but this likewise failed.

There was some question as to why the distance distinction between proved and wildcat territory should have been ten miles, and Mondell thought that a distance of five miles from proved territory was enough to establish territory as "wildcat." Others, however, thought ten miles not enough. Raker explained that it had originally been twenty-five to fifty miles, but that a group of California oil men had appeared before the committee and had secured the reduction to ten miles.

One of the objections to the mixture of sale and lease provisions was that those who had a patent to the land could sell oil cheaper than their neighbors who had to pay a royalty. La Follette thought the fee owner might thus even lower his price and prevent the sale of the oil which had to pay a royalty.

Most of the discussion involved the general question of conservation, and especially the question whether resources should be privately or publicly owned. Thompson, of Illinois, one of the leaders in support of the bill, argued that the public lands belong to the United States, and not to any particular state; that the United States government had bought the Louisiana Territory, and much other territory, with money from the general treasury. He denied that the federal government had given the earlier states the public lands within their borders, pointing to Ohio, which had been admitted to the Union with a reservation to the government of all the unappropriated public lands within the state. Thompson cited Webster and Clay as having favored federal control of the public lands.[13]

Mondell pointed out that the withdrawal of the coal and oil lands had resulted in a complete tie-up of these resources, and that it was becoming more and more clear to western people that the only way out was to adopt a federal leasing system; but he declared that he had always favored public retention of coal and oil lands, not by the central government, but by the local government, because he feared that federal control meant "centralization, bureaucracy, absentee control," and other evils, and because the federal government would get the cream of the royalties. Mondell gave a clear expression of the view that the western people were entitled to some of the benefits of the public land disposition, after they had gone into the West.

Taylor, of Colorado, represented an extreme western view, in his declaration that the bill was "in violation of the moral, legal and constitutional rights of the western states; in contravention of the enabling act by which they were admitted into the union, and to that extent unconstitutional." In the minority report which accompanied the bill he announced: "I look upon this bill as absolutely taking from the people of the arid West some of the most sacred property and political rights they have, not only reversing the traditions of this government for over a hundred years, but violating the very constitutional guarantee upon which these states were admitted into the union."[14] Taylor frequently recounted the history of the early lead-mine leases as proof of the impracticability of a leasing system. He cited Secretary Ballinger's report for 1910 on the same point. He used a number of the characteristic western arguments against the principle of reservation: that it would result in the development of a huge bureaucracy which would absorb the royalties in expensive administration, would create a host of government employees who could never be pried from their jobs; that tenants would never take as good care of the property as owners; that the leasing system involved a heavy loss of taxes to the states and counties of the West; that it deprived the western people of their freedom, compelling them to "surrender the sovereign right of American citizens to local self government and become permanently helpless, if not servile, tenants under federal tyrants and autocratic predatory bureaucrats, . . . an outrage upon a free people." Like Mondell, Taylor seemed to feel that his arguments had little effect on the House but that did not lessen the warmth with which he gave them.

There was some complaint of the loss of taxing power, because of the reservation of government lands, the claim being that the states had to pay heavy taxes to support courts of justice on the reserved lands. "Why," Taylor asked, "should we supply the government and its agents and tenants with modern civilization on a silver platter without any expense, and moreover, pay the government a royalty on our own resources for the privilege of doing so?"

Much of the debate turned on the question of the disposition of the royalties. The bill provided that the amount of the royalty paid under each lease should be fixed by competitive bidding, should be payable in money or in oil, and that the royalties should go to the reclamation fund, and when repaid, half should go to the community from whence they originated. Mondell was particularly opposed to this provision, which he considered a "joker," since reclamation funds had seldom been repaid. He had introduced a bill himself,[15] which gave all of the royalties to the states, and he offered an amendment to this bill giving half of the royalties to the states and half to the reclamation fund. Taylor admitted that the royalties charged at first would be low, but feared that they might be greatly increased in the future. Lessees, according to the bill, were required to pay $1 per acre rent on the land in addition to the other royalty.

This contest over the distribution of the royalty or income derived from public resources is one which had been waged for years in regard to public timber lands, and is destined to become an increasingly important question, if the reservation policy is applied as fully as it should be. The western men, finding themselves unable to secure the resources themselves, occupy themselves with plans for securing as much as possible of the income, and are often willing to trade votes on almost anything else for concessions on this point.

There were, in the debates, some echoes of the Taft-Roosevelt fight and the Progressive defection of 1912. Humphrey, of Washington, took time to accuse Gifford Pinchot of a number of official blunders whereby millions of acres of land were lost to the government, and criticized the ex-chief forester so caustically as to elicit a remonstrance, even from some of the Democrats in the House.

Mondell, acting for Church, of California, introduced an amendment giving special rights to the California oil men who had been the objects of so much solicitude in Congress—those who were on the naval reserves at the time they were withdrawn in 1910. The amendment provided for them a preference right to lease not over 2,560 acres for any individual or corporation, for a royalty of not over 8 per cent. Church wanted a further proviso that the amount accruing from this royalty should be given to the navy, offering this sop to draw support for the rest of the amendment.[16] Mann opposed this strongly, as he had several times opposed similar proposals in Congress, and he succeeded in having the royalty provision changed from "not over one-eighth" to "not less than one-eighth."

The provision of Mondell's, or Church's, amendment allowing leases of 2,560 acres was discussed somewhat. Ferris thought 640 acres was enough, and Lenroot thought so too, but Mondell stated that, while 640 acres might be enough in California it would not be enough in Wyoming; and Hulings, of Pennsylvania, who was a wildcat oil operator himself, declared that no oil man would go out to an untried field and explore even for a possible 2,560 acres. He pointed out that oil companies needed larger tracts, and that if they were limited to 640 acres, or even 2,560 acres, they would not go upon the land at all; but this he thought no great misfortune, since too much oil was already being produced. Hulings pointed out that in almost any untried territory, where the land was privately owned, the owners were usually anxious to have their land tested, and would offer any operator far better inducements than the chance of a fee to 640, or even 2,560 acres, under the Ferris Bill. As he expressed his own view, "I am not in favor of giving away the public domain, but when you are legislating, I should like to see it done in a reasonable way." Hulings was without doubt correct in his contention that larger tracts should be provided for— such tracts as could serve as the basis of an efficient and economical oil industry, but few men in Congress were educated to that point in 1914. Even an entry of 640 acres showed a great step in advance of the Placer Law.

A provision indicating some consideration for the exigencies of oil extraction was that forbidding lessees to drill wells within 200 feet of the outer boundaries of the tract leased, unless the neighboring land was owned by private parties, without definite agreement with the

Secretary of the Interior. This provision was designed to eliminate the wasteful "offset" wells so common in most privately owned oil fields.

Raker offered an amendment giving a preference right to applicants who had developed wells on adjacent claims, and Mondell supported this, but it failed; and after several days of desultory debate, the bill passed the House.

When the bill came up in the Senate, Clark, of Wyoming, remarked that it would not pass anyhow, and offered a bill of his own—a bill for the relief of the California oil men, which passed with scarcely a comment. The House would not agree to the substitution, so nothing further was done.

Representative French, of Idaho, tried to get a bill through to provide for surface entries of withdrawn lands, reserving to the government the particular mineral supposed to be contained therein. Mann, of Illinois,—the watchdog of the public lands during these years,—objected strongly to this bill, because it did not reserve all of the minerals. He pointed out the possibility that entrymen might take up lands classified as coal, reserving only the coal to the government, and find out later that the land contained potash, or some other valuable mineral which the government should also have reserved. Mann's objections stopped consideration of this bill.[17]

Thus the general question of the disposition of the public oil lands was unsolved. The discussion of the Ferris Bill had, however, brought out clearly the vast complexity of the problem, and the very great difficulty of securing any agreement among the various factions in Congress; and it is not surprising that in the Congress which convened in December, 1915, another "Ferris" Bill of similar provisions, after being buffeted back and forth in both houses of Congress, failed again to settle the question. This bill was approved by the Secretary of the Interior, by the Geological Survey, and by the Bureau of Mines, was debated at great length, and finally passed the House, but the Senate tried to substitute another bill and so defeated it.[18]

During these years, the question of relief for operators on withdrawn lands was always the rock on which legislative proposals were wrecked. It would probably have been possible to get the House and Senate together on most other points, but on the question of relief for these California operators—most of them were in California—it was impossible to secure any agreement. These men—there were perhaps a hundred of them—seemed for several years to be permanent parts of the legislative machinery at Washington. They were said to have formed an association to conduct a campaign to get the withdrawals rescinded, and to have contributed $100,000 to a "slush" fund.[19] At Washington they were so intent on securing special concessions of some kind that they were willing to block all attempts at legislation that did not meet their demands. One of the Wyoming newspapers once complained of the manner in which the California men blocked all efforts to secure general legislation because it did not concede enough to them. The State Oil and Gas Supervisor of California pointed out that there were some honest claimants and some others, and that it was impossible to grant proper relief to the former because of the persistence with which the others pressed their claims.[20] The Operating Agreements Act of 1914, which left the adjustment of their claims to the Secretary of the Interior until other legislation was secured, was not final, and these men hung around the committee rooms at the Capitol in 1915 and 1916, and later, seeking further concessions in the general leasing bill. An amendment in the last of the Ferris bills—the so-called Phelan amendment, offered by Senator Phelan of California—provided leases and preferential rights for them, but Mann objected very strongly to further concessions of this kind, and was warmly applauded for his stand. It was Mann who had stopped consideration of Church's relief bill shortly before.

As already pointed out, there were several different classes of operators on the withdrawn lands. There were, in the first place, a great many who had gone upon the lands and who had made discovery before the time of the withdrawal order of September 27, 1909. These men were able to secure title to the land upon which they had made discovery, and so were not considered much in the discussion of relief measures, although some of them were subjected to loss by the order of withdrawal, because they had planned more extensive development than was possible after the land near them had been withdrawn. On the other hand, these men, along with all other operators in California, found the withdrawal order highly beneficial in another way, in that it tended to reduce the vast over-production of oil. About this time, some California oil was selling as low as 10 cents a barrel.

The second class of operators included those who had gone upon the public lands in a proper and lawful manner, and had tried with due diligence to find oil but had not yet made discovery at the time of withdrawal. The men of this class were caught in all stages of development. Some had merely filed their claims, and were drumming around in search of capital to drill with, when the withdrawal order came. Others had secured capital, had just brought their drilling rigs upon the land, but had not actually started to drill. Still others had begun drilling, and some had drilled almost to the oil sands. Some had drilled industriously enough, but, because of bad luck, had been unable to complete their wells in time. Some operators of this second class were clearly in a situation demanding consideration.[21]

A third class of operators included those who had filed upon the public lands properly, before their withdrawal, but had not pursued their work with due diligence, and had not made discovery. The degree of diligence, or lack of diligence, here varied widely. A fourth class had gone on the lands prior to September 27, 1909, but had not been, and at the time of withdrawal were not, in diligent prosecution of their work, but after withdrawal had started drilling and effected a discovery. A fifth class included those who had assumed that the withdrawal order was invalid, and had gone upon the lands after their withdrawal, but before the confirmation order of July 2, 1910. It will be recalled that after Congress passed the law of June 25, 1910, confirming the withdrawals and the right of the President to make them, President Taft had issued an order, on July 2, 1910, confirming all withdrawals previously made. Some operators maintained that the original withdrawal order was not valid, and that entries were legal and proper, until the land was properly withdrawn, after authorization by Congress. A sixth class may be designated to include those who had entered withdrawn lands after the confirmation order of July 2, 1910; and there was even another class, who entered lands still later, after the Supreme Court had declared the original withdrawal order legal and valid. Then of course there were some operators who claimed merely through dummy entrymen.[22]

Many of the men who went on the land after the confirmation order of July 2, 1910, undoubtedly did so in the hope or belief that if they were able to find oil, they could at least exploit it for a time, while Congress was debating, and in the final settlement would somehow be able to square themselves, or at least get some concessions. Wildcatting on ordinary oil lands would not have been attractive on such a chance as this, but drilling on some of these reserved lands, even with a very modest chance of securing part of the production, looked like a fair sort of speculation.[23]

The claims of these California operators were largely on land which had been set aside as naval reserves, and which were under the jurisdiction of the Secretary of the Navy, Josephus Daniels; and Daniels was always at loggerheads with Secretary Lane, of the Department of the Interior, on the question of dealing with these claims. In his position he had the support of Attorney-General Gregory, of many of the geologists in the Geological Survey, and probably of a majority of the officers of the navy. Secretary Lane favored granting relief, in the form of a lease on the land claimed, not only to the second class of operators—those who had entered the land prior to the time of withdrawal, but also to the fifth class—those who had entered the land after the date of the withdrawal order, but before the confirmation decree of July 2, 1910. Secretary Daniels was willing to accord relief only to the first of these classes. Those who had gone on the lands after the withdrawal, in direct violation of the order of the government, seemed to him entitled to no special dispensations. From this position he steadfastly refused to budge, and he appealed to President Wilson for support in his stand.[24]

In every session of Congress for several years, when the oil and gas bill was brought up, the California operators would demand relief; Secretary Lane would second the motion, and Daniels would promptly refuse to have anything to do with the proposal. In that way the entire bill would be defeated. Early in the year 1917, when one of the oil-land bills was up before Congress, a conference was called which included not only Lane and Daniels, but representatives from the Department of Justice and from the Public Lands committees of both the House and the Senate; but there was no way of securing an agreement between the two factions. Daniels was absolutely uncompromising in his stand, and the California oil delegation were equally so.

The record of Secretary Lane on the question of protecting the naval reserves, and on the general question of conservation, merits a little careful consideration. At the time he was appointed, he was heralded as a fortunate choice for the very important position of Secretary of the Interior. He was supposed to be a man of pure motives, of uncompromising integrity, and a thoroughgoing conservationist. It was thought that the public domain would be entirely safe in his keeping. While in office he managed to maintain this reputation with the general public, by writing articles and making speeches on conservation, and by writing on other subjects with such style and effect as to get a strong personal following. He always preserved a tradition of eminent correctness and respectability which was almost impervious to criticism. Pinchot in an open letter to Lane written in August, 1916, declared, "There is a widespread impression that you have made yourself the champion of the conservation of our natural resources for the general benefit"; but he added, "That impression is less justly founded than I could wish."

Pinchot and other conservationists had welcomed Lane's appointment as Secretary of the Interior, but Lane's record in office, particularly on the oil-land question, presently alienated Pinchot and many others. For several years there was almost a continual struggle—between Lane, representing the California oil men, on one hand, and Pinchot, Daniels, and the officers of the navy, and Attorney-General Gregory, on the other hand. In this quarrel, Pinchot went so far as to publish, in August, 1916, an "open letter" to Lane on the subject of the naval oil lands, tracing in careful detail Lane's record on the naval oil lands and on the general subject of conservation. "In such matters as these," declared Pinchot, "you have sometimes been actively and openly against the public interest. In others you have let your subordinates speak for the Department and have kept in the background yourself. In others again, you have been silent when you should have been the champion of public rights. In every one, a duty and responsibility, which can neither be avoided nor evaded, attaches to you." Pinchot accused Lane of trying to abandon the naval reserves, by favoring the Phelan amendment: "The record shows beyond question that your program and the program of the Phelan amendment, for rewarding wilful trespassers with preferential leases at the expense of the Navy's reserves of oil, were one and the same. . . .

"Two Attorneys-General McReynolds and Gregory, and the Secretary of the Navy, have from the first preferred the Navy and the national defense to the claims of the trespassers. You have from the first advocated setting the claims of the trespassers ahead of the Navy and the national defense," declared Pinchot. "In the face of their solemn and repeated warnings, you have persisted in your advocacy of private claims (admittedly not valid before the law) as against the public right." Pinchot accused Lane not only of undue sympathy with the trespassers upon the naval reserves, but of trying to protect their interests by questionable methods, as, for instance, by approving a "joker" in one of the bills, the effect of which would have been to open the reserves to the exploiters. Pinchot pointed out the friendly attitude Lane took toward the proposals of attorneys for the oil operators; and accused Lane of misrepresentation in various matters—in short, of lying.

Pinchot's "letter" contained charges too serious to be ignored; and there is abundant evidence elsewhere that Lane was not consistently and sincerely interested in conservation, that his sympathies were largely with big business, and that he labored persistently to promote the interests of big business—particularly big oil business, to the neglect of the interests of the people. A fair argument might be built up to show that he was one of the most dangerous men that have ever held the office of Secretary of the Interior, because, while he was working persistently to promote the ends of exploiting interests, he was writing articles on conservation, and in general preserving an attitude of impenetrable sincerity and respectability. As Pinchot once said, "We could stop Ballinger, but we couldn't stop Lane."

Many public utterances of Secretary Lane prove clearly that he belonged in the camp of the big interests and the exploiters, and not with the conservationists. His attitude in regard to the California reserves was excellent evidence on this point. In this matter, year after year, he was always lined up with Senator Phelan, and the various representatives of the California oil interests.[25] He thought there was plenty of oil, that there was "no danger of the Navy being short of oil." In his discussion of one of the Phelan amendments, he complained that over 6,000,000 acres of oil lands were "locked up" because of the failure to agree upon relief questions which concerned only about 5,000 acres of land. In another letter he spoke of the 6,000,000 acres of oil lands which were "useless to the country today," because not being exploited. There has hardly been a time since 1859 when there was not too much oil on the market, and it would have been most fortunate if much of the oil lands could have been "locked up" for future use. Similarly, in his 1919 report, he spoke of the "years of struggle and baffled hope" which were to end with the passage of the new leasing bill, as if it were unfortunate that the draining of the withdrawn lands had to be delayed so long. Thus his views on this point were those of the oil exploiter.

Lane's final resignation from the Department of the Interior, and his acceptance of a highly remunerative position—reported at $50,000— with the Pan-American Petroleum and Transport Company, have been criticized a great deal, and with much justice, although it is not un- common for government officials to capitalize their official experience in this way. As a lawyer, Lane was probably not worth more than a frac- tion of $50,000 a year, but as Ex-Secretary of the Interior, his influence was apparently thought to be of great value. It is the guess of the writer that he was hired by the Pan-American company for the sake of the influence it was hoped he might have in Mexican affairs. The Pan-Ameri- can had immense interests in Mexico, and was having difficulty with the Mexican government over the Mexican land laws; and Lane had the "correct" attitude toward Mexico, that is to say, he saw only one side of the question, and that the American oil companies' side.[26]

Perhaps the truth is that Lane merely had a different attitude from that held by most strict conservationists. He was a lawyer by training, and, with his brother, George W. Lane, had been attorney for the Inde- pendent Oil Producers' Agency. The training of a lawyer often leads to an exaggeration of individual rights and an inability to see social rights, and the fact that Lane had actually worked for California oil pro- ducers, and that his brother represented them during the very time that he was Secretary, could hardly have failed to influence him. The troubles of the California operators were the troubles of his friends, and of his brother's friends. It should be added that his record in the management of the Indian lands reflects more credit on him than his management of the public domain.

Finally, it should be said that there was some justice in the demand of the California oil operators for relief. Some of them really believed that the withdrawals were illegal, in fact had been told by competent lawyers that they were illegal. Also in some cases what were at first wrongs had, by transfer, become in a sense vested rights in the hands of at least partly innocent purchasers.[27]

With the entry of the United States into the World War, there was little time for discussion of land legislation, although there were scat- tered efforts to make provision for opening the public oil lands. Senator

Walsh, of Montana, brought in a bill similar to the Ferris Bill of the previous administration, but after considerable debate, it was defeated, largely because of a disagreement between the House and Senate on the question of relief for the California oil men. There was some effort on the part of the anti-conservation forces to take advantage of the war to get a law providing for the sale of oil lands, the pretext being that the needs of the war were so pressing as to demand an immediate increase in oil production, regardless of the means of securing it. The California State Council of Defense carried on a campaign in 1917 to get the lands opened for exploitation, but this was without effect.[28]

There were, on the other hand, an increasing number of men who urged not only government retention of the oil lands but government drilling and operation of the oil wells themselves. Senator Borah—an active enemy of the conservation policy generally—announced, in the discussion of the Walsh Bill, that he had come to favor such a policy, on the ground that any other manner of operation, under a system of sale or leasing, was certain to result in monopoly. Borah saw that the oil industry was a natural monopoly, and a few other members of Congress began to see this. Pittman suggested that the President would have to take over and develop the oil lands if no law was passed providing for it, and he thought Congress was not likely to agree on any bill. Secretary Lane was said to have wanted such a provision in 1917, authorizing either the Department of the Interior or some other department to drill for oil on the government reserves.[29] As early as 1913, Secretary Daniels had recommended legislation which would "enable the (Navy) Department to refine its own oil from its own wells."[30] Senator Walsh once said that there was a sentiment growing up in favor of government exploitation of oil lands, and that there were a number of senators who favored that policy.[31]

The alignment of forces on the Walsh Bill above mentioned showed how complex and how difficult the oil question had become. It was not strictly a question of East versus West, as it had once been on the forest land question and some other public land questions. It was not entirely a conservation question. Many men seemed hardly to know whether the Walsh Bill was a conservation measure or not, it had been so patched and doctored in an attempt to effect a compromise. In the Senate, Smoot, who, on forest land questions, had been a conservationist, opposed it. Borah opposed the leasing provision, and asked for government operation. Phelan, of California, would have liked more relief for the oil men, but supported the bill anyhow. King, of Utah, opposed strongly, and asked that the "clammy oppressive hand of the government" be removed from the "adventuresome persons who are willing to go upon the public domain and hazard their money in the costly efforts to find producing wells." Penrose, of Pennsylvania, recalled that many Pennsylvania oil men had gone out to Wyoming and wanted more liberal treatment for them. La Follette thought it was too generous, and Kendrick, of Wyoming, thought it was about right. Kendrick pronounced it "the herald of a new era."

When the conference report on the Walsh Bill was made in the House, Ferris, who had the bill in charge, pronounced it satisfactory, but he was about the only one who had this view. Some men thought it was not liberal enough in its treatment of the California oil men and other "pioneers." Others thought it entirely too liberal. Stafford, of Wisconsin, called the conference report an "abject surrender of the principles the House has always stood for." London, of New York, declared it a surrender of the people's wealth to private interests, "an attempt at public plunder in the expiring days of a session." Frear, of Wisconsin, asserted that two men, chairmen of Democratic national committees from two western states, were hired to help get the bill through, and that they were in Washington all the time, working for the passage of the bill. Crampton, of Michigan, objected to the provision giving the President the right to compromise California claims, and Mann could see no reason for giving trespassers seven-eighths of the oil they could get in their illegal operations. Chandler, of Oklahoma, thought the provision requiring a certain depth of drilling was bad, and Juul, of Illinois, declared no prospecting would be done under the law. Thus the bill seemed to suit nobody very well, and yet everybody thought some sort of bill should be passed. Of course there was much political buncombe in the discussions, which tended to cloud the real issues.

One of the strong influences against the bill, as it emerged from the conference, was Gifford Pinchot. He was not at this time occupying any government position, but was following the oil-land legislation closely, and he wrote a letter to one of the congressmen, pointing out various faults in the bill. In the first place, he opposed granting any absolute patents, pointing out that the existence of numerous alienated tracts in an oil field would destroy that unified control which is the main purpose of a leasing system. In the second place, Pinchot opposed the granting of so much to the oil trespassers in California. The conference report had given each of these trespassers who had a flowing well the right to lease 31 acres surrounding it, with a preference right in the leasing of the rest of his claim. The government was forbidden to inquire into the fraudulent character of any claim unless it could be shown that the corporation which had bought it had actual notices of the fraud.

To all this, and to many other features of the conference bill, Pinchot objected strongly, and he pointed out that the conference report had falsely represented the attitude of a number of government officials as to some provisions of the bill. Pinchot asserted that under the Operating

Agreements Act of 1914, giving authority to make arrangements for continuing the operation of the wells of the trespassers during the suits against them, the Interior Department had abused its discretion in giving to every trespasser seven-eighths of the oil wrongfully taken by him. Pinchot sharply criticized Secretary Lane for his action, and declared he "should not be trusted again." There is no doubt that Pinchot's attitude had great influence in Congress, and was one of the reasons for the defeat of the bill. Frear explained that in his opposition to the bill he was largely influenced by Pinchot.[32]

NOTES

1. S. 5485, S. 5492, 61 Cong., 2 sess.
2. S. 9011, S. 9817, 61 Cong., 3 sess.
3. S. 1584, S. 1587, H. R. 777, 62 Cong., 1 sess.
4. Stat., 35, 844.
5. Stat., 36, 583.
6. Stat., 37, 105.
7. S. 3045, 62 Cong., 2 sess.; Stat., 37, 496.
8. H. R. 26812, H. R. 27831, 62 Cong., 3 sess.; Stat., 37, 687; 38, 509.
9. H. Res. 108, S. 60, S. 475, H. R. 6203, 63 Cong., 1 sess.; *Cong. Rec.*, Appendix, p. 473.
10. *Rept., L. Off.*, 1916, 64.
11. H. R. 16136, 63 Cong., 2 sess.
12. *Eng. and Min. Jour.*, Aug. 18, 1917, 300.
13. *Cong. Rec.*, p. 14945-14950.
14. *Ibid.*, p. 15053-15064.
15. H. R. 12246.
16. This amendment was essentially the same as a bill introduced in the same session by Church, of California. (H. R. 15661.)
17. H. R. 15036, 63 Cong., 2 sess.
18. H. R. 406, 64 Cong., 1 sess.
19. *Oil and Gas Jour.*, Nov. 11, 1915, 3.
20. *Ibid.*, March 7, 1919, 3.
21. Hearings before the House Committee on Public Lands on H. R. 24070, May 13 and 17, 1910, p. 75, 78.
22. *Cong. Rec.*, Sept. 3, 1919, 4746.
23. *Oil and Gas Jour.*, July 11, 1918, 73.
24. *Ibid.*, June 22, 1916, 32.
25. Perhaps Secretary Lane should be allowed to state his own view on the question of relief for the California operators. In his 1915 report appears the following:

There was doubt at the time of these withdrawals as to their legality, there being no specific statute on the books authorizing the action. So serious was this doubt that as a precautionary measure Congress at its next session passed an act authorizing such withdrawals, and the same lands were subsequently, in July, 1910, withdrawn again. It was the opinion of many of the most competent members of the bar that the withdrawal of 1909 was void, and the operators proceeded to act in accordance with this advice. The result was that when the second withdrawal, that of 1910, was made, there were a number of operators engaged in drilling and some of them had already found oil on these lands. The government insisted upon the validity of the 1909 withdrawal, and after failing to have its view sustained in the lower courts, was at last successful before the Supreme Court. . . . If the full measure of the government's right is acted upon as a basis of our policy in dealing with these lands it will bankrupt many oil companies and do what appears to me to be an injustice, and an unnecessary injustice, to those who have invested many millions of dollars under a mistake as to the law. I shall not assume to say what policy should be followed as to the naval reserves, but as to the other withdrawn lands I believe Congress (which is the one forum wherein relief can be sought) should so act as to recognize the equitable rights of these operators. This might be done by saying that those who would today be entitled to patent were the land not withdrawn may have leases under which they will pay a liberal royalty to the government. This plan will doubtless be urged. I am of the opinion that it is too liberal. We might draw a line at the time of the second withdrawal. If this were done, leases could be made to all who were actually operating upon the land at that time. And if it is thought advisable, there could be imposed a higher royalty than would be called for under the general development bill. I feel that this is one of those situations often arising in the life of the individual and of the state when it is not wise to exact all that the law allows even as to those who are in the wrong. (*Rept.,* Sec. of Int., 1915, 16.)

26. See *Cal. Oil World,* May 26, 1921, Second Section, p. 29.

27. U. S. Geol. Survey, *Bul. 705;* Morrison, *Mining Rights,* 422; Open Letter to the Honorable Franklin K. Lane, Secretary of the Interior, by Gifford Pinchot, Concerning the Navy's Oil Lands, and Other Conservation Legislation, August 12, 1916; Bulletin to the Members of the National Conservation Association, Jan. 1, 1917, National Conservation Association, Washington, D. C.; *Cal. Oil World,* April 26, 1923; *Oil and Gas Jour.,* Aug. 24, 1916, 3; Feb. 20, 1920, 2, 78; May 27, 1921, 2; *Outlook,* Jan. 3, 1914, 13.

28. S. 2812, 65 Cong., 1 sess.; *Sunset Magazine,* Sept., 1917, 8; *Oil and Gas Jour.,* July 26, 1917, 29.

29. *Oil and Gas Jour.,* March 8, 1917, 33.

30. Hearings before the Senate Committee on Public Lands, on S. Res. 282 and S. Res. 294; Oct. 22, 1923, p. 156.

31. *Cong. Rec.,* Aug. 21, 1919, 4112; Jan. 4, 1918, 648; Sept. 17, 1918, 10382.

32. *Ibid.,* Feb. 18, 1919, 3705.

CHAPTER XXIV

THE PUBLIC OIL LANDS: THE
LEASING LAW

THUS the Sixty-fifth Congress passed into history without doing anything to open up the reserved oil and coal lands. The interests involved were too important to permit the question to sleep long, however, and, early in the next congress, in August, 1919, Senator Smoot introduced a bill for the Committee on Public Lands, which he declared to be unanimously approved by that committee; and, after days of debate and discussion, this was finally enacted into law. This bill, with its 38 long sections, covering coal, oil, oil shale, gas, phosphate, and sodium, presents a wilderness of intricate and complicated provisions, but the important sections relating to oil must be examined.[1]

The Smoot Bill divided oil and gas lands into wildcat or unproved, and proved, and provided regulations for each. For wildcat prospecting, it authorized the Secretary of the Interior to grant prospecting permits, running for two years, and covering not over 2,560 acres. The permittee was required to begin drilling within six months, and must drill 500 feet within two years, unless he effected a discovery in less time, or unless, for a good reason, he was excused by the Secretary. In Alaska these prospecting permits ran for four years, and were more liberal also in their other provisions. On discovery, the permittee was allowed a lease covering one-fourth of the land in the permit, the lease to run for 20 years, at a royalty of 5 per cent, and a rent of $1 per acre, with a right of renewal. Thus the Smoot Bill, unlike the bills presented in earlier years, was a pure leasing bill, with no provision for the granting of any fees whatever. For a period of six months, the permittee also had a preference right to the remainder of his 2,560-acre prospecting permit, at a royalty of from 12½ per cent to 25 per cent.

In proved territory, leases were to be awarded to the highest bidder in tracts of not over 640 acres, the leases to run for 20 years at not less than 12½ per cent royalty, with a privilege of renewal for successive periods of 10 years, on terms to be prescribed by the Secretary of the Interior.

There were various provisions to prevent waste of oil or gas, to prevent flooding of the oil sands, and to prevent the wasteful offset wells, no wells were to be drilled within 200 feet of the outer boundaries of any lease, unless necessary to offset wells of adjoining private owners. There were also various provisions designed to prevent monopoly. No individual or corporation was allowed to hold more than one lease within a given field, or more than three leases in any state, or, as a stockholder of another corporation, to hold any interest in more than that number of leases. Aliens were forbidden to own any interest in a lease, by stock ownership or otherwise.

Leases for the exploitation of oil shale were limited to a maximum area of 5,120 acres, and might run for an indeterminate period, subject to a readjustment of terms at the end of 20 years. Royalties might be fixed by the Secretary of the Interior at any point between 5 per cent and 25 per cent, or the Secretary might even waive the royalties for a term of 5 years.

Of the royalties arising from all leases, except those in Alaska, 45 per cent was to go to the reclamation fund, 45 per cent to the state for schools, and 10 per cent to the Treasury of the United States. Thus the bill was exceedingly generous to the states, as might have been expected of a committee composed almost entirely of western men.

Once more the much debated relief provisions appeared, in several sections of the bill. Operators on land withdrawn on September 27, 1909, who had drilled one or more wells, were permitted to relinquish the interest which they had prior to July 2, 1910 (the day when the President reissued his withdrawal order, after being specifically authorized by Congress), and on the payment of a royalty of one-eighth for the oil which they had taken, were allowed a 20-year lease at from $12\frac{1}{2}$ to 25 per cent royalty. In case the claims were in the naval reserves, only the producing wells were to be leased. Persons who at the time of any withdrawal order, or who on January 1, 1919, were bona fide occupants or claimants of oil lands not withdrawn from entry, and who had previously performed all required acts but had not made any discovery, were allowed prospecting permits, or if they had made discovery, were allowed leases. The relief provisions were reasonably generous.

In the debates on the Smoot Bill, there seemed to be no well-defined division of opinion in the Senate. Some men from the West favored the bill, and others opposed, and the men from eastern states presented no united front. There were so many different provisions in the bill that most senators found some provisions that pleased them and some that did not; and the debates proceeded in rather desultory fashion. At times there were only about a dozen senators present at the discussions.

Most of the western men disliked the leasing policy, but felt constrained to vote for the bill, because it represented the only way of opening the reserved lands to exploitation. Smoot explained that he did not favor a leasing policy, but realized that no other policy would get the approval of Congress. King, of Utah, opposed the "paternalism, the bureaucracy, the autocracy, the un-American system that the leasing bill entails," and suggested that "private ownership" was "the basis of the prosperity of the American people," but made no further suggestion, except the entirely impossible suggestion that the lands be ceded to the states. Ashurst, of Arizona, bewailed the fact that "president after president seem to fall under the same baleful view that we must conserve the resources of the country by taking them from our people"; but he admitted that he might "hold his nose and vote for the bill." Walsh, of Montana, was one of the few western men actively favoring the principle of leasing. Thomas, of Colorado, saw a need of safeguarding the "patrimony of posterity"; but evinced deep "consideration for the interests and welfare of the people who are now on earth, and who are suffering from discontents and difficulties innumerable"; and, although he did not like the bill, he was willing to support it, in order to relieve some of these difficulties. Fall stated that he preferred the Placer Law, but would vote for this bill, since the conservationists had been "successful in fastening upon the people of the United States their theories of governmental reservation and control."

Walsh, of Montana, asserted that there was a growing sentiment in favor of actual government development and operation in the coal and oil lands, and urged the passage of the bill to forestall any such procedure. Smoot agreed with him as to the possibility of government operation; and Walsh, of Massachusetts, actually advocated it strongly. The latter pointed out that they were dealing with "the last assets in lands and minerals that the people of the United States possess"; and in conclusion declared: "I personally think we have reached a stage in our public affairs where the development of public lands and public properties should be done solely and alone by the government. We have seen the mistake and the failure of private enterprise." Kirby, of Arkansas, offered a substitute for the bill, authorizing the President to mine and develop coal, oil, and gas on any of the public lands, when in his discretion the public exigencies should require it. Of course men like Fall and Jones, of New Mexico, and some other western men criticized this policy severely, and Kirby's substitute was promptly voted down. La Follette's amendment authorizing the government to fix the prices of any products coming from leased lands was likewise voted down, although there were ten votes in favor of it.[2]

The views of many of the western senators represented typical business men's views, such as might have been voiced at a chamber of commerce dinner or a convention of manufacturers. There was much praising of "American initiative and American energy, and American industry," and much distrust of government activity. Fall was convinced that "individually we are the greatest people on the face of the earth in all commercial and industrial matters"; but he was also convinced that "collectively as a government in business matters we are and always have been and always will be a failure."[3] King, of Utah, declared, "It is an axiom now, as it has been in the past, that that government is best which governs least." King voiced a very great distrust of the "commissions, agencies, governmental instrumentalities, departments, bureaus, so numerous as to be unnamed." "These bureaus and agencies and departments and commissions and legislative and executive instrumentalities," he declared, "are provided to promulgate more orders and more regulations, and to bring confusion and discord to communities and a despotic paternalism to individuals and to states." With this view went a characteristic business man's distrust of experts, Fall declaring that there had "never been a mine opened in the United States by any geological or other experts."

La Follette of course represented a radically different attitude. He was intensely interested in the bill, and spoke for many hours, in fact he occupied a considerable portion of the time for several days, in his discussion of various aspects of the bill. His account of the manner in which the conservation movement arose is worth quoting:

Wisconsin was one of the finest timber states in the United States. We went through a similar struggle in Wisconsin in regard to our vast timber resources. We had the same argument made to us there that you hear day by day from the western Senators here. It is all an old chapter to us. We did with our timber what you propose to do here. They swept it off; they organized their lumber trust; they exploited the people of Wisconsin. We got very little benefit from their "development." They devastated the northern half of the state by leaving conditions there which produced annual fires that swept over it and made a large portion of it a charred and blackened waste year after year. These great lumber organizations, enriched by the despoliation of our natural resources in Wisconsin, became the dictators of the public policy of the state. They took charge of legislation and exempted themselves from taxation up in their timberlands. They were not content with that. They became interested, of course, in the transportation problem. They went in with the railroads of Wisconsin, and became a great force, a great corrupting force, in that state. I undertake to say that senators from the West fully realize how the Amalgamated Copper Company, a great mining organization, has gone into your western country and under the name of development acquired control, by corporations and by syndicates, of your great natural resources, amassed great wealth, dominated your politics, elected your governors, controlled your state legislatures, and entered contests for United States senatorships.

Those high-handed methods finally resulted in the education of the public until out of it there grew a widespread demand for the conservation of these natural resources. It had its beginning along about 1906.[4]

As usual, the relief sections of the bill came in for heated discussion. La Follette declared that these sections as drawn would destroy the naval reserves, and he read letters from Secretary Daniels, and Assistant Secretary Franklin Roosevelt, of the Navy Department, and from officers in the Department of Justice, opposing similar clauses in a previous bill; and he finally offered an amendment providing that the bill should not apply to naval reserves, but this was rejected.

Those who argued for the relief provisions based their argument largely on the fact that the withdrawal of September, 1909, was thought by many lawyers to be unconstitutional, and that therefore the oil operators who went upon these lands after they were withdrawn were not ordinary trespassers. In other words, the argument was that the man who sought competent legal advice, and then followed the advice, was not guilty of any great offense, even though he presently found himself taking oil from government lands. On the other hand, La Follette argued with considerable logic that these men, in trespassing upon reserved land, took a chance, and knew that they were taking a chance. When their legal advice later proved to be faulty, that was no occasion for government relief. The government had warned them to stay off the land, and was certainly under no obligation to relieve them of the consequences of their violation of its own edict. Advocates of the relief sections quoted liberally from a report of Joseph A. Phelan, who, as special "oil examiner" for the United States Shipping Board, had made an examination of the naval reserves, and who had reported that it would be very difficult to retain any oil in them, anyhow, because of the draining of the oil by private owners scattered through the reserves. La Follette, however, asserted that Phelan was really an oil man himself, had been an employee of various oil companies most of his life, and as a matter of fact had been employed by the Shipping Board only for a couple of months while he was making this report. La Follette accused him of being part of the Standard Oil lobby working for the bill.

In connection with the discussion of the relief section, the old quarrel between the Secretary of the Interior and the Secretary of the Navy was again brought forward for an airing. Thomas, of Colorado, told the story of the assistant in the Attorney-General's office who had criticized Secretary Lane so severely that he was dismissed by the succeeding Attorney-General. This particular assistant had accused Lane of trying to push some fraudulent locations to patent. La Follette declared the bill as drawn would destroy the naval reserves, and praised Secretary Daniels for his staunch stand against any such action. La Follette quoted from Secretary Daniels to show how powerful were some of the oil interests involved in this relief question. The list of companies included the Standard Oil Company of California, the Union Oil Company, the Associated Oil Company, the Honolulu Consolidated Oil Company, the General Petroleum Company, the American Consolidated Oil

Company—some of the largest oil companies in the West. The size of some of these oil companies was significant only as it threw light on the influence of the lobby that was always at work when the bill was under consideration.

After several days of not very spirited debate, the bill passed the Senate with little opposition; but the House, as in the case of many previous land development bills, promptly substituted another bill. The House bill differed in several respects from the bill the Senate had passed. In the first place, it eliminated the maximum limits which the Senate had imposed on royalties. In the second place, it changed the alien ownership provision so that it applied only to aliens of those countries that discriminate against American oil companies abroad; and in the third place, it was less generous in its relief provisions for oil-land trespassers. In general it was less liberal to the oil operators, and more liberal to the government.

In the House, there were about the same general arguments about "pioneers," "American enterprise," "development of the West," "monopoly" and the "Standard Oil Company," alien ownership and the like; but, as in the Senate, there was no definite party of opposition. Some men thought the provisions of the bill too liberal. Griffin, of New York, called it a "bill to surrender the coal, phosphate, oil, gas, and sodium on the public domain to the coal trust and the Standard Oil Company." Other congressmen, largely from the West, thought it not liberal enough. Elston, of California, thought that the provision giving the wildcatter a lease covering only one-fourth of his tract was unfair to him, and left the government all the "velvet," but he nevertheless favored the bill. Other western men, as in the Senate, opposed the general policy of leasing. There was, however, a hint that a considerable number of people wanted not only government reservation of the land, but also government development.

The question of alien ownership of leases proved a delicate and baffling question. Peace negotiations were going on in Europe, and it seemed to be thought necessary to avoid any open criticism of our late allies; but there was much veiled suspicion and hinted criticism of the policy of Great Britain with respect to the world's oil resources. Some of the senators resented the "closed door" that American oil companies were reported to have found in some of the British dominions; while at the same time they wished to adopt a similar policy for our own public lands, but were fearful of retaliation from friendly countries, particularly from Mexico. There was something interesting, if not actually laughable, about the manner in which some men tried to figure out a way to maintain a "closed door" at home, and at the same time preserve the "open door" for American oil companies in foreign countries. In trying to reconcile these two aims, some of the senators were hardly less logical, however, than they were in calling for the open door abroad, so the United States could be assured her future supplies of oil, at the same time that they asserted that the domestic supply was "inexhaustible," so that there was no need for conservation.

Anything in the nature of "pork" always stirs up interest in Congress, and the division of the receipts was debated with some spirit. The Senate bill had given 45 per cent to the states, 45 per cent to the reclamation fund (this also went to the West of course), and 10 per cent to the federal treasury. Sinnott, of Oregon, representing the Public Lands Committee, offered an amendment reducing the states' share to 20 per cent, and increasing the share for the reclamation fund to 70 per cent—of the past production from reserved lands. Nearly $20,000,000 worth of oil had been produced on these lands, and the Secretary of the Interior was holding this sum for division among the oil companies, the federal government, the states, and the reclamation fund. Ferris, of Oklahoma, declared that the western states were too greedy in their demands for these receipts, and he offered an amendment giving all of the receipts to the reclamation fund, whence it would finally go, as funds are supposed to revolve out of the reclamation fund, one-half to the states and one-half to the federal government. The strong protest called forth by this proposal indicated that the western men did not value their reclamation fund very highly. In the debates, the opinion was several times expressed that money once put into the reclamation fund would remain there a long time. Mays, of Utah, declared that money put into the reclamation fund would probably not come back "in the lifetime of any member here"; and Mondell expressed a similar view. The western men wanted their money now, to use as they pleased.

In this connection, several pointed to the division of forest reserve receipts, and particularly to the division of the receipts from the forests which had, only a few years before, been recovered from the Southern Pacific Railroad Company, in Oregon and California. Mays, of Utah, pointed out that the state of Oregon had received the lion's share of these receipts, and now, since Utah was rich in coal instead of timber, he wanted the states to receive a large share of the royalties from the coal and oil lands. Of course the men from California and Wyoming were anxious to get for the states a large share of the oil receipts; the men from several western states were interested in phosphates and some were interested in oil shale. Thus most of the western men wanted a large share of the receipts from at least one kind of natural resource, and by trading among themselves they were able to secure a very large share of the receipts—undoubtedly too large. The eastern men were generally less interested in securing for the national government a fair share of these revenues than they were in raiding the treasury in their own interest, on some other proposition. Ferris got almost no support for his amendment, and finally withdrew it to "give it the dignity of a burial."

With the question of division of the royalties came the question of the loss of taxing power in the West through the reservation of lands. This had been a frequent cause of complaint, in discussions of the forest reserves, and it was here brought up in connection with coal and oil. Mays put the western view clearly:

If this bill brings the operations which its friends predict, great enterprises will be established, towns will be built, roads must be constructed, schools must be maintained, local government must be supported, the states must police and protect the properties. Titles to all the lands are to remain in the federal government exempt from taxation. Where are the states and the counties to get the money? Gentlemen say the improvements are to be taxable. But we must realize that the tipple of the coal operation or the derrick of the oil well represent but a small value compared to the coal or oil resources of the land itself.

Mondell expressed his opinion even more strongly:

If the federal government can retain title to extensive areas of very great value in states and communities and hold them thus indefinitely, free from taxation, the federal government can through that means absolutely arrest development, bankrupt a community partly developed, and paralyze the operations of local government. The doctrine that such a policy is tolerable is monstrous.

There was some point to the taxation argument, but most of the talk on the subject was political buncombe. The reply of Ferris to one of these arguments was not inapt. Evans, of Nevada, in discussing the plan to turn royalties over to the reclamation fund, asked Ferris who would build the roads in Nevada, where 90 per cent of the roads and hundreds of miles of territory were government owned; and Ferris promptly replied: "Who will buy me a Packard automobile? Who will furnish the money to do for me the things I cannot do for myself? The gentleman would go down into the government treasury and take money to spend locally that ought to be spent by the people locally." If the government builds up oil towns or coal towns on its own domain, it would not fail to provide such schools and roads as would be necessary for decent living. It would be impossible for the government to secure development of its property, without these. As to roads on the western government forest reserves, they are built by the federal government, and are generally far better than the roads maintained in those communities where the land has been patented. Probably this would also be true in coal and oil reserves.

Much of the reasoning regarding this loss of taxing power was superficial, or worse. In figuring the revenues that would have been enjoyed if the lands had been subject to taxation, western men usually assumed that the resources in the land—the coal, oil, etc.—would be taxed like any other property. Mays stated this clearly in his argument. But no rational system of taxation would include any considerable tax on undeveloped resources. A heavy tax on coal or oil deposits, like a heavy tax on standing timber, would encourage rapid and wasteful exploitation, and would therefore be most unfortunate. There are already far too many factors favoring rapid exploitation of our natural resources, and the tax on undeveloped resources should always be very light. The tax should fall mainly on the current production, and of course the government royalty does this.

Much of the reasoning on this subject took only a short-time view of the matter. If all natural resources were turned over to private exploitation, there can be no doubt that they would be exploited more rapidly than they ever would in federal reserves, and that some of the communities involved would enjoy an era of what Americans commonly regard as prosperity; but, if such a policy resulted in the speedy and wasteful destruction of the resources, such prosperity would be short-lived. The government, in its reservation policy, is aiming at long-time results.

It was commonly assumed in the complaints about the loss of taxing power, that if the reserved lands were turned over to private ownership, and coal and oil towns were built there, the only effect would be an increase in tax revenues. No attention has usually been given to the fact that with an increased population would come an increased expense for enforcing justice, providing schools, etc. These expenses might not increase in the same proportion as the revenues, or perhaps not as fast, but they would certainly increase somewhat. As a matter of general reasoning conservation does not involve any general sacrifice, for present or for future generations, even in the West. It is true that it may result in a slower rise in rents and land values, and may at first involve all the disadvantages that arise from sparseness of population, but it must be remembered that sparseness of population has its own advantages. If the conservation policy means a wiser and more economical use of resources, the standard of comfort in the West will in the long run be higher than it would have been under a policy of exploitation.[5]

The House finally reduced the share of receipts going to the states, and passed the amended bill by a vote of 391 to 169. The Senate refused to accede to the House amendments and a conference committee was appointed, but the conference committee allowed most of the House amendments, including the amendments eliminating the upper limits on royalties, the amendment offering less generous relief provisions, and the amendment refusing leases only to aliens from those countries that discriminated against Americans in their own oil development. On the question of the division of the royalties, the conference committee effected a compromise, giving 52½ per cent to the reclamation fund, 37½ per cent to the states, and 10 per cent to the federal treasury. The conference report was accepted by the House almost unanimously, and by the Senate without a word of comment, and so became law.

The Leasing Law, as finally passed, is a rather long and complicated piece of legislation, providing for the disposition of coal, phosphate, sodium, oil, oil shale, and gas deposits. To oil prospectors or wildcatters it offers a prospecting permit covering 2,560 acres of unproved land, on which the prospector must begin drilling operations, within six months, and must drill wells to an aggregate depth of 2,000 feet within two years, unless he finds oil sooner. In Alaska prospectors are allowed as many as five permits, with more liberal terms as to time of drilling. On discovery of oil or gas, the prospector is entitled to a 20-year lease of one-fourth of the land, or to 160 acres if the permit covers as much as that, at a royalty of 5 per cent and a rental of $1 per acre. The permittee is further entitled to a preferential right to a lease of the remainder of the land in his prospecting permit, at a royalty to be determined by the Secretary of the Interior, but not less than 12½ per cent.

Proved oil lands on the public domain, under the Leasing Law, may be leased by the Secretary of the Interior to the highest bidder, in tracts not larger than 640 acres, at a royalty of not less than 12½ per cent, and a rental of not less than one dollar per acre. These leases run for twenty years, with a preferential right to a ten-year renewal, on terms set by the Secretary of the Interior.

Certain provisions are designed to prevent waste. No wells are to be drilled within 200 feet of the outer boundaries of leases, unless the adjoining lands are privately owned. To encourage complete extraction of oil the law authorizes the Secretary of the Interior to reduce the royalty on wells of ten barrels daily production or less. The Secretary is given the general power to draw up regulations to prevent waste.

No person or corporation is permitted to hold more than one lease "within the geologic structure of the same producing oil or gas field," or more than three leases in any state. Monopoly in the sale of oil is strictly prohibited. Receipts from government royalties, rentals, and bonuses, as already stated, are to go, 10 per cent to the United States Treasury, 52½ per cent to the reclamation fund, and 37½ per cent to the states in which the lands are situated, for roads and schools.

The relief sections of the Leasing Law are too complicated to receive careful treatment, but they may be noted. Claimants on the public land,

outside the naval reserves, were permitted to relinquish to the government whatever "right, title and interest" they possessed prior to July
2, 1910 (the time when Congress validated the Taft withdrawals), and
secure a lease of their land, for 20 years, at not less than 12½ per cent
royalty. Claimants on the naval reserves might secure leases only of
their producing wells, unless the President, at his discretion, chose to
permit leases of the remainder of their claims.

The bill as passed was clearly regarded as a victory for the conservation forces. Letters were read in the House, from Gifford Pinchot, from
Secretary Daniels, and from Lane, praising the House bill, which was
substantially the same as the bill passed; and Representative Sinnott,
of the Public Lands Committee, was warmly commended by some of his
fellow congressmen, for his work in securing so good a law. The action
of the Senate, in adopting the conference bill without a word of comment, indicated that the Senate was not enthusiastic about the bill, but
felt that it must take it or nothing. The following article, appearing in
a Colorado state publication soon after the passage of the bill, probably
indicates fairly the attitude of many western people:

The bill is a complete reversal of the original land policy of the United
States. . . . The rights of the West, and particularly state rights, have
been jeopardized by the enthusiastic federal control school which seemed to
take the lead in proclaiming federal over state rights. . . . There has never
been a measure more bitterly fought out along technical lines with apparent
determination on the part of the Administration to give no quarter and to
allow no liberality in phrasing, and the bill as now completed, while a complete defeat for states' rights advocates, practically results in confiscation
of many properties. . . . It will help the West though, to open some of
this, even under a lease.[6]

The administration of the new law was placed in the hands of the
Secretary of the Interior, the General Land Office, the Bureau of Mines,
and the Geological Survey coöperating in the work.

The outstanding feature of the new law, and the feature which con-
trasts most decisively with the laws of most foreign oil-producing countries, is the smallness of the leases, suggesting, of course, our reliance
upon competition. In Persia, India, the East Indies, and in most South
American countries, at least a degree of monopoly is taken for granted,
and the land is leased in large tracts.

Development or exploitation of the public oil lands has proceeded
with sufficient rapidity under the Act of 1920. During the consideration
of the bill, scouts and geologists had studied the promising areas of
reserved lands; and even before the bill was signed, claimants were
camped within striking distance of the reserves like "crows on a fence."
It is said that some oil prospectors had aeroplanes placed at the tele-

graph stations, awaiting the word of the approval of the legislation, to carry the news to waiting representatives near where claims had been spotted. Applications under the act began to come in by telegraph on the day it was signed. In some sections, particularly in Wyoming and Montana, important and unexpected discoveries were made just about the time of the passage of the act, and the entire areas for many miles around these discoveries were plastered with applications. At the end of the year, the Commissioner of the Land Office reported that 5,000 applications for prospecting permits had been filed, not only in the western states but in Alaska. Many conflicts appeared among these applications, in fact some applications were reported involving as many as 20 conflicts.[7]

Thousands of prospecting permits are issued each year. On June 30, 1924, a total of 32,103 applications had been filed under the Leasing Act, and most of these were applications for oil permits. More than 4,000 applications for oil permits were received in the year 1924. Only a small proportion of these permits have resulted in discovery and in leases. In the fiscal year 1922, 10 leases were issued, covering 4,860 acres of land; in 1923, 4 leases were issued, covering 1,604.8 acres; and in 1924, 21 leases were issued, covering 3,674.93 acres.

These figures do not include operations within the naval reserves, which by executive order were placed under the administration of the Bureau of Mines. The oil and gas wells on the public domain are already so numerous and productive that their supervision constitutes a large business. In the fiscal year 1922, over 18,000,000 barrels of petroleum were produced from these wells; and of this amount the government received 3,616,852 barrels, worth $4,768,397 as royalty. In addition, the government was receiving at the end of the year, $14,000 a month from gas leases in Wyoming. In the Rocky Mountain field alone, there were on government land 336 wells, of which 267 were in the Salt Creek field, and 100 more were being drilled. The largest number of producing wells belonging to the government were in the Salt Creek field of Wyoming, where the government holds title to almost all of the land.[8]

The Leasing Act has not proved to be of very great importance in the discovery and exploitation of new oil fields, although it has contributed to that end. It has been more important in quieting the title to lands which were already proved. Almost all the holders of oil placer claims in tested areas, such as Salt Creek, even of claims dating back to the early days, applied for government leases, instead of pressing their placer claims.[9]

The government has adopted various measures to prevent waste. Not only have the government officials in the Bureau of Mines established careful rules and supervision to assure the protection of the oil sands against inflow of water, and to prevent outright waste of oil, but they have tried, as far as possible, to prevent the waste involved in over-production, or production at low prices. During the period of depression and low prices in 1921, for instance, they granted lessees relief from drilling requirements, except as to offsetting wells and as to wells on lands which, under the relief provisions of the law, had been leased at 5 per cent royalty. In this last respect, government ownership has demonstrated its superiority to private ownership, for during the time when the government was allowing operators to cease drilling, many private owners were compelling operators on their lands to drill regardless of the market.

In the adjustment of royalties, the Bureau of Mines adopted an important and sensible innovation, in the sliding scale of royalties, the rate of royalty depending upon the production of the wells, and upon the quality of the oil. This scheme is based upon recognition of the fact that the operator of a larger well, or of one which yields petroleum of higher grade, can afford to pay a higher royalty, whereas the operator of a small well, or a well yielding low-grade oil, must have a low rate or he must close down. The abandonment of a well means the waste of all the oil remaining in the ground, as well as the end of all government royalties, and in adopting this sliding scale the government has again done better than have private owners of oil lands.

NOTES

1. S. 2775, 66 Cong., 1 sess.
2. *Cong. Rec.,* p. 4733.
3. p. 4290.
4. p. 4742.
5. See the author's *United States Forest Policy,* 273-281.
6. *The Colorado School of Mines Magazine,* March, 1920, 48.
7. *Rept.,* L. Off., 1920, 79. The following account of the excitement in Wyoming following the passage of the leasing act appeared in a leading oil journal a few days afterward:

When the word arrived in Casper, just seven minutes after the President had affixed his signature to the leasing bill, there was a scurrying and much telephoning to the remote districts, and then everyone took a half hour off to watch the stock boards and gossip over the successful outcome. Credit was given Mondell, and also the senators from Colorado and Wyoming. The expected skyrocketing of oil issues failed to materialize tho.

Out in the oil fields was excitement aplenty. The hurried news that the bill was a law started guards out in every direction, and in a few instances claim jumpers were also in evidence. The guards reached their previously allotted stations in some of the outlying districts at dusk and took up the vigil that was to last through the night. At Salt Creek, as soon as the daylight faded so that objects became indistinct, at a short distance, a corps of autos patrolled the trails and roads, carrying armed men who scrutinized any strange car or traveller closely. Luckily there was no gun play recorded, but the sight of a validation stake in a strange car or a traveller carrying an axe and stakes would have precipitated a shooting affray immediately, and would-be claim jumpers seemed to realize this.

When Thursday morning dawned, however, there were several tracts in the productive area that had blossomed over night with a new set of stakes and validation notices. (*Oil and Gas Jour.,* March 5, 1920, 44.)

8. U. S. Bu. of Mines, *Rept.,* Director, 1922, 10; Bul. 232, *Manual for Oil and Gas Operations,* by Swigart and Beecher; *Rept.,* L. Off., 1922, 22; 1923, 24; 1924, 29.

9. *Oil and Gas Jour.,* Sept. 17, 1920, 38; Jan. 28, 1921, 2.

CHAPTER XXV

THE NAVAL RESERVES AND THE TEAPOT DOME INVESTIGATION

THE naval reserves in California and Wyoming have been the subject of so much discussion recently that they must be considered in somewhat more detail.

After the withdrawal order of September 27, 1909, the Navy Department set about to secure definite reservation of part of these withdrawn lands for the future use of the navy, and finally, with the help of the Secretary of the Interior, secured the creation of two petroleum reserves: number 1, created on September 2, 1912, covering 38,969 acres (gross) in Elk Hills, California; and number 2, established on December 13, 1912, covering 29,341 acres (gross) in Buena Vista Hills, California. These were the naval reserves which received so much attention in the debates on the general leasing bills. On April 30, 1915, naval reserve number 3, the Teapot Dome reserve, covering 9,481 acres near Salt Creek, Wyoming, was set aside by executive order; in 1916 two oil shale reserves were set aside in Colorado and Utah, covering 45,-444 acres and 86,584 acres respectively; and in February, 1923, President Harding set aside a reserve, immense in area if not in oil content, in Alaska.[1]

These naval reserves included some very valuable oil lands, and enterprising oil operators cast many a longing glance at them. Some operators even went farther, and asked for exploiting privileges; and some even went still farther and hung around the Capitol at Washington for some ten years, demanding recognition of all sorts of "rights" and equities.

The Leasing Act of February 25, 1920, was not extremely generous with the claimants for land in the naval reserves. The act provided merely that such claimants might have a lease on any producing *wells* within a reserve, with enough land for the operation of their wells, on payment of not less than 12½ per cent royalty to the government. The *President* was authorized to lease to such claimants the remainder or any part of the claim on which the wells were drilled, on any terms he might consider proper.

The reader who bears in mind the persistence with which the California oil men demanded consideration of their various equities from 1910 to 1920, will find it easy to understand the later history of the reserves. The oil men did not get all they wanted out of the Leasing Act, and they were about as persistent afterward as before; and when Warren Harding came into office in 1920, inaugurating perhaps the most corrupt and capital-ridden administration the United States has ever seen, their opportunity seemed at hand. The election of Harding, by a tremendous majority, was itself a rather ominous and disturbing circumstance, since he was known to be much under the influence of the business interests dominating the Republican party machine. When President Harding announced his cabinet appointments, there was occasion for further uneasiness. His Secretary of the Interior, Albert Fall, formerly Senator from New Mexico, was a lawyer of poor repute in one of the "rotten boroughs of the West," and had been known for years as one of the most aggressive enemies of the conservation policy. His Attorney-General, Harry Daugherty, was one of the leading politicians in the "Ohio Gang." These two men had much to do with the public lands, and it was clear from the first that they would probably favor an exploitation policy. The naval reserves, however, were in charge of the Secretary of the Navy, Edwin Denby, and there was at first no reason to believe that he wished to exploit the reserves.

The general political situation, and the atmosphere of the times, were favorable to the operations of the exploiters. The election of Harding by such an overwhelming majority put his administration in a position where it seemed that criticism could almost be ignored; and most people were at first so well satisfied with the "normalcy" program that they about ceased to do any thinking on political questions. Finally, it must be noted that the psychology of the post-war period was favorable to corrupt machinations in government, and in business as well. The United States had "won the war," and the people were so busily engaged in shouting over their victory, and in congratulating themselves on being such a great people, that they had little time or stomach for the drab task of watching their government. Such patriotic excitement as we were enjoying always makes an excellent cover for fraud in public office. Following the hilarity of the victory celebration came an orgy of fanatical "red hunting," up and down the "sweet land of the free," and this took much of the public attention for a while, offering further opportunity for official malfeasance.

The Harding administration was only about a year old when rumors began to circulate that the Teapot Dome reserve had been leased. Early in April, 1922, urgent requests for information began to pour into the office of Senator Kendrick of Wyoming, and, in response to these, he made inquiry at the various offices of the Departments of the Interior and Navy, but could get no information. On April 16, he introduced a resolution into the Senate calling upon the Secretary of the Interior for information as to whether a lease of naval reserve No. 3 had been made or was in contemplation. During this time, representatives of the press had also been trying to get information, but without success, until April 14, when the *Wall Street Journal* announced that the reserve had been leased. On April 18, a definite statement came from the Department of the Interior, to the same effect; and on April 21, Assistant Secretary Finney, of the Department of the Interior, transmitted to the Senate official notice that a lease had been executed to the Mammoth Oil Company, a Sinclair company, two weeks before.[2]

The circumstances of this lease were such as to arouse some suspicion in the minds of the public. The secrecy with which the arrangements had been carried on; the fact that no official notice of the lease had been issued until two weeks after it was signed; the fact that a lease for the entire area had been issued to one company, when it was generally understood that the Leasing Act contemplated smaller tracts; all these circumstances looked suspicious. Furthermore, some people could not understand how it happened that the Secretary of the Interior was leasing one of the naval reserves which had been placed in the custody of the Secretary of the Navy.

A few days after the Senate received official notice of the Teapot Dome lease, Senator La Follette introduced a resolution calling upon the Secretary of the Interior, Albert Fall, to send to the Senate copies of all leases, orders, papers, and correspondence relating to the three naval reserves, and authorizing the Committee on Public Lands and Surveys to investigate the entire subject. In response to this resolution, Secretary Fall sent to President Harding a report which merely stated some of the conditions which he considered justification for the making of the lease, entirely ignoring the demand for a copy of the lease and various other papers. Fall treated the Senate's instructions with discourtesy in two ways: in ignoring the request for certain papers, and in sending his letter to the President, rather than to the Senate. He declared that his letter was not "written in the slightest degree as an attempt at defense of actions or of policies. The writer recognizes no necessity for such defense."[3]

President Harding sent Fall's letter on to the Senate with the further information that the policy adopted by Fall and Denby had been submitted to him prior to its adoption, and with the assurance that this policy and all subsequent acts had at all times had his entire approval. This assurance had a great deal of influence in allaying public suspicion, because President Harding still had high personal prestige with a large majority of the people.

In February of the next year, the Senate Committee on Public Lands selected two geologists, J. O. Lewis and Frederick G. Clapp, to make a geological examination of the Teapot Dome reserve; and, late in the summer, these men made their reports, Lewis definitely commending the policy of leasing, and Clapp taking a more cautious attitude.

The Committee on Public Lands, charged with the investigation, was composed of Smoot of Utah, chairman, Lenroot of Wisconsin, Norris of Nebraska, Ladd of North Dakota, Stanfield of Oregon, Norbeck of South Dakota, Bursum of New Mexico, Pittman of Nevada, Jones of New Mexico, Kendrick of Wyoming, and Thomas J. Walsh of Montana. The personnel of this committee was not exactly promising, since a majority were western men, and the West had generally been hostile to the reservation policy. The Senate Committee on Public Lands was, in almost all Congresses, hostile to conservation. The Chairman, Smoot, was a conservative, a friend of Fall's, and had a record of hostility to the reservation policy, as far as it applied to oil lands. He was hostile to the investigation from the first, as were several other members of the committee. The entire administration at Washington was also hostile to the investigation. President Harding died before the committee began its work, and Coolidge was not in office long before he showed his hostility to the probing that was going on. At one time there were eleven investigations going on in Washington, and of course the administration could see no campaign thunder in any of them. Furthermore, President Coolidge and Smoot and Lodge and the rest of the standpatters knew enough about affairs at Washington to see that some of their friends were likely to get burned before it was over.

Fortunately, there was one man on the committee who had the tact and shrewdness, the ability and the persistence required to overcome the obstacles that were constantly being placed in the way. Thomas J. Walsh, of Montana, was the real power in the committee, and it was he who did almost all of the work. He was the one feared by the men implicated. He was the one they were constantly trying to "call off," and it was he who finally brought the investigation to a successful close.

The committee began its work in October, 1923, and after some reading of documents and reports, called Secretary Fall to the stand. Walsh tried to get Fall to admit that the President's order transferring the reserves to the Secretary of the Interior was illegal, but with no success. He asked Fall why he had called for no bids in leasing Teapot Dome, but Fall insisted he had got a better contract than he could have got by competitive bidding. He asked what legal advice the Secretary had relied upon in his policy, and Fall replied he was himself a lawyer, and that he had relied upon his own judgment. Fall was generally far from conciliatory in his answers, and several times showed fits of temper when cornered. He did not make a good appearance before the committee. Perhaps the most damaging part of his testimony was his refusal to name the two oil companies he claimed he had consulted before he leased Teapot Dome. He admitted having made a business trip to Europe in the interests of Harry Sinclair, after leaving the office of Secretary of the Interior, and confessed that he had given business advice to Doheny, the lessee of the California reserves, but denied that he had received any compensation for either service. Here were suspicious circumstances: leases to Sinclair and Doheny, without any competitive bidding; then, after resigning from office, employment by these two men, stated to have been undertaken without any compensation.

After Fall had been grilled for two days, Edwin Denby, Secretary of the Navy, was called. Denby opened his testimony by stating that the transfer of the reserves from the Navy Department to Fall's jurisdiction had been at his own (Denby's) initiative. He admitted, however, that there was strong opposition in the Navy Department to any such transfer, that few of the officers of the navy had been consulted about it, but he denied the charge that any of those officers who opposed the transfer had been removed or shifted because of their opposition.

Several naval officers were called next. Admiral Griffin testified as to his opposition to the transfer, and told how he had tried to get a clause into the order, requiring the Secretary of the Interior to refer any proposed leases to the Secretary of the Navy, and how this clause was eliminated by Fall, in spite of the efforts of Assistant Secretary Roosevelt to get it inserted. Foster Bain, Director of the Bureau of Mines, Theodore Roosevelt, Assistant Secretary of the Navy, E. C. Finney, Assistant Secretary of the Interior, were called, but no very important evidence was elicited. H. F. Sinclair told about his contributions to the campaign funds of both parties, and about the organization of the Mammoth Oil Company—a very complicated form of organization, with a peculiarly close connection between the affairs of the company and the personal affairs of Sinclair. Robert Bell, a special assistant to the Attorney-General, testified to the worthlessness of certain claims in the Teapot Dome reserve which Sinclair had bought out at a high figure.

Days were consumed in the testimony of geologists as to the amount of drainage from the reserves. Commander Stuart, of the navy, told an interesting story of the manner in which Fall had tried to have him detached and sent out of Washington because of his hostility to the lease. Amos L. Beaty, President of the Texas Company, told of his efforts to get Fall to entertain a bid from his company when it was rumored that Teapot Dome was to be leased. Some of this testimony was sufficiently scandalous to be fairly interesting, but when Carl Magee, the young editor of the New Mexico *State Tribune* appeared, his story was the "best yet." Magee told of his first visit to Fall's home at Three Rivers, New Mexico, in February, 1920, where he had gone to purchase Fall's interest in the Albuquerque *Morning Journal*. He described the dilapidated condition of Fall's ranch at that time, and recounted what Fall had told him as to his straitened financial situation. He then told of a later visit at Three Rivers, and of the great change that had been made at Fall's ranch; of the "pillars built up to this road, and beautiful woven wire fence put along, and trees planted, and beautifully concreted gutters, and a very expensive road," as far as he could see to the ranch house. Magee told how he had exposed, through his paper, certain irregularities in the management of the New Mexico state land office, in which Fall had a great deal of influence, and recounted the financial difficulties he had met thereafter, because of the refusal of one of the banks to renew loans on his newspaper.[4]

Magee's testimony as to the sudden rise of Fall from financial ruin to affluence was presently corroborated by other evidence. A certificate from the County Treasurer of Otero County, New Mexico, disclosed the fact that Fall had not paid any taxes on his property between 1912 and 1922, but that in June, 1922, he had paid up his back taxes in full, amounting to $8,000. He had bought a neighboring ranch for $91,500 in November, 1921, making the first payment of $10,000 in one hundred dollar bills which he had in a small tin box. He had, through an agent, bought other small ranches about the same time, paying for them a total of $33,000. He had bought some blooded stock from a stockman in Kentucky, and some from H. F. Sinclair, whose farm was in New Jersey. He had built an irrigation reservoir and hydroelectric plant costing some $40,000.

There seemed to be no doubt that Fall had suddenly come into possession of a large amount of money, and there was some reason to believe that this money had come from Sinclair. Certainly social and business relations of these two men were highly improper, considered in connection with the lease which Fall had secretly given to Sinclair. Sinclair had gone down to Three Rivers, in his special car, and with his family had visited Fall at his ranch house, the visits being returned by Fall and his family in Sinclair's special car. Sinclair later sent the foreman of the Fall ranch a blooded horse as a gift, and sent Fall himself a few head of blooded Holstein cattle and two hogs. These were shipped from Sinclair's New Jersey farm in February or March, 1922, at an expense to Sinclair of $1,105. There was no discussion of any price to be paid for the animals, and it was not until after the investigation of the Teapot Dome lease had been started that someone, believed to be Fall's son-in-law, appeared at the office of Sinclair in New York to make payment. The auditor thereupon called up the foreman of the Sinclair farm, and got a bill for the cattle and hogs, the bill amounting to $1,100, which was paid by the stranger. Thus, for the livestock he sent to Fall, Sinclair had received $5 less than nothing. There is hardly a reasonable ground for doubt that these cattle and hogs were meant as a gift from Sinclair to Fall, and that no payment at all was intended, or would ever have been made, had not the investigation been ordered. Fall's payment of a little less than the amount of the freight charges did not make the consignment much less of a gift, anyhow.

The question now was: Where did Fall get so much money? The cattle business had been very unprofitable in 1921 and 1922, yet here was a cattle man rolling in wealth. While the committee was probing at this question, Fall apparently got nervous, and wrote a long letter to Chairman Smoot, in which he discussed at considerable length his private circumstances and transactions, and named Edward B. McLean, of the *Washington Post,* as the man who had loaned him $100,000. This was on December 26; and on January 3, A. Mitchell Palmer, as attorney for McLean, who was in Florida, wrote to the chairman of the committee corroborating Fall's story of the $100,000 loan, and indicating McLean's willingness to make any written statements or affidavits that might be asked for. Palmer insisted, however, that McLean would not be able to leave Florida because of the condition of his own and his wife's health. Fall, by the way, had left Washington a few days before, because of his health.

The committee was here at the end of a promising lead; and some of the members were clearly discouraged. The chairman, Lenroot at this time, could see no reason for calling Fall before the committee, and indicated that he wished to close the hearings as soon as possible. Walsh thought he must follow the matter a little farther, and after much discussion, he finally decided to go to Palm Beach and examine McLean in person.

About two weeks later, on January 21, Walsh came back with real evidence. He reported that McLean, after being questioned a few minutes, admitted that he had given Fall no money, although he asserted that he had given him checks for $100,000, which Fall had not cashed. Walsh promptly wrote a letter to Fall, who apparently had a room at the same hotel, acquainting him with the nature of McLean's testimony—but Walsh's story may well be given as he told it:

This letter having been written, and signed by me, it was enclosed in an envelope addressed to former Senator Fall, which I sealed. I took it to the desk in the office at the hotel, and accosted the gentleman at the desk, whom I assumed to be the manager of the hotel. I asked him if he would not have the kindness to present the letter to Senator Fall. He said, "Senator Fall is not registered here." I said, "I understand he is not registered here, but I know as well that he is in the hotel, and I very respectfully ask you to hand the letter to him." He said, "Well, I do not want to hand him the letter." I then said to him, "Now, I do not care to place this in the hands of the sheriff and have him go rummaging through your hotel to find Senator Fall, but Senator Fall must get this letter." He thereupon said, "Well, I will give this letter to Mr. McLean and let him deliver it to Senator Fall." I said, "That will be satisfactory!"[5]

All this happened on Friday, January 11, and within a day or two the newspapers were full of it. Many of the Sunday papers, on January 13, made special feature stories of it. Fall had lied; and further disclosures were inevitable. Immediately other sources of information, opened up.

As soon as the story of McLean's testimony came out, Archie Roosevelt, who had a position with the Sinclair Company, began to foresee a possibility that the distinguished name of Roosevelt might be smutched by impending revelations. He had for some time entertained suspicions that all was not right in the Teapot Dome transaction, and, to strengthen his suspicions, Sinclair, on Monday, the day after the story appeared, called him to his office and told him to go and buy him a ticket to Europe on the first boat that sailed, but to see that his name was not put on the passenger list, and to tell no one in the office about his departure. Sinclair sailed on Wednesday. That same day, Senator Caraway made his speech on the Teapot Dome lease, and Archie became even more suspicious. The following day, G. D. Wahlberg, Sinclair's private secretary, came to see him and advised him to resign his position. Wahlberg explained that he himself was very unhappy in his own position. Archie Roosevelt's account of his interview with Wahlberg is perhaps worth quoting:

Then I asked him if he thought that Mr. Sinclair had bribed Fall. Mr. Wahlberg hesitated—it is a nasty word—and he said, "I think somebody lent Mr. Fall money." No, he didn't; he said, "Somebody might have lent Mr. Fall money." I think that is how he put it. I want to get his words as nearly accurate as possible. Then I asked him why he thought Mr. Sinclair was leaving the country. He shook his head, and he said, "Well," and he said that "it must be, of course, on account of the findings of Senator Walsh's trip down at Palm Beach." He then said to me that he was extremely worried; that leaving him over here all alone, with Mr. Sinclair away, he was afraid that he would be forced to explain certain things; that he would undoubtedly be expected to lie about certain things; that one of the things that he was worried about was a payment which was made to a foreman of Mr. Fall's; that that payment was $68,000 and that he had the cancelled checks. Now this was the main reason. This, of course, was the thing that took my breath away at first.[6]

At this time young Archie Roosevelt seems to have become very "unhappy," and he called up his older brother Theodore—the Assistant Secretary of the Navy—and explained his perplexity. He was promptly advised to resign and to appear before the committee and tell about his suspicions. His testimony, as recited above, tended to make the Fall-Sinclair transaction look even worse than before.

Wahlberg, when called to the stand, told a strange story of 3,500 shares of Sinclair Consolidated stock (worth about $70,000 at the time) which he had taken from Sinclair's vault and had loaned to Sinclair's attorney Zevely, and of $25,000 worth of liberty bonds which he had sold, giving the cash to Zevely, both transactions completed without any receipts or papers of any kind.

The trail was getting fresher all the time, but the most sensational news of the entire investigation suddenly came from an unexpected source. The evidence had been pointing more and more to Sinclair as the source of Fall's money, and Walsh was following that lead when E. L. Doheny, who had secured leases in the two California reserves, asked that he be permitted to make a statement before the committee, and, on taking the stand, admitted that he had loaned Fall $100,000 on November 30, 1921. He told a touching story of early friendship for Fall when they were poor miners together, and assured the committee that it was this early friendship, and not the awarding of leases on the California naval reserves, which had induced him to make the loan. He testified that his son had taken the money to Fall in cash in a satchel.

Disclosures followed fast after this. The next day after Doheny testified, Sinclair's attorney Zevely testified that Sinclair's private secretary Wahlberg had sent $25,000 or $30,000 in bonds to Fall—to a bank in El Paso—in June, 1923. This, like Doheny's $100,000 was a "loan," although the note for it was not made out until several weeks later, and was made out, not to Sinclair, but to Zevely.

It seemed necessary to call Fall to the stand again, but his physicians reported that he was too ill to come to the committee room to testify. The committee sent three other physicians to examine him, however, and, on their report that he was able to appear, a subpoena was issued for him. When he appeared, however, on February 2, he refused to answer any questions, on the ground that the committee was without the authority to conduct the investigation, and on the ground that he might incriminate himself. Sinclair likewise refused to testify, after some of the above evidence had been presented, on the ground that the Senate had no power, under the Constitution, to require an unwilling witness either to attend or testify before it or any of the committees.

The main work of the committee—perhaps it would be more accurate to say the work of Senator Walsh—was done, however. Later probing brought out further unsavory details of the Teapot Dome lease, and of other transactions of Secretary Fall, and, most unhappily for the Democrats, brought out the fact that William McAdoo, a very promising candidate for the presidency, had been hired by Doheny soon after he left his cabinet position, at a fee or salary of $25,000 per year, in addition to a retainer fee of $100,000 paid to McAdoo's law firm of McAdoo, Cotton, and Franklin.

Altogether the story of the leasing of the naval reserves constitutes one of the most scandalous and disgraceful stories in American history, a story of official corruption or incompetency which involved to some extent the President and several members of the cabinet.

Whether the looting of the naval reserves was deliberately planned before the convention of 1920 is a question which cannot be answered definitely either way. There were persistent rumors that, before the nomination of Harding, a group of oil operators and others entered into a conspiracy to make Senator Fall Secretary of the Interior, in consideration of his promise to permit them to exploit the naval reserves. It was even specifically charged that Harding had been nominated after he had agreed to some such arrangement. Al Jennings, the Oklahoma ex-convict, testified that Jake Hamon, a well-known Oklahoma oil operator and politician, had once told him he spent a million dollars on the election, and that he (Hamon) was to be the Secretary of the Interior. Later developments indicate that there was something wrong with part of this story, but a number of other witnesses testified as to the amount of money Hamon spent on the election, and as to the deals whereby either Hamon or Fall was to be the Secretary of the Interior. One oil operator told the committee that Hamon (who was then dead) had told him that "it would be worth $500,000 to Fall to be Secretary of the Interior, payable approximately $150,000 at the time and the balance annually for four years." Several testified that General Wood had been approached with a proposition that he give certain interests three cabinet members, in consideration of their support, but that Wood had declined.

Testimony on this point was criticized severely; and some of it could not be given a high rating as evidence. Nevertheless, some of the testimony fits in so perfectly with later actual events that it is not to be ignored. While it was not possible to prove that there was a conspiracy, there is some reason to believe that there was such a thing; and there is hardly any doubt that certain powerful oil interests worked hard to have, if not Fall, at any rate a man of Fall's general attitude, put into the office of Secretary of the Interior.

If we ignore the possibility of a conspiracy in the campaign of 1920, we may say that the first step on the part of the looters of the naval reserves was the selection of Albert Fall for Secretary of the Interior. The Secretary of the Interior had charge of immensely valuable natural resources of all kinds, even before President Harding turned the naval reserves over to him. The position called for a man with a strong appreciation of public rights and interests. Fall was condemned as absolutely unfit for such a post by every detail of his record in the Senate. He had been an exploiter, and a friend of the exploiters. He had always opposed the conservation movement, and was known to be hostile to the oil reserves before he took office. President Harding had been in the Senate with him for years, and was entirely familiar with his record. Harding and Fall had long been friends.

Fall had not been in office many months before he announced his hostility to the conservation policy in unmistakable terms. In a speech at Colorado Springs, in September, 1921, he declared:

The conservation policy of the last 20 years has merely resulted in giving the corporations who control the market what you Colorado people call "a lead pipe cinch" on prices. The same is true of our grazing lands, our oil and mineral lands, our homestead lands, and our fisheries. All natural resources should be made as easy of access as possible to the present generation, the prospector, the homesteader, the lumberman, the miner, the fisherman, the sheep herder, and the cattle- and horse-man. The more these resources are developed, the greater the progress, not only of the present generation, but of the generation still to come.

The second step in the exploitation of the reserves was the order of President Harding, transferring the administration of the reserves from the Navy Department to the Interior Department. Harding took office on March 4, 1921, and the order for the transfer was issued on May 31. This was almost unseemly haste. The excuse offered for this order was that since the Secretary of the Interior had charge of the oil lands on the public domain, he could administer the naval reserves best; but the naval reserves were set aside, not for the purpose of immediate exploitation, but as a future oil reserve for the use of the navy, to be used when fuel from other sources should not be available. Furthermore, there was much reason to believe that the order was illegal. By an act of June 4, 1920, the Secretary of the Navy was "directed" to take possession of the naval reserves, "to conserve, develop, use and operate the same"; and the idea that the President could, by his fiat, transfer powers reposed in one cabinet officer to another cabinet officer, seems indefensible. No opinion was sought from any of the law officers of the government. The Attorney-General was not consulted; and when Senator La Follette addressed an inquiry to Secretary Fall, as to the legal opinion on which the transfer was justified, Fall addressed a not very courteous reply, stating, among other things, that the details as to the handling of naval oils could not be given out without permission of the President. President Harding addressed a letter to "Dear Secretary Fall," in which he stated, "I quite approve of the manner in which you responded to his [La Follette's] inquiry."[7]

The transfer was apparently resolved upon as early as April 1, 1921. About this time—less than a month after the inauguration of Harding —Secretary Denby, of the Navy Department, announced his intention of surrendering control of the reserves to Fall. Denby apparently believed that he himself had originated the idea of transferring the reserves to Fall's jurisdiction, but there is evidence that Fall wrote the order of transfer. In signing the order, Denby did not consult any of the naval officers who had been charged with the administration of the reserves; and some of the evidence tended to show that when Admiral Griffin, who had charge of the reserves, objected to the transfer, he was replaced by Admiral Robison, who knew nothing about the matter but had been talked into a favorable attitude by the oil magnate, E. L. Doheny. Two young naval officers who objected to the transfer were apparently detached and sent away from Washington, at Fall's suggestion.

The original draft of the executive order of transfer was drawn by Fall, who sent it to Denby with a draft of a letter to be signed by the latter and transmitted to the President, explaining the occasion and necessity for the order. Amendments to the order, drafted by Admiral Griffin, designed to necessitate the sanction of the Navy Department for any lease made, were taken by Colonel Roosevelt to Secretary Fall, but Fall declined to adopt them. On his return from this errand, Roosevelt explained to Griffin that he had done the best he could with Fall. A formal protest against the proposed order, written by Admiral Griffin, was never sent to the President.

Thus, by two moves, the oil reserves were placed in charge of Secretary Fall. The next move was to get them into the hands of Sinclair and Doheny. It was not many months after President Harding's order was issued when Sinclair made his visit to Three Rivers, New Mexico, where he stayed several days. Fall later pretended that Sinclair came to see him about some leases on Indian lands in Oklahoma, but it is more likely that he came expressly to open up negotiations for Teapot Dome. While at the ranch, he arranged to send the foreman the horse already mentioned, and to send Fall the cattle and hogs, from his farm in New Jersey. Since there was no discussion of price for the cattle and hogs, and since no bills were made out for them, it is to be assumed that they were gifts, just as the horse was. These were shipped some time in February or March, and on April 7, the Teapot Dome lease was signed.

Before the lease was signed, however, it was necessary to make some disposition of a great many claims to land in the Teapot Dome reserve. The act of June 4, 1920, had given the Secretary of the Navy authority only over those portions of the reserves on which there were no "pending claims or applications for permits or leases" under the Leasing Law.

Now the entire Teapot Dome reserve, excepting about 400 acres, was covered by "claims," practically all worthless, yet pressed with much persistence; and, under some arrangement or understanding between Sinclair and Fall, the former acquired all these so-called claims, paying handsomely for them.

As a matter of fact, all of the claims were held or controlled by the Pioneer Oil Company, closely related to the Midwest Refining Company, which was in turn a Standard subsidiary. One or more of these companies had for several years been trying to secure a lease on Teapot Dome. They were, of course, amply able to undertake the development of the territory, and were potential competitors for the lease. A short time before the execution of the lease, Sinclair paid, or agreed to pay, a million dollars for these worthless claims. Although Sinclair denied it, there is good reason to believe that this payment was not made in fact for the transfer of the title to the worthless claims, but was made to remove from the field a formidable competitor for the lease, and there is reason to believe that Fall understood this clearly.

In shadowy affairs of this kind, the "hush money," once given out, often has to be distributed rather widely. The million-dollar payment to the Pioneer did not end the claim adjustment; and this brings in another interesting story. In the latter part of the year 1920, an adventurer by the name of Leo Stack associated himself with E. L. Doheny in an effort to secure from Secretary Daniels the lease of a row of offset wells along the boundary line between the Teapot Dome reserve and the Salt Creek field. The effort was fruitless, and by some kind of an arrangement, Doheny retired from the agreement, and the Pioneer Oil Company was substituted for him. The Pioneer Company entered into an agreement with Stack under which he was to assist, presumably in Washington, in securing a lease of the reserve for the Pioneer, in return for which service he was to have an interest in any lease which might be secured. Stack was supposed to be an experienced lobbyist; but he was never called upon for assistance. The Pioneer Oil Company, after entering into its agreement with Sinclair, ceased all efforts to secure a lease. Stack felt that he had been treated badly, and presently he got the interest of F. G. Bonfils, publisher of the *Denver Post*, and entered into a contract by which he and Bonfils were to divide whatever they could get out of the Pioneer Company. Immediately upon the execution of this contract, the *Post* began the publication of a series of articles denouncing the "corrupt" Teapot Dome lease, sending copies of the paper to members of Congress and of the Cabinet. Stack and Bonfils also started an investigation in New Mexico, touching the question of Secretary Fall's finances, securing evidence of a damaging character; and they started a suit in Colorado against the Pioneer Oil Company and Sinclair, alleging a conspiracy against Stack. Before this suit had gone far, however, Sinclair settled the suit, by paying $250,000 cash and by agreeing to pay $750,000 more. The attacks of the *Denver Post* promptly ceased.

The editor of the *Denver Post* was not the only man who knew how to make the newspaper business pay. John C. Shaffer, publisher of several newspapers, including the *Rocky Mountain News* and the *Denver Times*, got $92,500 from the Pioneer Oil Company, for absolutely no consideration, according to his own testimony. Such princely generosity is hard to explain; and Shaffer never really did explain it.

Troubles came, "not single spies but in battalions," to the lessee of the Teapot Dome. When the Mammoth Oil Company had taken possession and had begun drilling, the Mutual Oil Company entered the Dome also, under agreement with the owners of a worthless placer claim, and began drilling. In this situation Secretary Fall persuaded Assistant Secretary Roosevelt, of the Navy Department, to send a squad of marines all the way out from Washington to eject the Mutual Oil Company as a trespasser. The reason for this use of armed forces was stated to be that if the ordinary civil remedy of an injunction had been resorted to, it would probably have resulted in a judicial inquiry into the validity of the Sinclair lease. If Sinclair had filed a civil suit, he would have had to establish his own title, and in trying to do so might have started an embarrassing inquiry.

The Teapot Dome lease was a fairly well-concealed betrayal of the rights of the public. The lease gave to the Mammoth Oil Company the right to occupy the Dome practically until the oil should be exhausted, and to exploit the oil on payment of royalties graduated from $12\frac{1}{2}$ per cent to 50 per cent, according to the size of the wells. (The average royalty from wells drilled at the time of the investigation would have been between 16 and 17 per cent.) The government did not get this royalty in oil, however. It sold its royalty to the Mammoth Company, receiving oil certificates which it might exchange for fuel oil, gasoline, lubricating oil, or other petroleum products, or which it might redeem in cash or might exchange for oil storage tanks, all exchanges to be with the Mammoth Oil Company.

This last mentioned provision, like the President's transfer order, was apparently illegal, since "all purchases and contracts for supplies and services in any of the departments of the government, except for personal services," must, according to statute, be made by advertisement and competitive bidding. Sinclair ventured the opinion that the contract called for personal service, in the extraction of the oil, and that therefore competitive bidding was not required, but this was absurd. According to the contract, the government was to buy these at the market price—a most generous provision for the Mammoth, since the government ordinarily secures a large part of its supplies considerably below market price under a system of competitive bidding.

Two other statutes were apparently violated by the provisions of this lease. One statute required all purchases and contracts for supplies for the army and navy to be made under the direction of the chief officers of the department concerned. Another statute provided that no contract for the erection, repair, or furnishing of any public building or any public improvement should bind the government to pay a larger amount of money than was in the treasury, appropriated for the specific purpose. The Teapot Dome contract provided for the erection of tanks beyond the limits of congressional appropriations, in fact one of the grounds on which the contract was defended was that the navy needed more storage tanks than Congress had provided for.

Thus the Teapot Dome contract provided that the oil reserve which had been set aside for the use of the navy should be exploited, at a time when there was already a general over-production of oil from private lands, and that the royalty oil from this reserve might be traded for storage tanks. A little calculation will show just how much oil the government was finally to get out of this reserve. There were, according to the latest estimates, perhaps 26,000,000 barrels of oil in the reserve. The government royalty, as far as drilling had gone, amounted to less than 20 per cent of this, or a probable total of about 5,000,000 barrels, for the reserves. Two-thirds of this royalty oil was to be given for the storage to hold the other one-third, since it was estimated that it would cost two barrels of oil to build one barrel of storage. In other words, the navy was to have, finally, one-third of 5,000,000 barrels, or only 1,666,666 barrels of oil, of its original reserves of 26,000,000 barrels. Surely this was an extraordinary way of providing for the future fuel needs of the navy.

In the discussions regarding the Teapot Dome lease, much was made of the fact that the Sinclair Company agreed, as one of the considerations for the lease, to build a pipe line from the Salt Creek field to Kansas City, the argument being that this pipe line would provide Salt Creek oil with a better market, in which the government would share. Walsh pointed out, however, that since the government had no oil—having sold it, under the contract—it was in a poor position to profit from the building of this pipe line. Furthermore, some of the testimony tended to prove that this pipe line would have been built by some company, anyhow, regardless of the Teapot Dome contract.

Perhaps the most important reason given for the leasing of the Teapot Dome reserve was that the Dome was being drained by wells on adjoining lands in the Salt Creek field. In his letter to President Harding, Secretary Fall cited reports by C. H. Wegemann and C. K. Heald, of the United States Geological Survey, and by F. P. Tough, of the Bureau of Mines, as expert evidence that there was danger of drainage, although, as a matter of fact, Heald had reported that drainage was not taking place and would not begin for some time, had recommended only leases covering a few sections of the reserve, and had advised these only "if appreciable loss becomes imminent." Wegemann had urged the development of the property "as a unit"; and Tough had reported that a large area in the northern part of the reserve was subject to drainage. A. W. Ambrose, of the Bureau of Mines, indicated that there were many uncertainties in the matter, but that serious drainage would occur only after some years. While Fall included the Ambrose report in his letter to the President, as an exhibit, he did not quote it in his report; and the Heald report, which was exactly what he did not want, was stressed less than the reports of Wegemann and Tough.

The drainage of the reserve was generally regarded as less important by the Geological Survey than by the Bureau of Mines. It was conceded that the loss had not been very great yet. Furthermore, even if the entire reserve had been drained through wells in Salt Creek, the government would have got more out of the oil than it did under the Mammoth contract, because the wells in Salt Creek were on government land, and yielded a higher royalty than the Mammoth contract called for. No matter how serious the drainage, there was no way that the government could fail to lose by the Sinclair lease.

It is often claimed that the Teapot Dome lease would have been an unprofitable venture for Sinclair, even had the circumstances of its making never been divulged. It is doubtless true that the lease would not have proved as profitable as it was at first thought to be. Geologists originally estimated the oil content of the reserve at more than 100,-000,000 barrels, and that estimate was later scaled down to 26,000,-000 barrels, or less. Sinclair at one time testified that he expected to make $100,000,000 out of his lease; and later developments certainly did not justify any such estimate of possible profits. Yet the terms of the contract were so favorable to him that it should have been reasonably profitable.

The Doheny contracts, covering Reserve No. 1 and part of No. 2 in California, were similar to the Sinclair contract covering Teapot Dome, not only in their general provisions, but in the reprehensible manner in which they were obtained. As already indicated, the two reserves in California were subject to more or less drainage from adjoining wells. The drainage from No. 1 was not considered very serious, but No. 2, with a very large oil content, was so interspersed with private holdings that the drainage was very serious. Secretary Daniels recognized this at the time he was in the Navy Department, although he stubbornly refused to permit any drilling in the reserves, even to protect them from drainage. It was claimed by certain men in the Bureau of Mines that the failure to drill necessary offset wells prior to March 4, 1921, resulted in a loss to the government of 6,800,000 barrels of royalty oil, worth $8,800,000. This estimate can be little but a guess, but, at any rate, Secretary Daniels admitted that the drainage was serious, and before his retirement on March 4, 1921, had called for bids for the drilling of 22 offset wells in Reserve No. 1. The proposals were not submitted, however, until after the new administration came in, and the contract was awarded to Doheny by Secretary Fall, on July 12, 1921. There was no particular reason for criticism of this contract. The contract was awarded to the highest bidder, upon open competitive bidding.

On April 25, of the following year, another agreement was signed between the government and Doheny—Doheny acting for the Pan-American Petroleum and Transport Company—by which the government, through Fall and Denby, agreed to exchange its royalty oil for tanks constructed at Pearl Harbor, Hawaii. The contract called for the construction, not only of tanks, but of docks, wharves, and loading machines, and the dredging of the channel. This contract was the nucleus of a larger plan developed by the Navy Department, involving the expenditure of approximately $102,000,000, the expense to be met by the delivery of the royalty oil accruing to the government under leases of the various oil reserves. This contract, like the Teapot Dome contract, left only a very small fraction of the contents of the reserves for the fuel uses of the navy.

Here was another extraordinary piece of executive usurpation. The question of the legality of exchange of oil from the reserves for tanks and wharves and dredging was apparently considered in a casual way, but an opinion of the Judge Advocate General was deemed sufficient to justify the transaction. Now, the Judge Advocate General was not a lawyer at all, and he did not write the opinion relied upon. The opinion was written by a clerk in his office. The Judge Advocate General testified that in his opinion the word "exchange," as contained in the statute, authorized the Secretary to exchange oil for anything—a battleship, ordnance, or an addition to the State, War, and Navy Building.

Some of the strongest oil companies in the country declined to bid on the contract for the construction of storage at Pearl Harbor, on the ground that the statute did not authorize payment for work of that character in oil. The attorney for the Standard Oil Company, Oscar Sutro, after careful investigation, advised the officers of the Standard not to have anything to do with the contract, because it would be illegal. This information was conveyed to Fall and he was urged to get an opinion from the Department of Justice. Nevertheless, no opinion was sought from the Department of Justice, nor from the Solicitor of the Interior Department, nor from any of the law officers of the government, except the clerk above mentioned. Testimony tended to show that Fall did not want any legal opinion on the question.

An important provision of the agreement of April 25 was that, if future leases should be given covering any part or all of Reserve No. 1, Doheny's company should have a preference right in the leasing; and the following December 11, a contract was secretly awarded to Doheny to exploit all of Reserve No. 1, although some of it was not to be exploited until the government asked for the work to be done. In the April contract appeared one of the irregularities that characterized so many of the transactions of Secretary Fall. Several oil companies bid on this contract. Doheny, for the Pan-American Company, submitted two bids, one of which was in strict conformity with the proposal of the government, and the other which was not. The second proposal offered to do the work at slightly less cost, but it secured for Doheny the preference right to a lease of the eastern half of the reserve. The second bid was the one which was finally signed. In other words, Fall, after getting bids from several companies, awarded the contract on terms which the other companies had no chance to bid on, and gave Doheny a preference right to the rest of the reserve. This was about five months after Doheny had lent Fall the $100,000. Doheny insisted that his loan had nothing to do with his securing of the lease since Fall had nothing to do with the lease personally, but evidence showed that Fall had a great deal, if not everything, to do with the lease.

Aside from the folly, and probable illegality, of trading reserve oil for navy equipment, this Pearl Harbor contract was a strange transaction, in that it required an oil company to do things that an oil company was obviously not equipped to do. The building of wharves, the dredging of the channel, even the construction of storage tanks, had to be contracted to construction companies, and the Pan-American had to have something for its services in contracting the work, although Doheny insisted that he did not enter the contract for profit. This peculiar method of getting construction work done stirred up a great deal of criticism, and brought Fall's transactions before Congress. Officers of the Chicago Bridge and Iron Works could not understand why they should not be permitted to bid on such work, and wrote to Fall and to their congressman, Medill McCormick. When they failed to get any satisfactory reply from Fall, they employed attorneys to take the matter up. The Graver Corporation likewise made inquiries and the entire matter was given considerable publicity.

The facts regarding the leasing of Naval Reserve No. 2 are somewhat different. This reserve was supposed by some to contain much more oil than either of the others, but so many tracts within the reserve were held by private owners, particularly by the Southern Pacific Railroad— later the Pacific Oil Company—that there was no possibility of keeping the land as a government reserve, and this fact was generally recognized. Leases of producing wells and of placer claims within the government sections were freely made during the Wilson administration, after the passage of the Leasing Act, and this policy was continued by the Harding administration until all of the lands were leased. There was no apparent impropriety in the leasing of these lands, except in the case of the Honolulu Oil Company.

During the Wilson administration the Honolulu Oil Company applied for patent to 17 placer claims in No. 2. This application was approved by the Commissioner of the Land Office, Clay Tallman, as to 13 of the claims, and was denied as to the other four. His conclusion was approved by Secretary Lane, but, before the patents were issued, President Wilson directed that the case be reheard. Testimony was again taken before the local land officers, who advised against the issuance of patents, but their decision was reversed by Commissioner Tallman, whose decision came before Assistant Secretary Payne for review. Payne reversed Tallman's decision and denied the application. An application for rehearing was filed, in accordance with the practice of the Land Office, and this appeal was pending when Secretary Fall took office. Fall filed an opinion in which he asserted that were the matter one of first impression, he would grant the application for patents as ruled by the Commissioner, but that he would not disturb the decision of his predecessor. He accordingly denied the petition for a rehearing, but directed that leases be granted to the Honolulu company under the provisions of the Leasing Act of February 25, 1920. The leases were accordingly executed, covering over 3,000 acres.

Now the only power reposed in the Secretary of the Interior, under the Leasing Act, was the power to lease producing *wells*. The Honolulu leases covered *tracts of land*. The Leasing Act authorized the *President* to grant leases of claims within which producing wells had been leased, but no producing wells had been leased in the Honolulu claims, and the leases were neither signed by President Harding nor authorized by him, so far as the record disclosed. There was no written evidence that President Harding had any knowledge of the leases. They were thus awarded without any authority of law. The Honolulu claims had been for years the subject of much discussion and disagreement in official circles at Washington. In Wilson's administration, Daniels and Lane had a rather bitter quarrel over the proper disposition of the claims, Lane taking his characteristic attitude of friendship for the oil interests. President Wilson apparently sided with Daniels on this question.

Yet another detail of the record of Secretary Fall must be noted. In the very heart of Naval Reserve No. 1 was a section of very valuable oil land—section 36—which had been granted to the state of California for educational purposes and later sold to the Standard Oil Company. Lands known to be mineral could not be selected by the state. In the year 1900, reports sent to the Land Office indicated that some of the land in this region was oil-bearing, and the Land Office withdrew an extensive area, including section 36, and all of Naval Reserve No. 1. The surveyor who surveyed the lands two years later, after giving careful attention to the land in question, noted on his plats, and reported in his field notes, that section 36 was mineral in character. Meanwhile, an agent of the Department of the Interior, E. C. Ryan, was sent into the field to examine the land, and in March, 1904, he filed a report indicating that this land was non-mineral, whereupon the order of withdrawal was recalled. The state of California, relying on the surveyor's report that section 36 was mineral, asked to be permitted to select other lands instead, but this application was never finally disposed of. In 1908, however, the state issued certificates of purchase to this section to parties apparently representing the Standard Oil Company.

There was assuredly plenty of evidence to warrant a careful investigation as to the real character of this land. The report of the surveyor has been mentioned. Some years afterward, furthermore, the government instituted suit against the Southern Pacific for the recovery of lands in the immediate vicinity, and the government investigation instituted in connection with this suit revealed much information regarding section 36. As a result of such information, the government ordered hearings to be held to determine whether section 36 was known to be mineral at the time the survey was approved, and thus whether the title of the Standard Oil Company was good. The local land officers at Visalia, California, and the chief of the field service at San Francisco, were required to report to the General Land Office at Washington every six months as to the progress made in the proceedings. Notwithstanding all this, the whole affair was overlooked for a period of seven years, until, on February 2, 1921, the chief of the field service at San Francisco wrote the Commissioner of the Land Office that the papers in the case had been accidentally found in the "closed files," where they had been inadvertently placed, instead of being deposited in the "open files." Thereupon the subject was considered by the Commissioner of the Land Office, Clay Tallman, Secretary Payne of the Department of the Interior, who succeeded Lane when the latter resigned, and the Assistant Secretary, E. C. Finney, with the result that the local officers were ordered to proceed with hearings as before directed. During all this time the Standard Oil Company was producing large amounts of oil from this section; and one of the assistant attorneys-general advised injunction proceedings to prevent the further extraction of oil.

At this time a new Attorney-General was in office—Harry Daugherty—and he directed his assistant not to do anything until he had a chance to confer with him. As the Walsh report states, "It may or may not be of significance that the desire of the Attorney-General in the matter was carried to his assistant by an officer or attorney of the Standard Oil Company." This assistant to the Attorney-General testified before the committee as follows:

I heard nothing from my memorandum (urging the injunction), but three or four days after it was sent, Mr. Loomis, the vice-president of the Standard Oil Company of California, came to my office and asked me what action had been taken on the recommendation of the Interior Department for proceedings in this matter. I told him that the case had been formally sent to the Attorney-General's office by me, and I had no information as to what his attitude would be. Three or four days after that, Mr. Loomis again came to my office and brought a letter from the Attorney-General, written in his own handwriting, asking me to see that no action was taken in the case until he could have a personal conference with me. And I never had any conference with the Attorney-General during my stay there, up until May 11. (The memorandum was dated March 30.)

At this juncture, the attorney for the Standard Oil Company filed a motion before Secretary Fall to dismiss the entire proceedings. This motion was argued before Fall on June 8, 1921, by the attorney for the Standard, Oscar Sutro, representatives of the navy and of the Department of Justice attending. Upon the conclusion of the argument, Fall asked those present whether the facts were as stated by Sutro. The response was that the facts were substantially as represented, whereupon the Secretary announced that the motion for dismissal of the proceedings was granted. No argument was asked for from those representing the government, although the Assistant Secretary of the Interior, E. C. Finney, a man of wide knowledge of public land matters, was at variance with Sutro in respect to both of the legal propositions upon which he based his argument. Finney stated before the investigation committee that such an extraordinary exercise of the supervisory power of the Secretary of the Interior as Fall assumed was wholly unknown in the practice of the Land Department.

Thus the exploitation of the oil reserves was going along merrily until Walsh stopped it. Altogether the transactions of Fall, Daugherty, Denby, and of President Harding himself, constitute a most extraordinary record of corruption or of incompetence. The record of Fall is an amazing story in its details of illegal assumption of authority, of special favors and concessions to his personal creditors, of high-handed disregard, not only for the public rights and interests, but for legal limitations as well. Perhaps some will believe that Doheny's advance of $100,000 was really a loan, notwithstanding the fact that Doheny, according to his own testimony, tore the signature from the note, and notwithstanding all the other questionable circumstances surrounding the entire transaction. To some minds, there will be nothing suspicious about the fact that the money, in cash, was placed in a satchel and carried to Fall's room in Washington by Doheny's son, since a note was given for the money. The oil magnate tore the signature from the note, and gave the signature to his wife, for fear, in case anything happened to him, his executors or someone might ask Fall to pay it. That was certainly a thoughtful and neighborly turn, but it was poor business, for Mrs. Doheny lost the signature and could not find it until after the hearings were over. At the later trial of the Elk Hills case, a signature was produced which was claimed to belong to the note.

Some distrustful minds will incline to a suspicion that the $100,000 was not really meant as a loan at all, but as a gift, a little token of appreciation for some of the favors that Secretary Fall had shown his old friend, just as some will believe that the $25,000 or so from Sinclair, and the shipment of cattle and hogs from the same man, were gifts. Some will say, too, that even if these contributions had been made as loans, with the understanding that Fall should work them out in the service of Doheny and Sinclair, at $50,000 a year, they would not have represented a very high type of business.

Fall was not the only one with a discreditable record in the naval reserve affair. President Harding's part in the business has been indicated. Had he lived a few years longer, he would have faced deep embarrassment and humiliation in trying to explain his record. President Harding had long been a friend of Fall's; he appointed him to his cabinet with a full knowledge of Fall's previous record of hostility to these reserves and to conservation generally; and he certainly stretched the law to the limit in his efforts to put the naval reserves in Fall's control. His motives in doing all this need not be discussed here, although the writer attributes to him nothing worse than extremely poor judgment, and lower ideals of public service than should be expected of the President of a great and enlightened nation.

Daugherty's record on "section 36" has been mentioned; and his general official record—certainly one of the worst in American history —need not be traced here. The Wheeler investigating committee has done that sufficiently. As for Denby, nothing was proved but almost unbelievable incompetence. In his examination before the committee, he revealed almost complete ignorance of every essential fact regarding the naval reserves and the plans for their exploitation.

Certain prominent men who cannot be charged with complicity in the actual rape of the reserves deserve credit for their efforts to shield the guilty ones. Among these may be mentioned Edward McLean, A. Mitchell Palmer—Ex-Attorney-General of President Wilson's administration, law partner of J. W. Zevely, and attorney for Sinclair and Fall, as well as for McLean—C. B. Slemp, a friend of McLean's, and President Coolidge's private secretary, and, finally, President Coolidge himself. McLean went to extraordinary lengths in his efforts to prevent the truth from coming out. He not only lied to the committee, in stating that he had loaned $100,000 to Fall, but, during the time when the committee was hot on his trail, he leased a private wire from Palm Beach, where he was, to Washington, in order that he might keep in close touch with developments at Washington; and he left no stone unturned in his effort to prevent the committee from calling him to testify under oath. To aid him in this effort, he secured the legal and political assistance of A. Mitchell Palmer, whose Democratic affiliations were assumed to be of value, of Francis McAdoo, a son of William McAdoo, and of other influential politicians, not all Republicans. Palmer, as attorney for McLean, represented to the committee that McLean had loaned Fall the $100,000, and did his best to prevent McLean from being examined.

A number of circumstances throw a serious cloud over the record of President Coolidge in this matter. His secretary, C. B. Slemp, left Washington for Palm Beach about two days after McLean left, and at Palm Beach Slemp was associated much with McLean and somewhat with Fall, dining with McLean frequently, and a few times with Fall. On the evening after McLean confessed that he had not made any loan to Fall, Slemp and Fall and the latter's wife and daughters dined with McLean. Thus, during the days when the burning question was whether McLean would be compelled to testify, Slemp was in close touch with him. Confidential relations are indicated by the fact that McLean told Slemp of his intention to confess the truth about the Fall loan; although Slemp denied that he knew the real truth—that McLean had not made the loan—until he read it in the papers. Just what was the nature of Slemp's correspondence with President Coolidge during these critical days he refused to divulge, but he was in touch with Coolidge all the time. If there was nothing suspicious in the correspondence, it seems unfortunate that it was not laid before the committee.

Two telegrams that were sent to McLean by his associates in Washington suggest at least a mild possibility that, notwithstanding his denial, Slemp went to Palm Beach to keep in close touch with McLean and to keep President Coolidge informed as to developments. On the day of his departure from Washington, a telegram went forward: "Mr. Slemp and Mr. Whaley leave tonight." On the same day the doorkeeper at the White House sent the following telegram to McLean: "The secretary leaves today at 9.40." Thus McLean apparently was kept informed as to Slemp's movements, by men in the White House. A further peculiar circumstance was that McLean and President Coolidge at the White House employed the same telegraph operator—a man named Smithers.

Other telegrams found by the committee point to the same McLean-Slemp-Coolidge cabal. A telegram to McLean from his "confidential employee," John F. Major, dated December 22, stated that McLean's attorney had recommended the leasing of a private wire for various reasons, among which was that it would give "*easy and quick access to the White House.*"

After the truth had come out, after McLean had admitted his earlier lie to the committee, his relation to Coolidge remained as friendly as ever. Walsh took McLean's testimony on January 11. The very next day President Coolidge sent the following telegram to McLean: "Prescott is away. Advise Slemp with whom I shall confer. Acknowledge. Calvin Coolidge." The answer to this telegram is not available, but a month later Coolidge sent McLean another telegram: "Thank you for your message. You have always been most considerate. Mrs. Coolidge joins me in kindest regards to you and Mrs. McLean." A little before this, on January 29, one of McLean's employees had sent McLean the following telegram: "Saw principal. Delivered message. He says he greatly appreciates, and sends regards to you and Mrs. McLean. There will be no rocking of the boat, and no resignations. He expects reaction from unwarranted political attacks." Now the "principal" referred to was probably President Coolidge, although one witness, an employee of the *Washington Post*, tried to make it appear that Senator Charles Curtis was meant. There is no reason to assume that it was Senator Curtis. The rest of the telegram evidently refers to President Coolidge. Its similarity to Coolidge's telegram of February 12 is noticeable, and the promise of no "rocking of the boat and no resignations" could only have come from the President.

President Coolidge's close friendship with McLean has been the basis of some criticism. Senator Norris once commented on this as follows:

Men will perhaps condemn me for making the statement, but down in your own hearts, right down deep, you know that you felt humiliation when a President of the United States, just elected to the office, came to Washington and made his home in this man's house—this man McLean, who has lived a life of debauchery and whose only right to a place in respectable society is one that has been purchased by his money inherited from an indulgent parent.[8]

While absolute proof cannot be presented, there is strong reason to believe that the truth about Fall's "loan" transactions was known to President Coolidge, and probably to other influential politicians some time before it was exposed by Walsh. Could McLean have been ignorant of it? What did he think and say when Fall asked him to make the false statement that he had loaned him $100,000? Did he agree to tell that lie without inquiring as to the reasons for it—without finding out the real truth? It is impossible to believe it. No man would be so simple, so innocent, as to do that. McLean knew of the Fall "loans."

And if McLean knew of them, how could Slemp have failed to know of them? He and McLean were old friends, personally, both good Republicans, both greatly concerned over the threatened stain upon the record of their party. McLean was busy many hours each day sending and receiving telegrams—trying to avoid being called before the committee. That he was extremely anxious cannot be doubted. How could he have avoided telling his old friend Slemp the whole story? In a sense it was his party duty, as a good Republican, to tell the President's secretary, in order that the party leaders might arrange the defense to the best advantage. Slemp surely must have been intensely interested. It seemed likely at the time that such a scandal would wreck the party, and the President's secretary could not properly have been indifferent to such a contingency. There were many opportunities for the two men to talk about the affair, for they were together much, dined together, and Fall and his family dined with them several times. As Walsh suggested to Slemp in his examination: "I do not know anything about any of your conversations, but it did occur to me that it would be the most natural thing in the world that you would have conversations with these gentlemen [McLean and others] about this matter." The most natural thing indeed.

Slemp testified that he first learned the truth from McLean on the same day that Walsh arrived at Palm Beach. His story is perhaps worth quoting:

Senator Dill: You said that you were walking along the beach on that day that Senator Walsh arrived?

Mr. Slemp: That is correct.

Senator Dill: And Mr. McLean said that "I am going to tell all." You and he had been talking about this previously then?

Mr. Slemp: No, sir.

Senator Dill: Well, why would he say that "I am going to tell it all"?

Mr. Slemp: Well, that was as much of a surprise to me as it could be because I was amazed.

Senator Dill: He did not explain what he meant by saying that he was going to tell all?

Mr. Slemp: He did not, because I did not know that he was going to do it, and I urged him to do it, to tell the truth about it.

Senator Dill: Well, there was not any suspicion then, was there?

Mr. Slemp: Not the slightest in my mind.

If the reader will try to draw a mental picture of this episode, as Slemp told it, he can judge for himself as to the probabilities. It is difficult to believe that Slemp did not know the truth regarding Fall's transactions long before Walsh pried it from McLean.

And if Slemp knew the truth, how could Coolidge have failed to know it? Slemp was in close communication with the President; and it would have been a most extraordinary circumstance if he had failed to transmit what he knew to his chief. We might go further and say that if President Coolidge knew the truth, other party leaders did too—but it is unnecessary to go so far.

President Coolidge has been criticized for carrying his party loyalty too far; and his record in the Teapot Dome scandal must be considered in connection with his record in regard to the investigation of his Attorney-General, Daugherty. Senator Wheeler, chairman of the committee which was investigating Daugherty, openly charged President Coolidge with responsibility for the fact that Daugherty was not caught. It is no wonder that Senator Walsh betrayed some indignation at President Coolidge's attitude. The following quotation from the examination of Slemp seems justified:

Senator Walsh of Montana: Just another question or two. Now, you must have regarded that situation as a highly remarkable one when so eminent a citizen as Mr. McLean confessed that he had been misrepresenting so important a matter to this committee, and an ex-Cabinet minister (Palmer) was found in exactly the same situation with reference to this exceedingly important matter. Now, your duties, as you say, are confidential to the President of the United States. Did you not conceive that any duty whatever devolved upon the President of the United States in connection with that matter?

Mr. Slemp: Well, I would not like to pass on that, Mr. Senator.

Senator Walsh of Montana: Well, I would.

The investigations of the committee revealed the fact that certain oil interests have had remarkably close connections with the government. The policy of Doheny was apparently to keep on his pay roll as many influential ex-officials of the government as he could afford, unquestionably for the influence they could exert on the policy of the government, particularly the policy of the government in its Mexican affairs. He thus employed, not only Ex-Secretary Lane, and McAdoo, but Ex-Secretary of War Garrison, and, on one occasion, Ex-Attorney-

General Gregory, according to his own (Doheny's) testimony. He testified also that he was willing to hire Fall at any time, in fact, that he intended to offer him employment. The spectacle of millionaire oil operators, of the moral standards shown by Doheny and Sinclair, controlling the policies of the various departments at Washington through the influence of these ex-officials, is not one that thoughtful Americans can view with indifference.

Doheny's money was used at other national capitals than Washington. It is said that he offered the United States Government a plot of ground in Mexico City for embassy purposes, and that President Harding accepted the gift. He has loaned vast amounts of money to various Mexican governments. It is not recorded whether any of the loans were of the nature of his loan to Fall. He testified that he had loaned $5,000,000 to the Obregon government.

The above story, as revealed by the investigations of the Walsh committee, is by no means the whole of the oil scandal. Later court proceedings brought out further details, involving high officials of other oil companies in the United States. Several of the most influential oil men in the United States fled to Europe until some of the proceedings were over. Since the committee was unable to secure the testimony of many of these men, the details of their operations cannot be learned definitely. Only enough is known to justify the belief that there have been more large-scale crooks in the oil industry than has generally been supposed.

For a week or two after Senator Walsh had opened the Teapot Dome cesspool, President Coolidge made no move in the matter, and some of the Democrats in Congress introduced resolutions "directing" him to institute and prosecute suits to cancel the oil leases, and to dismiss the members of the cabinet who had been connected with the oil scandal. In response to such resolutions, President Coolidge on January 27 issued a letter stating, "When facts are revealed to me that require action for the purpose of insuring the enforcement of either civil or criminal liability, such action will be taken"; and explained that under his direction the Department of Justice (Daugherty's department) had been observing the course of the evidence at the hearings. He further explained that, since men were "involved who belong to both political parties," he would appoint two special prosecutors, one from each party, to institute appropriate judicial proceedings.[9]

There was much criticism in Congress, and even in a few of the newspapers, of the attitude of the President. For a while he hesitated to dismiss Daugherty and Denby from his cabinet—Fall had already resigned—and many thought that the situation demanded a real housecleaning. As to proceedings against the looters of the reserves, his first plan was to have some of Daugherty's associates follow the proceedings of the Walsh committee to see if "anything serious" were uncovered; and when he finally acceded to the idea of special prosecutors, among his first appointments was included Ex-Attorney-General Gregory, who had been attorney for an oil company, and had, on one occasion, received a small fee from one of Doheny's companies for services connected with Mexican oil properties. Another of his first thoughts was a lawyer named Strawn, who was said to have connections with a bank which had helped to float some Sinclair stocks, and which had Standard Oil connections. It was even stated in Congress that President Coolidge had seriously considered the appointment of two of Harry Daugherty's leading henchmen in the Department of Justice. These candidates were finally abandoned, however, and the names of Atlee Pomerene, a former Democratic senator from Ohio, and Owen J. Roberts, a lawyer of Republican faith, were announced.[10]

Nothing particularly uncomplimentary was said of Pomerene, except that he was a lawyer of no very wide reputation. It was said that he had never had a case in a federal court, and some men in Congress objected to him on the ground that he was unfamiliar with public land law. He was a railroad attorney, and had done some publicity work for the National Transportation Institute—an association of railroad companies organized for propogandic purposes. It seemed to be thought by some men in Congress that the President had picked out the most conservative Democrat he could find. In discussing the circumstances of his appointment, Senator Wheeler did not mince words:

The thing that nauseates me is the fact that the President of the United States, in selecting these men, has apparently not given any consideration at all to their qualifications. We find him first selecting Mr. Gregory, without ascertaining whether he was employed by any oil company. We find him then selecting Mr. Strawn, without inquiring whether or not he was connected in any way with any oil interests. Then we find him, without any hesitation at all, coming over into the Democratic Party and picking out, not a man qualified in law, but a man qualified as a politician. Why, the President of the United States knew when he appointed Mr. Pomerene that it would be difficult for many members in this body who had served with him here to vote against him. It seems to me that he was deliberately taking advantage of that fact.[11]

In discussing Pomerene's appointment, Senator Stanley asked:

Does the Senator [Dill] think, however, understanding the situation as it exists, that there is any very great probability that the present Chief Executive of the United States will, under any circumstances or under any pressure, name a man more progressive than Mr. Pomerene?

In similar spirit, Jones, of New Mexico, declared:

I have no hope whatever that we would get a better man than Pomerene named by the President of the United States.[12]

Criticisms uttered regarding Roberts were of such a nature as to indicate that he was far from an ideal appointee. The prosecution of these cases was a big job, one that called for a great deal of enthusiasm as well as ability. Sinclair and Doheny and the Standard Oil Company of course had capable and highly paid legal departments—Sinclair had nine lawyers and firms of lawyers at the Teapot Dome trial—and there was every reason to believe that the suits would be fought bitterly; yet Roberts was declared to have been hostile to the Walsh investigation, at least in the beginning. Extracts from a speech he had made before the trust company division of the American Bankers' Association sounded strongly reactionary for one who was to act as the champion of the public interests in a great legal battle with private capitalists:

Everywhere you turn judicial and semi-judicial administrative commissioners, investigating bodies, inspectors of every known variety, are found. The result is that the business man in America today feels that he is doing business with a minion of the Government looking over his shoulder with an upraised arm and a threatening scowl.[13]

Perhaps this speech did Roberts an injustice, but it sounds strikingly like many that have been made in Congress by men who objected to that government control over public resources which it was hoped these prosecutors might succeed in reëstablishing. A final suggestion as to Roberts' conservatism is found in the fact that his name was suggested by one of the most conservative men in Congress, Senator Pepper of Pennsylvania.

Doubt was expressed as to whether the President really wanted to recover the reserves and punish those who had taken them; and Walsh and others who had shown positive interest in these matters were much discouraged. Not only had the President been slow to act. When he finally began looking for men to carry on the work that Walsh had begun, he apparently consulted first, last, and all of the time, men who had been either indifferent to the investigation or actually hostile to it—such conservatives as Senator Lenroot (who as chairman of the investigating committee, seemed actually opposed to the investigation until Walsh had made his discovery), Senators Curtis, Pepper, and Lodge, and men in the Department of Justice who were avowedly partisans of Daugherty. The old line conservatives were "all honorable men," to be sure, but they could hardly be expected to have any interest in prosecuting cases against their friends—such men as Fall, Daugherty, and Sinclair. The whole affair, as exposed by the Walsh committee, was the worst scandal in the history of the Republican party, and it was hardly reasonable to expect good Republicans to be zealous in establishing such scandalous matters in court.

In all these appointments, President Coolidge never consulted Senator Walsh, until after the names had been selected, although Walsh was probably better qualified than any other man in the country, either to give advice or to take actual charge of the prosecution. He had a wide knowledge of public land law, and of course an intimate and accurate knowledge of the law and facts involved in these cases; and his ability and energy and pertinacity had been sufficiently shown in the investigation. Had the President really wished to prosecute these cases vigorously, he would doubtless have put Walsh in charge personally, or at least would have consulted him often. Walsh was opposed to both Pomerene and Roberts.[14]

Secretary Daugherty's record in regard to section 36, in Naval Reserve No. 1, was not such as to recommend him strongly for further inquiry into the status of that valuable piece of Standard Oil property, but the man for that job, Samuel Knight, was apparently selected by Rush Holland, one of Daugherty's avowed partisans in the Department of Justice.[15]

The new prosecutors started suits immediately for the cancellation of the leases and for the conviction of Fall, Sinclair, and Doheny on bribery and conspiracy charges. Chief Justice McCoy, of the District of Columbia Court, promptly quashed the indictments of Fall, Sinclair, and the Dohenys on a technicality. A few weeks later, however, on May 28, 1925, Federal District Judge Paul J. McCormick, of Southern California, handed down a decision, ordering the cancellation of several of the Doheny contracts and leases, on the ground that the payment of $100,000 to Fall was against "good custom" and against public policy, and that there was a conspiracy between Doheny and Fall. In general the McCormick decision followed closely the conclusions of the Walsh report, even to the extent of holding invalid President Harding's transfer of the reserves to the Department of the Interior. Of course Doheny appealed from this decision.

The government seemed to be getting the range of the exploiters, but only a few weeks later Federal District Judge Kennedy, of Wyoming, handed down a decision upholding the validity of the Teapot Dome lease. Contrary to the decision of Judge McCormick, the Wyoming judge held that the validity of President Harding's transfer order was immaterial; that the secrecy surrounding the negotiations was desirable in the navy's interest; and that there was no indisputable evidence of fraud. The government promptly appealed from this decision.[16]

It is impossible to predict what may be the final decision of the Supreme Court. The Supreme Court is at present a strongly conservative body—it has been purposely made so by Presidents Harding and Coolidge. Yet it is difficult to see how the leases and the transactions of Fall, Sinclair, and Doheny could be finally upheld; and, as a liberal journal has expressed it:

No amount of judicial whitewash will ever convince the American people that the episode of the oil leases is not one of the most sordid and humiliating in our entire national history.[17]

The unearthing of such a mass of corruption and official misfeasance, and the saving of valuable public property, might well have seemed a great public service, to be recognized and acclaimed by the government, the press and by the public; but strangely, it may seem, the revelations were scarcely out before a persistent campaign was inaugurated, by some of the administration in Washington, by the Republican leaders, and by most of the Republican press—and that includes most of the great newspapers of the country—to discredit the investigation and its findings. One does not need to be a Marxian, or even a "Bryanite," to see that the same interests that put Harding into the President's chair, and Fall into the cabinet, were operating through the press to befuddle the people as to what had happened. Insinuations were constantly made in the press that the investigating committee was delving into all kinds of vile and silly rumors, and was trying to besmirch the characters of innocent men. An effort was made to "frame" Senator Wheeler of Montana, to stop his investigation of the record of H. M. Daugherty; but Wheeler was acquitted twice, once by the Borah investigating committee, and once in court.[18] When Walsh made his report, many of the newspapers and journals promptly attacked this piece of "petty politics," although the report was extraordinarily judicious in its tone and in its conclusions.

One of the strangest aspects of the whole affair was the manner in which the business journals, particularly the oil journals, defended the transactions of Fall, Doheny, and Sinclair. Business men generally are not of such obtuse moral sensibilities that they would openly defend bribery and corruption of public officers, yet it is significant that one may search far in business literature without finding a word of condemnation of Doheny or Sinclair, or even of Fall. No business men's organizations, as far as the press informs us, ever repudiated either of the first named. The chambers of commerce, the associations of oil producers and refiners, have said nothing about the betrayal of the integrity of business men. It is a fair bet that Senator Walsh, who was responsible for the disclosures, has received ten times as much condemnation from business publications as either Doheny or Sinclair.[19]

Perhaps this account of the recent history of the naval reserves will seem overlong and tedious, but it shows how strong and insinuating are the forces working for the exploitation of those reserves, and of other valuable resources still belonging to the government, and how vigilant must public opinion be if our reserves of oil, oil shale, coal, and timber are to be retained for the benefit of all the people. The Teapot Dome scandal is not the first incident of its kind in American history; and it is not likely to be the last. Throughout our history we have had strong forces favoring exploitation, and the means by which they have attained their ends have often been similar to those employed by Fall, Doheny, and Sinclair. Secretary Teller, of the Department of the Interior, was said to have worked the clerical force of the Land Office overtime during the last days of his administration, in 1885, in order to get certain unearned patents out to land-grant railroads. The history of our forest lands contains many similar stories. The Oregon land frauds in the early years of the twentieth century involved several men in Congress, and a great many in the Oregon state legislature. The Ballinger-Pinchot controversy, ten years later, and Roosevelt's defection from the Republican party, were concerned with the public coal resources. The struggle between the forces favoring conservation and the forces favoring exploitation is one that goes on all the time.

NOTES

1. U. S. Bu. of Mines, *Rept.*, Director, 1922, 11; U. S. Geol. Survey, *Bul. 623;* S. Doc. 210, 67 Cong., 2 sess.

2. Information here is taken largely from the Hearings before the Senate Committee on Public Lands and Surveys, on S. Res. 282, and S. Res. 294; and from the Walsh Report on those hearings, S. Rept. 794, 68 Cong., 1 sess. Ravage, *The Story of Teapot Dome,* is an excellent summary of the evidence brought out by the hearings.

3. S. Doc. 210, 67 Cong., 2 sess., p. 26.

4. Magee's account of the origin of this feud with Fall runs as follows:

About four months after I took over the paper I returned from Las Vegas, and my wife met me at the station and told me Senator Fall was waiting for me at the office. I went to see him, and he had a copy of the *Journal* containing a criticism of the public land office, and he held it up and asked me whether I wrote it. I told him I was responsible for anything that was in the paper, and I put on my glasses, and then I said, "Yes, I wrote it." He said, "Didn't I tell you that the public land office was organized to suit me?" I said, "Yes; I think you did tell me something like that." And I said, "It isn't organized to suit me." He said, "I want you to back away from it." And he said, "Will you do it?" And I said, "I should say not." And he said, "I am going to break you." I said, "Put on your hat and wade in; the water is fine." And we have not been on good political terms since that time. (Hearings, p. 842.)

5. Hearings, p. 1699.

6. Hearings, p. 1715. It was here that one of the peculiar incidents of the investigation occurred. Roosevelt insisted that Wahlberg said "68,000." Wahlberg always insisted, on the stand, that he had never said anything about any $68,000 or about cancelled checks, and insisted that he said "six or eight cows," and that Roosevelt misunderstood this for $68,000. Wahlberg denied much that Roosevelt testified to. (Hearings, 1717, 1721.)

7. Hearings, p. 1291.

8. *Cong Rec.*, Jan. 30, 1924, 1671.

9. *Ibid.*, Jan. 28, 1924, 1520.

10. *Ibid.*, Jan. 28, 1924, 1537; Feb. 7, 1924, 1957, 1958; Feb. 26, 1924, 3169.

11. *Ibid.*, Feb. 18, 1924, 2637.

12. *Ibid.*, Feb. 16, 1924, 2550, 2564, 2565.

13. *Ibid.*, Feb. 16, 1924, 2563.

14. *Ibid.*, Jan. 29, 1924, 1612; Feb. 16, 1924, 2548-2559.

15. *Ibid.*, March 18, 1924, 4425.

16. U. S. vs. Mammoth Oil Co. et al., 5 Fed. Rep. (2d series), 330.

17. *New Republic*, July 1, 1925, 135.

18. S. Rept. 537, 68 Cong., 1 sess.

19. *New Republic*, June 11, 1924, 63.

CHAPTER XXVI

INDIAN OIL LANDS

THE history of the Indian lands is an extremely confused and tangled story, taking in, as it does, all the various laws and treaties dealing with each of the various tribes; and no detailed account can be included here. Nevertheless, some account must be given, for the Indian lands have at times been the dominating influence in the oil industry. Most of the great oil fields in Oklahoma have been found on Indian lands.[1]

In general the Indians have been declared to be wards of the United States government, and, as such, their lands have been governed either directly by federal statutes or by treaties and agreements drawn up under the provisions of federal statutes. In February, 1891,[2] a law was passed authorizing the Indian councils to lease for mining purposes certain lands not needed for agricultural purposes, for not over ten years, on conditions fixed by the agent in charge of the reservation with the approval of the Secretary of the Interior.[3] It has been reported that a number of wells were drilled in the Indian Territory as early as 1891.

Previous to 1893, the avowed policy of the United States, in dealing with the Five Civilized Tribes (the Five Civilized Tribes included the Creek, Cherokees, Choctaw, Chickasaw, and Seminole tribes) of the Indian Territory, was to maintain an Indian community away from the influence of white people. Every treaty from 1828 to 1866 was based on this idea of exclusion of the Indians from the whites and non-participation by the whites in the political and industrial affairs of the Indians. During the eighties and early nineties, however, it became evident that the government effort to preserve the tribal organization was failing, for a great many white people were going into the Territory, and were living there, with the approval of the Indian authorities. In fact, the whites presently greatly outnumbered the Indians, and they got possession of some of the most valuable of the lands. The theory of the government, when it turned the title to the lands over to the Indian tribes, was that the title should be held for all of the Indians of the tribe. All were to be the equal participators in the benefits accruing from the land. It developed that a few enterprising citizens of the tribe, frequently not Indians by blood but by intermarriage, became practically the owners of the best lands. As a result of these conditions, a law was passed in March, 1893, providing for the allotment of the lands to the Indians in plots of not over 160 acres, the Indians to become citizens of the United States.[4] In executing this policy, Congress aimed to

conserve the interests of the Indians and to fulfill the national obligation, not simply by assuring an equitable apportionment of the land,
but by safeguarding the individual ownership of allottees through
suitable restrictions which were designed to secure them in their possession and prevent their exploitation.

The Curtis Act of June 28, 1898, was one of the important land-
marks in the history of Indian lands.[5] This law provided for the enroll-
ment of members of the tribes and for the allotment of the surface of
the lands to the Indians, and for the leasing of mineral lands by the
Secretary of the Interior. The Curtis Act had to be ratified by each of
the various tribes before it went into effect, but it was ratified by all of
them within a short time.

It will be unnecessary to trace all the various laws and rules and
regulations that have been adopted dealing with each of the tribes, but
some consideration of the more important tribes must be given.

The Osage Indians have received more publicity than any other
equally numerous tribe of Indians, because of the phenomenal wealth
they have received from their oil lands. Some of the richest of these
oil lands the Osages bought from the Cherokee Indians in 1883, for
$1.25 per acre. The Cherokees, it is said, thought they had driven a
very good bargain in this sale. As early as 1896, the Osages granted
a lease covering the entire area of the Osage nation, nearly two million
acres, to Edmund Foster and his associates, the lease to run for 10
years. Foster had difficulty getting enough money to do any drilling,
but finally managed to get a couple of successful wells down. In the
meantime, however, he had considerable trouble with his associates,
and died before he had achieved any particular success with his lease.
His brother took up the work, but he soon died too, and a nephew
took over the job, and after a hard struggle, in January, 1903, succeeded in organizing the Indian Territory Illuminating Oil Company to take over the lease. The Indians had cancelled the lease in
1898, on complaints as to the manner in which it was being managed,
but had reinstated it the following year. The Indian Territory Illuminating Company carried on exploitation largely through sublessees,
drilling some 783 wells during the three years ending January 1, 1906.[6]

In 1906 the Foster lease was renewed for another ten years, but its
area was cut down to 680,000 acres, and the royalty was raised from
10 per cent to 12½ per cent. About the same time, a lease covering
350,000 acres, part of it released from the Foster lease, was granted
to J. S. Glenn, who had done the first drilling for Foster. The rest of
the original Foster lease was restored to the Osage tribe, the minerals
to be held by them as tribal property for a term of 25 years.[7] The exploitation of the Foster lease during the decade 1906 to 1916, just as

before, was carried on almost entirely through sublessees.[8] The work of the Indian Territory Illuminating Oil Company, in subleasing these lands, must have been fairly profitable, for it paid the Indians a royalty of 12½ per cent, and subleased at one-sixth royalty, with an additional bonus of one dollar to five dollars per acre.

As the time approached for the expiration of the second term of the Foster lease, the Indian Territory Illuminating Oil Company applied for a renewal. There was considerable discussion of the matter, and in March, 1915, a series of hearings were held to give all interested parties a chance to be heard. At some of these hearings, the Osage tribal council was present and the council, after some days of deliberation, adopted a set of resolutions, asking that the lease be not renewed, but that all future leases be made direct to the companies doing the actual work, in tracts of not over 4,800 acres each.[9] The council also petitioned that Congress be asked to extend the trust period on their other lands, which otherwise would expire in April, 1931. All of these resolutions the Secretary of the Interior promptly approved,[10] and in 1921, Congress passed an act extending the trust period, reserving the mineral deposits, until 1946.[11]

The termination of the Foster lease meant increased revenues for the Osages, for the royalties were generally raised. Previously the tribe had received 12½ per cent royalty, with no rentals. The royalties were now raised to one-sixth for each quarter-section of land producing an average of less than 100 barrels per well per day, and one-fifth for such units producing 100 barrels or more, while a rental of $1 per acre annually was also charged.[12]

Since November, 1912, more than 24 public sales of Osage oil and gas leases have been held, and a total of 547,156 acres of land have been leased, for a total in bonuses alone of over $76,000,000. These sales, where men often bid above a million dollars for a single quarter-section of land, are very interesting events.[13]

The policy of the Osage nation in dealing with its mineral lands has been different from that followed by the Five Civilized Tribes. The Osages reserved to the tribe all the mineral deposits when allotments were made, "nationalizing" their deposits after the manner of many foreign governments. Thus the vast oil revenues did not accrue to a few fortunate individuals but to all members of the tribe equally. On the other hand, the Indians of the Five Civilized Tribes did not reserve the minerals to the tribes when allotments were made, and, as a result, some Indians received tremendous sums in oil royalties, while others received nothing.

The Choctaw and Cherokee nations became interested in the oil possibilities of their lands in the early eighties. In October, 1884, the

Choctaw Council passed a law creating the Choctaw Oil and Refining Company, to explore for petroleum; and the Cherokees made provision for charters for exploration a few weeks later. There was considerable interest in the matter at the time, but the following year the Cherokees repealed the charter they had granted, and this action scared capital out of the field for many years.

The Choctaw nation was generally much more closely affiliated with the Chickasaw tribe than with the Cherokee. On August 24, 1898, the Choctaw and Chickasaw nations by special election ratified the Curtis Law with some amendments. Under the regulations issued, some of the oil lands of these tribes were to be leased, the leasing to be under the control of two trustees appointed by the President on recommendation of the executives of the two nations, each of whom was to be an Indian by blood, of his respective tribe. Leases were to cover not over 960 acres, on a royalty of 10 per cent, and must be approved by the Indian Inspector. Not all of the oil lands were to be leased under this act, however, and on July 1, 1902, another agreement was made by the tribe whereby about 440,000 acres of land were to be set aside, to be sold to the highest bidder, by the Secretary of the Interior. The Secretary offered these lands in 1904, both the mineral and the surface, but did not approve of the price offered, and so refused to sell. These lands were withdrawn by the act of April 26, 1906, which provided for the sale of the surface only.[14]

Like the Choctaws and Chickasaws, the Creeks and Cherokees are sometimes considered together. As early as 1892, Michael Cudahy, the meat packer, who had been in the Ohio and Indiana oil fields, secured from the Indians a large number of oil leases, aggregating nearly 200,-000 acres, in the Cherokee and Creek nations, but a number of the Indians were opposed to the leases, and the Secretary of the Interior, under the provisions of the Curtis Law, refused to approve more than one, which happened to include the townsite of Bartlesville.[15] James Givens, of a law firm in Muskogee, later got leases covering about 300,000 acres of Creek and Cherokee lands, and several other individuals and companies tried to get large leases, but the Secretary of the Interior refused to approve most of them.[16]

Thus the protectorship of the United States government has been maintained at all times. During the year 1902, agreements were made with the Creek and Cherokee tribes providing that Indian citizens might lease their lands, for mineral purposes only, with the approval of the Secretary of the Interior.[17]

Leases have been offered for other reservations than the Osage and the Five Civilized Tribes. Considerable tracts have been leased in the Kiowa and Otoe reservations, in Oklahoma; in the Wind River and

Shoshone reservations of Wyoming; in the Blackfeet and Crow reservations, in Montana; in the Quinaielt reservation, in Washington; and in others. The discovery of oil near Farmington, New Mexico, led to great interest in the nearby Navajo reservation, and in October, 1923, the government held an auction of oil and gas leases in the Navajo reservation in New Mexico and Arizona.[18]

Various criticisms have been aimed at the administration of the Indian lands. During the years following the passage of the Curtis Law, there was much complaint among oil operators of the restrictive and illiberal policy pursued by the Secretary of the Interior, and of the delay and the uncertainty attending all negotiations.[19]

Within the last few years there has been complaint from the oil operators on Indian lands that the royalties were too high, and even that the bonuses paid at the auctions were higher than the bidder could really afford. It has often been stated that more money has been spent for the Indian lands than was ever secured from operations on them. One oil journal complains of the "senseless bidding" at the auctions, and of the unreasonably high prices paid. It is difficult to see how bidders in general could justly complain that the prices they set are too high.

During some of these years, determined efforts were made in Congress to secure a law providing for the sale of some of the Indian lands, minerals and all, especially some of the Choctaw and Chickasaw lands. It was claimed in Congress that the Commissioner of Indian Affairs had recommended the policy of sale, and of course many of the Indians were anxious for anything that looked like immediate income. While some men were trying to provide for the sale of such lands, La Follette and a few others tried to nail the leasing policy down more firmly. Of course President Roosevelt favored the leasing policy.

There have been, in the past, many allegations of fraud in the management of Indian lands. During the old days of tribal management, rank frauds were unquestionably perpetrated on the ignorant Indians, sometimes by the shrewder of their own people, and sometimes by the white people who were always hanging around the reservations. The reports of the Dawes Commission indicate the general situation that existed for many years. Even later, at times somewhat the same situation existed in various tribes. The Indians, many of them lacking in understanding of business tricks, were always more or less the victims of crooks and tricksters; and, of course, when they became wealthy, they were more valuable prey. It is said that some of them became so suspicious, as a result of being so often victimized, that they were afraid to take packages or mail out of the post office, or to sign their names on an ordinary hotel register.

Many reports of the Commissioner of Indian Affairs mention the hordes of shyster lawyers who always infested the Indian reservations and the adjoining sections of Oklahoma, trying to find flaws in land titles, or stir up business in some way. Of course, business in later years, with the numerous millionaire Indians, has been what lawyers call "good picking." Million-dollar lawsuits have been fairly numerous among the Indians, and not infrequently they have involved some allegation of fraud or misrepresentation.[20]

In another case, there was a suspicion of some kind of improper influences in the Osage Council. In November, 1912, the Secretary of the Interior opened the bids covering 42,410 acres of Osage lands and prepared to grant the lands to the highest bidder, as required by law. In the meantime, however, the Uncle Sam Oil Company was trying to secure a large lease covering some of these offered lands, and the Indian council rejected all the bids received by the Secretary. The Secretary ordered an investigation of the entire matter, and on receiving the report, issued an order removing the entire Osage Council. One of the deposed councilmen sued the Secretary of the Interior, questioning his right to remove him, but the Supreme Court of the District of Columbia upheld the right and the tribe elected a new council, which accepted the auction bids. This was not the end of the matter, however, for several years later the Uncle Sam Oil Company announced that it was going to sue the Standard Oil Company for one billion dollars, for having taken its lease away, through "underground influence" on "several prominent former government and state officials." This suit soon fizzled out, however.[21]

There is every reason to believe that the Office of Indian Affairs, at least in recent years, has been honest and conscientious in its administration of the Indian oil and gas lands. Unfortunately, however, the general policy of the government has been extremely unwise. It has been particularly unwise in two respects: first, in opening lands for exploitation far in advance of any real need for oil, and more rapidly than the market could absorb the production at any reasonable price; and second, in opening these lands in too small tracts for efficient and economical exploitation.

Indian lands have been opened to exploitation far more rapidly than there was any possible reason or excuse for. At times when oil was so cheap as to be hardly worth barrelling, new fields of Indian lands were opened up, to add to the over-production and waste. This sort of thing has happened even very recently. In 1921, for instance, a year of extreme depression and low prices, when nothing was less needed than more oil, the Secretary opened up over 60,000 acres of the rich Osage lands. During the next two years there was no lasting recovery, for oil was constantly being produced in larger quantities than it could be consumed, and was being stored by millions of barrels almost every month; yet the Secretary held two sales of Osage lands in 1922, and two more in 1923, adding over 100,000 acres to the territory under lease, thus helping to increase the general demoralization of the industry. He was required to do this, it should be noted, by the terms of the law dealing with Osage lands.[22]

To some extent the responsibility for this unfortunate policy has been with the various Secretaries of the Interior, for they have sometimes put Indian lands on the market when there was no legal requirement for such action. On the other hand, in some instances the provisions of the law have required the Secretary to lease a certain amount of land each year, and the Secretary had no alternative but to follow the law. The law of March 3, 1921,[23] for instance, provided that the Secretary of the Interior and the Osage Council should offer for lease all the remaining portion of the unleased Osage land before April 8, 1931, offering it annually at the rate of not less than one-tenth of the unleased area each year. The present Secretary of the Interior, Hubert Work, and Assistant Secretary Finney, have tried to get this law changed to authorize the Secretary to withhold such lands from leasing during times of over-production; and there is some evidence that Secretary Lane, of the Department of the Interior, and Cato Sells, Indian Commissioner, would have been glad to see more of the Indian oil saved for the future. The Indians have generally favored a policy of rapid exploitation, and of course the oil interests have always wanted more land to exploit.

Indian lands have played a very important part in the oil production of the country. In the Five Civilized Tribes a great many oil fields have been found and exploited: Bird Creek, Bartlesville, Copan Wann, Hog-shooter, Nowata, Delaware, Chelsea, Inola, Bald Hill, Bixby-Leonard, Cushing, Boynton-Cole, Glenn, Hamilton, Henryetta, Morris, Muskogee, Mounds, Haskell-Stone Bluff, Tiger Flats, Healdton-Fox, and a number of lesser fields. The great Cushing field is located in the Creek Nation, and the more recent Burbank field is on Osage land. The fact that these fields were found in Indian territory does not mean that Indians derived a royalty from all the production, except in case of Osage land. Some of the tribes had allotted the land with rights, and titles had gone to individuals, whites and Indians. In the fiscal year 1915 more than 27,000,000 barrels of oil were produced from government leases. In the fiscal year 1924, the production was 49,640,458; and the next year it was 48,138,278 barrels.[24]

The leasing of so much land during years of over-production would not have been quite so unfortunate in its results if the land had been leased in large enough tracts to permit economical exploitation, but the general policy of the government has been to lease in small tracts, on the false theory that competition could thus be insured and on the unfortunately valid theory that such a scheme would force rapid "development." In sales of Osage leases, for instance, the unit for years was 160 acres, no lessee being permitted to have more than a total of 4,800 acres. Perhaps 4,800 acres, in a contiguous tract, would have been enough to permit efficient and economical exploitation, but in scattered tracts of only 160 acres, it frequently meant only hasty and wasteful extraction of the oil, over-production, and all the forms of waste that go with it.

To make matters worse, the Secretary of the Interior has usually inserted in the leases certain requirements as to the number of wells to be drilled within certain specified periods, these requirements applying regardless of market conditions. Such drilling requirements were intended to prevent speculation in leases and to force prompt exploitation, but they only tended to increase the chronic demoralization of the oil market. In a few instances these requirements have been relaxed, but at other times they have been enforced, to the very great injury of the operators as well as of the public. During the depression that gripped the oil industry in 1905, these drilling requirements were said to have been the most important factor in forcing drilling and in increasing the over-production. As Professor Haworth, of the Kansas Geological Survey, pictured the situation in 1905:

With the entire area controlled by one good business head, development would be stopped until the stocks had decreased greatly and the price of oil advanced; but with the Secretary of the Interior insisting on the "development clause," no one cares to jeopardize the title of valuable leases by declining to drill. Therefore, the price goes downward and production continues to increase.[25]

The limitation of 4,800 acres to each lessee was imposed by the Secretary of the Interior in his regulations issued under the act of June 28, 1906, the act which provided for the reservation of the Osage lands not included in the Foster lease. This limitation was a characteristic American attempt to preserve small units where small units were economically indefensible. As long as the limitation was observed, it prevented the large companies from engaging extensively in the exploitation of Indian lands, and the large companies of course conducted their operations more efficiently than small companies. In other words, this limit was an attempt to keep the oil business in the hands of small and inefficient units.[26]

The limitations as to size of tracts leased have been changed in recent leases of some of the Osage lands and in the leases of lands belonging to some other tribes. In 1920, the Secretary of the Interior issued a circular raising the limitation for any one lessee to 20,000 acres on what is known as the East Side of the Osage reservation, and entirely abolishing the limitation on the West Side. The next year, the limitation of 9,600 acres was abrogated as to Indian reservations in Oklahoma, except the Five Civilized Tribes, the Osages, and the Kiowas. In leasing the Navajo lands, the regulations issued provided that the secretary might, in his discretion, grant leases covering as much as 4,800 acres; and the same provision applied in the leasing of the Ute lands in Colorado and New Mexico.[27]

As a result of the too rapid exploitation of Indian lands, some of the Indians have become phenomenally wealthy. In the Osage nation, where the oil deposits were held in common, is to be found perhaps the highest per capita wealth to be found anywhere in the world. Each Osage Indian has his homestead and allotment of 160 acres and in addition about 500 acres of surplus land. In the fiscal year ending June 30, 1922, the bonuses paid for Osage leases amounted to over $22,000,000, in addition to which they received royalties running as high as one-fifth on a total oil production of 29,000,000 barrels. The average income of the enrolled Osage Indians during that year was approximately $10,-000, and some families received as high as $80,000, largely from royalties and bonuses.

In spite of their vast incomes, many of the Indians are in debt much of the time. Many of them have married whites, who are just after money, and of course the type that seek money in this way are not usually a very high type. Many of the Indians have developed very extravagant habits. The following description of the Indian life in Pawhuska is taken from the 1917 *Report of the Commissioner of Indian Affairs*:[28]

What is most impressive is the fact that it seemed as though the community was filled with Osage Indians who were having series of continuous joy rides. Automobiles were everywhere filled with them, and so confirmed has the motor habit become that many of them will not walk two blocks to go from one store to another, without hiring a motor for the purpose.

The Indian village is about a mile from the business section of Pawhuska. When the jitney craze was at its height an enterprising liveryman thought that he saw a good business opportunity and put in jitneys to run out to the Indian village. The undertaking was a quick failure. No Osage Indian would demean himself by riding in a cheap public vehicle of that sort. He would far rather pay a dollar or more and ride in state alone, or with a friend or two, each of whom paid the same, than to travel in the same company but with the stigma of paying only a dime for the round trip. The fact that the automobiles were identical had no influence. I was informed that it was not unusual for parents to give their children $20 notes at a time for spending money. I saw one little girl go into a drug store and buy a glass of soda water, handing the clerk a $20 bill in payment and, with her hands full of change, leave the store headed straight for the moving picture show across the street. Expensive jewelry is also bought by the children without regard to its cost or appropriateness, while it is stated that more expensive French ribbons are sold in Pawhuska than in any town in the Southwest.

Automobile rides and movies have not been the only indulgence of the Indians. For generations Indians have had a strong liking for the white man's "fire water"; and those with plenty of money and no work to do have often found it possible to get a poor grade of "fire water" in spite of the prohibition laws, although the prohibition laws were fairly well enforced. When unable to get any whisky from bootleggers, the Indians have often bought large amounts of lemon and vanilla extracts. Grocers in Pawhuska kept large stocks of such extracts on hand. The Indian Bureau finally had to adopt the policy of using no extracts preserved in alcohol.

By no means all of the wealthy Indians are Osages, although the Osages have received more oil money than any other tribe. In some of the other tribes, where the oil and other minerals were the property of the individual surface owner, there have been some dazzling illustrations of unearned incomes. Jackson Barnett is only one example of many that might be given.[29] In the Cushing field, on the Creek lands, there were the other famous cases of Barney Thlocco, Tommy Atkins, and Emma Coker, all supposed to be worth millions.[30]

A fair share of some of these large fortunes was dissipated in litigation over titles. The Tommy Atkins case was an example of this. Tommy Atkins, the full-blooded Indian allottee to a very valuable piece of land in the Cushing field, died in infancy. A fierce legal battle was waged in an attempt to prove him a myth. After his existence had been legally determined, a second and more intense contest arose. Charles Page, a wealthy oil operator, held a lease under Minnie Atkins, who claimed to be the mother of Tommy. Others held a lease under Nancy Atkins, who also claimed to be his mother. After a long trial, the federal judge decided in favor of Page. This decision was affirmed by the circuit court of appeals, and the case was considered settled. Presently, however, a new suit was brought for possession of the estate by a Sapulpa lawyer and others, who contended that they had found a third woman, a negro named Sallie Atkins, who was the real mother of Tommy. They claimed to have obtained a lease of the land from her. The judge before whom this trial was held decided in favor of Sally; but soon the judge himself was arrested on charge of accepting a bribe for his decision. This case was finally appealed to the Supreme Court of the United States, where the title of Minnie Atkins was held to be good, and the Page lease was confirmed.[31]

The extraordinary fortunes that have fallen to many of the Indians seem fair and just in one way, at least. The government established the Indian reservations in Oklahoma because the land there was thought to be of little value; and when oil was later discovered there, it almost looked as if there were some sort of justice in the turn of fortune's wheel.[32]

In other ways, however, the situation is far from happy. Such wealth as many of the Indians have does not contribute either to their usefulness or to their happiness. In many respects it is said to be a demoralizing influence. Furthermore, while they are smothering in their wealth now, it is likely that the time will come when oil royalties will cease; and the decline to poverty again will be a more painful change than was the elevation to riches. It is greatly to be regretted that the Indian lands could not have been closed to exploitation entirely, many years ago, or at least exploited much more slowly than they have been. Had they been so, we would have had less distress in the oil industry, less waste of oil, less demoralization of the Indians; and we should have more oil left for the future, and a longer future of financial competence for the Indians.[33]

NOTES

1. For a list of the oil fields, see the *Report* of the Commissioner of Indian Affairs, 1915, 399, 406.

2. Stat., 26, 795. See Snider, *Oil and Gas in the Mid-Continent Fields,* 208; and Ok. Geol. Survey, *Bul. 2,* p. 166.

3. Several years before this, grazing leases had been authorized, and a few had been made. (*Rept.,* Sec. of Int., 1886, 94.)

4. Stat., 27, 645. For a general account of legislation relating to Indian Lands, see S. Doc. 690, 62 Cong., 2 sess.

5. Stat., 30, 495.

6. *Independence Daily Reporter,* History of the Mid-Continent Oil and Gas Field, p. 19, 26; *Mineral Resources,* 1904, 705; 1905, 855; *Rept.,* Sec. of Int., 1905, 61.

7. Stat., 34, 539.

8. Ok. Geol. Survey, *Bul. 2* (1911), p. 192.

9. The Indians do not seem to have been consistently favorable to the policy of smaller leases, for early in 1912 the Osage council had authorized the chief to execute four leases of about 200,000 acres each, covering the entire unleased part of the reservation. These leases the Secretary of the Interior refused to approve. (*Rept.,* Comr. of Indian Affairs, 1913, 31.)

10. *Rept.,* Comr. of Indian Affairs, 1915, 27; 1921, 18; *Mineral Resources,* 1915, 660; *Oil and Gas Jour.,* March 11, 1915, 12; April 15, 1915, 2; May 13, 1915, 3; July 1, 1915, 2.

11. Stat., 41, 1249.

12. *Rept.*, Comr. of Indian Affairs, 1916, 59.

13. The following table shows the record of Osage lease sales, to April, 1923:

HISTORY OF OSAGE LEASE SALES

No.	Date	Acres sold	Bonuses	Av. per acre
1	Nov. 11, 1912	24,541	$ 39,436.00	$ 1.60
2	Sept. 29, 1913	10,132	498,182.00	49.17
3	April 30, 1916	14,377	2,057,600.00	143.12
4	June 20, 1916	2,482	1,169,280.00	471.10
5	May 31, 1917	8,160	1,947,600.00	238.68
6	Nov. 12, 1917	19,869	1,677,200.00	84.41
7	Feb. 14, 1918	25,440	1,275,500.00	50.14
8	May 18, 1918	38,980	1,180,575.00	40.55
9	Nov. 9, 1918	34,720	3,350,200.00	30.28
10	March 5, 1919	20,520	2,790,925.00	136.01
11	June 6, 1919	38,451	3,884,925.00	101.03
12	Oct. 6, 1919	34,670	6,056,950.00	174.41
13	Feb. 3, 1920	30,173	3,056,950.00	101.31
14	May 18, 1920	31,303	2,851,900.00	91.10
15	Oct. 12, 1920	36,877	3,993,750.00	108.30
16	June 14, 1921	25,918	4,559,100.00	175.90
17	Dec. 12, 1921	35,665	7,261,600.00	203.60
18	March 2, 1922	33,467	3,949,950.00	118.02
19	June 28, 1922	33,220	10,889,700.00	324.80
20	Jan. 18, 1923	24,231	6,215,700.00	256.52
21	April 5, 1923	23,960	8,029,100.00	335.10
	Total	547,156	$76,736,123.00	$140.25

14. *Cong. Rec.*, Jan. 11, 1912, 868; *Rept.*, Sec. of Int., 1905, 122; 1906, 57.

15. The townsite was said to have been approved the day after Cudahy's lease was approved. For some time there was a bitter fight between the Cudahy Oil Company and the town, as to his drilling rights, but it was eventually compromised.

16. *Rept.*, Comr. of Indian Affairs, 1900, 127.

17. *Rept.*, Sec. of Int., 1903, 224, 245, 380.

18. Comr. of Indian Affairs, Various *Reports,* especially 1913, 30; 1918, 53; 1921, 16, 17; *Oil and Gas Jour.*, May 3, 1917, 33. On the Navajo leasing, see the *Wyoming Oil Index*, May 5, 1923, and July 14, 1923, and the K. C. *Star,* Feb. 11, 1923. For a discussion of the leasing of oil lands in the Seneca reservation in New York, see *Rept.*, Sec. of Int., 1900, 300, and *Eng. and Min. Jour.*, Feb. 6, 1897, 146.

19. *Mineral Resources,* 1904, 705.

20. *Oil and Gas Jour.*, July 30, 1914, 14; Sept. 26, 1919, 73.

21. *Rept.*, Comr. of Indian Affairs, 1913, 31.

22. *Rept.*, Comr. of Indian Affairs, 1914, 33, 35.

23. Stat., 41, 1249.

24. Below is given a short summary of activities on the various reservations for the fiscal year 1925:

	Acres leased	Production	Revenue received
Osage	20,175	33,662,179	$12,141,621
Five Civilized Tribes,	80,001	13,532,857	4,214,100
Ponca	8,920	413,890	221,638
Pawnee	3,185	149,578	96,264
Shawnee	13,458	39,283	85,228
Kiowa	3,477	75,906	47,014
Shoshone	891	54,610	25,861
Navajo	....	210,175	55,794
Crow	1,100	....	9,625
Consolidated Ute Agency	3,477	....	40,876
Seger	....	....	832
Blackfeet	....	....	676

25. *Mineral Industry*, 1905, 482. For an account of over-production a few years later, see *Rept.*, Sec. of Int., 1907, 26; 1908, 30.

26. *Stat.*, 34, 539; *Oil and Gas Jour.*, Jan. 7, 1915, 4; May 20, 1915, 3; April 26, 1917, 2; Dec. 5, 1919, 2; Aug. 20, 1920, 76.

27. *Rept.*, Comr. of Indian Affairs, 1921, 18.

28. p. 339-344.

29. See above, p. 208.

30. *Rept.*, Comr. of Indian Affairs, 1915, 32.

31. *268 Fed.*, 923; *260 U. S.*, 220.

32. See an article by Francis E. Leupp in *R. of Rs.*, Oct., 1910, 468.

33. *Oil and Gas Jour.*, June 21, 1917, 2; July 23, 1920, 34; Nov. 12, 1920, 94; July 8, 1921, 84.

CHAPTER XXVII

RESULTS OF OUR OIL POLICY: THE PRESENT SITUATION

IN what situation, we may finally ask, has our oil policy left us? Where will the oil come from to satisfy our mounting requirements? How long can we keep up the pace? As David White has said:

These are plain, common sense business questions, predicated on our present oil requirements as an established fact, and on the suggested possibility that our prodigal spending of our petroleum heritage may cause its too rapid depletion if not its early exhaustion in the midst of our spendthrift career, and at some untoward moment send us as beggars to foreign countries for the precious fluid necessary not only to satisfy our extravagance but even to sustain our industrial prosperity, our standards of living, and our civilization.[1]

According to the best estimates available, the United States has now used over one-third of her original supply of oil, and is consuming the remainder at the rate of about 8 per cent each year. The most recent estimates have placed the remaining reserves at about 9 billion barrels, which are being used at the rate of about 700,000,000 barrels per year.[2]

As early as 1908 David T. Day, then in charge of the petroleum statistics of the United States Geological Survey, in connection with Roosevelt's conservation movement, calculated the total quantity of oil originally available in the ground and recoverable, at between 10 and 24.5 billion barrels, with 15 billion barrels as a reasonable guess.[3] This estimate was based upon data accumulated by extensive field investigations, but at that time some of the country's richest pools had not been discovered. Ralph Arnold calculated that there were between five and eight billion barrels in California alone;[4] and in 1915 the same authority revised Day's estimates, in the light of additional data, and placed the original supply at approximately 9.1 billion barrels, thus reducing Day's estimate considerably.

The following year geologists in the oil and gas section of the United States Geological Survey worked out a third estimate, placing the original recoverable supply at 11.2 billion barrels;[5] and the next year, 1917, the same men reconsidered their figures and reduced their estimate from 11.2 billion to 10.1 billion barrels. Late in the year 1918 David White and his associates in the United States Geological Survey recalculated the entire question, and placed the original supply at 11.3

billion barrels, as a conservative estimate.[6] Early in 1920 David White placed the figure at between 11 billion and 13 billion barrels, with 12 billion barrels as a conservative estimate. This estimate was frequently cited on the point. George Otis Smith declared it a liberal estimate, although he suggested that an increase of 25 per cent in the figure might be necessary.[7] Arthur Little considered it liberal.[8]

In March, 1921, at the meeting of the American Association of Petroleum Geologists, David White asked President Pratt of the association to coöperate with the Geological Survey in making a new survey. The joint committee thus formed, in coöperation with a number of state geologists, consulting specialists, and company geologists especially familiar with the stratiography, structure, and mode of occurrence of oil in the different fields, worked for nine months, and in January, 1922, published their estimates, which gave the United States an original reserve of about 15 billion barrels, "recoverable by present methods," with slightly over 9 billion barrels still unused. Thus the most recent estimate coincides very closely with that of Day in 1908, fifteen years before. It was pointed out in the estimate that nearly half of the oil remaining in the ground, over 4 billion barrels, was of the heavy group containing the fuel oils; while on the other hand, perhaps these estimates might be raised from 40 to 80 per cent by new methods of extraction.[9]

The detailed estimate of this committee, for oil remaining in the ground in January, 1922, was as follows:

	Millions of barrels
New York	100
Pennsylvania	260
West Virginia	200
Ohio	190
Indiana and Michigan	70
Illinois	440
Kentucky, Tennessee, Alabama, and Mississippi	175
Missouri, Iowa, North Dakota, Wisconsin, and Minnesota	40
Kansas	425
Oklahoma	1,340
Northern Louisiana and Arkansas	525
Texas, except Gulf Coast	670
Gulf Coast, Texas, and Louisiana	2,100
Colorado, New Mexico, and Arizona	50
Wyoming	525
Montana, Nebraska, and South Dakota	100
Utah, Nevada, Oregon, Washington, and Idaho	80
California	1,850
Eastern Gulf Coastal Plain and Atlantic Coast States	10
	9,150

There have been other estimates, by individual authorities. E. W. Shaw, of the United States Geological Survey, thought David White's 1919 estimate of 11.3 billion barrels too low, and ventured an estimate of between 12.6 and 14.6 billion barrels. Shaw had ventured the prediction in 1915 that while it was "within the range of possibility that the number of undiscovered pools is as great as that of the discovered pools," that was "not at all probable."[10] Professor John Bownocker, State Geologist of Ohio, estimated the remaining reserves in June, 1921, at between 6 and 7 billion barrels. Professor David B. Reger, of the West Virginia Geological Survey, ventured the prediction early in 1917 that 36 per cent of the country's petroleum resources had been used. This suggested an original reserve of over 11 billion barrels. Dr. Van de Gracht, geologist with the Marland Oil Company, recently ventured the opinion that the 1922 estimates of the Geological Survey needed little modification.[11]

In 1925 a Committee of Eleven, appointed by the board of directors of the American Petroleum Institute, issued a report in which the known oil reserves remaining at the present time, and recoverable by present methods of extraction, were placed at 5,300,000,000, while more than 26,000,000,000 barrels of oil would be left in the oil sands when the wells ceased to flow, a "considerable portion" of which could be recovered by better methods of extraction; and over a billion acres of land were yet to be prospected.[12]

The recent vast and sustained production of California, and of the entire country, has undoubtedly come as a surprise to most geologists, and it may be assumed that many of them would raise their estimates of probable future production somewhat in the light of this new development. However, the California fields were already developing rapidly at the time the last estimate was made, and probably this had something to do with the fact that the 1922 estimate was considerably higher than any of the earlier ones, except that of Day, in 1908.

Many predictions have been made as to the time required to exhaust remaining oil supplies, and as to the time when the decline in production should set in. Day, in his report for the National Conservation Commission, in 1909, calculated that if the rate of production continued to increase as fast as it had previously, the supply would be exhausted about 1935, while if production did not increase at all, the supply, estimated at a minimum of 15 billion barrels, would last 90 years. Day saw, however, that as the supply and the rate of production declined, and as the price increased, the consumption would be reduced and so the supply would last "a long time." He was in error in his estimates as to the time when the decline should set in. This he thought would happen within a very few years. Ralph Arnold, in 1915, calculated that the supply would last 22 years at the rate of production then prevailing, but he saw, as Day did, that as the production decreased, the price would rise, consumption would decline, and the oil supply would last from 50 to 75 years. Arnold predicted prices as high as $5 a barrel.

In 1916, with the war demand for petroleum, supplies seemed inadequate, and many writers became pessimistic as to the possibility of great increases. One writer in the *Oil and Gas Journal* ventured the view that there was "little of record to show any real territory to add to the former fields of development," and that the production was destined soon to decline.[13] Van H. Manning, of the United States Bureau of Mines, in the same year, calculated that the store of oil "in sight at present," would not last over 30 years at the prevailing rate of consumption, although he saw that the rate of consumption would decline. Four years later, the same authority calculated that, at the prevailing rate of consumption the supply would last only 20 years.[14]

In February, 1919, David White stated that "many well informed geologists and engineers" believed the peak in the production of natural petroleum in this country would be reached at least as early as 1921, and perhaps in 1919 or 1920, and he offered this opinion also as his own. In September, 1920, David T. Day declared that a decline in oil production could be "only a matter of a few months." The same year White revised his prediction as to the date when the peak of production should be reached, this time allowing until 1925 for that to happen, but declaring that the annual production would probably never go as high as 450,000,000 barrels. This latter estimate proved too low, for in 1923 the production reached 735,000,000 barrels. In discussing the question of the duration of the supply, however, White saw clearly that the annual production would decline for a "long time," and that oil wells would still be producing "at least 75 years hence." He even thought it might be 75 years before some of the salt domes would be revealed.

Mark Requa declared that at any rate it was a problem for the present generation, and not altogether one for posterity:

Those of us who believe that posterity must settle these problems of heat, light, and power are living in a fool's paradise, and must inevitably awaken within the next few years to face, subdued and chastened, the real truth.[15]

F. G. Cottrell, Director of the United States Bureau of Mines, predicted in October, 1920, that our oil should last 18 to 20 years.[16] Dr. W. R. Jillson, of the Kentucky State Geological Survey, ventured the prediction in 1922 that Kentucky would be producing oil in 1980.[17] About the same time, R. P. McLaughlin, California State Oil and Gas Supervisor, became much concerned about the average output of California wells.[18] George Otis Smith saw that high prices of crude oil would result in an active drilling campaign, but he thought this would afford only temporary relief, and pointed out that "energetic pushing upward of the production curve only hastens the coming of the year when that curve must turn downward."

About this time, Arthur Little published an interesting pamphlet on the *Petroleum Outlook*, in which he took a gloomy view of the future of oil production. Referring to the 6½ billion barrel reserve which the Geological Survey had estimated was still in the ground, he wrote:

With the estimate itself there is no fault to be found, but the choice of term is misleading, for its sarcasm may be overlooked. What remains underground is there, not as a reserve, but in defiance of the wildest extravagance of effort to get it out.

After a careful consideration of the situation in each of the main oil-producing regions, he declared:

From the foregoing it develops in résumé that, of the seven domestic sources, three, the Appalachian, Lima-Indiana, and Illinois fields, are in a state of hopeless decline; one, the Gulf Coast field, is an unknown quantity; the two greatest, the Mid-Continent and California fields, are at their best; and just one, the Rocky Mountain field, is assuredly still in its youth. . . . The present boom, with its wild orgies of extravagant competitive effort, indicates what is to be expected with every successive advance in the utilization of oil, and as the successive evidences of shortage accrue. This outrageously wasteful forcing of production can only drive the resource into its early decline.

Conditions during the year 1921 were not such as to direct attention to the question of the exhaustion of our oil reserves. A severe financial and industrial depression caused a drastic decline in the demand, while the opening and rapid exploitation of several new pools—Mexia, Haynesville, Tonkawa, Burbank, Huntington Beach, Signal Hill, Santa Fé Springs, and Eldorado—greatly augmented the supply. Charles N. Gould, of the Oklahoma State Geological Survey, writing in October, predicted that the oil supply would last at least a century.[19] Edward Butler, in his book on *Oil Fuel*, published in 1921,[20] quoted from Day, Engler, Lewes, and Sir Boverton Redwood, and concluded that "the available store of petroleum will have become so depleted toward the end of the present century that long before then, other sources of supply will become necessary." Butler was writing of the world's supply, but it is generally conceded that the reserves of the United States will be among the first to be exhausted. Writing in a somewhat different humor, Professor S. H. Leslie, of the University of Michigan, denounced the optimism of the oil barons and Wall Street publications, and predicted that another 13 years would exhaust the supply if the 1920 rate of production was maintained.[21]

During 1922 the over-production of oil continued, and even grew

worse, nevertheless prophets of impending shortage appeared, some
of them practical oil men. Professor Ralph McKee, of the Columbia
University School of Mines, in an address before the Toronto section
of the Society of Chemical Industry, in December, declared that the
petroleum production of the United States was at its peak and would
probably begin to decline within three years.[22] A writer in the *Magazine
of Wall Street*, Barnabas Bryan, discussed at considerable length the
question of our oil reserves, and predicted that our supply would meet
present demand for only 20 to 30 years.[23] J. Koster, of Dutch oil circles, predicted an oil shortage within two years.[24] James E. O'Neil,
president of the Prairie Oil and Gas Company, of Independence, Kansas,
told the Senate investigating committee in December, that he thought
the United States had seen her peak of crude oil production, and that
within ten years would have to import at least half her supply.[25] The
policy of the Prairie Oil and Gas Company, in buying and storing immense quantities of crude oil, paying very good prices for some of it,
seems to indicate that O'Neil was willing to back his judgment of the
situation.

E. W. Marland, president of the Marland Oil Company, has for
some time been pessimistic as to future reserves of oil, and has often
pointed out the necessity of rapid amortization of the capital invested
in the industry. As Marland has recently stated:

It matters little whether the unmined reserves are nine billion barrels or
twice that amount. In either event they are distressingly small when it is
realized the security of our industry and of our nation are both builded
thereon.[26]

In some ways even business men have evinced unmistakable uncertainty as to the future of the oil supply. The price at which some of
the stocks of motor companies have sold probably indicates that Wall
Street is uncertain how long we shall have the fuel for our automobiles.
Likewise the slowness of some of the railroads and shipping lines and
other industrial concerns, as well as domestic consumers, to turn from
coal to oil fuel has been due in large measure to the uncertainty of the
fuel supply. In several instances, railroads have turned from coal to
oil fuel, and later have been forced to return to coal in periods of high
oil prices, of course at considerable expense. C. V. Kettering, of the
General Motors Company, speaking before the American Petroleum
Institute in November, 1920, declared that, "The only cloud on the internal combustion engine field today is the fuel supply"; and there can
be no doubt that the business public is apprehensive on that point.[27]
Not long ago a number of business men in Tulsa, Oklahoma, were discussing with considerable seriousness the question as to what Tulsa
would do when her oil supply was gone.[28]

During the last ten years or more, it has been observed that when there is an imminent shortage of oil, prices go up promptly, while, when there is an oversupply, prices go down slowly. This was seen even at the time of the development of the great Cushing field, and has been even more noticeable in recent years. For instance, in December, 1923, and January, 1924, even though there was an unprecedented and constantly increasing accumulation of oil in storage—over 300,000,000 barrels—prices of crude oil were raised throughout the country, in evident anticipation of a coming shortage.[29]

Many oil men and correspondents of oil journals have made fun of the estimates of geologists, and of all predictions of exhaustion of the oil supplies. A writer in the *Engineering and Mining Journal* in January, 1904, C. T. Deane, declared that "The permanence of the supply (of California oil) need not trouble this generation."[30] *The Oil and Gas Journal*, in August, 1916, decried the pessimism of the conservationists in the following vigorous fashion:

The fact that our forefathers were always equal to every emergency does not seem to weigh one particle with them. The suggestion that our children may be perhaps as smart as ourselves and fully as capable of attending to their business does not appeal for a minute to such prophets of disaster as Pinchot and his faction. They want to legislate for all times. They desire to tie the hands of future generations with useless statutes which but hinder the efforts which practical and common sense men are putting forward every day to develop the resources which we now have for the benefit, comfort and happiness of the men, women and children who are now on earth.[31]

The *Petroleum Refiner* carried on an energetic campaign during part of 1921 and 1922 to discredit the geologists' predictions of oil scarcity. In October, 1921, that journal published an article by H. G. James, secretary of the Western Petroleum Refiners' Association, in which the central theme was that, "In all probability, the full needs of the people will undoubtedly be satisfied indefinitely"; and in the same issue appeared an editorial, "Plenty of oil for our children's children."[32] In December, this journal again raised the question: "Is oil actually being exhausted?" and concluded after a discussion of the long-enduring Pennsylvania fields, that "consumers die and are forgotten," but the oil would always be available. In the same issue the editor, Charles B. Marsh, demanded: "Where are the evidences of shortage? Why not cut out all these groundless dire predictions? They have been dished up to us annually now for the past four years. And all the time we have gone on increasing our stocks."[33]

When Professor Ralph McKee made his prediction, early in 1922, that petroleum production would probably begin to decline within three years, the *Petroleum Refiner* immediately came to the rescue: "The good professor's declarations are an awful strain upon his intelligence. He is just too reckless with his reputation for anything. Evidently he has not been reading the late news from Arkansas, Oklahoma and Texas. When Ralph is dead and forgotten the 'peak of production' will be a petroleum promontory still unscaled." The *Refiner* did not think we would be near an oil famine for a "long, long time to come."[34]

The *Oil News*, in November, 1921, published an article, "Prophets of Gloom," written by H. G. James, making much fun of an article which had appeared in *Hearst's Magazine*, predicting an oil famine. "From the very beginning, like the poor, we have always had the oil prophet of gloom with us," wrote James, "and we have never failed to meet every requirement for oil. . . . Is it not significant that the dire predictions are coming from the press, while oil men have no fears of their ability to provide."[35] James wrote for other oil journals a great deal, and always in a very cheerful vein.[36] Recently he has ventured another prediction that: "There will be plenty of oil for decades to come, and when there isn't oil enough, something else will take its place—possibly cars will be run from power extracted from the air. Who knows?"[37] An editorial in the *Oil and Gas News*, in September, 1921, took Arthur Little to task for his predictions of coming scarcity: "That's the same nonsense so called experts peddled through the public prints only a few months ago."[38]

The resolutions adopted by the Western Petroleum Refiners' Association at its meeting May 9, 1921, represent fairly characteristic business men's views:

Whereas, the general public through the public press and periodicals and trade journals has been led to believe there is danger of a shortage of petroleum products and the exhaustion of nature's crude supply within a given period, and

Whereas, this discussion has caused the manufacturers of internal combustion engines and users of fuel oil much concern, deterring many prospective consumers from the adoption of oil as a fuel, and

Whereas, we believe this has proved a serious detriment to the industry, now therefore be it

Resolved, that we, the Western Petroleum Refiners' Association, . . . call the attention of the oil consuming world to the fact that the exhaustion of crude petroleum in 15 to 25 years has been constantly predicted by alleged experts for the past 30 years, and that notwithstanding we have just emerged from the greatest war of all times, making enormous demands upon the petroleum industry, that the industry not only met every demand, but today, in spite of the marvellous expansion of the automobile industry, holds in reserve storage larger quantities of gasoline, kerosene and fuel oil than at any other time in its history. . . .

We do not share in the alarming predictions of so-called experts and hereby publicly express our confidence in the ability of the petroleum industry of the world to supply the needs of the world indefinitely. Yet we anticipate there may be periods of temporary shortage, even as there have been in the past, but these have only marked the way for expansion. These periods have always been of short duration, while operators exploited for new fields, and we confidently believe, in view of present prospective future fields, it will prove so in the future.

Therefore, the Western Petroleum Refiners' Association places itself on record as saying to the oil consuming world that it need have no more fear of a supply of oil in the future than it has of a supply of coal.[39]

In April, 1924, James again entered the lists against the prophets of future oil scarcity, the particular object of his sarcasm being a prediction of George Otis Smith that our oil resources would soon be exhausted. "There is no more probability of exhaustion of petroleum within the next 20 years now than there was 20 years ago, when the same predictions were being made," declared James; and he pointed out that he had "this advantage of Mr. Smith in the argument: While he [Smith] has been predicting exhaustion in 20 years for nearly 20 years, my side of the argument has been consistently sustained by the increase of production faster than consumption could take care of it. . . ." James even expressed his opinion that a government official "has no right to prejudice an industry by making such statements."[40]

When the Geological Survey issued its last estimate of reserves, in January, 1922, a great many of the oil operators and oil journals immediately began a campaign to offset the iniquitous effects of this last publication of the "experts." The *Oil News* devoted considerable space to a symposium on the question; and a number of optimistic contributions were offered, one of them by H. G. James.

Most practical oil men have been unreceptive to any predictions of coming scarcity. At the annual meeting of the American Petroleum Institute in November, 1920, R. D. Benson, president of the Tide Water Oil Company, after casting some ridicule at the "Geological Survey and other learned and scientific gentlemen," ventured the prediction: "I believe that production of the United States will be entirely governed by the price paid for crude oil, for a longer period than anyone, no matter how learned, can safely predict." With a religious faith which hardly seemed appropriate at an oil men's convention, Benson asserted his "belief that kind Providence never limits the supply of anything so valuable as petroleum."[41]

Thomas O'Donnell, always one of the most optimistic of all the oil men, regretted that the public had been so "frequently alarmed by statements of well-meaning and learned scientists," and pointed out that "Agitation by government officials, . . . is just as dangerous as governmental regulation and interference." O'Donnell had "no apprehensions as to the future petroleum supply of either this country or the world at large."[42] At the next annual meeting of the Institute, he again spoke of the "over-production of super-scientists," and of the "very able writers with a great command of the English language and a free use of mathematics," who gave an "exaggerated viewpoint to the public mind"; and predicted plenty of oil in the United States for 100 years.[43] Henry L. Doherty likewise has at times expressed his confidence in the future;[44] and Harry Sinclair once declared: "There is plenty of petroleum, and always will be. Exhaustion of the world's supply is a bugaboo. In my opinion, it has no place in practical discussions."[45] Thomas F. Galey was even more optimistic: "Personally," he declared, "I believe there will never be a long continued oil shortage. I believe that domestic production can and will be increased and that the anticipated peak will be indefinitely advanced. . . . The search for petroleum is being carried into all parts of the world, and it is inconceivable that nature, in the abundance of her resources, should not provide the crude oil."[46]

It is difficult to conceive of our oil supply as lasting longer than "indefinitely," but Representative Weaver, of Oklahoma, declared in 1913 that: "The United States has well-defined oil areas already prospected, tested, and exploited, that show enough oil to supply the world's fuel, illumination, and lubrication for at least 5,000 years. . . . Recent developments in fair Oklahoma alone but indicate the boundless scope of this one natural resource. Its development by the sturdy explorer has converted cow camps into cities, cow tracks into highways, poor men into rich men, and hovels into homes. It is and has been a beacon light pointing the pathway of hope, obliterating skepticism and doubt."[47] Surely this is as much as should be expected of any sort of resource.

At the 1922 meeting of the American Petroleum Institute, there was little pessimism as to the future. A. C. Veatch, chief geologist for the Sinclair Consolidated Oil Corporation—formerly of the United States Geological Survey—spoke of the "exhaustion bogey" raised by those "sincere but misguided patriots" (in the Geological Survey), and asserted that "far from facing world oil exhaustion, the oil industry is merely on the threshold of a glorious future." A. L. Beatty, president of the Texas Company, was, if possible, even more cheerful:

Perhaps it is not a part of the Divine plan that man should know the limits of provision made for him. . . . Profligacy is to be condemned at all times, but rather than fail to take advantage of opportunities in sight, should we not believe that somewhere protected for us are such resources as we shall need? The world is very wide, and despite the teachings of today, there may be more petroleum than mankind will ever require.[48]

The "Committee of Eleven" appointed by the American Petroleum Institute, in its recent report, did not criticize the geological estimates of oil reserves, as far as those estimates went. This committee, however, emphasized the rapid development of knowledge as to the occurrence of oil, and the probability that wildcatters would become more and more successful in finding it. They pointed to the vast amount of oil left in the ground when wells ceased to flow, and ventured the hope that a large proportion of this 26,000,000,000 barrels will be brought to the surface by improved methods of recovery, such as mining, flooding, and other means yet to be devised. The committee also included in its estimate of oil resources the shale oil, and the liquid products contained in the coal and lignite deposits of the country, and drew a very cheering conclusion as to the future of the oil supply. Their first two conclusions were: (1) There is no imminent danger of the exhaustion of the petroleum reserves of the United States; and (2) it is reasonable to assume that a sufficient supply of oil will be available for national defense and for essential uses in the United States beyond the time when science will limit the demand by developing more efficient use of, or substitutes for, oil, or will displace its use as a source of power by harnessing a natural energy.[49]

Oil operators and oil journal correspondents and editors are not the only men who distrust all estimates of reserves and all predictions of scarcity. The California State Mining Bureau figures in 1916 credited the state with double the reserves that were given by Secretary Lane's figures.[50] The Oklahoma Geological Survey was for many years unreceptive to any pessimistic estimates of the state's oil resources. In 1908, the Oklahoma Survey reported: "It is the opinion of the Survey staff that not one-tenth of the probable oil field of the state has been touched, and that not one-tenth of what has been proved productive is yet developed. According to this estimate, not more than 1 per cent of the state's oil and gas has been developed."[51] This estimate would have given Oklahoma an original reserve of about 4½ billion barrels, which is twice as much as the United States Geological Survey has estimated for the state. Four years later State Geologist C. N. Gould estimated that less than 10 per cent of the oil in the state had been taken, and predicted that the development of new fields in Oklahoma would continue for 50 years, and that they would continue to produce for another 50 years.[52] Even in very recent years, occasional adherents of the inorganic theory of the origin of oil appear, and, as in early Pennsylvania days, generally believe that the supply of oil is inexhaustible.[53]

The question of the justice of these criticisms will be best determined by the events of the next decade or two, but even now a few conclusions seem justified.

In the first place, it is to be admitted that the events of the past seemed to justify the optimism of the "practical" men in the field, or perhaps more accurately, have thus far failed to justify the pessimism of the geologists. It is now over half a century since the first predictions of scarcity were made, and during that half century, scarcely a year passed without its crop of predictions of approaching exhaustion of the fields. Almost every year, prophets have ventured to predict that the beginning of the decline was imminent and could not possibly be postponed much longer.

Such predictions were freely made when the total production of the country was 2,000,000 barrels; when it had grown to 5,000,000 barrels; when it reached 10,000,000 barrels; when, in 1882, it reached the grand total of 30,000,000 barrels; when it doubled that figure in 1896; when it doubled again ten years later; when it doubled again only seven years later; when it reached 300,000,000 barrels in 1916; when the production failed to increase very much during the Great War; when it increased to 443,000,000 barrels in 1920; when it rose to 472,000,000 barrels the next year. When the production grew nearly 100,000,000 barrels in 1922, predictions of approaching decline were voiced by dozens of writers, scientific and otherwise. In spite of the deluge of oil, even many practical oil men began to grow apprehensive and to wonder where their future supplies of crude were to come from; and when, early in 1923, rumors of insufficient gasoline supplies were whispered among those in the trade, a veritable scramble for remaining sources of supply resulted. So apprehensive about the matter seemed some of the big purchasing companies that prices jumped 100 per cent in a few weeks. But the logic of events was again on the side of the optimists, for the great fields of southern California and Oklahoma, instead of declining as some of them were scheduled to do, mounted steadily week by week, and presently the refiners were swamped with the crude oil which they had feared would be insufficient for their needs. And yet, at this very time, geologists were predicting an approaching decline in production, and a not far distant "oil famine."

Some of the recent predictions of approaching decline in oil production sound strangely like similar predictions made nearly a half-century ago. In June, 1879, J. C. Welch, in his *Views of Future Production*, expressed the hope that oil production would presently begin to decline. In January of the next year, he said, "Great hopes are entertained that in six months the production will necessarily show a very important falling off." Six months later he wrote: "The next point is for the production to show an appreciable falling off. This point has not arrived yet, although producers, on account of the falling off of wells throughout the district, expect it to do so pretty soon." For six months there was a slight decline in production, and in January, 1881, he wrote, "I cannot understand, in view of the facts, how there can be an increase in the production and, in plain words, don't believe it." Six months later he announced: "The sanguine hopes for an important decrease in the production have been postponed for some months at least. Bradford is expected to decline rapidly at some time, and it was confidently hoped the time was near at hand; but the figures on the May production have been disappointing, and any marked decrease in the production is still a matter of the future." The production did not decline, and six months later he wrote: "For the time being, the increase at Allegheny equals the loss at Bradford, but this relation is likely to change soon, and not only Bradford will decline, but Allegheny will accelerate the decline by declining herself."

Shortly afterward, in February, 1882, a correspondent of the *Oil and Drug News* took up the question, "How far off is the date when the production of petroleum will not be in excess of the demand?" which he declared was the "great question of the hour to all parties concerned"; and came to the conclusion, after an exhaustive analysis of production and consumption figures, that prices would soon go higher. Not long afterward, the large wells of the Warren district brought production to unprecedented figures, and prices went still lower. S. F. Peckham, writing on Petroleum for the Census, commenting on these prophecies of future decline in oil production, declared:

When prophecy, indulged even by the most sagacious producers of longest experience, proves so futile, I think I am warranted in expressing the opinion that, as regards the future supply of petroleum, the drill alone gives valid testimony. I wish to emphasize the futility of prophecy, and the abundance of the present supply.[54]

Thus for decades, geologists and others have been predicting a decline in oil production, yet that production has steadily and rapidly increased all the time. In many instances, the men in the industry have been somewhat influenced by these predictions and have loaded up with more reserve lands and more crude stocks than they otherwise would have been willing to carry; and when the anticipated scarcity failed to appear, or when the market broke, as it frequently did, under unexpected new floods of oil, these men of course felt that predictions of scarcity had no place in practical business, and that geologists were more or less of a detriment to the industry, at least as far as their estimates and predictions were concerned.

Much of the criticism of the 1922 estimate of oil reserves, as of other estimates, was wholly lacking in fairness and in intelligence. There was criticism of the inaccuracy of such estimates; yet the geologists were careful to make no claims of exactness, and they allowed four billion barrels, of the total of nine billion, for oil regarded as prospective and undiscovered. As to this yet undiscovered oil, the press bulletin was careful to state: "The estimates of oil in possible territory are absolutely speculative and hazardous," and "although they represent the best judgment of the geologists, they nevertheless may be, at least in part, wildly erroneous." Criticism was offered on the ground that these "experts" were merely theoreticians, with little knowledge of the actualities of the oil business; yet, in addition to the 15 men responsible for the estimate, perhaps 60 more aided in the compilation, who were largely men of wide practical experience. Some of these were in the employ of large oil companies, and had done notable work in locating oil fields. It seems strange that oil men hire geologists to help them find oil, yet often have so little confidence in some geological work.[55]

With the disbelief in all predictions of approaching exhaustion of oil has gone a great deal of criticism of the geological profession, and of what are sometimes called the pretensions of the geologists. "Practical" men have often pointed to Professor Whitney's early declaration that there was no oil in California, and some of them have denied that the geologists know anything at all about the geology of oil.

It is only rather recently that practical oil men have been willing to accord geologists much respect. Only a few years ago President J. C. Donnell, of the Ohio Oil Company, was heard to say, "When the geologist comes into the oil industry, I go out"; yet the geology department of the Ohio Oil Company is now one of the most important departments in that company. Only about 10 years ago the man in charge of the producing department of the Standard Oil Company of California declared that "oil is where you find it"; but he soon found it profitable to build up an extensive geology department.

Geological predictions, as already suggested, have frequently turned out unfortunately. Professor Whitney's predictions on California are not the only ones that have been proved false. All predictions of exhaustion or even of decline in the total production have been falsified by the logic of events, and inevitably many practical men have assumed that it will be true indefinitely. During the last few years it has seemed each year that it was impossible production could go higher. Most people like to pose as prophets, and knowing that the oil production must finally decline, many have from time to time been tempted to rush out and announce the inevitable. Perhaps geologists have sinned more than most others in this respect. Ralph Arnold once called the estimation of oil reserves the "king of indoor sports" for petroleum geologists.[56]

On the other hand, as already stated, the men who have worked out the estimates of our oil reserves have generally been among the first to admit the impossibility of exactness or finality in estimating the extent or probable duration of our oil resources. They would doubtless say also that the question as to whether the present reserves are 9 billion barrels or twice that amount, or even three times that amount, or whether the supply will last, at present rate of consumption, 10 years, or 20 years, or 30 years, is a relatively unimportant question. As Day expressed it in 1909: "With the certainty of exhaustion of the present fields by the present generation, it is not a matter of vital argument whether such exhaustion comes in 10 years or 40."[57] The point is that *soon*, from the point of view of the life of a nation, our oil will be insufficient for even the most essential industrial uses, as, for instance, lubrication.

The individual well gushes forth, declines and finally ceases to produce; the oil field has its period of flush production, its peak, and its long or short decline to extinction; and there is little sense arguing that the United States contains an infinite number of undiscovered fields. The geologist may well say that the United States faces the absolute certainty of final shortage of oil; and a fair probability, judged from all the data at present available, that this shortage will be in evidence within the next few years. In constantly bringing before the people the certainty of approaching scarcity, and the necessity of conserving remaining supplies, geologists have performed a service which will be better appreciated as the future brings to us a clearer realization of the problems involved. The only unfortunate circumstance seems to be, not that geologists should have made predictions of scarcity, but rather that their predictions have received less attention than they merited.

It is worth noting, too, that the geologists themselves are entitled to much of the credit for having staved off the inevitable decline in production so long. Two or three years ago, when there was a very serious overproduction of oil, and stocks were accumulating rapidly, the suggestion was made in an oil journal that, if the more important oil companies would raise a fund and give the geologists a three-year vacation, the over-production would soon be a thing of the past; and there can be no doubt that the suggestion had merit. As a matter of fact, most of the major producing fields of recent discovery have been drilled on geological advice. No important producing company tries to get on without a staff of geologists.[58]

Some practical oil men even object to the very idea of making estimates of our oil reserves, regarding it as a useless, or worse than useless, waste of labor. Yet business men cannot proceed intelligently without coming to some sort of conclusion as to the amount of raw material they have to work on. A lumber manufacturer who built an expensive lumber mill without first assuring himself of about a 20 years' supply of timber to work on, would be regarded as a poor business man; yet we have the peculiar spectacle of some oil men investing millions in the oil industry, with no apparent desire to know whether the raw material for their business is sufficient for 10 years or 20 years, or longer, or even with a decided hostility toward the work of those who are trying to make such information available.

While the geologists have made their mistakes, we must not fail to note one occasion on which they have had the opportunity to laugh last. In the early months of 1921, Ralph Arnold, one of the leading geologists of the United States, made the prediction that the proved oil fields of Mexico were approaching exhaustion. This Mexican "salt water bugaboo" was the butt of much ridicule from oil men for a few months, but less than a year afterward all had to concede the accuracy of the prediction.

Perhaps it will be fair to point out that the "experts" and "theorists" and "scientific gentlemen" have had the last laugh on another question similar to this. Not many years ago, Gifford Pinchot and other scientific foresters were warning the people of the United States of the approaching exhaustion of our supplies of cheap virgin timber. Many practical lumbermen made much fun of these predictions, and declared that the supply of lumber was practically inexhaustible. For some years they seemed to have the better of the argument; but the tide has turned and the same lumbermen who, ten years ago, were talking about the inexhaustible supplies of southern pine are now getting their lumber from the Far West, and paying two or three times as much as they did. Cheap wood shingles are gone from the market, wooden frame houses are passing, and gradually, as prices rise, the consumption of wood is declining, and the uses of wood are being reduced. For a decade or two the practical lumbermen seemed to be correct in their view: for the next ten thousand years the scientific foresters will seem to have been correct.

In a certain sense, practical oil men and geologists have both been correct. Their points of view have differed. Perhaps business men can see farther than "the end of their noses"—a great economist once said that most of them could not—but their view is generally short. They are not generally greatly concerned as to what will happen in ten or twenty years; in fact, as business men they would sometimes hardly dare be so. The oil men who based their business calculations on the theory of an impending shortage of oil, in the years from 1870 to 1924, have mostly been eliminated from the business field. They were acting on the wrong theory. There have actually been, at various times during the past half-century, men so ill advised as to gamble in oil stocks in the belief that production was presently to decline and prices to rise correspondingly. Some geologists have had enough confidence in geological predictions to do that; but it is unnecessary to state that this has usually been poor business. The business men who have succeeded in the oil game have necessarily been, in general, those who had a very optimistic view of the future oil production.

On the other hand, most geologists, and to some extent economists as well, have a long view of human affairs. Men who talk in terms of millions of years naturally do not regard a period of 20 or 30 or 100 years as a very long time. Thus an "oil famine" due in 20 years—nothing for a business man to consider very seriously—is to a geologist a very immediate matter. This difference in point of view is undoubtedly responsible for much of the mutual distrust of practical oil men and geologists, or for that matter business men in general and academic men in general.

In yet another respect, however, business men and all academic men differ in their point of view. Business men look only from the individual point of view; academic men, including geologists, from the social point of view. Business men are interested in getting money, in accumulating wealth for themselves, and not often can they be induced to take a serious and intelligent interest in the social effects of their activities. Even if they do take such an interest, their personal bias is generally too strong to enable them to form judicious or far-reaching opinions. On the other hand, most academic men are trained to study questions from the social point of view, to deliberate much upon social consequences.

Thus, when geologists foresee an impending scarcity of oil, they look upon it as an extremely important matter, for it involves the prosperity and happiness of countless millions of people, yet unborn, people whose interests are, from a broad and humane social viewpoint, nearly as important as the interests of those now living. But few practical oil men see the question from this angle. Most of them see that conservation will mean decreased profits for themselves, and that consideration outweighs all others. Answering this indictment, they would say, of course, that they really do not believe there will ever in any reasonable time be a scarcity, and in a sense that is true; but that illustrates the extent to which personal interests often color judgments. It is the exceptional business man who ever holds economic opinions contrary to his own private interests.

A fair share of the talk and writing about the "inexhaustible" supplies of oil has, without a doubt, been directly inspired by the personal interest of the oil operators. Conservation talk was bad for "business." It made the oil industry, the automobile industry, and all industries dependent upon them, seem temporary and unworthy of permanent support. It tended to discourage the adoption of oil-burners, and the purchase of gasoline trucks and other gasoline driven vehicles, perhaps even made some people hesitate about buying automobiles. If the people had been certain of an inexhaustible supply of cheap oil, the oil industry would have expanded more rapidly. Furthermore, talk of future scarcity always led to talk of conservation, and talk of conservation to consideration of government regulation, or at any rate government interference; and that is of course the very last thing that any oil man wants. One of the tasks of men like H. G. James was that of convincing the people that there was plenty of oil.

Not a few of the leading oil men have tried to espouse two contradictory views, have declared that there was no danger of a scarcity of oil in the United States, yet have insisted that the United States should bestir herself to secure supplies abroad, or, to be more accurate, should help and protect those oil companies that are trying to secure foreign reserves. The logic is that while the United States has enough oil to last indefinitely, she is in serious danger of finding herself in grave need of foreign supplies; that although she will always have plenty, she will soon need more. Some of the officials of big oil companies with holdings abroad find their personal interest leading them to hold such contradictory views. They must never admit that there is not enough oil in the United States, for that is poor business; and on the other hand they always want the backing of the government in their activities abroad. They are about as logical as another set of big business men who used to call for a ship subsidy, to encourage the building of ships to haul goods in, and at the same time urge a tariff, to prevent goods from being shipped. Logic and self-interest often refuse to dwell together.

As to the future of oil, it may safely be assumed that sometime, probably within the next few years, production will reach its peak and begin to decline. Perhaps it has already done so, in the summer of 1923. How rapid this decline will be, it is impossible to predict. There are too many unknown and unpredictable factors in the problem: the actual reserves of oil still in the ground; the work and expense necessary to its extraction; the cost of extracting oil from shale; the availability of substitutes; the cost of oil imports; and many other factors that could not even be named at this time. It is impossible to base any predictions on the history of the older Appalachian fields. The decline of the Pennsylvania fields since 1891 has been very gradual, but that has perhaps been due to the fact that there have always been new fields to the westward to care for the growing demand. If new fields had not been available, the price of oil would have risen very high, and this would have stimulated more rapid exhaustion of the old fields. Perhaps in the future, when new fields are no longer available, a higher price will stimulate very rapid exploitation and thus presently bring such a scarcity of reserves as will necessarily cause a rapid decline in production. There are some reasons for believing this possible. On the other hand, however, if new sources are available, in the form of shale oil and imports, these may perform the service that has heretofore been rendered by new fields, and the decline in our domestic production may be gradual, somewhat as that of Pennsylvania has been.

As to the question, "When will our oil be finally exhausted?" there is no doubt that the answer is, "Never." Production will reach its peak and will decline through years and through centuries, but oil will always be available, in smaller and smaller quantities, and used for constantly higher and more important purposes. There is no reason to doubt that it may rise to a hundred times its present value, or even much more. Oil has sold as high as $250 a barrel, and there is no reason to doubt that it might, under some circumstances, rise to such a figure again. But even if it should, it will still be available for commercial use, and at that price would supply many important commercial wants.

The estimates of our oil supply mentioned above refer to the actual oil reserves recoverable under present conditions. It is now necessary to point out that there are various ways in which this supply will be increased or supplemented as prices rise. Methods of extraction will be improved; refining processes will be perfected, so that more of the valuable products can be obtained from the crude; shale will be mined and forced to yield its treasures; substitutes will be devised; and foreign supplies will be brought in wherever they can be secured.

As the price of oil rises, various means will be used to increase the recovery. In the first place, more care will be used to protect the oil sands from injury, by more care in drilling, in adjusting casing, in sealing the oil strata, and in various other ways. In the second place, perhaps the flooding of the oil sands with water, as already practiced in the Bradford field, may be tried in other fields; but this has seldom been successful, except in the Bradford field; in fact, it has frequently done more harm than good, and in some instances where a property has been benefited it has been at the expense of adjacent holdings. Since flooding practically ends all further possibilities of increasing recovery by other methods, its employment, until other methods have been tried, seems a short-sighted policy.

In the third place, gas or vacuum pumping will perhaps be done more in the future. By this process, the pressures at the wells are reduced, to permit further expansion of the gas and the expulsion of more oil from the oil-bearing sands. This method is one of the oldest in use, but it has only slightly increased the total recovery from most fields, and the expense has usually equalled the value of the additional production, so that perhaps it will not play a very important part in the future.

In the fourth place, what is known as the Smith-Dunn process, or the Marietta process—because first used extensively at Marietta, Ohio—will be used more. By this process, compressed air is forced into the oil-bearing formation through a few wells, thus forcing the oil underground toward the other wells, from which it is pumped in the usual way. This process was known as early as 1865; in fact, a patent for it was obtained in that year, and it has been used for a number of years, particularly in the Appalachian fields, where it has been successful in most cases where used. It is perhaps the most promising method for increasing oil recovery, and will doubtless be used a great deal in the future. Natural gas can be used instead of air, but because of its cost, it will doubtless be used less in the future. J. O. Lewis, formerly of the United States Bureau of Mines, has ventured the prediction that it may be possible to go to exhausted fields and recover perhaps one-half as much, or, in some cases even fully as much, as was obtained in the first place.[59]

In the fifth place, probably oil that cannot be extracted otherwise will to a limited extent be "mined," when the price rises high enough to justify it. In some of the shallower fields, it will perhaps be possible to sink shafts, as in coal mining, and bring the oil-laden sands to the surface, run them through a retort and extract the oil. E. L. Doheny claims to have suggested the mining of petroleum in California thirty years ago; and it is said to have been considered seriously in the Mid-Continent field in the spring of 1923, when oil prices were high. Such a process would be very expensive, but when the price of oil rises to $5 or $10 a barrel, or even higher, perhaps mining may be profitable in some fields. The French are said to be carrying on operations of this character in the old Alsace-Lorraine fields; and several American engineers and technologists are studying the question.[60]

Two great difficulties in the way of oil mining are apparent from studies already made. In the first place, the element of danger will be very serious, for even in very old fields, there is enough gas left to kill the miners when it is released, unless they are isolated from it by means of some sort of protective apparatus. In the second place, the expense will certainly be very great. It is difficult to realize the tremendous amount of labor and capital that would be required to secure even a small fraction of our present production of petroleum.

Deeper drilling will, of course, be a very important means of increasing the oil recovery in many fields. Wells a mile deep are not uncommon in California, and doubtless even greater depths are possible, when the price of oil once justifies the expense. In some fields it is probable that oil horizons lie untouched below the strata that have been drained, and deeper drilling will bring this oil to the surface.[61]

Refinery methods will be improved in the future, just as in the past, and one result of this will be an increase in the proportion of the more valuable oil products, and a decline in the proportion of the less valuable products. A great change has already come since crude oil was burned as fuel without even being run through the refinery at all. For instance, in the ten years from 1911 to 1921, the proportion of gasoline extracted from crude oil increased from 6 per cent to 22 per cent, while increasing quantities of gasoline were being taken from natural gas, which previously went to waste.

In like manner, the tremendous demand of our millions of automobiles has made necessary an enlarged supply of lubricants, and has led to the manufacture of lubricants from the asphaltic crudes of California and Texas, which were not long ago regarded as unfit for such use, the paraffin crudes having previously met all lubricating needs. On the other hand, the production of kerosene, a relatively unimportant product in these days of automobiles, has fallen from 58 per cent of the crude in 1899 to 12.7 per cent in 1920. The production of gas oil (the product just above the fuel oil residuum) has in similar fashion been reduced by the expansion of the cracking process in an effort to secure more gasoline; and of course the amount of fuel oil has declined since the day when most of the California product was used entirely as fuel. In other words, the production of the more valuable gasoline and lubricants has increased at the expense of the relatively less valuable kerosene, gas oil, and fuel oil; and this change may be expected to continue in the future until some of the least valuable products will no longer be made at all.[62]

NOTES

1. *Mineral Industry*, 1921, 506.

2. Pogue, *Economics of Petroleum*, Ch. 2; *Oil and Gas Jour.*, Nov. 28, 1919, 70. On the general oil situation in recent years, see: U. S. Geol. Survey, *Bul. 599*, 13; *Natl. Geographic Mag.*, Feb., 1920, 181-. *Oildom*, Nov., 1921, 20; *Sci. Am.*, June 22, 1918, 394; address of Geo. O. Smith on "Industry's Need of Oil," before the Am. Iron and Steel Inst., May 28, 1920; Am. Petroleum Inst., *Proceedings*, Ann. Meeting, 1920, 8; *R. of Rs.*, Sept., 1920, 291; *Magazine of Wall St.*, Feb., 1922, 517-; *The Petroleum Outlook*, Arthur D. Little, pub. by the author, Cambridge, Mass., 1920; Requa, *Petroleum Resources of the U. S.*, S. Doc. 363, 64 Cong., 1 sess.

3. U. S. Geol. Survey, *Bul. 394*, p. 35; *R. of Rs.*, Jan., 1909, 49.

4. *Mineral Resources*, 1910, 413.

5. S. Doc. 310, 64 Cong., 1 sess.

6. S. Documents 310 and 363, 64 Cong., 1 sess; *Econ. Geol.*, Vol. 10, 1915, pp. 695-712; *Mineral Industry*, 1915, 528, 529, 578; *Oil and Gas Jour.*, July 12, 1917, 34; *Eng. and Min. Jour.*, Aug. 4, 1917, 196; *Smithsonian Rept.*, 1916, 273-287; *Bul.*, Geol. Soc. of Am., Vol. 28, Sept., 1917, 603-616; *Proceedings*, Am. Mining Cong., Vol. 19 (1916), 473-490; Federal Trade Comm., *Rept. on the Advance in Price of Petroleum Products*, 1920, p. 12. See particularly a paper read by David White, on "The Unmined Supply of

Petroleum in the United States," at the annual meeting of the Society of Automotive Engineers, Feb. 4-6, 1919, appearing in *Jour., Society of Automotive Engrs.*, Vol. 12, 1919, 361-363.

7. *Annals Am. Acad.*, May, 1920, 111; *Oil and Gas Jour.*, June 18, 1920, 54.

8. *The Petroleum Outlook*, Arthur D. Little.

9. *Oildom*, Oct., 1922, 50; *Oil and Gas Jour.*, March 25, 1921, 64; *Natl. Petroleum News*, Jan. 25, 1922, 59; *Oil News*, Feb. 5, 1922, 31.

10. *Proceedings*, Second Pan-American Scientific Congress, Sec. III, p. 198.

11. *Oildom*, Sept., 1921, 64; *Oil and Gas Jour.*, June 10, 1921, 3; March 1, 1917, 3; *Oildom*, Dec., 1923, 18.

12. *American Petroleum, Supply and Demand*, A Report to the Board of Directors of the Am. Petroleum Inst., McGraw-Hill, 1925.

13. December, 1916, p. 21.

14. *Oil and Gas Jour.*, Sept. 21, 1916, 30; *Petroleum Mag.*, June, 1920, 30.

15. S. Doc. 363, 64 Cong., 1 sess.

16. *Oil and Gas Jour.*, Oct. 15, 1920, 72.

17. *Petroleum Age*, June 15, 1922, 92.

18. *Cal. Oil Fields*, July, 1920, 13.

19. *Oil and Gas Jour.*, Oct. 7, 1921, 56.

20. *Oil Fuel, Its Supply, Composition and Application*, London, 1921.

21. *Wyo. Oil News*, July 9, 1921.

22. *Oildom*, Dec., 1922, 64.

23. *Magazine of Wall St.*, Feb., 1922, 517.

24. *Oil and Gas Jour.*, Nov. 23, 1922, 11.

25. K. C. *Star*, Dec. 22, 1922, p. 1.

26. *Natl. Petroleum News*, Dec. 19, 1923, 33B.

27. Am. Petroleum Inst., *Proceedings*, Ann. Meeting of 1920, p. 36.

28. *Oil and Gas Jour.*, March 19, 1920, 87.

29. *Mineral Resources*, 1915, 560.

30. p. 56.

31. Aug. 24, p. 2.

32. Oct. 20, 1921, 12, 30.

33. *Petroleum Refiner*, Dec. 1, 1921.

34. Issues of Jan. 5, 1922, 12; March 2, 1922, 15.

35. Issue of Nov. 20, p. 21.

36. *Oil and Gas Jour.*, April 23, 1920, 53.

37. *Petroleum Age*, Aug. 1, 1923, 14.

38. Issue of Sept. 22, 1921, 13.

39. *Oil and Gas Jour.*, May 13, 1921, 58.

40. *Petroleum Age*, April 15, 1924, 15.

41. *Proceedings*, p. 32.

42. *Ibid.*, p. 35.

43. *Petroleum Refiner*, Dec. 8, 1921, 26.

44. Am. Petroleum Inst., *Proceedings*, Ann. Meeting, 1920, 45.

45. *Petroleum Age*, Dec. 15, 1921, 19.

46. *Natl. Petroleum News*, Nov. 22, 1922, 47.

47. *Cong. Rec.*, 63 Cong., 1 sess., Appendix, p. 471, 472.

48. *Natl. Petroleum News*, Dec. 20, 1922, 73; *Oildom*, Jan., 1923, 26.

49. *Am. Petroleum, Supply and Demand.*

50. *Oil and Gas Jour.*, April 27, 1916, 2.

51. "Mineral Resources of Oklahoma," Ok. Geol. Survey, *Bul. 1*, 1908, p. 23.

52. *Econ. Geol.*, Dec., 1912, 730.

53. *Oil and Gas Jour.*, June 24, 1921, 64; *Petroleum Age*, July 15, 1922, 84.

54. *Tenth Census*, Vol. 10, p. 267-.

55. For a sane and judicious discussion of this subject, see an article by J. Elmer Thomas, in the *Oil News*, March 5, 1922, 31.

56. *Oil Weekly*, April 5, 1924, 27.

57. *R. of Rs.*, Jan., 1909, 49, 50.

58. *Inland Oil Index*, Dec. 16, 1922; *Mineral Resources*, 1889 and 1890, 287; 1910, 383; *Econ. Geol.*, Oct.-Nov., 1905, 1; Vol. 9, 235; *Mineral Industry*, 1913, 547; *Oil Weekly*, Oct. 13, 1923, 33; Pa. Geol. Survey, *Ann. Rept.*, 1885, 55.

59. Bacon and Hamor, *Am. Petroleum Ind.*, 428, 431; *Oil and Gas Jour.*, June 29, 1916, 35; Feb. 22, 1917, 28; Feb. 6, 1920, 64; Oct. 15, 1920, 74; June 1, 1922, 12; July 6, 1922, 14; *Econ. Geol.*, Dec., 1915, 704; *Mineral Industry*, 1908, 687; 1913, 545; *Natural Gas*, Jan., 1922, 17; *Natl. Petroleum News*, March 28, 1923, 75; April 4, 1923, 91; Feb. 6, 1924, 37; *Oil Weekly*, Oct. 10, 1924, 24; Oct. 24, 1924, 19; *Am. Petroleum, Supply and Demand*, 102-115.

60. *Petroleum Age*, June 15, 1923, 18; *The Lamp*, June, 1925, 18; *Natl. Petroleum News*, Oct. 25, 1922, 69; *Oil and Gas Jour.*, April 30, 1920, 2; Feb. 1, 1923, 10. On the general question of the amount of oil left in the ground, see references on p. 154.

61. On the general question of increasing the oil recovery, see: *Sci. Am. Supp.*, April 6, 1918, 15; Panyity, *Prospecting for Oil and Gas*, Ch. XVII; U. S. Bu. of Mines, *Bulletins 134, 148, 177, 194, 195;* Tech. Papers, 32, 38, 42, 45, 51, 53, 66, 68, 70, 72, 130; Bacon and Hamor, *Am. Petroleum Ind.*, Ch. IX; *Eng. and Min. Jour.*, Aug. 4, 1917, 198; *Standard Oil Bulletin*, Oct., 1922, 2.

62. Pogue, *Economics of Petroleum; Natl. Petroleum News*, June 20, 1923, 37; *Oil and Gas Jour.*, Dec. 30, 1921, 6; Sept. 20, 1923, 108.

CHAPTER XXVIII

THE QUESTION OF SUBSTITUTES FOR OIL PRODUCTS

SUBSTITUTES will be available for some of the oil products; and the oil shale deposits of the country will probably yield very large quantities of oil which will furnish many of the essential products now secured from petroleum.[1]

The extraction of oil from shale has been a commercial industry for many years. Such an industry was started in France as early as 1838, and in 1839, at the Paris industrial exposition, an inventor by the name of Selligue exhibited samples of shale oil products, light oils, burning oils, heavy oils, and paraffin. The French industry grew in importance until 1864, when it was set back by imports of cheap American oils. For some years the French government imposed a heavy import duty, but in 1893 this was reduced, and the result was another heavy blow to the industry. The introduction of the Scotch type of retort about 1903 increased the yield of oils, and in recent years the industry has given indications of recovering some of its former importance, its development being stimulated by the high price of imported petroleum oils during the World War and the years 1919 and 1920, and by a high import duty on petroleum.

In Scotland, also, the oil shale industry is of long standing. In 1850, James Young erected a plant for the production of oil from Boghead coal. In 1862, the deposits of Boghead coal were exhausted, and since then the oils obtained in Scotland have been almost entirely produced from oil shales. The industry developed rapidly, although many of the companies organized were unsuccessful. As in France, the industry suffered very greatly from the competition of American oils, which caused the failure of many companies and the consolidation of most of those that remained. This consolidation resulted in the development of more economical methods, so that the industry survived a half-century of American competition, and in 1919 was credited with a production of 2,758,555 long tons of shale. The consolidation movement, which started in the sixties, has ended in a complete monopoly. In 1871, there were 51 active companies, in 1910, only six, and in 1919, the four largest companies were consolidated into the Scottish Oils, Ltd., which practically controls the field.[2]

Australia has had considerable experience with shales, since as early as 1865 or 1866; yet, in spite of the assistance of a government bounty, under the Act of 1917, the production was only 25,453 long tons in

1919, enough to supply the present consumption of the United States for a little over a half-hour. Some of the Australian shales are reported to be very rich, 80 gallons per ton, and some even richer. The industry of New Zealand and Canada has been less important.

Shale deposits are reported to exist in many other parts of the world: England, Jugo-Slavia, Spain, Sweden, Germany, Switzerland, Italy, Bulgaria, Esthonia, Morocco, Angola, Belgian Congo, Brazil, Argentina, Chili, Uruguay, Manchuria, Mongolia, and elsewhere. In recent years many experiments have been carried on in the extraction of oil from some of these shales, and a great many companies have been formed for their exploitation.[3]

Oil shales are known to exist in many parts of the United States, some of the most extensive, and probably the most valuable, being in Colorado, Nevada, Utah, and Wyoming. The richest of these are in the Rocky Mountain region, and belong to what is known as the Green River formation. Also there are the black shales of the eastern states: Illinois, Missouri, Indiana, New York, Kentucky, Ohio, Pennsylvania, and Tennessee, and some in Texas and some in California, which, strictly speaking, are not true oil shales.[4]

According to Winchester, the Green River formation contains persistent beds up to 49 feet thick that will yield at least 35 gallons of crude oil per ton. More thorough surveying and sampling may discover thicker seams of equal or greater richness. A recently discovered 20-foot seam, sampled across the face at one point, yielded over 60 gallons of oil per ton. Some beds, most of them too thin to be worked commercially, have yielded as high as 90 gallons, but the average for the Green River formation is probably 35 gallons, or nearly a barrel. Kentucky shales have yielded as much as 16 gallons of oil per ton.

Before the development of petroleum, there were a number of companies engaged in the production of oil by distillation of various kinds of bituminous substances. In 1860 there were 53 companies in the business, mostly in the eastern part of the country, many of them operating under licenses from the Young Company of Scotland. The methods were crude, and the materials used ranged from bituminous and cannel coals to some true oil shales. The product was kerosene or "coal oil," which latter term survives the time when most of the kerosene used in this country was really derived from coal. Some of these companies were just getting started when the petroleum industry began its career, and the low cost of petroleum products soon drove the shale oil or coal oil companies out of business.

About 1913, the United States Geological Survey began to take a serious interest in the oil shale deposits, and within the next three or four years published several bulletins on the subject. These bulletins

A MOUNTAIN OF OIL SHALE, NEAR DeBEQUE, COLORADO

attracted wide attention, and led to a great deal of public interest in oil shales and their exploitation. Many retorts of various designs have been proposed, several have been patented, and a few have been erected; but most of them have been so small that they could only be considered as experimental or demonstration plants. Two or three plants had been erected in Hoboken, New Jersey; one or more in Illinois and Kentucky; one in Kansas City; several in Denver; a half dozen or more in the region between Denver and Salt Lake City; two or three in Salt Lake City; and at least two in Nevada, and an equal number in California. There are two retorts at Elko, Nevada, one of which is financed by the Southern Pacific Railroad, and is placed in charge of David T. Day. The other, the Catlin Shale Products Company, has been in successful commercial operation for several years. The chief product seems to be paraffin wax, of which it turns out an excellent quality, but the average yield of oil is over a barrel to the ton. Martin Gavin, of the Geological Survey, reported that he had found in the neighborhood of 75 proposed processes for the treatment of oil shale. *The Shale Review*, a monthly magazine devoted to the interests of the industry, has been published for several years.[5]

Unfortunately the first steps in the development of the new industry have been characterized by much fraudulent promotion. Over 100 companies have already been organized—one authority says over 150 companies, and another over 250 companies—ostensibly to develop oil shale in some way or other, but these have mostly been formed to fatten the bank accounts of conscienceless promoters, rather than to exploit oil shale. Dr. Raymond F. Bacon, of the Mellon Institute, reported in May, 1921, that there were over 250 companies listed as shale oil companies, only 19 of which had erected experimental plants, many of the rest being frauds.[6] Advertisements have recently appeared in the *Inland Oil Index*, asking for valid oil shale placer claims in Colorado, Wyoming, or Utah. The oil boom of 1919 and 1920 of course contributed to the formation of such companies.

A few conclusions seem justified regarding the oil shale industry and the availability of oil from this source to supplement our oil supplies. In the first place, there seems to be no doubt as to the vast quantities of shale in the United States. Winchester estimated that there were about 896,000 acres in Colorado, 2,696,000 acres in Utah, and 460,000 acres in Wyoming classified as chiefly valuable for its oil shale. As to oil content, estimates of 100 billion barrels have been made, and, whether this happens to be correct or not, there is no doubt that the reserves are very great, probably many times as great as the original petroleum supplies of the country. Winchester has estimated that there are at least 92,159,000,000 tons of 15-gallon shale in the Uinta Basin of

Utah, perhaps 40 to 60 per cent of which can be recovered; 7,176,000,-000 tons in Wyoming, at least 15-gallon shale, of which 60 per cent could be recovered; 7,219,000 tons of somewhat richer shale in the Elko region of Nevada; and a large amount in Colorado. Winchester was very careful, however, to disclaim any pretense of accuracy in his estimates.[7]

One estimate by men in the United States Geological Survey—Dean Winchester and A. R. Schulz—gave Colorado, Utah, Wyoming, and Nevada at least 75 billion barrels, in 25-gallon shales three feet thick or more. Dr. Ralph McKee, of Columbia University, estimated that there were 64 billion barrels in shales of 42-gallon richness, in the Green River section of Colorado, Utah, and Wyoming alone, and large deposits elsewhere.[8] Willard Jillson, State Geologist of Kentucky, estimates a possible 12,308,217,065 barrels in Kentucky. John R. Reeves, of the department of Geology of the University of Indiana, has estimated that there are approximately 45 billion tons of oil shale in the 500 square miles of the outcrop district of southwestern Indiana, and a much larger amount west of this which is covered with other rocks, available only by underground mining. This shale has an oil content of from 6 to 16 gallons per ton, with an average of perhaps 10 gallons. Floyd Parsons somewhere got an estimate of 23 billion barrels for Colorado, and another writer put Colorado's reserves at 20 billion barrels.[9]

A committee of the American Petroleum Institute has recently made the following estimate of the oil shale resources of the country:[10]

State	Oil shale in 1,000 tons (short)	Shale oil in 1,000 bbls. (42 gals.)
California ...	13,939,200	5,575,680
Colorado	113,792,000	40,640,000
Indiana	69,696,000	13,939,200
Kentucky ...	90,604,800	28,993,536
Montana	6,969,600	1,393,920
Nevada	7,219	2,165
Utah	92,159,000	16,588,620
Wyoming ...	7,176,000	1,291,680
Total ...	394,343,819	108,424,801

While the reserves of oil shale are undoubtedly very extensive, the extraction of oil from these shales will be a very expensive business. M. L. Requa has stated that the oil shale industry, developed to the scale of the present petroleum industry, would require a mining activity comparable in size to the coal mining industry. In Scotland, in 1908, 8,300 men, 4,000 of whom were miners, were required to produce something over 1,000,000 barrels of oil. It is to be hoped that American labor will do somewhat better, for at this rate over 6,000,000 laborers would be required to produce our present petroleum output of over 2,000,000 barrels a day.[11] Gavin estimates that it would require somewhere near 750,000 miners alone; and George Otis Smith makes the point that "our country cannot afford to support another such army of workers, . . . until we reach another stage in our industrial development."

Gavin estimates that the industry will require an investment of $3,000 per barrel of shale oil daily capacity. To produce 400,000,000 barrels of oil—a little over half the present annual consumption—would, according to Gavin, require nearly 1,100 shale retorting plants, each putting through 1,000 tons of shale daily; and this would mean an investment of over $3,000,000,000 for retorts and refineries alone. This does not include the cost of lands, the cost of opening up and developing mines, or transportation and marketing facilities; nor does it include the cost of developing subsidiary industries, without which the shale oil industry could not exist. Probably the industry will not soon, if ever, attain such a magnitude, for it can develop only when prices of oil products are far higher than they are now, and we will not consume so much of this expensive oil. Nevertheless, the production of enough oil for our absolutely essential uses will require a vast amount of labor and capital; will require the establishment of great plants, the building of towns, and the housing and feeding of a great industrial population numbering hundreds of thousands of men.[12]

Many serious difficulties will be encountered in establishing the shale industry. The handling of tremendous quantities of heavy shale will be a great engineering problem.[13] Assuming an annual consumption equal to that of 1923, 735,000,000 barrels, and assuming shale of a richness of 30 barrels, more than 1,000,000,000 tons of shale would have to be handled. This is twice as much as the total amount of coal now being mined. To secure even half our present annual oil consumption from shales would require about as extensive an industry as the present coal mining industry. Moreover, shale may prove more difficult to mine than coal.

In mining the western shales, the problem of securing an adequate and reliable water supply for the retorts, and refineries, and for the general use of the communities engaged in the work, will be a tremendous problem, for some of these shale fields are in regions where water is very scarce. Retorting and refining operations will probably require from 150 to 250 gallons of water per ton of shale treated, and in part of the shale country the expense of developing an adequate and uniform surface water supply may be prohibitive. There are few large water courses, and these are far apart. Many of the smaller tributaries are periodic in their flow; they are generally dry part of the year and are raging torrents during the rainy months. To secure a uniform supply of water from such streams, it would be necessary to construct expensive dams, which, in some sections would soon fill with mud and silt. In much of this country it will be impossible to secure underground water. Many rich shale beds lie in high cliffs and on mesas where the cost of pumping water from the valleys will be very great. Furthermore, the water on many streams is always appropriated for agricultural or other uses, perhaps lower down the stream, and the amount of such appropriation may increase, since agricultural demands ordinarily take precedence over industrial requirements. In some sections of the shale country, it would probably be very difficult to get even enough water for the domestic use of the communities engaged in the industry. So great are the difficulties in the way of an adequate water supply in the West that some authorities think eastern shales may be exploited about as cheaply, in spite of their relatively low oil content.[14]

It may be necessary to build power dams to develop power for various mining purposes. There will be need for a great deal of power in the industry, to mine and crush the shale, and haul it to the retorts. Where water power is not available, there will be a considerable demand for coal, and thus the shale oil industry will be a factor in hastening the exhaustion of our most valuable mineral resource.

There are other difficulties. The mere question of disposing of spent shales will be hard to solve in some places. It has been a rather difficult problem in Scotland. There will be difficulty in transporting shale oil. On account of the high percentage of solid paraffins in most shale oils, they are said not to flow well at a temperature below 85 degrees Fahrenheit; and it might be necessary either to heat the oil at intervals along pipe lines and force it through at high velocity, or to transport it entirely in tank cars. Either way would of course be far more expensive than the transportation of crude petroleum through a modern pipe line. The refining of shale oil, as well as its extraction and transportation, will probably be expensive, more expensive than the refining of petroleum, because the oil must be distilled many times in order to separate the different ingredients, which have varying degrees of volatility and specific gravity. The average refining loss for shale oil in Scotland is about 25 per cent, as compared to about 7 per cent in refining petroleum in the United States.

Whether development will be cheaper in the West or in the East will depend upon a number of considerations. The western shales are generally richer in oil content—perhaps more than twice as rich—but offsetting this advantage are several advantages in favor of the eastern shales: the fact that some of them can be cheaply mined by steam shovels, rather than through tunnel mines; the fact that many of the eastern fields are more accessible to good transportation facilities, to markets, and to water supplies; and the fact that the oil produced is apparently of higher quality than that produced from many western shales. It is also stated that lean shales can be retorted more rapidly than rich shales. Reeves claims that the New Albany shales of Indiana, lying immediately on the surface, can be mined so much cheaper than either western or Scotch shales, that oil can be extracted much cheaper, in spite of the low oil content of the Indiana shale. According to his estimate, from 8 to 10 tons of New Albany shale can be secured as cheaply as one ton of shale in regions where it must be secured by underground methods.[15]

The question of by-products will of course help to determine the expense of producing shale oil; and there is no doubt that many by-products can be made. In Scotland, oil shale operations have been successful largely on account of the high recovery of ammonium sulphate, which has a ready sale as a fertilizer; in fact Pogue states that is the main product of the Scottish industry, and oil the by-product. Many American shales contain at least as much of the nitrogen from which this is derived as do the Scotch shales. Various other substances are mentioned as possible by-products of the industry: vanillin, salicylic acid, pepsin, rubber, perfumes, medicinal preparations, drugs, and dyestuffs. Paraffin will almost certainly be produced in large quantities. It is too early to say much about this question, however, because the output of by-products is in some measure determined by the demand for such products. Furthermore, the character of the American shales differs widely from region to region.

Estimates as to the actual cost of producing oil from shale differ considerably, and they are not much more than guesses. One authority estimated in 1921 that it would cost at that time from $1.50 to $2.50 a ton to mine, break, and deliver shale to the retorts.[16] David T. Day, who has devoted much time to this problem, has estimated that the Colorado shale oil will cost about $3 a barrel. Johnson and Huntley make a similar estimate; and Victor Alderson has put the cost at $2.50. Experts in the United States Geological Survey estimated in 1916 that oil could be secured from the black shales of the eastern states at a cost of perhaps $4.20 per barrel. In Australia, the total cost of getting shale oil onto the market was recently reported to be about £15 per ton (of oil), while imported crude oil could be bought for £3 per ton.[17]

Not only will shale oil be a relatively expensive product; but it is not yet certain that it has all the important qualities of petroleum. It is not yet known, for instance, whether lubricating oils of high viscosity and durability can be made from American shales. None that have yet been made are suitable for steam cylinders, internal combustion engines, or heavy duty bearings. Gasoline recovery may not be high—9.9 per cent, according to one estimate, although this can be changed more or less by using different processes. This brings out once more the criminal wastefulness of burning as fuel petroleum products which still contain a large fraction of lubricating oil.[18]

Unfortunately the government has lost the title to much of its shale lands, largely under the Placer Law. Dean Winchester, of the United States Geological Survey, when he visited the shale region around De Beque, Colorado, in 1918, reported:

So-called "assessment pits," many of which are in reality nothing more or less than small excavations dug at random on the hillside, and are neither essential or even of assistance in the ultimate development of the shale, have been dug on most of the claims in order to comply with the law, which required that $100 be spent on each claim each year. A large part of the land underlain by valuable beds of oil shale and within the reach of the railroad has now been filed upon as mineral land and is therefore covered by the preferential rights recognized by the Leasing Law of 1920. Many applications for patent to oil shale land have been filed with the General Land Office, and by July 1, 1922, the government had issued final patent to about 29,000 acres of oil-shale land, most of it near Grand Valley.[19]

In Utah, "lands in nearly every part of the state known to contain oil-yielding shale have been surveyed, staked, examined, and covered with such so-called development or assessment work as was thought necessary to meet the requirements of the laws relating to mining claims. Assessment work consists in digging prospect pits along the shale outcrop, building roads, trails, and houses to be used in the development of the oil shale and in some places, the construction of tramways, etc."[20] One of the western mining journals, in March, 1918, reported a "rush to locate the more accessible outcrops in Colorado, Utah, Wyoming, Nevada and California."[21]

It is very doubtful whether the present law governing oil shale lands will permit the most economical methods of extraction. Some shale lands, as pointed out, were entered under the Placer Law before the passage of the Leasing Act of 1920 and are now in private ownership. Those which still remain the property of the federal government, except the naval reserves, are now subject to lease in tracts of not over 5,120 acres; and this is probably too small a unit for the establishment of an efficient industry. It is far in advance of the Placer Law which governed previously, not only in being a leasing act, but in providing larger tracts; but probably still larger tracts should be provided.

It is conceivable that the exploitation of these shales in some regions should be carried on in tracts of 50,000 acres, or even more, in order that efficient and well-planned railroad connections, water supply, pipe lines (if pipe lines are used), refineries and other plants may be possible. A tract of 5,120 acres would hardly cover a good sized mountain in Colorado or Utah, and it might be difficult for two companies to divide a mountain in such a way as to exploit it most cheaply and efficiently. The plant of the Utah Copper Company at Bingham, Utah, and many other great mining plants, show that efficient mining must often be on a very large scale; and there can be no doubt that the shale oil business will be a business for large corporations, and nothing for the "small man" to play with.

Estimates by various engineers as to the cost of a complete retorting plant, handling 1,000 tons a day, run from $1,000,000 to $5,000,000. To justify the erection of a first class plant, perhaps a company should have more than 5,120 acres of land in sight. It seems possible, or perhaps even probable, that the new industry will be best developed by the great oil companies, if they are given the proper opportunity.[22]

It is to be hoped that the government will realize from the start the exhaustibility of the shale deposits, and will provide adequately for the avoidance of waste, in mining and in refining. It is to be hoped, and expected, that no such crime as the past and present petroleum industry will ever be committed again in the United States. The history of the coal industry shows some of the wastes that are likely to develop in the new shale industry. It may be profitable for the individual operator to leave a large share of the shale in the mine, in the form of supporting pillars, etc. It may be profitable for him to waste entire seams of shale, of slightly inferior quality or thickness, in securing the richer and thicker deposits. It may be profitable for him to throw aside all deposits of inferior richness. One authority in 1918 reported that no American company expected to use shales of less than 30 gallon content per ton— twice as rich as the Scotch shales that have been used. Obviously a considerable proportion of our oil will be in poorer shales, and the government will either have to establish certain regulations as to the kind of shales that may be wasted, or we shall presently be put to the very great expense of working some of the shale deposits a second time. Private interest cannot be trusted to guard our future national interests in this matter.[23]

It may be profitable for the individual operator to waste some of the less important elements in refining. The methods used in some of the small and inefficient oil refineries show what may be possible here. There are many ways in which operators might profit by wasting this irreplaceable resource, and the government will have to be alert in guarding against such methods, as far as possible. Fortunately the system of leasing leaves the title to lands in the government, and so provides a means of regulating the methods of mining; and when the oil production has begun its inevitable decline, and the shale oil industry has arisen to supplement it or take its place, the recollection of the iniquities of our oil policy and of its disastrous results will be so fresh and clear that conservation may be the order of the day. Of course the high prices which must prevail before the shale oil industry can prosper will always be an influence against waste.

While shale oil is probably the nearest to a complete substitute for crude petroleum, there is no doubt that substitutes will be devised from various sources, for many of the 300 refined products of crude oil. These have not been widely used during the past half century of cheap petroleum, but will be available at the higher prices which will prevail when petroleum production begins to decline.[24]

There have been several "epidemics" of substitute invention in recent years, when men in many countries of the world turned their attention to the quest for satisfactory substitutes, particularly for gasoline. During the Great War, for instance, the price of gasoline went so high that a great deal of attention was directed to the problem of finding a cheaper motor fuel. Many foreign governments took up the question officially, and appointed commissions to study the problem of substitutes. In Great Britain, no less than 50 substitutes for gasoline were offered to the government by experts in oil blending. In Germany, the problem was far more acute, since oil could not be secured from abroad, but the German difficulties were not published to the outside world.[25] Even in the United States, on our entry into the war, Secretary Lane, of the Department of the Interior, issued an appeal for a better aeroplane fuel. In response to this appeal, some scores of fuels were submitted and examined, but none were found any better than high-grade gasoline.[26]

Again in 1919, with a very high price for oil products, came another wave of interest in substitutes, and official commissions in several countries began work on the question. The British Society of Motor Manufacturers and Traders not long ago offered a prize of $10,000 for the discovery of a volatile fuel suitable for use in combustion engines in place of gasoline, and a number of other prizes have been offered by the International Association of Recognized Automobile Clubs, which represents the motoring interests of about every civilized country on the globe. The French government not long ago offered cash prizes amounting to $50,000 for the invention of a gasoline substitute, and a government subvention of $80,000 for laboratory experiments. The French government has devoted a great deal of attention to the question of finding a satisfactory substitute for gasoline, largely, no doubt, in the hope of becoming independent of imports in time of war.[27]

For gasoline, there is no doubt that there will be a wide variety of available substitutes, in small quantities. In the first place, there are substitutes derived from petroleum. Kerosene, for instance, now a relatively cheap and unimportant product, selling for a few cents a gallon, will be used more in the future. The supply of kerosene will probably always be greater relatively to the demand, than the supply of gasoline, and its use is likely to expand. Nearly one-third of the stationary combustion engines in the United States are run today on kerosene instead of gasoline; and a number of freight and passenger boats use it, because of its low cost. The chief objections to it as a motor fuel are: the offensive odor from the exhaust; the impossibility of starting the motor without first heating it with gasoline or some other substitute; the difficulty of vaporizing; the great heat production in the combustion chamber; and the tendency to create carbon deposits. If all the automobiles used in a large city now burned kerosene, the odor and the smoke would probably be almost intolerable. All these difficulties combine to make kerosene a far less desirable fuel for automobiles, but its use in stationary engines will likely increase as the price of gasoline goes up.[28]

There is the possibility also of utilizing still lower petroleum products, gas oil or fuel oil, or perhaps unrefined petroleum. In the Diesel engine, this has been successfully done in marine propulsion, and some authorities expect a wider use of this type of engine, perhaps even for automobiles. There are objections to it, particularly its great weight and size per horse power, and its high initial cost; but it is very economical of fuel. One recent report was to the effect that the Bethlehem Ship Building Corporation had perfected a miniature Diesel engine which would propel automobiles a distance of 50 miles at a fuel cost of $2\frac{1}{2}$ cents. This engine was given considerable publicity by enterprising journalists, but the manufacturers report that it is not yet being manufactured for commercial purposes.[29]

A second source of gasoline substitutes is to be found in coal, lignite, and peat deposits. Pogue estimates that the potential yield of the entire coke industry, when by-product practices are used entirely, will be about 25,000,000 barrels of oil products annually, on the basis of present technological methods, which will of course be improved. Coal has the possibility of yielding a barrel or more of oil, and from 20 to 30 pounds of ammonium sulphate per ton, the residue being fuel more valuable than the raw coal; in fact, Pogue and many other authorities see the use of coal in its natural solid form as a gradually diminishing use. Many authorities predict that in the future, more and more of the coal will be converted into gas and oils before being used.[30]

The Committee of Eleven, in its report to the American Petroleum Institute, estimated for the United States a total of 525 billion barrels of liquid products from our deposits of coal, yielding perhaps 92 billion barrels of motor fuel; but of course it will take some centuries to mine this coal and extract these liquid products. With coal production of 500,000,000 tons annually, a liquid yield of 20 gallons or one-half of a barrel per ton would amount to 250,000,000 barrels annually—of liquid products, not of petroleum. This is estimated to yield nearly 20 per cent of motor fuel, but it would probably yield only a small fraction of lubricating oil.

At the present time, gasoline blended with commercial benzol (recovered at by-product coke ovens) and blends of gasoline, alcohol, and benzol are being used to a limited extent as satisfactory motor fuels. Benzol is used in considerable quantities in England, and Germany used it extensively during the war and since. Officers of the United States Steel Corporation reported in 1920 that they had under construction plants with a capacity of 95,000,000 gallons of benzol per year—nearly a week's supply for the automobiles of the United States, at the present rate of gasoline consumption. As by-product coke ovens replace the wasteful beehive ovens so much used in America, more and more products can be made from coal, but coal by-products can be of only minor aid in supplying substitutes for petroleum products. It has been estimated that if all the coal mined in the country were coked in by-product ovens, the motor fuel thus secured would amount to only about 20 per cent of the annual domestic consumption of gasoline. Some authorities place this estimate far below 20 per cent—as low as 3 per cent.

In Czechoslovakia, the Ministry of National Defense have urged the extensive use of what is termed dynalkol, a composite of 60 parts of benzol and 40 parts of alcohol, which is claimed to be a more efficient aeroplane fuel than gasoline. This fuel is being produced in Czechoslovakia at a lower price than gasoline can be brought in, and it is being used in the automobiles and motor-driven vehicles which are in charge of the Ministry of Defense.[31] Doctor Friedrich Bergius recently gave to the Birmingham University Mining Society an account of the discovery of a process for producing gasoline from coal. He stated that experiments with this process have been successful at Mannheim, Germany, where a plant with a capacity of 60 tons a day has been installed. The conversion of coal into petroleum is said to be achieved by introducing hydrogen into the coal, thus completely changing its chemical character, and converting about 90 per cent of it into a liquid similar to fuel oil. This oil, by another process, is then transformed into light oils and gasoline, the latter totalling about 40 per cent of the volume of the fuel oil treated, there being also an equal percentage of Diesel engine oil.[32]

Some attention has been given to the manufacture of motor fuel from peat. The governments of several countries have investigated this question, and some encouraging reports have been issued. It has been claimed that a ton of peat will yield 30 gallons of alcohol, or even more of other forms of motor spirit. A committee of the American Petroleum Institute recently estimated that there were 70 billion barrels of liquid products in the country's lignite deposits, 12 billion barrels of which could be used as motor fuel.[33]

Even if benzol or other coal products were satisfactory and adequate at the present time, they are, for the long run, defective in being derived, like gasoline, from an exhaustible and irreplaceable natural resource; and for that reason a third class of gasoline substitutes must be considered, a class of substitutes that can be produced from inexhaustible sources for an indefinite time in the future. The most important substitute of this class is alcohol, and there is little doubt that alcohol will play an important part in the future of automobile transportation; in fact it is already an important source of power in some parts of the world.[34]

Alcohol, as a motor fuel, has practically all the essential qualities of gasoline; in fact, investigators in the United States Bureau of Mines have claimed that "denatured alcohol more nearly approaches the ideal fuel than does gasoline, for at any one compression it shows greater efficiency."[35] Alcohol has a number of advantages over gasoline: its thermal efficiency is greater; it is safer in storage, in transportation, and in use; it is cleaner in all ways, and lacks the unpleasant exhaust of gasoline; it does not carbonize, and therefore is not so likely to cause the engine to "knock"; and it is supposed to be more economical of lubricants. Alcohol, owing to a slow rate of flame propagation, gives a slower speed and lower piston velocity, which means steadier running. Its outstanding disadvantage is its low vapor pressure. Compared with gasoline, alcohol is so stable at normal temperatures that it is difficult to start a cold engine with it.

Alcohol is produced from three general kinds of vegetable sources:

1. Sugar-containing products, such as sugar cane, molasses, mahua flowers, sugar beets, mangolds, and grapes.

2. Starch or inulin-containing products, such as maize or corn and other cereals, potatoes, artichokes, etc.

3. Cellulose-containing products, such as sulphite, wood and wood pulp, straw, cornstalks, lyes, etc.

The essential characteristic of all these products is that they contain fermentable sugar or some component which can be converted into a fermentable sugar. It will be seen immediately that the future demands for refined sugar and for alcohol may be strongly competitive.

An Australian report of 1921 gave the following yields of alcohol from various sources:[36]

	Imperial gallons of alcohol (95%) per ton of 2,240 pounds
Sugar molasses	65
Sorghum stalks	12½
Wheat	83
Barley	70
Maize	85
Sorghum grains	87
Potatoes	20
Sweet potatoes	35
Sugar beets	18
Artichokes	22
Cassava	39
Apples and pears	12
Apricots and peaches	11
Grapes	18
Bananas	13
Watermelons	3
Zamia palm	18
Grass tree	12
Sawdust	20

Some of these sources may be considered in detail. Sugar molasses offers possibilities, and is even used in some countries to produce alcohol. In Australia, about a million gallons a year are produced from this source—enough to last the United States a half-hour or so—and it is reported that this could be increased considerably. It was estimated that, with an average yield of 15 tons of sorghum stalks per acre, one acre would produce 187½ gallons of alcohol. According to these figures, it would take about 40,000,000 acres planted to sugar cane to produce enough alcohol to equal the present gasoline production of the United States. A report from South Africa, however, states that the South African Railways have found "natalite"—a sugar cane distillate made largely from waste materials—to be cheaper than gasoline, and have signed a six months' contract with the Natal By-Products (Ltd.), the manufacturers of natalite, for motor fuel to be used throughout their transport service. This report stated that natalite was offered to the railways at 31 cents per gallon, while the lowest tender of gasoline was about 47 cents; and the natalite was claimed to be 90 per cent as efficient as gasoline. The government of South Africa has long been keenly interested in substitutes for gasoline, because of the high price of oil products there.[37]

In January, 1923, a Brazilian Congress of Coal and Other National Fuels, established in October, 1922, made a report in which it urged increased use of alcohol in motors, and recommended the establishment of alcohol coöperative societies in important centers, in order to distribute the product at a low price. A proposal was made to organize a company at Pernambuco for the manufacture of motor fuel from alcohol, sulphuric ether, and pyridine, the alcohol to be produced from sugar finals. It was estimated that Brazil could produce an amount equal to about 80 per cent of her gasoline imports of 17,000,000 gallons per year (1921).[38] A bill was recently introduced into the legislature of Uruguay, providing for the granting of a government monopoly for the manufacture of alcohol, and authorizing the government to require the use of a mixture of alcohol and gasoline, half-and-half, in all motor vehicles.[39] Some of the Cuban sugar manufacturers have experimented with alcohol production for many years, as a means of utilizing their vast quantities of sugar refuse.[40] Small quantities of Cuban alcohol have been sent to the Canary Islands for experimental purposes, and it is claimed that it could be retailed there at 20 cents a gallon, if it were not for the heavy tariff on it.[41]

The cassava plant holds out some promise in the production of alcohol. The value of the plant lies in its large tubers, which have an average weight of from 8 to 10 pounds and contain a quantity of poisonous juice. The poison is, however, very volatile, consisting of hydrocyanic or prussic acid. The Bureau of Science at Manila, Philippine Islands, estimated that one acre of cassava would produce 10 tons of roots, containing 5,000 pounds of extractable starch, in addition to from 4 to 6 per cent of fermentable sugars. The Australian commission estimated that a ton of roots should produce about 39 gallons of alcohol, at a raw material cost of slightly over one shilling. If these calculations are accurate, cassava might yield as much as 390 gallons of alcohol per acre, at a moderate cost.[42]

In Germany, large quantities of industrial alcohol have been made from potatoes. This is an old industry in parts of Germany, but its growth must be attributed in considerable measure to government subventions and encouragement. Thus in 1887, all alcohol used in Germany for industrial purposes was exempted from taxation; and in 1895, a progressive distillation tax was imposed, and the revenues arising from this were used for paying a refund or bonus on the industrial alcohol used in Germany. In various ways the industry was fostered by the government. To assist in building up a market for the product, a "Central Association for the Disposal of Alcohol" was formed, on a coöperative basis, and by various means industrial uses of alcohol were stimulated so that the industry was more successful in Germany than in any other country.[43]

Our large crop of corn, or "maize," as the English call it, has been mentioned as a source of future alcohol supplies. One bushel of corn, however, will produce only about two gallons of alcohol, and our corn crop averages about 3 billion bushels, so that it would require a larger crop than the United States has ever produced to yield the equivalent of our present annual gasoline output of something like 7 billion gallons. The Secretary of the Kansas State Board of Agriculture, a few years ago, estimated that a ton of wheat straw would produce the equivalent of 40 gallons of gasoline, and that Kansas, with nearly 12,-000,000 acres of wheat, could produce 46,560,000 gallons of alcohol from the straw alone. Other enthusiasts estimated that the wheat straw produced west of the Mississippi River could be made to furnish about one-fifth of our gasoline requirements, while cornstalks would also help out materially. Popular writers took up this interesting and cheering theme, and drew pictures of the farmers plowing their land and running their machines of all kinds with the product of their straw piles—a sort of a new agricultural Elysium. Government investigators, however, presently reported that these pictures were overdrawn; that the cost of getting the straw transported to extracting plants was in many cases prohibitive. It must also be noted that the straw piles are needed on the land, and that the best farmers do not burn them, but scatter at least part of them back on the land.[44]

Other waste substances of various kinds are often mentioned as valuable sources of alcohol: wood waste, sawdust, sugar-mill refuse, etc. Henry Ford announced several years ago that he was going to convert all the abandoned breweries in Michigan into alcohol distilleries, to furnish motor fuel for the automobiles of the country. Very recently he has made another threat of similar nature. Professor Ralph McKee announced, in 1920, a gasoline substitute made from wood waste which, mixed with kerosene and benzol, was in some respects superior to gasoline and could be made even cheaper. This mixture, he stated, had been used successfully in Norway and Sweden. A French chemist claims to have been successful in extracting motor spirit from linseed oil.[45] There is no doubt that some of these substances will be used, as gasoline becomes more expensive, but it seems reasonable to believe, with J. O. Lewis, of the United States Bureau of Mines, that the total product from such sources will be small. The chief obstacle to the use of such waste materials is the fact that they are widely scattered, and the cost of transportation of such low grade materials will always be high, compared with the value of the products secured. Such materials as cornstalks and beet tops contain only about 2 per cent of sugar, and it would not generally pay to transport the other 98 per cent to a central distillery for the sake of so small a recovery.[46]

A British commission reported in 1919 that "so far as vegetable sources of raw material for the manufacture of power alcohol are concerned, we must rely mainly, if not entirely, on increased production in tropical and sub-tropical countries."[47] Nevertheless, the production of denatured alcohol in the United States totalled nearly 100,000,000 gallons per year during two years of the Great War—nearly 2 per cent of our present annual gasoline consumption—and it could be raised above that figure. Since 1918, the production has, however, decreased very greatly. Much of the industrial alcohol manufactured today is made from sugar molasses and waste sulphite liquor.

Various patented and unpatented mixtures and concoctions, actually hundreds of them, have been advertised as gasoline substitutes. Scarcely a week goes by without the announcement of a new discovery of this kind, and seldom is anything ever heard of it again. If the newspapers and magazines were to be believed, there should now be as many substitutes for gasoline as there are cures for cancer.

Some writers see interesting possibilities of substitution. One such declared that a gallon of liquid hydrogen would run an automobile 250 miles. A competent chemist replied that, even if liquid hydrogen could be produced cheaply enough and kept long enough, it would run an automobile only about two miles per gallon.[48] Professor E. H. S. Bailey of the University of Kansas, suggests the possibility of getting some of our powerful explosives under control, so that they may be harnessed to our automobiles. Sir Oliver Lodge has predicted a time when "atomic energy" will supersede both coal and oil as a source of power.[49] Other writers see radium as the successor of coal and oil; in fact some writers have grown very enthusiastic when discussing the possibilities of radium.

Recently a patented fuel called Ethyl, a product of the laboratories of the General Motors Corporation, has been much discussed in the newspapers. It is advertised as an "anti-knock" compound, to be mixed with gasoline, but it has been claimed that with motors built for its use, it would nearly double the mileage per gallon of gasoline. Officers of the General Motors Chemical Company have not made this claim, however. There are a number of similar anti-knock compounds on the market, and considerable success is claimed for some of them.[50]

As already suggested, a number of governments have become greatly interested in the development of substitutes for gasoline, particularly in the production of alcohol. The efforts of Great Britain, Australia, Germany, Brazil, and Uruguay have been mentioned. France has perhaps paid more attention to this question than any other country.

During the war, the French government was so anxious to secure an adequate supply of motor fuel that it provided for government purchase of all the alcohol offered, the price and qualities being fixed in advance. In this way, the government accumulated a large surplus of alcohol, during and after the war. In order to get rid of its surplus and stimulate the production of alcohol, so as to lessen the dependence of France on foreign supplies of gasoline, the government passed a law early in the year 1923, requiring all importers of gasoline to purchase from the government one-tenth as much alcohol as they imported of the other fuel. The government of Switzerland has also made persistent efforts to find a successful substitute for gasoline, but thus far has failed to discover a motor fuel that was either better or cheaper.[51] President Coolidge's Oil Conservation Board has evinced much interest in the question of substitutes for oil products.

On the whole, prospects for the invention of a cheap and satisfactory substitute for gasoline do not seem very bright. Unquestionably substitutes will be available, when the price of gasoline rises high enough to justify their production, but they will probably be expensive.

One substitute will not have to be invented by any chemist. "Dobbin" has done yeomanlike service for some thousands of years and can be recalled to the harness whenever other forms of motive power become too expensive. The extent to which horses will displace automobiles is interesting to speculate upon, even if difficult to predict. The writer entertains little doubt that within the next ten or twenty years, there will be an important displacement of gasoline power by horse power, for some purposes. This will perhaps come first on the farms, where in many instances tractors have already proved more expensive than horses.

It is likely that horses will begin to replace automobiles first on the farm, for several reasons. In the first place, few farmers have high enough incomes so that they could ignore a 100 per cent increase in their gasoline expense. In the second place, farmers have a passable substitute at hand, for they can use their draft horses occasionally if they wish to, or, if they wish to keep driving horses, they can do so at far less expense and trouble than would be necessary in the city. On this whole question, there is some experience to guide us, for during the severe agricultural depression of 1921, many farmers put their cars up and drove their horses where they needed to go. They already had the automobiles, and the questions of initial outlay, interest charge, and depreciation did not require consideration, but the mere operating expense was more than they could afford.

How fast, or to what extent, horses may be expected to replace gasoline power in the city, is a more difficult question, but a guess may be ventured. Horse power seems a possible competitor of gasoline power for heavy, slow-moving trucks, for in pulling heavy loads, the bulk of a heavy draft horse is used to reasonable advantage. When it comes to high-speed, light-load service, however, such as is needed in taxi service, in some types of retail delivery, or, even more particularly, in doctors' cars or in fire-fighting vehicles, horses seem at so great a disadvantage that it is to be hoped they may never have to be used again. A team of good horses, pulling their own weight in coal or lumber or sand or cement, even at slow speed, seem reasonably well employed. They are moving themselves and the load, two equal weights, let us say ; and, if we assume that the moving of the load is the valuable service sought, we may conclude that only half of the energy of the horses is wasted— the half which is used in self-propulsion. On the other hand, a team of horses hammering over the brick or cobblestone pavements with a man or two, or with a few baskets of groceries, is wasting most of its energy. Two 1,250-pound horses hauling a 250-pound man, are really moving 2,500 pounds of horse and only 250 pounds of man. Ignoring the carriage, for the sake of simplicity, we may say that ten-elevenths of their load is themselves, and only one-eleventh is the real object of the trip— the man. In other words more than nine-tenths of the energy of the horses is wasted in self-propulsion.

Not only is too much horse energy devoted to self-propulsion, but it is applied in a very clumsy and inefficient way. Without any reflection on the wisdom of the Creator of all living things, we may say that the horse is a rather inefficient machine for using energy to effect a change of situs. This will be evident to anyone who will observe the manner in which the horse moves forward—constantly starting to fall down and catching himself. The horse uses a tremendous amount of energy, particularly when moving rapidly. On the other hand, the automobile moves on the principle of the wheel, which, once started, would move forever but for friction. This is true, not only of the wheels of the automobile, which are just like the wheels of a horse carriage in this respect, but of the engine itself. It is mechanically demonstrable that but for friction an automobile engine, once started, would turn forever.

Even with expensive alcohol as fuel, rapid long-distance transportation would probably be cheaper by automobile than by horse power, at least if the automobile were equipped with the very best type of motor. Such a motor should yield about 30 miles per gallon of fuel. Two gallons of alcohol would carry four or five men 60 miles over good roads. This alcohol could be made from one bushel of corn. Now, it would require two horses to haul the same load, and these horses would require an average of two days to cover the same distance. In two days they would have to be fed about 24 quarts of corn each (assuming a feeding of four quarts three times a day for each), a total of 48 quarts or a bushel and one-half. This would be enough corn to run the automobile 90 miles, in the form of alcohol. Even with a motor efficiency of 20 miles per gallon, the "fuel" cost of driving the automobile is no greater than that of driving the horses, and the wear and tear is likely to be less. Thus, even if alcohol, at a price of 35 cents a gallon, had to be made from 70-cent corn, it would be as cheap a means of rapid long-distance transportation as the corn, fed to driving horses. Perhaps the horse is permanently out of commission for this sort of service.

But there is another substitute for gasoline which will almost certainly be used widely in the cities and even somewhat in the country. Electricity has heretofore been a less satisfactory motive power than gasoline, for many purposes; and as long as gasoline is cheap, electricity will probably not be called much into use. In the future, however, gasoline will almost certainly become much more expensive, while on the other hand electric cars may be expected to improve in efficiency and fall in price. Perhaps we shall not soon have electric cars that will draw power from the air as they go, but it would be a courageous man who would venture to place limits to the possibilities of electric power. We shall almost certainly see great developments in this field, and it is entirely possible that some electrical wizard will invent an electric automobile which will do many of the things that the gasoline automobile now does for us.

None of the available substitutes mentioned above, and perhaps none that will ever be invented, will do all the work of cheap gasoline. Probably none of them will ever have the dominant position in the motor world that gasoline now has. Perhaps they will all be used more or less. In other words, we shall have, with the remainder of our gasoline, shale oil gasoline, kerosene, gas oil, fuel oil, benzol, alcohol, and various combinations of some of these, in varying proportions as time passes. We shall have benzol and similar spirits not only from coal, but perhaps from lignite and peat. We shall have alcohol from many different sources and regions and climes, and perhaps other materials that we have not yet dreamed of; and, with all these, will reappear horses in places from which they were only recently driven, and electricity in forms that cannot now be visualized. As the supply of gasoline dwindles, some of these substitutes will be used to supplement it, and thus the gasoline supply may be prolonged for many years. On the other hand, an unfortunate probability, perhaps it may be termed a certainty, is that with the decline of gasoline will come increasingly heavy demands on the other hydrocarbon, coal, and for unimportant uses we shall hasten the exhaustion of another indispensable natural resource.

Even with a considerably reduced supply of oil and oil products, we should be able to fare reasonably well, for the automobile engines of the future will certainly be far more economical of fuel and of lubricants than the engines of the present day. Gasoline and lubricating oil have always been so cheap as to be hardly worth saving, and most engines have been built with very little regard for fuel economy. It is entirely possible that within the next decade or two automobiles will be improved 100 per cent in this respect. Some authorities declare that there is no reason why automobiles should not go 50 miles on a gallon of gasoline. Pogue states that a mileage of from 30 to 50 miles per gallon should be possible. Perhaps the improvement in the use of lubricating oil will be even greater than in the use of gasoline. As these products become more expensive, consumers will become more saving and careful in their use. It has been estimated that careful carburetor adjustment alone would save some 600,000,000 gallons of gasoline a year.[52]

The problem of motive power is unfortunately not the only problem, or even the main problem, connected with the exhaustion of petroleum. The problem of lubrication looms more portentous. Substitutes for gasoline can be produced, in quantities sufficient for the essential work of the country; but equally satisfactory and plentiful substitutes for petroleum lubricants have not yet been found.[53] M. L. Requa put the future lubrication problem clearly:

We are prone to think of our coal supply as limitless, and that if it ever is exhausted the hydroelectric power will supply the deficiency. We forget that the machinery must be lubricated, and that there is no known substitute for petroleum as a lubricant. This problem of the source of our heat, light and power is dependent on the problem of solving the question of lubrication.[54]

George Otis Smith, in an address before the American Iron and Steel Institute in May, 1920, stressed the same general idea:

In our attention to the generation of power to meet the needs of industry and transportation, we give too little thought to one unique function of oil—that of saving power. Machinery without lubrication is unthinkable; adequate lubrication saves energy and makes it available for use, as well as adds to the life of the machine. . . . On second thought, we realize how universal is the use of lubricating oil, in the home as well as in the largest steel plant, in the motor cycle and locomotive, in the electric fan and the power station—everywhere oil is needed in the bearings, a single drop or many gallons. And in looking to industry's future needs of this petroleum product, large plans should be made, for our use of power and machinery is increasing faster than is generally appreciated.

Edward Prizer, of the Vacuum Oil Company, a practical oil man, spoke in similar terms:

As long as metal revolves on metal, no lack of a separating and friction reducing fluid is thinkable. Today the wheels of all industry the world over revolve upon petroleum. The day may come when the lubricating qualities of petroleum will be the real vital need of our civilization, and will have to be conserved with judgment and care, lest mechanical operation comes to a halt for lack of a lubricating substance. The vast expansion of mechanical operation is calling for larger and larger quantities of lubricants, and there is nowhere in sight any substitute for petroleum in this particular.[55]

The significance of the lubrication question appears most clearly when we consider the vast future power development, and the vast demands for lubrication that may be expected within the next century. Hydroelectric development on a vast scale is to be expected. The Federal Power Commission, during the first two years of its existence, reported applications for the development of over 20,000,000 horse power. As the country grows in population, more railroad equipment will be required, more engines and cars and more extensive shops and machinery for their construction and repair. Shipping will probably increase. Urban electric utilities will expand, and domestic power equipment will grow rapidly as housewives demand more labor-saving devices. From a hundred different sources will come increasing demands for lubricants, while the source from which good lubricants are derived is dwindling in its daily runs.

There are substitutes for some of the work of petroleum lubricants. For some purposes, animal oils will serve; and chemists state that animal fats, if treated in a certain way, will yield a hydrocarbon oil which is just as efficient a lubricant as petroleum, but oil secured in this way would be extremely expensive. Castor oil is an excellent lubricant, but far more expensive than petroleum oil, and if we had to produce enough castor oil to serve all the purposes now served by petroleum oils, it would require the use of a large area of our best agricultural land. Furthermore, castor oil will not serve all the lubricating purposes that mineral oils serve. Perhaps some lubricant will be found by chemists, a lubricant that will be cheap and as effective as the petroleum lubricants. Some chemists assert their absolute faith that when this problem arises, the chemists will be able to solve it. This is a cheerful view, and there is some justification for it, but we must remember that optimism and wisdom do not always live together. Certainly the very efficient German chemists—probably about as efficient as those of America— found no substitutes for petroleum lubricants during the war. The American army, following the retreating Germans in 1918, found many trucks abandoned, ruined by lack of proper oil; and many locomotives were rendered unserviceable in the same way. In May, 1917, 8,000 German locomotives were reported to be laid up at Essen for repairs, through inadequate lubrication, and the conditions that were found after the war indicate that such a report was probably no exaggeration.[56]

It may be worth while to speculate on the question as to how soon the shortage of lubricating oil will be felt. Our present consumption of such oils, including exports, runs well above a billion gallons annually. If the average lubricating content of the crude oil supply of the country is placed at 25 per cent, it is evident that a crude oil production of four billion gallons, or about 100,000,000 barrels, would be necessary to supply lubrication requirements. If exports be reduced or eliminated, this estimate may be considerably reduced, since we have been exporting more than one-third of our lubricants. Most authorities on the subject would doubtless be willing to predict that the crude oil production of the country will fall below 100,000,000 barrels within 10 or 20 years. In other words, present indications are that we shall feel the shortage of lubricants within two decades, at most.

The question of substitutes for the other petroleum products need not concern us here. When kerosene becomes too expensive to burn in lamps, farmers will have to turn to other means of illumination, many of which are better, even if more expensive. In most cities and towns, electricity has already displaced oil lamps. If gas oil becomes too expensive for the production of city gas, cities will have to turn to coal. They should have done so long ago, for much of what has been burned as gas oil should have been saved for lubrication. Likewise, when fuel oil becomes too expensive to burn, other fuels will serve. Far too much petroleum has been burned as fuel. There will always be a residue from petroleum refining which will serve as fuel, but the smaller this fraction can be made, the better.

Thus the question of substitutes for petroleum products is a very complicated question, quite too difficult for definite solution in this brief chapter, or for any definite solution whatever at the present time. There are too many unknowns in the problem. Nevertheless, a few tentative conclusions seem to be justified by the evidence at hand. In the first place, substitutes, for gasoline, lubricating oil, and other petroleum products, will generally be more expensive than the petroleum products now used, in fact many of them will be far more expensive. In the second place, substitutes will generally be inferior to the products now used. The mere fact of expensiveness will tend to force the use of inferior products and inferior grades of products. The fact that these substitutes cannot now be sold in competition with petroleum products suggests that probably, although not certainly, they are either inferior in quality or more expensive to produce, or both.

In the third place, as far as substitutes are made from coal or other hydrocarbons, they merely represent a substitution of one exhaustible natural resource for another, and this only postpones a serious situation for a few generations. The really critical situation will come when our coal supply approaches exhaustion. In the fourth place, as far as substitutes are made from agricultural products, from sugars, grains, animal fats, etc., they will necessitate a subtraction from the area that can be devoted to the production of food. When our population has doubled or tripled, as it easily may within a time which thoughtful men must consider, there will be heavy enough demands on our agricultural land for food purposes alone, and the additional demands for petroleum substitutes will involve almost inevitable hardship. As far as these demands are imposed upon tropical agriculture, they will of course be less felt, but the competition with food supplies will go on nevertheless. Finally, it must be pointed out that the production of these various substitutes will require a vast amount of capital, doubtless far more than is now used in the production of oil.

NOTES

1. U. S. Bu. of Mines, *Bul. 210;* U. S. Geol. Survey, various bulletins, especially 705 and 729; Alderson, *The Oil Shale Industry* (1920); *R. of Rs.,* Sept., 1920, 293 (article by David T. Day); Pogue, *Economics of Petroleum,* Ch. XXVII; Colo. Bu. of Mines, *Bul. 8* (1919); *Natural Gas,* July, 1921, 19.

2. *Mineral Resources,* 1899, 177, 242.

3. For an account of the industry in New Brunswick, see *Mineral Resources,* 1915, 729.

4. U. S. Geol. Survey, *Bul. 729.*

5. *Colorado School of Mines Magazine,* Feb., 1920, 20; *Mineral Industry,* 1923, 519.

6. *Oil and Gas Jour.,* May 6, 1921, 62.

7. U. S. Geol. Survey, *Bul. 729.*

8. *Oil and Gas Jour.,* Feb. 24, 1922, 6.

9. *Oil News,* Feb. 5, 1922, 36; *Sat. Evening Post,* March 20, 1920, 34; *Sci. Am.,* April 22, 1916, 429; July 13, 1918, 29.

10. *American Petroleum, Supply and Demand,* by the Committee of Eleven of the Am. Petroleum Institute.

11. U. S. Geol. Survey, *Bul. 641,* 139-.

12. *Min. and Sci. Press,* May 14, 1918, 613; U. S. Geol. Survey, *Bul. 641,* p. 141.

13. Gavin, Hill and Perdew, *Notes on the Oil Shale Industry,* U. S. Bu. of Mines, Serial No. 2256, p. 7.

14. Jakosky, *Uses of Water in the Oil Shale Industry,* U. S. Bu. of Mines, Tech. Paper 324.

15. Reeves, *An Economic Study of the New Albany Shales,* U. S. Bu. of Mines, Serial No. 2466.

16. Gavin, Hill and Perdew, *Notes on the Oil Shale Industry,* U. S. Bu. of Mines, Serial No. 2256.

17. *Commerce Reports,* March 19, 1923, 753.

18. The United States Bureau of Mines has given the following figures as to the products made from Scotch shale oil: (Pogue, *Economics of Petroleum,* 338.)

Products	Percentage yield
Naphtha (gasoline)	9.9
Burning oil (kerosene)	24.7
Gas and fuel oil	24.4
Lubricating oil (low viscosity)	6.6
Wax	9.5
Loss (including still coke, 2 per cent)	24.9

19. U. S. Geol. Survey, *Bul. 729,* 72; *Mineral Industry,* 1921, 531.

20. U. S. Geol. Survey, *Bul. 729,* 119; *Eng. and Min. Jour.,* Aug. 14, 1920, 324.

21. *Min. and Sci. Press,* March 30, 1918, 433; *Mineral Industry,* 1921, 531.

22. U. S. Bu. of Mines, *Bul. 210,* 140; *Eng. and Min. Jour.,* Feb. 1, 1919, 217.

23. *Min. and Sci. Press,* April 13, 1918, 509.

24. On the general question of substitutes, see *Gasoline and Other Motor Fuels,* Ellis and Meigs, N. Y., 1921. See also an article by Warren K. Lewis in *Oildom,* June, 1923, 58; one by J. O. Lewis in *Oil and Gas Jour.,* Oct. 15, 1920, 74; and one by Roy Cross in *Natl. Petroleum News,* March 22, 1922, 41. Consult also Cassier's *Engineering Monthly,* Vols. 43-44, 143-146; *Motor World,* Feb. 15, 1912, 845.

25. *Oil and Gas Jour.,* Dec. 7, 1916, 2; April 19, 1917, 3; *Commerce Reports,* No. 243, Oct. 16, 1916, 202.

26. *Oil and Gas Jour.,* Oct. 11, 1917, 2.

27. *Petroleum Refiner,* March 23, 1922, 9; *Oil and Gas Jour.,* March 19, 1920, 29.

28. *Motor World,* Feb. 15, 1912, 845; Cassier's *Engineering Monthly,* Vols. 43-44, pp. 143-146; *Sci. Am.,* July 3, 1920, 3.

29. *Lit. Digest,* March 4, 1922.

30. Pogue, *Economics of Petroleum,* 332; *Oil and Gas Jour.,* Nov. 28, 1919, 74. Address of Geo. O. Smith before the Am. Iron and Steel Institute, N. Y., May 28, 1920.

31. *Commerce Reports,* Aug. 27, 1923, 537.

32. *Ibid.,* Jan. 15, 1923, 163.

33. Cassier's *Engineering Monthly,* Vols. 43-44, 143; *American Petroleum, Supply and Demand,* by the Committee of Eleven of the Am. Petroleum Inst.

34. U. S. Bu. of Mines, *Bul. 43;* U. S. Geol. Survey, *Bul. 392;* U. S. Dept. of Agr., *Dept. Bul. 82; Farmers' Bul. 429; Natural Gas,* July, 1921, 19; Australia Inst. of Science and Industry, *Bul. 20,* "Power Alcohol." (This bulletin has a good bibliography.)

35. U. S. Bu. of Mines, *Bul. 43,* p. 236.

36. Australia, Inst. of Science and Industry, *Bul. 20.*

37. *Commerce Reports,* Jan. 29, 1923, 292.

38. *Ibid.,* Jan. 8, 1923, 89.

39. *Ibid.,* Jan. 28, 1924, 223.

40. *Eng. and Min. Jour.,* June 20, 1903, 938; *Oil and Gas Jour.,* Jan. 13, 1922, 6.

41. *Commerce Reports,* Aug. 20, 1923, 482.

42. Australian Bul., *No. 20,* as cited.

43. U. S. Dept. Agr., *Bul. 82;* Australian Bul., *No. 20,* as cited.

44. *Oil and Gas Jour.,* Oct. 15, 1920, 74 (article by J. O. Lewis), Oct. 22, 1920, 109.

45. *Ibid.,* Sept. 9, 1921, 3.

46. *Ibid.,* Nov. 23, 1916, 29; Feb. 8, 1917, 3; April 16, 1920, 64; Oct. 15, 1920, 74.

47. *Petroleum, Executive Report of the Inter-Departmental Committee on Various Matters Concerning the Production and Utilization of Alcohol for Power and Traction Purposes,* London, 1919.

48. *Oil and Gas Jour.,* June 25, 1920, 62.

49. *Ibid.,* Jan. 2, 1920, 3.

50. Ethyl has been found so poisonous and so dangerous to handle that it has been taken off the market.

51. *Commerce Reports,* Feb. 5, 1923, 357; March 19, 1923, 753; Dec. 31, 1923, 887.

52. Pogue, *Economics of Petroleum,* 346.

53. *Oil and Gas Jour.,* May 3, 1917, 37; March 5, 1920, 84.

54. Hearings before the Senate Committee on Public Lands, S. 45, 65 Cong., 1 sess., Pt. 2, p. 84.

55. *Natl. Petroleum News,* May 9, 1923, 45.

56. *Oil and Gas Jour.,* May 31, 1917, 3.

CHAPTER XXIX

THE QUESTION OF FOREIGN SUPPLIES OF OIL

IF the United States can secure an adequate supply of oil from foreign countries, after her own supply is gone, the situation will not be desperate. The possibility of doing this hinges upon two general questions: first, have other countries a supply sufficient to meet our needs as well as their own, when our reserves shall be gone; and second, if they have, will it be possible for us to secure a reasonable share of that foreign supply?

As to the reserves of oil in foreign countries, no estimate worthy of the name can be made. The best that can be offered is a scientific guess carefully formulated on the basis of the data now available, and necessarily subject to a wide margin of error. Of the important producing regions, only Roumania, Galicia, and the Baku and Grosny districts of Russia have been sufficiently exploited to offer criteria comparable to those of the United States, for the estimation of oil reserves. In none of the producing or prospective oil regions of other countries have the geologic data been published and the stratigraphy and structure described sufficiently to serve as the basis of accurate estimates.[1]

In many other countries, as for instance in Mexico, the detailed geological examinations, sometimes accompanied by drilling tests, have been confined to restricted areas, with very indefinite data as to remaining regions, which, from surface indications or other data, are believed to be oil-bearing. In some regions, we have only the evidence of oil and gas seeps or pitch or asphalt deposits scattered over great areas, in which general geologic conditions, similar in essential characters to those of producing districts, are reported to prevail. In other regions, as for instance in the Near East, including Mesopotamia and Persia, a tremendous potential value is assumed on the basis of the character, relative abundance, and widespread distribution of well-recognized surface indications of the presence of petroleum, although geological details are meager, and actual oil production restricted to comparatively few areas. In countries like the Philippine Islands, Madagascar, and Australia, the indications suggest a possibility of the existence of oil deposits of some importance, yet the Standard Oil Company has expended more than a million dollars in prospecting in the Philippines, without any success; and the other countries named have not been sufficiently explored to justify any conclusions as to their reserves.

As early as 1915, several eminent geologists devoted considerable attention to the question of the world's oil reserves. In 1916, Ralph Arnold published several articles regarding the oil and gas resources of the Americas, and the following year Frederick G. Clapp wrote a *Review of Present Knowledge Regarding the Petroleum Resources of South America.* Neither of these authorities, however, ventured any definite estimate of the reserves of the various countries.[2]

According to Arnold, Mexico apparently possesses almost unlimited potentiality; the conditions in Central America were little known; the West Indies, Dutch Guiana, French Guiana, Ecuador, Bolivia, and Chile offered little promise; Argentina and Brazil only moderate promise; and of all the South American countries on which Arnold reported, only Colombia and Peru were said to justify hopes of extensive reserves. Arnold did not report on Venezuela and Trinidad because he had investigated them professionally for certain private oil interests.[3]

Conceding the impossibility of accurate estimates, and the necessity of frequent revision from time to time, as more adequate information becomes available, Eugene Stebinger, Chief of the Foreign Mineral Section in the United States Geological Survey, worked out provisional estimates of the resources of various countries in 1920, which he regarded as "conservative":[4]

Country or region	Relative value	Millions of barrels
United States and Alaska	1.00	7,000
Canada	.14	995
Mexico	.65	4,525
Northern South America, including Peru	.82	5,730
Southern South America, including Bolivia	.51	3,550
Algeria and Egypt	.13	925
Persia and Mesopotamia	.83	5,820
S. E. Russia, S. W. Siberia and the region of the Caucasus	.83	5,830
Roumania, Galicia and western Europe	.16	1,135
Northern Russia and Saghalien	.13	925
Japan and Formosa	.18	1,235
China	.20	1,375
India	.14	995
East Indies	.43	3,015
Total	6.15	43,055
Total eastern hemisphere	3.03	21,255
Total western hemisphere	3.12	21,800
Total north of equator	5.20	36,400
Total south of equator	.95	6,655

These estimates, as Stebinger and White were careful to admit, are the most general approximations only. Nevertheless, they are the best estimates at present available, and must serve until additional data are secured. Taking into consideration the "geology of other regions in which oil has not yet been proved to be present in commercial amounts," David White later raised the estimate to 60 billion barrels, which he regarded as conservative, although he was very careful to point out that these estimates were "highly speculative." Two years later, in September, 1922, he revised his calculations again, in the light of additional information that had become available, and raised his estimate to 70 billion barrels.

Assuming world reserves of 70 billion barrels, and an annual production of approximately 700,000,000 barrels a year, there should be about a hundred years' supply of oil in the world. This corresponds with Redwood's dictum of only a few years ago that, "The available store of petroleum will have become so depleted toward the end of the present century that long before then other sources of supply will become necessary." As in the case of the oil supply of the United States, however, the matter is not quite so simple. There is a possibility that world production may increase above its present average, and so hasten the exhaustion of the world's reserves. There is, on the other hand, a possibility, perhaps a strong probability, that when the production of the United States begins to dwindle, no other country will rise to take her place. Certainly it seems improbable that any of the other great oil-producing countries could exploit their oil as rapidly as the United States has drained hers, because none of those countries have the transportation facilities, or the industrial skill and equipment for such a tremendous effort. Furthermore, it is doubtful whether the people of any foreign country, even the most backward and uncivilized, would be so lacking in sense as to wish to drain their oil as rapidly as we have drained ours. Whatever may be the rate of exploitation during the next few decades, it is certain that for the world, as for the United States, the time is coming when higher prices will reduce consumption and so conserve the remaining supplies. It is altogether likely that the world's petroleum reserves will last, not for one century, but for many centuries, in fact, practically forever.

Thus the world supply seems reasonably adequate for a considerable time in the future. The disquieting feature of the situation is the tremendous rate at which the United States is placing herself in a position of dependence upon foreign supplies. In 1920, Van H. Manning estimated that the United States was consuming her supply 14 times as fast as the rest of the world. In 1922, David White wrote forcibly of the rapidity of our drift toward dependency:

In the United States we have at the present moment produced a total of 5½ billion barrels of oil—more than ⅓ of our estimated original heritage of oil. Contrasted with our more than 5 per cent rate of annual depletion, the rest of the world withdrew, in 1921, not much over 280 million from its store of over 60 billion barrels, or less than ½ of one per cent of its reserve recoverable by present methods. In other words, the reserves of the rest of the world would stand the present rate of drain for over two centuries. From the standpoint of the world distribution of oil and the economic relations of our reserves to those of the rest of the world, an error of two or three billion barrels—four to six years' supply—in the estimate of the oil reserves of the United States is comparatively insignificant.[5]

The world has considerable amounts of oil, and the United States will soon need some of it. The next question is, will she be able to get what she needs? For half a century she has been the generous dispenser of oil to all the peoples of the world. She has mined her supply of the precious fluid and sent it out to every nook and corner of the world; has scattered tin oil cans over the deserts of Africa, the malarial swamps of India and South America, and the steppes of Siberia. She has lighted with her refined oils the house of the Bavarian peasant, the camp of the Bedouin herdsman, the hut of the Brazilian peon. For half a century she has poured out her matchless lubricants to oil the wheels of the world's factories, the cotton mills of Lancashire, the woolen mills of Saxony and Udine, the silk mills of Lyons, and the steamships that have plied every sea. She has sent her gasoline, billions of gallons of it, to drive automobiles wherever automobiles are driven. She has sold it all for a mere pittance; and now she faces the time when she must buy for her own needs. How will she fare in the world's oil markets?

During the past few years we have been fortunate enough. Imports of petroleum have increased with tremendous rapidity during the past decade or more, as the following table indicates:

Net Imports (Imports Less Exports) of Crude Petroleum into the United States, 1913-1924

Year	Net imports	Percentage of net imports to total consumption
1913	12,345,969	4.7
1914	13,943,227	5.3
1915	14,371,170	5.3
1916	16,700,963	5.2
1917	26,008,919	6.7
1918	32,834,918	8.0
1919	46,802,916	11.1
1920	97,418,197	18.5
1921	116,424,000	22.1
1922	114,117,000	19.5
1923	82,015,000	11.5
1924	77,775,000	10.4

Almost all our imports in recent years have come from Mexico, and Mexican production is now on the wane. The decrease in imports since 1921 has been due partly to the decline in Mexican production, and partly to the great increase in domestic production. It is possible that our imports may decline further, for there are no fields in sight which give promise of duplicating the performance of the Mexican fields for the past ten or fifteen years.

It is hardly possible that large amounts of oil could be secured elsewhere in the world as cheaply as oil has been sold in the United States, even if foreign states wished to encourage the exploitation of their deposits, because most of the great reserves of oil are supposed to be in relatively undeveloped regions. In Mexico, oil has been produced more cheaply than in the United States, but Mexican production is apparently already on the decline, and in most of the other regions supposed to contain great reserves of oil—Persia, Mesopotamia, South America, and Russia—lack of transportation, lack of other industries essential to oil exploration, hostile climatic conditions, lack of efficient labor, and remoteness from all important markets, will make oil development a relatively expensive matter. It has been estimated that in some foreign fields, the cost of drilling each well will amount to perhaps $500,000. In such fields only the strong and wealthy companies will be able to operate, and oil will not be cheap.[6] The situation prevailing in regions like this is well described in a recent article relating to Venezuela:

An estimate of possible oil to be produced from Venezuela would be grossly misleading without some estimate of what that oil will cost per barrel. . . . Costs of development are enormous, while to achieve success the investment must be carried for many years before the profits will appear. The best managed companies have not recovered a gross return equal to their investment, after 10 years of development. Any work in Venezuela based on a return of less than $2.50 per barrel for the oil, f.o.b. tanker, is dangerous and probably unjustified.[7]

Even in Mexico, where the greatest oil wells in the world have been found, near the coast, where transportation is a simple and inexpensive matter, some of the large producing companies claim to have suffered great losses with the recent decline in production. Some writers have estimated the losses of the Standard Oil Company at $250,000,000, and of the Royal Dutch interests at $100,000,000. Probably these estimates are grossly exaggerated, but it is doubtless true that Mexican exploitation has been less profitable than many people have assumed.

It is further to be noted that much of the oil produced in foreign countries is of relatively poor quality, heavy, and of small gasoline content. A comparatively small proportion of the world's production is of the quality found in the Pennsylvania, Mid-Continent, and Wyoming fields. Figures of exports and imports for the United States have not indicated our actual balance of trade in petroleum products, in any event, because our imports have been heavy Mexican oil, and our exports have included a large proportion of the more valuable products, gasoline and high-grade lubricants.

Even if we assume that foreign countries have sufficient reserves of oil for our needs, and that the physical obstacles to their exploitation can be overcome, the question still remains whether political obstacles may not be placed in our way.

One political obstacle that will always be encountered in many of the important oil regions is the instability of governments. Much of the future oil supply will probably be found in South and Central America, in Persia and the Near East, and in Russia; and in all these regions governments are so insecure and unstable that international agreements are subject to great danger of abrogation. The histories of Turkey and of Russia illustrate this very well.

Even where governments are reasonably stable, it is doubtful if oil exploitation in the interests of the people of the United States will generally be welcomed. Recent investigations by several different departments of the government at Washington have tended to show that many foreign countries are decidedly hostile to exploitation by American oil companies.[8]

In the British Empire, each dominion and colony has its own legislation on the subject of petroleum exploitation, but the general policy of the Empire has followed certain general lines:

1. The barring of foreigners and foreign nationals from owning or operating oil-producing properties in some of the most promising oil territory of the Empire.

2. Direct participation by the government in the ownership and control of petroleum companies. The controlling interest of the British government in the Anglo-Persian Oil Company is a case in point.

3. Arrangements to prevent British oil companies from selling their properties to foreign-owned or -controlled companies.

4. Prohibition of the transfer of shares in British oil companies to other than British subjects or nationals.

The restrictive policy of the British government has been in force in India for a long time. American oil companies are expressly excluded from doing business in Burma by a proclamation of Queen Victoria issued in 1884, and by a blanket concession for 99 years given the Burma Oil Company in 1885, protecting this company from all foreign competition. The Burma Oil Company, by the way, is partly owned by the Anglo-Persian Oil Company, in which the British government owns a controlling interest. When, in 1902, the Colonial Oil Company, a subsidiary of the Standard Oil Company, applied for a license to prospect in Upper Burma, the application was denied without any reason being given for the refusal. In the same year, another subsidiary of the Standard applied for a prospecting license, securing the aid of General Patterson, United States Consul General in India, in support of its application, but was informed that the government of India did not desire "to introduce any American oil companies or their subsidiaries into Burma." This opposition to American companies was not limited to the acquisition or development of oil lands, but extended to the ownership of refining and distributing facilities. For example, the Standard Oil Company of New York was not allowed to purchase a warehouse site in Burma.

This policy of exclusion has not been confined strictly to Burma but has been extended to other parts of India which showed promise of petroleum production. The alacrity with which the local government of India amends any deficiency in the law of exclusion is illustrated by the following letter from the Standard Oil Company of New York:

Early in 1917 one of our representatives visited Sylhet, in Assam, India, and obtained an option expiring November 15, 1917, to purchase or lease land and mineral rights from private owners of freehold property. It was discovered by our representative that in Sylhet most of the land had been settled under what is known as the "permanent settlement" tenure, and which conveyed absolute title to both the surface and underground rights. On July 6, 1917, our representative registered with the local authorities an agreement to lease the private property secured under the option referred to above for the purpose of prospecting for oil. We were blocked in this effort to secure a foothold by a new regulation of the government of India, No. 11917, dated October 6, and published in the official *Assam Gazette* on October 24, 1917. This regulation prohibits any owner from transferring his interests in a mine—which expression, it is notified, includes any mineral deposits, or land known or believed to contain a mineral deposit of commercial value. Thus before the expiration date—November 15—of our option on the Sylhet property the government of India, by its regulation No. 11917, stepped in to prevent the transfer to us of the petroleum or mineral rights on that private property.

One explanation that has been offered for the restrictive policy of India is that the government feared that the Standard Oil Company wished to secure control of the Burma fields, in order to shut off production there, so as to raise the price and keep the market for its American products. The tenacity with which the government has adhered to its policy, without any exception since 1884, indicates that this explanation is hardly adequate.[9]

In northern and western Australia, no mineral oil lease may be given to a foreign company, but some of the other states are apparently more generous. In the mandate of New Guinea, the Australian government has apparently adopted a policy of exclusion. In 1920 the United States Consul General at Melbourne reported that he was given to understand positively and distinctly that while the administration policy in regard to New Guinea had not been wholly formulated, it had been specifically determined that aliens would not be allowed to prospect the oil possibilities, and that any oil discovered would be promptly confiscated by the government. Again on November 29, 1921, he reported that the authorities had stated informally that Australian policy contemplated complete jurisdiction over all oil operations in the territory and the exclusion of other nationals. The government itself has planned to do the prospecting, although the Anglo-Persian Oil Company has done some work. In May, 1920, this company entered into an agreement with the commonwealth of Australia by which the company and the government together were to build a refinery. The Australian and British governments agreed, on July 7, 1919, to provide in equal shares a fund of not to exceed £500,000 for exploration work in Papua, a part of New Guinea, to be conducted by the Anglo-Persian Oil Company. This was apparently an exclusive right to prospect in that territory, but it is reported that the British government has withdrawn from this agreement.

In British African possessions, some restrictions appear. In Nigeria, Gold Coast Colony, and British East Africa, leases are granted only to British subjects, and British companies; and in Uganda and Somaliland, prospecting may be carried on only in certain defined areas and under special license of the governor. In Egypt, all the crude oil is produced by a single company, the Anglo-Egyptian Oilfields, although several other companies are prospecting. The government itself has done some prospecting, and has even planned to engage in refining. In British Guiana, none but British subjects are permitted to secure rights in oil lands, and in British Honduras, oil lands belonging to the crown are leased only to British subjects and to British companies. In Trinidad, leases on crown lands are given with the stipulation that the company receiving the lease shall never permit more than 25 per cent of its capital or voting power to be held by aliens, and that the chairman and managing director and a majority of the other directors shall at all times be British subjects. Trinidad, by reason of its location, has a great strategic importance for the British navy, and it is unlikely that foreign capitalists will ever be given an opportunity to secure petroleum concessions. In Brunei, a British province in Borneo, restrictions are similar to those of Trinidad.

In some of the remaining portions of the British Empire, there are no nationality restrictions. In the United Kingdom, there are no restrictions, perhaps because there is believed to be no oil there. In the Federated Malay States, grants of mineral rights are subject to the approval of the British government. In New Zealand, in British North Borneo, and in Sarawak, there are no restrictions; but in the latter two provinces, all land is already covered by concessions. A British company, the British North Borneo Petroleum Syndicate, holds an exclusive concession covering British North Borneo; and the Royal Dutch-Shell interests control all the production in Sarawak.

In Canada, by a law passed in March, 1920, Dominion lands are leased only to companies "registered or licensed in Canada, and having their principal place of business within His Majesty's Dominion"; but in February, 1921, these regulations were amended to meet the reciprocity provision of the leasing law passed in the United States in February, 1920. The provision requiring registration or license in Canada probably was not meant to shut out foreign companies, for there was nothing to prevent the registration of foreign companies. Perhaps the most active oil company in Canada is the Imperial Oil Company, a Standard subsidiary.

Thus it appears that foreigners are barred from exploiting oil deposits in the British Empire, either by specific laws, or by the fact that all the valuable lands are already leased. About the only provinces in which lands are open to exploitation without restriction, are those which are not supposed to contain oil, or those in which promising lands are already taken.

Persia, while not a part of the British Empire, has for years been within the British "sphere of influence." Persia is credited by the United States Geological Survey with a possible oil reserve of nearly 6 billion barrels, and Great Britain has tried hard to get control of this reserve. Through the Anglo-Persian Oil Company, she long ago got control of southern Persia, and has claimed title to three and one-half of the five northern provinces, a territory covering about 500,000 square miles and reported to contain the most promising oil lands in Persia. The Persian government, however, has denied the validity of Great Britain's claim to the northern provinces and has even negotiated with American companies regarding concessions there. Early in 1921, the government awarded a concession for the exploitation of its five northern provinces to the Standard Oil Company of New Jersey, but this concession had to be ratified by the Persian parliament, and while this ratification was pending, Sir John Cadman, a director of the Anglo-Persian Oil Company, secured for his company the right to share the concession with the Standard Oil Company on a fifty-fifty basis, when the concession should be ratified. When this agreement was published, the Persian parliament refused to ratify the concession, and in February, 1923, had awarded no concessions in the northern provinces. The Sinclair Consolidated has been active in seeking concessions there, and has accused the Standard of New Jersey and the Anglo-Persian of working in opposition.[10]

Within recent years there has been much talk about the possibility of British domination of the world's oil supply, the talk emanating to some extent from English sources. The efforts of the Royal Dutch Company to absorb several of the American companies, and its final success in absorbing the Union Oil Company of Delaware; the failure of American companies to secure concessions in the British dominions and elsewhere; the San Remo Conference, in which England and France tried to divide between them some of the important undeveloped oil regions of the world; the adoption of oil fuel by the chief navies of the world; the numerous predictions of exhaustion of the oil supply of the United States; and the indiscreet boasts of some prominent Englishmen, that England has the future supply cornered, have tended to stir up considerable apprehension.

It is said that Admiral Lord Fisher saw the importance of oil fuel for the British navy at least as early as 1880, and for years he preached the necessity of having oil stored for this purpose. He gradually won converts, until, in 1914, the British government, under the leadership of Winston Churchill, bought a controlling interest in the Anglo-Persian Oil Company, to secure a future supply of fuel for the navy. About this time, there was considerable careless talk in England about the necessity of securing some of the world's oil reserves, and about the extent of England's ambitions in this direction; and some of this talk was heard across the Atlantic.

In 1914, H. W. A. Deterding, directing head of the Royal Dutch-Shell interests, stated that, unless unforeseen events occurred, within ten years the Dutch-Shell companies would dominate the fuel oil supply of the world, and that no oil-burning vessel could sail the seas without Dutch-Shell oil in its bunkers. In January, 1916, Walter Runciman, president of the Board of Trade of Great Britain, stated in the course of a speech in Parliament, that the future policy of Great Britain would be to control the oil supply of the world.

In 1919, Sir E. Mackay Edgar attracted wide attention with this startling announcement:

America has recklessly and in sixty years run through a legacy that, properly conserved, should have lasted her for at least a century and a half. . . . Just when Americans have become accustomed to use twenty times as much oil per head as is used in Great Britain; just when invention has indefinitely expanded the need for oil in industry; just when it has grown to be as common and as true a saying that "oil is King" as it was twenty years ago that steel was king; just when the point has been reached where oil controls money instead of money controlling oil—the United States finds her chief source of domestic supply beginning to dry up and a time approaching when instead of ruling the oil market of the world she will have to compete with other countries for her share of the crude product. . . . America is running through her stores of domestic oil and is obliged to look abroad for future reserves.

They (the Americans) are diligently scouring the world for new oil fields— only to find, almost wherever they turn, that British enterprise has been before them and that the control of all the most promising properties is in British hands. . . . The British position is impregnable. All the known oil fields, all the likely or probable oil fields, outside of the United States itself, are in British hands or under British management or control, or financed by British capital. We shall have to wait a few years yet before the full advantages of the situation begin to be reaped. But that the harvest will eventually be a great one can be no matter of doubt. To the tune of many million pounds a year America before very long will have to purchase from British companies, and to pay for in dollar currency, a progressively increasing proportion of the oil she cannot do without and is no longer able to furnish from her own stores.

According to Sir Edgar it was the deliberate policy of Great Britain to secure all the oil lands she could get, and then hold on while the United States exhausted her supply.[11]

A number of Englishmen, about this time, took pains to point out the unfortunate position of the United States, and to congratulate England on being more fortunately situated. A writer in the *Financial News* (London) of February 24, 1920, made the following comments:

At the commencement of the war we believed that the effective British share of the oil resources of the world was about two per cent. Careful admiralty calculations recently made have shown that it is now about 56 per cent. This figure includes the Persian and Burmah resources, but takes no account of the vast South American fields commanded by the British Controlled Oilfields. The exact amount of their contribution cannot, at the moment, be estimated with anything like precision. Probably a modest estimate might put it at another 19 per cent. If that be so, our present command of the world's oil resources runs to no less than 75 per cent of their entirety.[12]

In 1923, Sir Edgar apparently was still of the same mind:

As for oil, America has already reached the importing stage. Our business as Britons is to sit tight on what we have and exploit all the oil, cotton and metal possibilities in the non-American world. In that way we shall do more than safeguard our position. We shall be able to supply America.[13]

Beeby Thompson, in October, 1921, expressed similar views as to the British policy: "We have sat on our oil resources whilst America be-

stowed her richest mineral resources lavishly and generously, regardless of the future. In doing so she has deliberately wasted billions of
cubic feet of natural gas that no regrets can replace. We have scoured
the world for new sources of oil supply, whilst America has rested
content with her home supplies. She has skimmed the cream of her wealth
at a period when prices were low, whilst we enter the world's markets
with our flush production when the value of oil has appreciated, and
with the lesson of conservation before us." Beeby Thompson saw, however, that England owed a "greater debt to America than is generally
supposed by the layman," and declared: "The restrictive policy we are
adopting appears undesirable, and is liable to antagonize other nations,
and is likely to lead to reprisals sooner or later." Government participation in the oil business, Beeby Thompson thought particularly unwise.[14]

Some Americans expressed similar views. As early as 1917, Mark
Requa expressed some concern at the efforts of the British to "dominate
the oil industry," the immediate occasion of his alarm being the efforts
British oil interests were making to secure larger holdings in California.
C. C. Osborn, economist of the Marland Refining Company, in an article in September, 1920, declared that Great Britain had reserves of
21 billion barrels, while the United States had only six billion barrels,
and that Great Britain was using oil from the United States while she
conserved her own for future use.[15] David White, in his article in the
Annals, in May, 1920, devoted much attention to the question of political control of the world's oil fields, and concluded that,

If to the petroleum resources in the countries now held by Great Britain,
France and the Netherlands, there be added the concessions held by their
nationals in producing or prospective oil regions of other countries, the
total oil resources in the control of these nations will probably exceed three-
fourths of the world's oil reserves outside of the United States.

Some observers insist that it is the deliberate policy of the Royal
Dutch-Shell interests to exploit American resources as long as possi-
ble, while conserving their resources elsewhere. The acquisition of the
Union Oil Company of Delaware in 1919 and 1920, and the rapid expansion of Royal Dutch-Shell interests in the United States since the
war, are often pointed out as illustrations of this general policy. Some
have insisted that the same policy was shown by the action of the Shell
Company of California in shipping millions of barrels of oil out of
California while it was retarding development operations in the East
Indies; but it is unnecessary to go so far for an explanation of the exports of California oil. The Royal Dutch-Shell interests, of course, got
oil where it was cheapest—in California, and it was not profitable to
exploit resources elsewhere in competition with the California flood of
cheap oil.

The growth of some of the great British oil companies has been very remarkable. The Anglo-Persian Oil Company, formed in 1909, with a capital of £1,200,000, had increased its total assets to nearly £7,000,000 in 1922, and on the latter date was paying dividends of 20 per cent. The oil demands of the people of England are being supplied to an increasing extent by the Anglo-Persian Oil Company.

The Royal Dutch Petroleum Company was organized in the Netherlands in 1890 for the purpose of developing petroleum lands in the Dutch East Indies. The issued and outstanding stock at that time was 1,300,000 florins ($522,600 at normal exchange). In 1921, this had increased to 321,457,000 florins ($129,226,000). In the meantime, its earnings increased from about 44,400,000 florins ($17,850,000) in 1917 to 129,500,000 florins ($52,010,000) in 1920. In 1921, the Royal Dutch Company reported assets of 740,000,000 florins or nearly $300,000,000, and they have increased further since 1921.

Starting as a local enterprise in the Dutch East Indies in 1890, the Royal Dutch Company became an international factor in 1907, when it was affiliated with the Shell Transport and Trading Company. These two companies, together with the Rothschild Russian interests, had organized the Asiatic Petroleum Company in 1903, to act as a joint distributing agent for their products. When the Royal Dutch and the Shell interests became affiliated in 1907, two new companies were organized, the Bataafsche Petroleum Company, capitalized at 140,000,000 florins ($56,280,000) and the Anglo-Saxon Petroleum Company, with a capital of £8,000,000 ($38,932,000) which took over the properties of both the Royal Dutch and the Shell companies. The Royal Dutch Company then became a holding company, owning 60 per cent of the stock of Bataafsche and Anglo-Saxon companies, while the Shell Transport and Trading Company owned the remaining 40 per cent.

The Shell Transport and Trading Company was incorporated in England in 1897, as a transporting and distributing company, with an authorized capital stock of £1,800,000 ($8,760,000). As already stated, it became affiliated with the Royal Dutch Company in 1907. Its authorized capital stock increased from £1,800,000 in 1897 to £3,500,000 ($17,033,000) in 1907, and to £33,000,000 ($160,595,000) in 1921, with £21,365,000 outstanding. Its earnings, like those of the Royal Dutch, have increased prodigiously in recent years.[16]

The Royal Dutch-Shell group of interests own or control over 100 affiliated and subsidiary companies outside the United States. They control also several very important companies in the United States, including the Shell Union Oil Corporation, the Roxana Petroleum Corporation, operating largely in the Mid-Continent field, and a considerable number of smaller companies. The Shell Union Corporation resulted from the absorption of the Union Oil Company of Delaware by the Royal Dutch-Shell interests. The Royal Dutch-Shell interests have tried hard to increase their holdings in the United States. In 1921, they tried to secure control of the Pacific Oil Company, but were prevented from doing so, it is said, by the Standard Oil Company, which bought about 20 per cent of the Pacific Oil stock to prevent the Royal Dutch-Shell from getting it. When they bought control of the Union Oil Company of Delaware, they secured a 26 per cent interest in the Union Oil Company of California, which the Delaware company owned, but certain American interests immediately organized to prevent the "Dutch" from getting control of the Union Oil of California.

The Royal Dutch-Shell group of interests control important petroleum reserves, not only in the Dutch East Indies, but in Sarawak (British Borneo), Roumania, Egypt, Venezuela, Peru, Trinidad, Mexico, and the United States; and they control five refineries in the United States, with a daily capacity of 65,000 barrels; four in Mexico, with a daily capacity of 155,000 barrels; one in Venezuela, one in Trinidad, one in Curacao, one in Suez, and others in Europe and in the Orient, together with compression plants, storage facilities, and other equipment. They own 752 miles of trunk pipe line in the United States and about 240 miles of pipe line in Mexico; and they own or control about 1,144,000 tons of tankers, barges, and tugboats. They operate about 120 fuel oil bunkering stations in all parts of the world. There has recently been talk of a possible merger of the Royal Dutch-Shell interests and the Anglo-Persian interests, through purchase by the former of the British government's shares in the Anglo-Persian Oil Company, but it is unlikely that the British government will soon part with her Anglo-Persian stock.

It must be recalled, in all this discussion of the Royal Dutch-Shell interests, that they represent Dutch as well as British capital, so that their growing wealth and power are not significant of British ambitions only. It has also been said that the French have acquired a considerable interest in the Royal Dutch-Shell.

The British government has been shrewd and far-sighted in choosing oil reserves at strategic points. The reserves of Persia, Mesopotamia, and Egypt are fortunately located with reference to the Suez Canal and the vast trade with India and the Orient. The oil of Trinidad and Venezuela will always serve excellently in the traffic westward, especially the trade through the Panama Canal. In time of war, when oil supplies will be most important, the British navy can secure oil at convenient and accessible points to the eastward and to the westward.

There is no doubt that Great Britain has built up a consistent oil policy, aiming at control of large reserves for future use. The government has not been content merely to depend on private enterprise to secure needed reserves, but has supported her oil companies diplomatically and otherwise, and, as stated before, owns the controlling interest in one of the world's largest companies, the Anglo-Persian Oil Company. It has been stated that the British government bought 40 per cent of the stock in the Royal Dutch-Shell Company during the Great War, but government officials have recently denied that the government holds any interest in the Royal Dutch-Shell now. A year or two ago, it was reported that the government was considering the sale of its Anglo-Persian stock, but no such action has been taken. The Labor Government which went into power early in the year 1924 promptly announced its policy of retaining the Anglo-Persian stock, and all the oil reserves that Great Britain controlled. Great Britain is not trying to secure a world monopoly, but she is trying to secure all the reserves she can get.

As a result of her oil policy, Great Britain is rapidly becoming independent of American oil supplies. Within the past few years, imports from Persia have been displacing American oil. In 1920, the United States and Mexico supplied 84 per cent of the British imports, and Persia 4 per cent; while in 1923, in spite of the tremendous production and consequent cheapness of oil in California, the United States and Mexico supplied less than 58 per cent of the imports into the United Kingdom, and Persia furnished over 29 per cent.

It is not at all difficult to understand the policy of Great Britain. For much more than a century Englishmen have known that their country was safe only as long as the British navy controlled the seas; and the navy that controls the seas must burn oil. It was Winston Churchill, as First Lord of the Admiralty, who persuaded the British government to buy a controlling interest in the Anglo-Persian Oil Company. Walter Hume Long, a later First Lord of the Admiralty, also urged upon the government the necessity of acquiring additional oil lands. The control of adequate reserves of oil is an absolute necessity for England, as it is for no other country in the world. The United States, or France, or Germany, or any other continental power, might survive without access to the sea, but every Englishman knows that on the day when control of the seas passes to another power, Britain's days of greatness are numbered. From this point of view, it should be easy to understand the British oil policy, and even condone the apparent selfishness of it, but it should also be easy to understand that this policy will probably not soon be changed and that many of the British dominions will perhaps always be closed to exploitation by foreign oil companies.

It might logically be argued that the control of foreign oil lands is not so important, even to England, as some Englishmen have believed. Lord Fisher, when he talked about the necessity of having oil fuel for the navy, had in mind supplies of fuel oil stored in tanks at strategic points, these stores to be increased, whenever possible, by advantageous purchase in the open market; and there is much to be said for this scheme, as a means of national defense. A supply of 100,000,000 barrels could probably be constructed and filled for between $200,000,000 and $300,000,000—not a very large item in national war budgets—and this amount of oil fuel would last the British navy even for a very long war, assuming a consumption of 10,000,000 barrels per year. Such a supply would insure British control of the seas perhaps better than any number of square miles of oil lands in Turkey or Persia, because it would be immediately accessible. With such a reserve of oil stored, England could retain her control of the seas and of her over-seas territories, including oil territories; and without an adequate store, she might lose her outside territories, including the oil reserves upon which she is dependent. Oil reserves, such as those of Persia, may be the means of preserving British control of the seas, and therefore of British national life, but it must be remembered that British control of the seas is in turn necessary to the maintenance of any control over such reserves.

The British point out that their policy is not altogether unlike that of the United States, that in the Philippine Islands we prohibit any but Filipinos and Americans from drilling for oil. The British have also claimed that not long ago, after a group of British capitalists had spent several millions in prospecting in Colombia, the United States government intervened and compelled the abrogation of the concession, on the ground that it was contrary to the Monroe Doctrine. British officials have also claimed that in the so-called Amory case, the government of the United States had refused to support a concession given to British capital in Costa Rica, and in fact had endeavored improperly to procure its annulment. The latter assertion has been denied by the government at Washington, which insists that the Amory concession was secured from a revolutionary government, the Tinoco government, which had not been recognized by either the United States or Great Britain, and that the United States government had refused to support even her own citizens in attempts to secure concessions from Tinoco. The Amory company was stated to be an American company, although subsequent events revealed the fact that British subjects owned much of the stock.[17]

Sydney Brooks, in December, 1920, declared that he did not "like the prospect."

The Americans are not going to yield their old supremacy without a struggle, least of all in those Spanish-American Republics which they regard as their natural preserve. The concessions which British subjects have acquired in Venezuela, Costa Rica, Colombia, Ecuador, and so on, are looked upon at

Washington with peculiar jealousy. . . . Moreover, in one of the great organizations that is fighting out the battle of oil, the British Government is itself the principal stockholder, and an unavoidably official and national character is thus imparted to its operations.[18]

It is finally to be noted that the British policy of exclusion, if maintained for the indefinite future, may be as injurious to British interests as to American interests. If there is a real need for oil, there is no good reason for barring foreign capital from coming in to help find it. Hitherto there has usually been too much oil on the market, and foreign countries which did not encourage exploitation were wise, but in the future, when oil becomes scarce and expensive, foreign countries will probably find it a wise policy to welcome the operations of the most efficient oil companies in the world—those of the United States. Canada has probably been wise to allow the Imperial Oil Company—a Standard subsidiary—to spend its money prospecting for oil. Certainly no Canadian company could do the work as well. If the Imperial had found vast resources of oil in Canada, as Doheny did in Mexico, it would have been a somewhat different story. Even then, however, if oil had been needed, Canada would have gained more than the Imperial Oil Company would. A country with its oil supplies in the hands of foreign capitalists in a time of war is in a somewhat precarious position, but capital tied up in a foreign country in time of war, particularly if it be an enemy country, is in a far more precarious position.

There is no important oil production in continental France, but the general policy of France with respect to her colonies is highly restrictive. The Sinclair Consolidated Oil Corporation reported as follows: "In practice, it has been found that France and the French colonies are more completely closed to development by American companies than (in) any other part of the world." The Federal Trade Commission was definitely of the opinion that the French government would not grant concessions to any alien oil company, at least in Algeria, French West Africa, and Madagascar, unless two-thirds of the directors of the company were Frenchmen. An act was passed in France in September, 1919, providing for the nationalization of mineral resources, but it is not certain whether this applies to petroleum resources.

Since the war France has tried to get a foothold in oil fields in various parts of the world. In the Polish fields, French capital is reported to be dominant, with control of 85 per cent of the oil fields. A large portion of the French holding was secured as a compensation for French aid in the Polish-Russian war.[19] On February 6, 1922, a Franco-Polish accord was signed, which permitted certain French petroleum companies named in conjunction by the two governments, to export petroleum freely from Poland, with the sole restriction that the Polish government should have the right to fix the amount needed for domestic use. Special facilities were to be granted to these companies for the exportation of petroleum, and the export tax was not to exceed 40 per cent ad valorem. About this time a French "Groupe National de Pétrole" was formed, including the French government and all the leading French companies interested in the industry. This consortium was created for the purpose of conducting petroleum negotiations and exploitation desired by the French government.[20]

In Roumania, France has been trying to build up her control in various ways. One report is to the effect that the French Minister of Commerce has been engaged in a scheme to subsidize certain Roumanian oil companies, on condition that the companies concerned leave undisturbed the present proportion of French capital, grant priority to orders for the French market, reserve important posts for French engineers, and give preference to French manufacturers, the subsidies to be repaid the French government by indemnities recovered from the Roumanian government, or by deliveries of oil.[21] It is very recently reported that these negotiations between France and Roumania have been abandoned. When Czechoslovakia gave a concession to a Standard subsidiary in 1922, the French government is reported to have insisted on being given control of the company.[22]

To put her oil business on the best possible basis, France incorporated a great oil combination in 1919, known as the Société pour l'Exploitation des Pétroles, which was to monopolize the production, sale, and distribution of petroleum and its products in France and in her colonies and spheres of influence. This combination was reported to have wound up its affairs early in 1924. Within the past few years France has developed a comprehensive petroleum policy, however, designed to insure an adequate supply of liquid fuels and lubricants, largely with a view, no doubt, to the necessities of war. Not only has the government tried to secure reserves abroad in Poland, Roumania, and Czechoslovakia, but, under the law of December 16, 1922, it has offered more liberal terms to prospectors in France and her colonies, and has even done some prospecting on government account in the Pyrenees, Madagascar, and in the Sahara desert. The government has also tried to stimulate production in the only oil district of continental France, the Pechelbronn district. French participation in the Mesopotamian fields will be discussed later.

Along with these efforts to secure adequate petroleum production under French control, the government has constructed extensive storage facilities, with the purpose of keeping reserves of petroleum products. It was reported not long ago that the government had storage tanks for over 1,000,000 tons of petroleum products, and was building more.[23]

The government Service des Essences et Pétroles has also encouraged the construction of more tank ships, to carry oil to the French fleet. The French decree requiring 10 per cent of alcohol to be mixed with all gasoline put on sale in France is an effort to encourage the use and production of alcohol and so make the nation more independent of imported gasoline.[24] With this regulation has gone an active effort to stimulate the distillation of oil shale, of which there are considerable deposits in France.[25]

Dutch dependencies have been quite as illiberal as the British or French. In the Dutch East Indies, prospecting licenses and concessions are granted only to Dutch subjects, inhabitants of the Netherlands or Dutch East Indies, and to companies incorporated under the Dutch laws, having on their boards of directors a majority of Dutch subjects. Persons or companies not established in the Dutch East Indies must be represented in the island by a trustee. The Royal Dutch-Shell Company has almost a complete monopoly of production. A considerable part of the Archipelago, particularly the smaller islands, is still entirely closed to private exploration. American companies have for many years tried without success to secure leases in these fields. It is a matter of record that a number of years ago, when an American oil company was trying to purchase the controlling stock interest in an important Dutch company, the Dutch government intervened and prevented the transfer. This Dutch company was later taken over by the Royal Dutch.

The government of the Netherlands has denied that its policy is as restrictive as has been claimed, and has pointed out that some foreign oil companies, including the Standard Oil Company, have secured concessions. On the other hand, it has been asserted by foreign interests that the best lands, those known to be valuable, have been turned over to the Royal Dutch-Shell companies, while it is only the lands of little promise that are open to exploitation by foreign companies. There is reason to believe that in this respect the policy of the Dutch government is like that of the British government.

The restrictive policy of the Dutch government in regard to the Djambi fields of Sumatra has caused a great amount of discussion and some international friction. In November, 1920, a bill, sponsored by the government, was introduced into the Dutch parliament, authorizing the Minister for the Colonies to establish with the person or persons to be nominated by him, a registered corporation, the Netherlands Indies Mineral Oil Company, for the development of the Djambi concessions in central Sumatra. From an explanatory memorandum accompanying the bill, it was clear that the "persons to be nominated" by the Minister for the Colonies were the Bataafsche Petroleum Company and the Shell Transport and Trading Company. In view of the fact that American citizens and companies had not found it possible to acquire petroleum properties in the Dutch East Indies, except a few of relatively negligible importance, the American State Department sent a note to the Dutch government, objecting to the proposed bill. In this note, the Secretary of State pointed out that Dutch subjects had been accorded full rights of exploration in the United States, and intimated that if the Dutch insisted on a restrictive policy toward American citizens and companies, the United States would have to retaliate by refusing to permit any Dutch-controlled company to acquire leases on the public lands.

A rather heated diplomatic correspondence followed. The Dutch Chargé d'Affaires replied that a subsidiary of the Standard Oil Company held concessions in the islands, and that the fact that these concessions were inferior to those of the Royal Dutch, was "exclusively because other companies working in the Dutch East Indies started much earlier with the exploitation of oil fields and therefore obtained better fields." To this, Acting Secretary Davis replied that the proposed Djambi concession was clearly an act of discrimination against American oil companies; that the Standard subsidiary had been able to secure only inferior concessions; and that it had secured even such concessions, not from the government, but by purchase from third parties. Representatives of the Dutch government claimed that there was no discrimination against American citizens, since they could take up residence in the Indies and become citizens; but the American reply was that this was more than the United States required of foreign oil interests.

Finally, the Dutch Minister for the Colonies pointed out that under a new law to go into effect July 20, 1920, no more concessions would be granted at all, but that all future exploitation would be done by the government, or by private companies under a contract with the government, and that this applied to Dutch and foreign companies alike. The American reply was that this was not as liberal as the law of the United States, and it was hinted that in the administration of the law the Dutch government would probably still favor Dutch companies. The American Department of State concluded by announcing that the United States would grant no lease to any company "where the controlling or major portion of the stock of the corporation is owned, held, or controlled by citizens of the Netherlands or other non-reciprocating countries"; and the Dutch government concluded by granting the Djambi concession to the Dutch company, as at first proposed.

The provisions of this concession are worth noting. The Shell subsidiary receiving the concession was to be under control of the Dutch government, and the board of directors must all be Dutchmen, the president, vice-president, and one other member of the board of directors to be nominated by the Dutch Minister of the Colonies.[26]

The Djambi concession stirred up a considerable discussion, and not a little international ill will. American oil companies were supported by the Department of State in their claim that the Djambi concession was equivalent to a denial of the "open door" by the Dutch government, and a cry arose for retaliation under the Leasing Act of 1920. About the most obvious way to retaliate against the Dutch seemed to be to shut them out of the Indian lands. In March, 1922, the Secretary of the Interior, Albert Fall, amended the regulations governing the leasing of Indian lands, to prohibit the making or renewing of any leases to aliens; and on March 3, 1923, his last day in office, he refused to approve the assignment of two Indian leases to the Roxana Petroleum Corporation, a Royal Dutch subsidiary.

Attorneys for the Roxana company, in their plea before Secretary Work, Fall's successor, asserted that the report of the Federal Trade Commission, showing discrimination against American companies in Dutch territory, misrepresented the facts, because of a misunderstanding of the laws of Holland and England; that the Standard Oil Company had secured concessions in the East Indies, and was even operating on Djambi concessions; that the Department of State had misrepresented the Djambi affair to the American people, by withholding from publication three notes written by the Dutch government; that the Djambi concession had been awarded on a more fair and open basis than the Teapot Dome concession in the United States, in the awarding of which Secretary Fall had not even called for any competitive bids. It has been stated repeatedly that Secretary Fall did not conduct the Roxana hearing in a fair manner, and even that "he did not scruple to support his refusal" to grant the Roxana company a lease "by the false witness of two forged documents." One of the attorneys for the Roxana company ignored the whole question of discrimination against American companies in the Dutch East Indies, as not germane to the question, and based his argument upon the impropriety or illegality of refusing to approve leases of Indian lands under the retaliation clause of the Leasing Act of 1920. This act, he asserted, did not apply to Indian lands, but only to the public domain.[27]

Secretary Work, on May 16, 1923, rendered a decision reversing Secretary Fall's decision, and approving the transfer of leases to the Roxana company, basing his decision on the ground mentioned, that the Leasing Act of 1920 applied only to the public domain, not to Indian lands, which the Indians owned in fee, and that the Secretary of the Interior therefore had no right to restrict the leasing of Indian lands to American citizens. There can be little ground for criticizing the logic of this decision.

The reversal of Secretary Fall's decision did not allay the feeling of hostility which had arisen, and there was continued agitation for retaliation against foreign oil companies operating in the United States, particularly against the Royal Dutch-Shell interests, which were often claimed to be unethical in their methods of competition. An oil journal recently asserted that at one time in the Indian Territory, subsidiaries of the Royal Dutch-Shell syndicate held 29,000 acres of oil and gas leases, when American citizens were allowed only 4,800 acres. The legislature of Oklahoma has seriously considered the question of banishing foreign oil companies from the state.[28]

Even before Secretary Fall's decision refusing to lease Indian lands to the Royal Dutch-Shell interests, he had refused the Shell Company of California a lease on public lands in Utah. This was on September 12, 1922, and Secretary Fall allowed the Shell Company sixty days in which to file "definite and positive evidence" that the government of the Netherlands allowed American interests the privilege sought by that company. On February 21, 1923, the Shell Company filed a formal withdrawal of its application, which was forthwith cancelled.

Some idea of the ambitions of England and France, as well as of the international importance of oil, may be gained from a study of the various negotiations and treaties covering the oil deposits of the Near East. Shortly before the Great War, Great Britain and Germany divided between them the oil interests of Mesopotamia. The German government at this time recognized southern Mesopotamia and central and southern Persia as the exclusive field of operations of the Anglo-Persian Company, and agreed to the construction of a railway from Kut-el-Amara to Mendeli, for the purpose of facilitating petroleum shipments. An Anglo-German syndicate organized the Turkish Petroleum Company for the acquisition and exploitation of the oil resources of Mosul and Bagdad. Half the stock of the new company was assigned to the National Bank of Turkey (controlled by British capital), and to the D'Arcy group; one-quarter was assigned to the Royal Dutch Company; and the remainder was turned over to the Deutsche Bank. Upon joint representations by the British and German ambassadors at the Sublime Porte, the Sultan, in June, 1914, conferred upon the Turkish Petroleum Company the exclusive right to exploit the oil resources of the Mesopotamian valley from Mosul to Bagdad. The exact amount of royalty to be paid was left unsettled, the Turkish government demanding more than the British were willing to pay.

When the war broke out, this diplomatic liaison of Germany and Great Britain was broken up, but in the fiercest heat of the war Mesopotamian oil was not forgotten. Of all the spoils of war, Turkey was among the richest, and the British policy in the war was dictated to some extent by a desire to secure a dominating position in Mesopotamia.

In March, 1915, a secret agreement was reached among Great Britain, France, and Russia for the partition of the Ottoman Empire, and Italy was later admitted to the agreement and given her "share."

After the allied victory, the question of the division of its fruits— these "fruits" consisted largely of coal, oil, and mineral lands—engrossed the time and talents of European diplomats for several years. The French took over the German interests in the Turkish Petroleum Company, and the British acquired a mandate over Mesopotamia and fixed the amount of royalty interest which the Turkish government should receive at 10 per cent. In 1918, the Anglo-Saxon Petroleum Company sent geologists to prospect the country, in order to secure all possible information regarding oil deposits as soon as possible, and in December of the same year, Lloyd George and Clémenceau signed an agreement by which Mosul and Palestine were handed over to Great Britain.

The Treaty of Sevres, of August 10, 1920, practically broke up the Ottoman Empire and divided its valuable resources among the allies, Great Britain receiving the lion's share. Even before this, France and Great Britain had signed the famous San Remo pact, by which the British government was to retain exclusive control of the oil resources of its mandate territories, but agreed to allow France a share of the oil. Mesopotamia was to be exploited by a company to be established under the permanent control of Great Britain, France to receive 25 per cent of the oil, or 25 per cent of the shares of the company, if a private company were called upon to do the exploiting. The government of the country or other "local interests" were to have not more than 20 per cent interest in this company.

The San Remo agreement applied to other fields than Mesopotamia. France and England agreed to support each other in all negotiations for holdings of stock formerly belonging to German or Austrian interests in Roumania, or for concessions in the state-owned petroleum fields of Roumania. Whatever interests might be acquired in either of these two ways were to be divided equally between France and England. The two countries likewise agreed to give joint support to their dependents in their "common efforts" to secure concessions and other privileges "in the territories belonging to the former Russian Empire." The French government accorded "any British group or groups of good standing" which could "offer the necessary guaranty" the right to concessions in the colonies of France or in French protectorates or zones of influence, including Algeria, Tunis, and Morocco, such groups to be, however, 67 per cent French-owned. In the British Crown colonies, French oil interests were to have corresponding rights, "as far as the existing regulations will permit." The Royal Dutch-Shell and Anglo-Persian companies received substantial rights under the agreement.[29]

But Italy began to get thirsty for oil too, and in August, 1920, a tripartite agreement among Great Britain, France, and Italy was signed, covering oil rights in the Turkish Empire. By this agreement the three countries undertook to "render diplomatic support to each other in maintaining their respective positions in the areas in which their special interests are recognized"; although another section of the agreement provided that nothing in the agreement should prejudice the rights of non-signatory nations to free access to the various regions for commercial and economic purposes, whatever that may mean. Walter Teagle, president of the Standard Oil Company of New Jersey, once confessed his inability to interpret all the terms of this agreement, but considered that it was another effort to shut the Americans out of Mesopotamia.[30]

These European cabals, especially the San Remo agreement, were viewed with considerable hostility in the United States. The people of the United States thought they had "won the war," and were entitled to some voice in the disposition of the spoils of war, if there were to be any spoils. Vigorous protests were sent to Great Britain and France by the Department of State, and the relations between the United States and Great Britain were somewhat strained for a while. The Harding administration, entering in March, 1921, was strongly sympathetic toward the big interests of the country, and pressed the demands of the United States for some share of the Mesopotamian oil with much vigor. The British government replied that the San Remo agreement was based upon concessions granted to British interests by a former Turkish government, but Turkey denied this, and the United States finally received a share of the spoils. The *Washington Herald*, on October 19, 1922, announced that the share of the United States was not yet decided upon, but would probably be between 20 and 30 per cent, which was to be surrendered by the Shell and Anglo-Persian interests.[31]

The exact steps by which American interests were admitted into the Near East are rather hard to ascertain, since some of the agreements were not published, but negotiations were going on during 1922, and before the middle of the year an "oil peace pact" was concluded between the Standard Oil Company on the one hand, and the British government and the Royal Dutch-Shell interests on the other. The Standard obtained the "right" to prospect in Palestine. The agreement extended to the Baku fields, and provided also for the joint exploitation of northern Persia by a British controlled North-Persian Exploitation Company, a strong minority of whose stock was to be held by the Standard Oil Company. Later in the year, this "oil peace pact" was extended to the Mosul oil fields, and other American oil companies were drawn into the agreement. Under this last agreement the British oil interests were to have a 49 per cent interest in the enterprise of exploiting the Mosul oil, 25 per cent was to go to the combined American interests, while the remainder was to go to the Royal Dutch and to the French. Since much of the Royal Dutch stock was owned by the British, England really controlled the Mosul enterprise.

The United States was not the only power that looked upon the San Remo agreement with disfavor. Already, in 1920, a small handful of Turkish patriots, under the brilliant leadership of Mustapha Kemal Pasha, declared that they would refuse to submit to the terms imposed on Turkey in these various agreements and treaties, and, by a series of successful military campaigns, freed much of Turkey from allied occupation and established their claim to be treated as an independent country. The victory of the Turkish Nationalists called for a new deal on the Near Eastern situation, and when the first Lausanne Conference for Peace in the Near East assembled on November 20, 1922, there were some hopes of a just and lasting settlement, but as one historian has expressed it:

The conference was only a few days old, however, when the time-honored obstacles to peace in the Levant made their appearance: the rival diplomatic policies of the Great Powers; the desire of the West, . . . to maintain a firm hold upon its vested interests in the East; the imperialistic struggle of rival concessionaires, supported by their respective governments, for possession of the raw materials, the markets and the communications of Asiatic Turkey.[32]

It is unnecessary to discuss here all the diplomatic deals and schemes and machinations of the powers that had been fighting to "make the world safe for democracy." In his new military success and power, the "unspeakable Turk" soon found himself courted by those who had only recently been damning the Germans for their unholy Turkish alliance. England soon found France and Italy working hand-in-glove with Mustapha Kemal Pasha. French statesmen believed that England had been getting the lion's share of the spoils in the Near East, and negotiated a Franco-Turkish agreement in March, 1921. Italy was won over to a friendly attitude toward the Turks by the fear of a too influential Greece. The Angora Treaty of October 20, 1921, gave France valuable economic privileges, in return for which the French government was "to make every effort to settle in a spirit of cordial agreement all questions relating to the independence and sovereignty of Turkey." French ambitions seemed to be advancing swimmingly. The Turks presently began to suspect French imperialism, however, as during the war they had feared German domination, and at the first Lausanne Conference they refused to ratify many of the French claims. The result was the break-up of the conference.

While the English and French and Italians were haggling and scheming for concessions and advantages, the Turkish government turned over to a retired American naval officer, Admiral Colby Chester, an immense concession covering the Mosul oil fields, and much other territory that was already claimed by the other powers. This new concession granted the "right" to build some 2,400 miles of railroad in Turkey, with the right to exploit, for a term of 99 years, all minerals, including petroleum, within a zone of 20 kilometers (12 miles), on each side of the railroad. The concession, as thus defined, included not only a part of the Mosul oil fields, already claimed by the British and French under the San Remo agreement, but also the famous Arghana copper mines, to which France asserted some sort of claim. The concessionaires were to have also the "right" to construct several ports on the Mediterranean and Black seas, and to build many other public improvements. Profits— the terms of the concession suggested profits as a cheering possibility— were to go, the first 12 per cent to the stockholders in the Ottoman-American Development Company, and the remainder to the company and the Turkish government in the proportions of 70 per cent and 30 per cent. The amount required to fulfil the terms of the concession was estimated at from $100,000,000 to $1,700,000,000.

The Chester concession was really approved by the Turkish Minister of Public Works as early as 1909, but a succession of wars, including the World War, had prevented adjustment of all the details until recently.

It is not easy to extract from the tangled snarl of Turkish politics the exact significance of the Chester concession, but it is reasonably clear that it was largely a political move. The allied powers were squabbling about their respective rights and privileges, and the Chester concession was a move to serve notice that the Turks refused to recognize the other concessions claimed by the various powers. The concession was also shrewdly calculated to enlist American support for Turkey, in the parleys at the next Lausanne Conference. Doubtless the Turks really preferred exploitation by the United States, anyhow, because they believed that the United States was the only power which had the capital and expert knowledge necessary for the industrial development of Turkish resources, and yet did not aim at some sort of political domination.

The Chester concession finally turned out to be much of a hoax. A vast amount of publicity was given to the venerable admiral's achievement, but it was presently revealed that he had no financial connections at all adequate for such a tremendous project. Few capitalists seemed at all inclined to "buy into an international lawsuit" of this kind, and when the Ottoman-American Development Company tried to sell stock to the public, several of the directors in the company promptly resigned, among them George W. Goethals, builder of the Panama Canal, because the affairs of the company began to resemble too much those of an ordinary oil promotion scheme. After all the patient work spent in getting the concession, and after all the publicity accorded it, conscientious members of the development company dared not allow their names to be published in connection with the sale of stock. Thus the whole scheme came to naught, although it had served the political purposes of the Turkish government.[33]

Thus the oil of Mesopotamia has been for some years the subject of a vast amount of diplomatic dickering and international intrigue. Turkey used her oil resources as a trump card in a game that seemed never to end. Even the events in the Ruhr district of Germany seemed to hinge upon the Mesopotamian oil situation, for France apparently traded her support of Great Britain in the oil game for British support, or at least non-interference, in her Ruhr adventure. National honor and apparently every consideration of humanity was subservient to the greed for oil. The Turks were free to massacre all the Armenians they could catch, as long as they would dangle oil concessions before the eyes of the great European powers. This subject is entirely too large for adequate treatment in this book, but it is to be earnestly hoped that someone with a great deal of patience will sometime work it out fully.[34]

It is doubtful whether all the oil in Turkey will ever be worth what it has cost. Some Englishmen already rue the money and trouble they have expended, thus far to no result. Viscount Grey is reported as expressing some such views in November, 1922:

We have spent £100,000,000 to make the country (Mesopotamia) hostile to us. Well, why did we do it? There are people who say it was for oil. The oil that is there, I believe, is very far inland, and if oil was a consideration in the matter, valuable as oil is, I will only say that it is going to turn out the most expensive oil that any country has ever yet purchased. No; if only we had stuck to the plain, simple thing we promised—self-government—we should have saved our money, we should have had the friendship of the people.

Thus most of the leading powers of Europe are trying hard to secure all the oil lands possible within their respective spheres of influence; and the chance of securing control of valuable oil deposits outside of these political areas is not very bright. The expropriation of private property in Russia has not made Russian oil fields as attractive as they formerly appeared. Poland is at present so much under French control that it is not an attractive field for American capital. Roumania has generally pursued anything but a liberal policy. The oil production of Japan is almost all in the hands of two Japanese companies, and it is reported to be practically impossible for foreign companies to secure mineral rights. The oil supply of Japan is apparently inadequate even for domestic needs anyhow. China has disclosed no important resources, despite the expenditure of several million dollars of Standard Oil money in prospecting.

Many of the countries south of us do not promise great oil supplies for the future. Mexican production is falling off rapidly, and Mexico has clearly indicated her desire to keep some of her own oil for future use. We may be able to bully her into letting us have it, but such tactics are always likely to be expensive in the long run. Guatemala permits only native-born or naturalized citizens to acquire leases; Venezuela is already largely controlled by British-Dutch interests; Argentina is exploiting her own oil resources; the Dominican Republic, Costa Rica, Ecuador, Colombia, Peru, and Bolivia offer no bar to foreign exploitation, but in the most important of these countries, Colombia and Peru, most of the promising oil lands have been taken. American companies have large holdings in some of these countries, but the expense of securing oil in the South American jungles will certainly be high.

In this connection we must note the strong drift of almost all foreign countries in the direction of nationalization of oil deposits, for this movement shows very clearly that people in other countries realize the importance of oil and are determined to retain close control over it. There is not an important independent producing country that has not nationalized the oil on its public lands; and several countries have even gone so far as to nationalize the oil on privately owned lands. Among the latter are, next to the United States, the two most important producing countries in the world—Russia and Mexico; and several others, including Colombia, Argentina, Salvador, Santo Domingo, and Roumania, have gone far in the direction of nationalization. There is a strong sentiment for such a policy in Poland. In some countries the oil industry is specifically declared to be a public utility. This is true in Argentina, Guatemala, Bolivia, Roumania, and Spain.

The illiberal policies pursued in some foreign countries have been said to be in a measure inspired by what some foreigners consider the excessive bonuses and royalties exacted by our government on public and Indian lands. In most cases these high bonuses are on proved lands, but foreigners do not always differentiate clearly.[35]

If foreign oil-producing countries merely refused to permit exploitation of their oil resources by American companies, that would not be a fatal matter, for, at least in peace times, we could still secure as much oil as we were willing to pay for in competition with the rest of the world. The mere fact that our American oil capitalists could not secure employment for their capital abroad would not be so important if no obstacles were put in the way of our getting the oil we need and are willing to pay for. A number of foreign producing countries, however, have discriminated against all foreign purchasers of oil by levying customs duties on exports. The reason for this is doubtless largely fiscal, the need for revenues, but the effect is to impose a high price on purchasers. The desire to conserve natural resources has also been an important motive.

Mexico has for years had an export tax on oil, amounting in recent years to as much as 21 cents a barrel. Peru has had such a tariff for a long time, the amount of the duty varying with the price of oil. Venezuela now has what practically amounts to an export tariff on oil. Poland has had an export duty for several years, as have also the Dutch East Indies. The duty levied in the Dutch East Indies has been so high as to cause a great deal of remonstrance from companies operating there, and has even stopped some operations. The Roumanian government has at times prohibited the export of certain products. It is said that Argentina will probably impose a production and export tax if her production ever rises high enough to permit exports. It is evident that a production tax, or royalty, levied by a country which exports most of its production, is the practical equivalent of an export duty; and most countries have such production taxes.

There are good reasons for believing that tariffs on exports of petroleum will be resorted to more in the future than they have been in the past. The need for revenues is increasing almost everywhere; and a more eligible tax than this can scarcely be imagined. It has a tendency to conserve domestic reserves, and, with a demand as inelastic as the demand for oil will be when production is once on the decline, such a tax will be paid in large measure by foreign consumers. An export tariff on an irreplaceable natural resource approaches very close to the ideal of a burdenless tax, falling largely if not entirely upon foreign consumers. When production once enters upon its long decline and the supreme importance of the oil resource is better appreciated, producing countries will be likely to levy increasingly heavy export taxes, as a means of conservation and of raising needed revenues.

Retaliation against those countries which refuse concessions to American producers and against those which tax exports, will be urged, as a means of securing better treatment; but it is doubtful whether it will avail much; and, of course, the greater our dependence on foreign supplies, the poorer will be our position for retaliating. As long as we have valuable resources in the United States which foreign oil corporations are anxious to exploit, we may possibly employ retaliation to some advantage, but our concern should be with the time when our own resources will be largely gone, and at that time we shall not be in a position to retaliate. Furthermore, such a policy entails a great danger of stirring up international ill will, which in the long run is likely to cost far more than any oil secured would be worth. The history of international tariff disputes indicates how serious this danger is.

The Federal Trade Commission has recently called upon Congress to adopt a policy of restricting exports of oil and oil products. This suggestion would have been most timely and appropriate a few decades ago, but is hardly in order at the present time. Restriction of exports would seem an ungraceful gesture, and a most unfortunate example to set, just at this time when the United States must soon becôme an importer of increasing amounts of petroleum.[36] A more feasible proposal is that of government backing and protection of domestic oil companies in their efforts to secure foreign concessions.

Many writers on the subject advocate a "vigorous national policy." David White, in a speech before the Engineering Congress in March, 1922, mentioned the need of foreign supplies first of all, among his "constructive deductions."[37] George Otis Smith thought the government should "give moral support to every effort of American business to expand its circle of activity in oil production, . . ." even if that meant "world-wide exploration, development, and producing companies, financed by American capital, guided by American engineering, and safeguarded in policy because protected by the United States government."[38] Mark Requa urged government support for oil companies seeking foreign supplies, but he saw that the policy of the government would have to be a consistent policy, and not subject to change with each incoming administration. Requa even suggested that the United States and Great Britain should invite other interested nations to join them in formulating an international petroleum policy and program. Van Manning, formerly of the United States Bureau of Mines, has often urged a more vigorous foreign oil policy. Secretary Lane, of the Department of the Interior, strongly urged a policy of diplomatic support to our oil companies abroad, as did his successor, Fall, in the Harding administration; and many oil men, particularly those interested in Mexican fields, asked for stronger support from their government. Even the Federal Trade Commission recommended "all proper diplomatic support in obtaining and operating oil producing property."[39]

To some extent, support of American oil companies abroad has taken the form of a demand for the "open door." The International Chamber of Commerce has long worked for the "open door" in petroleum exploitation. At the London meeting of the Chamber, in the summer of

1921, a resolution was adopted declaring against government participation in industry and commerce, as discouraging to individual initiative, and opposing export tariffs and international rivalry. In the meeting of the previous year, Thomas O'Donnell and other American oil men had fathered similar resolutions.[40]

Naturally, most of the influential men in the large oil companies have called for more government encouragement and protection in their foreign adventures, and the government of the United States has at times been rather firm in its demand for the "open door." During the Wilson administration, and to a greater extent during the Harding administration, American oil companies had the support of the government in their foreign activities. In dealing with Mexican "confiscation" of oil rights, the Department of State generally adopted a very firm tone; and the diplomatic notes exchanged concerning the San Remo pact and the Djambi concession reveal a strong determination to back American oil interests in foreign fields.

It may be noted that in demanding the open door in the eastern hemisphere, the United States has demanded something which is not strictly consonant with the provisions of the Monroe Doctrine, as interpreted at Washington. There is reason to believe that British oil companies have found some obstacles in the way of securing rights in Central America, on at least one occasion. It is true that foreign companies are heavily interested in South America, but they have not been welcome in some Central American fields. Furthermore, Americans have not generally appreciated all that the "open door" involves. It has cost England a vast amount of treasure and some human life to protect some of the fields to which British interests lay claim. The United States has heretofore refused to participate in any such activity, or to assume any of the responsibility, except in Mexico, yet has demanded an open door to the resources. The United States refused to make war with Turkey, or to make peace with Turkey, or to take up any mandates, yet has claimed equal privileges in exploiting the oil fields in Turkey. As Ralph Arnold has said:

In all fairness, if America asks for participation in foreign fields, she will have to take up her share in the burden of protecting those fields. We cannot ask England to establish order and then go in and share the profits without assuming our share of the responsibility.[41]

Senator Phelan, of California, a consistent friend of the oil interests, has even proposed that the United States government go into the oil business as the British government has done. The Phelan Bill, introduced into Congress in May, 1920, provided for the formation of a corporation known as the United States Oil Corporation, to exploit foreign oil resources.[42] The nine directors of this corporation were to be appointed by the President of the United States, but the capital was to be subscribed privately. This bill was given considerable attention in business circles, but not in Congress.

The general policy of according oil companies strong government backing is fraught with many dangers, and it is often most unprofitable. As already suggested, Great Britain's oil interests have cost her very heavily, in fact, if England's safety did not hang so much upon oil, it might well be doubted whether these interests are worth what they have cost. Such rights represent a very uncertain and unstable asset, in any event. In not a few instances, oil concessions of "fabulous" value have proved to be less than worthless; and in other cases have been retaken by grantor governments, leaving the government which supported the concessionaire the alternative of withdrawing its supports, or at any rate limiting it to "proper diplomatic" exchanges, or of waging an expensive war to enforce the concession. The history of Turkish and Mesopotamian concessions illustrates dangers and uncertainties of this nature. In our own experience, we have once or twice come very near to a war with Mexico over our oil interests in that country, and a war would cost more than the oil has ever been worth, at such prices as have prevailed in the past. An association of American oil producers in Mexico has spent thousands of dollars in stirring up hostility to certain Mexican governments, with the purpose of protecting American interests in that country. This association has not been scrupulous in its methods of operation. The part played by concessionary interests everywhere in the world, in stirring up international ill will, would make an interesting study for some patient and painstaking scholar.[43]

The importance of concessionary interests is usually exaggerated. One economist estimated in 1914 that concessionary investments represented less than 4 per cent of the total capital of Great Britain, 2 per cent of that of Germany, and less than 1 per cent of that of the United States. Our concessionary interests in Mexico have been less than one-half of 1 per cent of our total capital, and yet they have occupied much of our national attention at times, and have threatened to drag us all into an expensive war.

Government support of rich oil companies operating in weak and backward countries gravitates easily into unfair bullying. We can see some of this sort of thing in the attitude of France toward Poland and some of her African dependencies, and for that matter in our own conduct toward Mexico. The oil companies operating in Mexico have used large amounts of money to influence the press and public sentiment in the United States, and even to influence the government, and while our attitude toward Mexico may not seem unfair, it has often been that of the big bully. The Department of State has addressed notes to Mexico that it would never have sent to a great power like England. We take this attitude in a spirit of patriotism and love of country, to be sure, for the interests involved are shrewd enough to play on the motive of patriotism. American citizens would not ordinarily care to risk their lives in the protection of the property of E. L. Doheny or the Standard Oil Company, but of course they will cheerfully "fight for their country"; and when they have read the inspired messages on the subject of Mexican atrocities for a few months, they are likely to want to fight, without knowing definitely whether they are fighting for their country or for the Standard Oil Company.

The government of the United States should doubtless offer some measure of encouragement to her oil companies in their efforts to get a footing in foreign fields. From examples set by Great Britain, France, and the Netherlands, we see clearly that oil interests without any government backing are at a serious disadvantage in many of the oil fields of the world. The oil companies of the United States should be put in a position of equality with those of other countries, as far as possible. There is no reason why English or French or Dutch companies should dominate the world's sources of this essential commodity. For the "open door" in this sense, the government should stand firmly. Just how far it should go in trying to secure it, whether its stand should be "slightly firm," or "reasonably firm," or "very firm," is another question and might easily be a delicate one. It is clear, however, that a hasty or domineering attitude might cause no end of trouble. It is also clear that the United States dare not insist on the absolute open door in Mesopotamia, Persia, and the Dutch East Indies, and at the same time hold a sort of Monroe Doctrine protectorate over parts of South and Central America.

While the United States might properly ask that she have the opportunity to secure a reasonable share of the oil that she needs, she obviously must not demand any exclusive privileges to insure her more than a reasonable share, for such a demand would naturally cause resentment abroad. The United States has consumed oil gluttonously in the past, not only her own oil, but vast imports from Mexico and Peru. When she seeks oil mainly from abroad, she will certainly have to moderate her Gargantuan appetite. Even if American oil companies had concessions abroad so rich that they could furnish the United States a lion's share of the world's oil, it would be doubtful wisdom for the government to insist upon having it. The open door is just as good for European countries as for the United States.

One of the liberal American journals has put this very clearly:

And for this difficulty there is apparently only one adequate solution, and that is that every concessionary be required to conduct his business in an international, not a nationalistic spirit. In the end there will have to be systematic arrangements for rationing raw materials among the nations, if there is to be a society of nations resting on a basis of fairness. But pending such an arrangement an approximation to fairness could be attained if every concessionary were required to refrain from all nationalistic discrimination. If an American secures a concession he has a right to the profits of promotion. But let him give all the world equal opportunity to subscribe the capital, to bid on supplies and to purchase the product. The richest nation would get more than its share, even under such conditions, but the poorer nations would not find their poverty aggravated by the danger of absolute exclusion.[44]

Mark Requa's suggestion of a world conference on the oil question is worth considering. It is not likely that such a conference would be able to work out any definite international policy at first, for representatives would come to it filled with holy zeal for their respective countries, and in their anxiety to protect national interests they would almost certainly ignore the broader question of the general good. Such a conference would, however, accomplish one thing. It would focus the attention of the world upon the international aspects of the oil problem, and might thus lay a foundation for its ultimate solution. The problem must be solved, and the first step in the solution will be to drag it out of the secret chambers of the diplomats and statesmen and into the clear light of publicity. An international conference would accomplish something.

Even with the moderate government support which has been accorded, American oil companies have some important holdings in foreign fields. American interests have in the past controlled about 85 per cent of the Mexican production. The International Petroleum Company, a subsidiary of the Imperial Oil Company of Canada, which is in turn owned by the Standard Oil Company of New Jersey, is said to be the largest oil company in South America, with large interests in Peru, and in Colombia, where it owns the de Mares concession. American companies control oil lands of more or less importance in Palestine, Roumania, Czechoslovakia, Dutch East Indies, Canada, and in South and Central America; in fact they occupy the dominant position in Central America. They seem to have an excellent chance of securing a reasonable share of the lands of North Persia and Mesopotamia.

Speaking of the political control of the oil lands of the world, Van H. Manning summarized the world situation in 1920 by stating that the Americans were the most important producers in the United States, Mexico, Peru, Canada, and Colombia, the latter with a very unimportant production; the British controlled India and Persia; British-Dutch interests dominated in Russia, Dutch East Indies, Roumania, Trinidad, Egypt, and Venezuela; French interests dominated in Galicia, Alsace, Italy, and Algeria; Japanese controlled Japan and Formosa; while

Portuguese and American interests dominated in Angola. According to Manning, while the United States was producing about 69 per cent of the world's total production, she controlled the ownership of only about 18 per cent of the estimated world's reserves—12 per cent in the boundaries of the United States, 7.5 per cent in Mexico, and only about 2 per cent in the rest of the world.[45]

Perhaps it is reasonable to conclude from the evidence now available that while the world has oil resources which would satisfy even the gluttonous appetite of the United States for a considerable time, there will be very serious physical and political difficulties in the way of securing them, and the political obstacles may increase as a world oil-hunger develops, and as the importance of this resource is more generally understood.[46]

Many foreign countries have shown a far clearer appreciation of the value of their resources than has the United States, and it is reasonable to believe that when the United States has definitely started on her long decline, that fact will sharpen the foreign appreciation of the value of oil. Many foreign countries have denied us the privilege of securing oil freely, even while the United States was generously supplying the world; and when the United States, in the rôle of the rich idiot child, shall have squandered her inheritance, there is no reason to believe that other producing countries will adopt a more generous attitude. On the contrary, there are many reasons to believe that they will be less generous than they have been in the past. Furthermore, when the United States has shifted from an exporting to an importing basis, foreign producing countries will be asked to meet, not only the needs of the United States, but also the needs of other consuming countries which have previously been supplied by the United States. When Russia and Persia and Venezuela, and a relatively few producing countries, must supply a part of the needs of the United States, and most or all the needs of the rest of the world, there will certainly be a "sellers' market" for oil, and producing countries will be in a strong strategic position.

Many foreign oil-producing countries, it is true, may not dare to impose too many restrictions in the way of the United States securing such oil supplies as she absolutely needs. The United States is by far the wealthiest and potentially the strongest nation in the world, and her relative wealth and power will probably increase; and the United States should be in a position to command "fair treatment," whatever that may mean, if any country is in such a position. As far as foreign supplies of oil can be secured, either by purchase or by force, the United States ought to fare about as well as any nation. If Venezuela, for instance, or Colombia, should develop great oil production, it seems likely that the United States, with her dominating influence in the Americas, would be able to secure a reasonable amount of that oil, even though it happened to be produced by the Royal Dutch or other European companies. The oil that is purchased in the market will probably be very expensive, however, and that which is secured by international intrigue or threats or actual force will be far more so. In the future, oil will be a costly commodity, and before the people of the United States are through buying their scanty gallons from Mesopotamia and Persia and Venezuela, they will think regretfully of the millions of barrels that flowed down the creeks of Cushing and Sunset-Midway, and of the hundreds of millions that were sold to the refineries for 10 cents to 50 cents a barrel.

NOTES

1. See an article by David White, on *The Petroleum Resources of the World,* published by the United States Geol. Survey, in 1920; also published in *Annals,* Am. Acad., May, 1920, 111.

2. *Econ. Geol.,* Vol. 11, 203, 299, April, May, and June, 1916; *Proceedings,* Pan-Am. Sci. Cong., 1915-1916, Vol. 3, 207; *Bul.,* Geol. Soc. of Am., Sept. 30, 1917, 603; *Transactions,* Am. Inst. Min. Eng., Vol. 57 (1918), 914-966.

3. *Proceedings* of the Second Pan.-Am. Sci. Cong., Sec. III, p. 207.

4. Address delivered at the meeting of the Am. Society of Mechanical Engineers, Sept., 1922, published in *Oildom,* Oct., 1922, 49; *Mineral Industry,* 1921, 506; *Annals,* Am. Acad., May, 1920.

5. *Mineral Industry,* 1921, 507.

6. *Oil and Gas Jour.,* Feb. 17, 1922, 6; *The Lamp,* Dec., 1924, 14.

7. *Transactions,* Am. Inst. of Mining and Metallurgical Engineers, Vol. 68, 1053.

8. S. Doc. 272, 66 Cong., 2 sess.; S. Doc. 11, 67 Cong., 1 sess.; Federal Trade Comm., *Rept. on Foreign Ownership in the Petroleum Industry,* Ch. V. The report of the Federal Trade Commission has been criticized a great deal, by foreign writers, on the ground that its hearings were not conducted with scrupulous fairness and that its conclusions were biased.

9. *Natl. Petroleum News,* May 2, 1923, 47.

10. *Oildom,* July, 1923, 29; *Oil Weekly,* April 3, 1925, 37; *Natl. Petroleum News,* June 27, 1923, 94; *Petroleum Age,* Aug. 15, 1924, 12; *The Lamp,* Feb., 1924, 4.

11. *Annals,* Am. Acad., May, 1920, 133. For an interesting, if not very scholarly discussion of England's position, see De La Tramerye, *The World Struggle for Oil.* The *Commercial and Financial Chronicle* of May 22, 1920, p. 2146- has an extended discussion of this question.

12. Quoted from *Annals,* Am. Acad., May, 1920, 133.

13. *Oildom,* Feb., 1923, 14.

14. *Oil and Gas Jour.,* Nov. 11, 1921, 80.

15. *Ibid.,* June 21, 1917, 32; Sept. 3, 1920, 80.

16. See De La Tramerye, *The World Struggle for Oil.*

17. *Independent,* June 15, 1914, 478.

18. *Oil and Gas Jour.,* Dec. 17, 1920, 72.

19. *Oil Weekly,* Jan. 20, 1923, 22; *Oil and Gas Jour.,* May 6, 1921, 2; *Mineral Industry,* 1920, 522.

20. *Commerce Reports,* Jan. 14, 1924, 109.

21. *Oildom,* July, 1923, 27; Jan. 22, 1924.

22. *Petroleum Refiner,* Jan. 5, 1922, 10; *Oildom,* Oct., 1922, 21, 23.

23. *Oil Weekly,* Dec. 8, 1923, 21.

24. *Oildom,* July, 1923, 27; *Commerce Reports,* Feb. 17, 1921, 976; Feb. 5, 1923, 357; March 19, 1923, 753.

25. *Commerce Reports,* Dec. 3, 1923, 620; *Oil Weekly,* Dec. 18, 1923, 21; *Petroleum Age,* Jan. 15, 1924, 72.

26. *Oil and Gas Jour.,* July 15, 1921, 2.

27. *Natl. Petroleum News,* May 2, 1923, 47.

28. *Ibid.,* March 21, 1923, 26; *Petroleum Age,* April 1, 1923, 88.

29. Federal Trade Comm., *Rept. on Foreign Ownership in the Petroleum Industry,* p. 103; *Oil Weekly,* March 17, 1923, 9.

30. Am. Petroleum Inst., *Proceedings,* Ann. Meeting, 1920, p. 5.

31. Federal Trade Comm., *Rept. on Foreign Ownership in the Petroleum Industry,* p. 16. It was announced in March, 1925, that the share going to American interests had been fixed at 25 per cent. These American interests included six companies: The Mexican Petroleum Company, the Gulf Refining Company, the Atlantic Refining Company, the Sinclair Consolidated Oil Company, the Standard Oil Company of New York, and the Standard Oil Company of New Jersey. (*The Lamp,* April, 1925, 11.)

32. Earle, *Turkey, the Great Powers and the Bagdad Railway,* 306.

33. *Natl. Petroleum News,* Nov. 14, 1923, 24; *Barron's,* June 18, 1923, 9.

34. *The Oil Trusts and Anglo-American Relations,* by E. H. Davenport and Sidney Russel Cooke, is an interesting new contribution on this subject.

35. *Oil and Gas Jour.,* March 10, 1922, 6.

36. Federal Trade Comm., *Rept. on the Advance in Price of Petroleum Products,* 1920, p. 56.

37. *Mineral Industry,* 1921, 506.

38. *Natl. Geographic Magazine,* Feb., 1920, 202.

39. Federal Trade Comm., *Rept. on the Advance in Price of Petroleum Products,* 1920, p. 56.

40. *Oil and Gas Jour.,* July 30, 1920, 69; Aug. 12, 1921, 3.

41. Davenport and Cooke, *The Oil Trusts and Anglo-American Relations,* p. 121.

42. S. 4396, 66 Cong., 2 sess.

43. See on this subject an article by Alvin Johnson, "The War; By an Economist," *Unpopular Review,* II (July-Dec., 1914), 411-429; adapted in *Readings in the Economics of War,* Clark, Hamilton and Moulton, p. 47. See also *The World Struggle for Oil,* by Pierre L'Espagnol de la Tramerye, and *The Oil Trust and Anglo-American Relations,* by E. H. Davenport and Sidney Russel Cooke; and *Southwestern Political Science Quarterly,* Dec., 1922; as well as the Hearings on the Teapot Dome lease, p. 2759. For an

interesting discussion of the strained diplomatic relations between England and the United States arising out of the oil interests of the two countries in Mexico, see *The Life and Letters of Walter H. Page,* by Burton J. Hendrick, Vol. I, Pt. I, Ch. 6. According to Page, it was President Wilson's belief that concession seekers and exploiters were responsible for many of the revolutions in Latin-American countries, these interests stirring up revolutions and backing adventurers—such as Huerta—for the sake of the financial profit they might make out of it. Page seemed to believe that President Wilson was persuaded that even the British policy with regard to Mexico was largely dictated by oil concessionaires; and his letters afford reasonable ground for such a belief, although British officials denied it.

44. *New Republic,* May 2, 1923, 254.

45. *Oil and Gas Jour.,* Oct. 22, 1920, 64.

46. *Transactions,* Am. Inst. Mining and Metallurgical Engineers, Vol. 68, 1056.

CHAPTER XXX

CONCLUSIONS

ON the basis of the facts of our history and present situation, it seems pertinent to formulate certain constructive proposals for a wiser oil policy in the future. There can be no doubt of the necessity for a wiser policy. As David T. Day has said, "In no line of national conduct is the progress so recklessly devoid of any thought-out policy";[1] and, as George Otis Smith has suggested, the motto on our coins should not be made our national policy in providing a future oil supply.

It will be unnecessary to repeat all the various conclusions and recommendations advanced in the preceding chapters. It may be assumed that oil recovery will be increased in various ways; that foreign sources of supply will be secured; that shales will be retorted for their treasures of hydrocarbon; that substitutes of various kinds will be used when it is profitable to use them; that waste of oil and of capital will be reduced; that more efficient and economical methods of use will be worked out; and that some of the least essential uses of oil will be eliminated. All these developments may be regarded as assured, when our oil supply is well on the way to exhaustion. The trouble is that will be too late. It should be our purpose to hasten these developments, in order that we may postpone, as long as possible, the time of exhaustion of our reserves.

The great problem is that of conserving our oil resources. The time for greedy indulgence of our every unimportant want, while ignoring the needs of the future, is long since past. If we are to pose as a considerate and generous people, or even as an intelligent people, we must raise our eyes above our own immediate selfish desires and take some reasonable thought of the future.

It ought not to be necessary to argue the justice and reasonableness of a more modest use of this irreplaceable resource. It ought to be clear to any mind above that of an orang-outang that the people of the present generation are only life tenants on the earth, and that they have no right to waste and destroy that which, as far as we can see, will be as essential to the happiness of future generations as to our own. Yet it is almost impossible to get the average American to see that there is anything but his own immediate wants, however trivial and unimportant, to be considered.

We remember that the problem is, after all, one for distant posterity, and that posterity should shift for itself, and we drowsily mutter "laissez faire" and forget the future in our supreme self-satisfaction in the present.[2]

A great many practical oil men, and many others as well, defend the speed with which we have used our oil; some because of a real disbelief in all predictions of future exhaustion; some in an equally ignorant faith that the Lord will provide, through the instrumentality perhaps of some inventor or chemist; some in the characteristic American pride in large figures—an oil consumption of 700,000,000 barrels would be a splendid thing to many quantitatively-minded Americans, even if the oil were used only to drown gophers; and many for the good old simian reason that all wants but their own, whether present or future, seem of small importance.

Before proceeding further with this question, it will be pertinent to point out just what we mean by "conservation," and on what philosophical basis we demand such a policy. This subject has been ably treated by Professor L. C. Gray,[3] by President C. R. Van Hise of the University of Wisconsin,[4] and by others, and must be briefly reviewed here.

As Professor Ely has said, "Conservation means a sacrifice of the present generation to future generations, whenever it is carried far, this conflict beginning far before the ideal is reached which conservationists are inclined to advocate." Professor Gray likewise finds "the real heart of the conservation problem" in "the conflict between the present and future." "The primary problem of conservation, expressed in economic language," he asserts, "is the determination of the proper rate of discount on the future with respect to the utilization of our natural resources." This seems incontrovertible, and the question remains, at what rate should the future uses be discounted? It is clear that there might be two extremes. In the first place, we might not discount future uses at all, but rate even the infinitely distant wants equally with present wants. In this way, "The period of utilization would be increased to infinity; and therefore the amount of present use would become infinitesimal." Concretely, if we wished to provide equally for present wants, and infinitely distant future wants, 100 barrels of oil per year for the United States would be quite too much. Thus this position is reduced to absurdity. On the other extreme, we might rate future wants at zero, and consume our resources as rapidly as we can get them out and can find ways to use them, with absolute disregard for the future. That is practically what we have thus far done, at least with our oil; and it is not less wrong than the first extreme.

The rate at which individuals discount future satisfactions may have some significance in this question, but it will be far from decisive. Individuals discount future satisfactions at an average of 4 or 5 per cent, or, considering various risks, perhaps 6 per cent per year. If we adopted such a rate for the conservation question, we should hold present wants about twice as important as wants twelve years from now, or perhaps four or five times as important as the wants of the next generation.

The rate of discount will not, however, be the same for individuals and for the nation or group, for several reasons. In the first place, there is less uncertainty as to the life of the nation, and as to the fact of the future enjoyment. As far as discount of future wants arises from uncertainty as to the satisfaction of those wants, the rate of discount should be far lower for the nation than for the individual, in fact it should be almost negligible. In the second place, the individual may discount a future want somewhat because of the possibility that the keen edge of enjoyment will be dulled by advancing age. At any rate, there is such a probability for many kinds of satisfactions. For the nation or group, there is no such discount. In the third place, the general point of view of nations, or of individuals considering national questions, takes in a longer sweep. Fifteen years in the life of a man seems a long time, while fifteen years in the life of a nation seems a very brief span. In short, the nation will not need to discount the future at as high a rate as an individual, because the nation is more likely to live to enjoy the future, more certain of retaining undiminished capacity for enjoyment, and more accustomed to take a long view.

As far as the use of a natural resource may be termed "productive consumption," however, it means a high rate of discount, whether for the individual or the nation. If the oil consumed from 1860 to 1960 should finally appear to have brought such an increase in man's productive powers as to enable him to develop other sources of energy and mechanical efficiency as good as the oil itself, it would justify the past heavy discount of future uses of oil products. In fact, such a contingency would go far to justify the absolute disregard for the future which has always characterized the American people. At the present time, however, it seems unlikely that any such justification can ever be offered.

We need not philosophize too much, however, as to just how much we may properly discount future uses of oil. It is entirely safe to say that when we refuse to consider the wants of future generations at all, we are discounting at too high a rate. We probably have a right to prefer our thousandth joy ride to the thousandth joy ride of our grandchildren, but whether we have the right to deprive them of their only ride in order that we may indulge ourselves with two thousand such rides is another question. Certainly no one will claim that we have the right to deprive future generations of every use of petroleum, even to the indispensable lubrication for the machines that make civilized life possible, in order that we may indulge in our thousands of miles of senseless joy riding, some of it to our own absolute detriment.

Many thoughtful men have seen and pointed out the duty of conserving our resources. The following expressions of this view seem worth quoting:

The forces of greed and selfishness are so entrenched behind corporate power and influence that to attack them may often appear to you as useless as the labors of Sisyphus; but as you love your states and country, I adjure you to take up this fight for the conservation of our fuel resources with the determination never to surrender until the forces of greed and avarice which are so rapidly sapping the very foundations of our country's greatness, capitulate and agree to end the wild riot of destruction that has characterized the past. (I. C. White.)

The present generation has no moral right to destroy these resources which were not created by man or given solely to us. (John Mitchell.)

We see that the greatness of our nation, based as it is upon its industrial, its commercial, and its financial supremacy, depends absolutely upon the conservation of its natural resources. (John Hays Hammond.)

The right of the present generation to use efficiently of these resources what it actually needs, carries with it a sacred obligation not to waste this precious heritage. (J. A. Holmes, of the United States Geological Survey.)

It is our duty as American citizens to guard these resources as sacred trusts, to preserve them, and to use them wisely and with moderation. (Cardinal Gibbons.)

The earth belongs to each generation, and it is as criminal to fetter future generations with perpetual franchises, making the multitude servants to a favored faction of the population as it would be to impair, unnecessarily, the common store. . . . As the parent lives for his child as well as for himself, so the good citizen provides for the future as well as for the present. (W. J. Bryan.)

Without the resources which make labor productive, American enterprise, industry, and skill would not in the past have been able to make headway against hard conditions. Our children and their children will not be able to make headway if we leave them an impoverished country. (W. H. Taft.)

We are nationally in the position of a large family receiving a rich patrimony from thrifty parents deceased intestate. . . . Now the first duty of such a family is to take stock of its patrimony; the next to manage the assets in such a manner that none shall be wasted, that all be put to the greatest good of the living and their descendants. (Andrew Carnegie.)

A writer in the *Scientific American* is not flattering in his comments on our use of our resources:[5]

No thoughtful mind, in its contemplation of the phenomenal growth of this country, will deny that it is due largely to the over-lavish use of our resources, and that the migrations of our pioneer people, in their depletion of the natural wealth of the country, have been not unlike the flight of a swarm of locusts across a fruitful land. We have eaten up our capital with the lavish prodigality of a youthful and thoughtless heir to some rich and suddenly inherited estate.

Professor J. B. Clark discussed the question of conservation in a very able article in the *Atlantic Monthly*:[6]

There is every reason why a state should make use of forethought. A century is as nothing in its life, and yet how many acts do legislatures, congresses and parliaments pass for the benefit of coming ages? We seem willing that the earth should be largely used up in a generation or two.

Gifford Pinchot, the greatest figure in the conservation movement in America, has, of course, written much, and the following seems pertinent:

What should any of us say of a man adrift at sea in an open boat with a little barrel of water, enough water to last him, with ordinary use, for five days, and he knew that the chances of his being picked up at sea were infinitesimal, he was out of the track of ships, and he knew with the best time he could make, it would take twenty days to reach land, what should we think of that kind of a man, if under these circumstances, he not only drank all the water he wanted, but used the rest of it to wash his hands?[7]

One of the severest excoriations the extravagance of the American people has ever received, has been at the hands of the Englishman, E. Mackay Edgar:

The time . . . is coming, is, indeed, well in sight, when the United States, partly through recklessly improvident exploitation and partly through natural processes of exhaustion, will be nearing the end of some of the available stocks of raw materials on which her industrial supremacy has been largely built. . . . The size and magnificence of the American inheritance and the rapidity and wantonness with which it has been squandered are an almost incredible commentary on human folly. On no country, perhaps, had "affluent Fortune emptied all her horn" in such varied and bountiful profusion, and no country could have shown itself more utterly ungrateful. The Americans have dealt with their resources, and deal with them today, in the pioneer spirit of sheer, unmitigated pillage.[8]

Four years later, in 1923, Sir Edgar again took up his tomahawk:

In all the years I have known America, I have never been so struck as during the past two months by her prodigality. . . . You can hardly name one of the big staples of industry which the Americans are not literally devouring. One hundred and fifteen million people are feverishly tearing from the earth its irreplaceable wealth, using it to maintain a rate of growth unprecedented in all human industry.[9]

Assuming the necessity, or, at any rate, the justice, of conserving our oil resources, of using them less lavishly, there are in general two ways in which this may be brought about: first, through higher prices, brought about by the unfettered action of economic law; and second, by government intervention. Government intervention might aim at regulating uses directly, or it might aim at conserving merely by making oil more expensive.

As pointed out repeatedly in the preceding pages, oil prices have always been far too low.[10] Even in periods of what were called prosperity for the industry, oil has been much too cheap. Low prices have been and still are a constant and irresistible invitation to waste. Oil should never have been sold for 10 cents, or 50 cents, or 75 cents, or $1, or even $2 a barrel. No substitute could be made by any chemical process now known, for less than $20 a barrel, in fact, no entirely satisfactory substitute could be made at any price whatever, except from other exhaustible hydrocarbons.

Oil and gas, and some other exhaustible natural resources should, as a general principle, be sold only at a price approximating the price at which satisfactory substitutes can be produced. For example, if it be assumed that oil can be made from shales for $5 a barrel, the price of petroleum ought to be brought somewhere near $5 a barrel. That is the price it will command eventually, and the sooner it comes, the more opportunity there will be for a painless adjustment to the inevitable. A price of $5 now would immediately stimulate the development of substitutes. For some petroleum products, adequate substitutes would be found. If for some other petroleum products no adequate substitutes could be found, it would be extremely fortunate that all the petroleum was not gone before that fact was demonstrated. If, for instance, the shale oil should be found to afford adequate supplies of gasoline, but no satisfactory lubricants, it would be fortunate if that were known before all the petroleum is gone, because it would be possible to enlarge the lubricant factor in the refining of the petroleum, while getting increased supplies of some of the other products from shale oil or other sources.

On the side of demand, higher prices would have the desirable effect of reducing the amount consumed, of conserving for higher uses. They would bring consumers face to face with the situation which is coming soon anyhow. Higher prices would prevent the building of a vast industrial structure which could be only of temporary service. The American automobile industry, as it is now developing, is only a transitory industry, which can hardly prosper when our small reserves of cheap oil are gone. Higher prices of oil would have the desirable effect of stopping, or at least retarding the growth of this industry, of keeping it down to a size at which it might hope for greater permanence. Furthermore, expensive oil would hasten the development of this industry in the direction which it must some day take—the direction of greater economy in the use of fuel and lubricants. Expensive oil will some day bring much greater efficiency in automobile engines. Perhaps it will double the gasoline and lubrication mileage of our automobiles. Obviously, the sooner this comes the better.

There will be some who will say, "But what if efficient substitutes should be discovered for all the petroleum products?" This sounds like the little boy who, upon being instructed to wash his face because his grandmother was coming for a visit, replied, "But if she shouldn't happen to come, then what?" If a higher price for oil resulted in the development of substitutes, no particular harm would have been done; and if, as seems far more likely, or almost certain, substitutes for some oil products were not forthcoming, the sooner it is known the better. Thus a higher price for oil could do no great harm at worst, and might do almost immeasurable good.

A fair question might arise as to whether shale oil should be considered a substitute for petroleum, since it is, like petroleum, an exhaustible resource, or whether we should go a step farther and consider some substitute that will be available when both petroleum and shale are exhausted. The latter would make a very different problem, for it is extremely doubtful whether any inexhaustible substitutes, even for the gasoline and lubricants in the barrel of crude oil, could be made for less than $20 a barrel, or perhaps much more than that. Also, for some of the 300 products of petroleum, no substitutes will ever be available at any price. This will perhaps be true of the most important of petroleum products—the lubricants. It will not be necessary to deal too extensively in the uncertainties of this question, however. The main point is that in taking shale oil as the substitute for petroleum, we are taking the cheapest possible substitute, and one that is itself exhaustible. In demanding that petroleum should not be sold at a price much below the cost of extracting shale oil, we have taken the most moderate position that the circumstances will justify.

But the question arises, how can prices of petroleum be made higher, in order that consumption may be reduced? Prices will, it is true, go higher as our oil reserves are further depleted. When it is once evident that oil production is definitely on the decline, prices of oil products will rise, perhaps very rapidly, and will eventually reach the level set by the cost of shale oil. In fact, if shale oil fails to furnish efficient lubricants, the price of petroleum may go far above that of shale oil. In other words, we shall get higher prices, regardless of what we do. All we need to do is to wait.

This is not an intelligent solution of the problem, however. It is a little like waiting until the house has burned down before calling the fire department, because the fire is easier to put out then. When the oil

496 THE UNITED STATES OIL POLICY

is about gone the price will go up indeed, but our purpose should be to postpone the exhaustion of the oil; and here arises the necessity of government intervention. It is only through government intervention that the vast production of oil can be reduced, or that oil can be conserved for its higher and more important uses. Professor Ely has said:

Adequate conservation in general means a course of conduct in economic affairs which is dictated by the common interest, but which will not be followed sufficiently by the private person under a system of laissez-faire, or non-intervention.

Francis S. Peabody pointed to the necessity of government intervention in a somewhat different way: "I do not like government paternalism, but I think we need it to preserve what we have got in the ground that God put there, that we did not have anything to do with putting there, and which is needed for the use of all the people."[11] Mark Requa, in a speech made at the 1920 meeting of the American Petroleum Institute, pointed out that if the private oil interests did not follow the path of "efficiency, coöperation and conservation," the government would surely intervene. Requa insisted that,

Ways and means must be found to place authority in the hands of experts who can and will plan in advance and see that these plans are carried into effect sufficiently in advance of the necessity, rather than wait and attempt to cure after the patient is moribund.[12]

Government intervention should take two forms: first, government ownership of as much as possible of the oil deposits remaining in the ground; and, second, government regulation of the extraction of the oil that is privately owned. As to the first, it must be admitted that the history of the United States oil policy, as traced in the preceding pages, is a strong indictment of the policy of private ownership of oil deposits. The institution of private property has often been credited with "turning sands to gold," and, with regard to most kinds of wealth, its operation has unquestionably been beneficent. Doubtless private ownership of ordinary agricultural land has been abundantly justified; but private ownership of forest lands, coal lands, oil and gas deposits, water power sites, and even to some extent arid and swamp lands, has been unfortunate in its results. In the case of oil and gas, at least, it has resulted in feverish and wasteful exploitation which will some day be regarded as one of the worst sins of all history.

Every influence under the régime of private ownership has worked for rapid exhaustion of the oil resources. The immeasurable stupidity of the system of offset wells, with the competitive scramble of adjoining owners for the oil in the ground; the system of leases which often require exploitation within a certain number of years; the fact that the private

owner must always charge a heavy annual interest against his investment; the heavy bonding of many producers, requiring heavy production regardless of price; the insane zeal of fortune hunters who prick
the earth full of holes in their search for quick wealth; all these factors
have worked for rapid exhaustion of the oil deposits. Only in regions
which are mainly controlled by a monopoly, or by a few large owners, or
by the government, is there any possibility of due regard being given to
future interests. Private ownership is resulting in far too rapid exploitation of our oil resources, and one of the most important lessons
to be drawn from the preceding chapters is the wisdom of the policy of
government ownership of oil resources.

It has been demonstrated many times that the government is the
proper agency to hold most mineral resources, or perhaps all mineral
resources. The government is the only agency that can afford to hold
such resources indefinitely in the promotion of the general good. Private
ownership has almost always led to wasteful use of the resource, and
not infrequently to bankruptcy of the owner. Private owners of timber,
coal, oil, and gas deposits have often found interest and carrying costs
compounding at such a rate that they were forced to get their resource
into the market, regardless of price or of the waste involved in the operation. When prices have been low, use has, of course, always been wasteful. Millions of acres of timberlands have been cut over carelessly and
wastefully, by owners who did not find it profitable to own timber and
were forced to cut and rip it into some salable product, regardless of
price. Hundreds of timber owners have gone into bankruptcy while committing such waste as this. Billions of feet of gas have been wasted outright or piped to market at a few cents a thousand, either because private owners did not wish to wait until it was worth a reasonable price,
or because scattered holdings made waiting impossible. Millions of
barrels of oil have been forced into the market because operators and
lease owners could not anticipate a price increase rapid enough to justify
them in holding it. A resource as plentiful as most kinds of bituminous
coal presents almost no considerable chance for appreciation in value,
and therefore private owners have almost no incentive for saving it for
the future. The result is that they are encouraged to exploit it as rapidly
as they profitably can, without any regard for future necessities. If a
single individual had bought all the coal deposits of the United States
at the time coal was first used as fuel, for even as much as 50 cents per
acre, he would have a fabulous investment represented long before the
coal supply would approach exhaustion. Even buying at such a price,
but for current sales to meet his carrying costs, he would probably have
lost on the investment. In other words, private individuals could hardly
afford to own coal lands for any long period of time, no matter how

cheaply they had purchased. This is, of course, not true of anthracite and some of the more valuable coking coals.

The history of private ownership of oil deposits in the United States is about the best possible argument for government ownership of deposits, but perhaps some of the positive advantages of that policy may be pointed out. In the first place, if the government owned the oil deposits and leased them to operators, it could in some degree reduce the instability of the industry by intelligent regulation of the amount of drilling in times of serious over-production. This the federal government has several times done through its control of Indian lands. During the over-production of 1922, the Secretary of the Interior ordered the cessation of drilling operations on Osage leases, and no one can deny that the order was sane and beneficial. Even when the government does not wish to go so far, it can at least suspend all drilling requirements on government leases, and thus allow operators to save themselves if they can. This the government has already done in some instances, while private landowners were doing their best to bring the industry to chaos by insisting on the drilling requirements of their leases. Thus if the government owned all the oil deposits it could secure greater stability in the industry, by preventing recurrent periods of over-production. This would not only conserve oil, but capital as well.

In the second place, the government could prevent much of the waste which has disgraced the history of the oil industry. The greatest saving would result from the avoidance of over-production, but oil could be saved in other ways: by better guarding of the oil sands; by stricter regulations regarding waste in transportation and refining and in the use of oil. It could secure more thorough draining of oil sands by allowing a lower royalty on small wells and on wells nearing exhaustion; in fact, it has already moved in this direction in some leases.

The government could, in the third place, greatly reduce the amount of capital expended in the exploitation of oil resources. Under a system of government leases, there would be far less speculation, less secrecy, and therefore more effective drilling, and in many fields fewer wells. There is no reason why, in some fields, at least half the present cost of wells could not be saved. Under a system of government ownership and leases, there would never have been any such debauchery as was found at Spindletop, Burkburnett, McKeesport, Signal Hill, Santa Fé Springs, and Huntington Beach; and the saving would have amounted to hundreds of millions. The waste represented by offset wells could have been reduced or eliminated by an intelligent form of lease. The government could permit more economical grouping of leases, or even lease an entire field to one company. The Teapot Dome lease illustrates a possibility. With all the scandal surrounding the Teapot Dome lease, there is no doubt that it represented a more sane and economical arrangement than has generally been possible on private lands.

Some will deny that it is any business of the government how much capital is wasted, as long as it is private capital. As Representative Smith of California once expressed it: "I really can not bring myself to see that it is any part of the Government's business, speaking bluntly, whether I drill twice as many wells as I ought to, or whether in a given field I pump twice as much oil as the market will take." To this George Otis Smith properly replied: "I think it is the affair of a good government to try to prevent waste. We want to encourage economy."[13]

In the fourth place, government ownership of oil deposits would save a vast amount of capital and energy by reducing speculation in leases, and by reducing the number of idle landowners. Under a system of government leases, instead of a few thousand millionaires, rich by no merit of their own, we might have much-needed public improvements, better roads and schools, and education. Along with this would go healthier social conditions, less inequality, and less waste of energy by those who try to keep up the pace of the "newly-rich." If the hundreds of millions that have been poured into the laps of fortunate landowners could have been used for public improvements, it would have been better for the public and better for the landowners.

Under a system of government leases, monopoly conditions might develop, in fact to some extent they ought to be recognized as inevitable in the oil industry, but they could always be easily controlled. Even where the government grants a lease to one company covering an entire field, where it confers a monopoly of a considerable area, as at Teapot Dome, it can secure for the public, part, if not all, of what would otherwise be monopoly profit. If the government had owned all the oil deposits and leased them, the Standard Oil Company would perhaps not be so rich, but the public interests would have been better guarded.

It seems rather strange how slow the idea of government ownership of oil deposits, or of any mineral deposits, was to appear in the United States. Economists generally either ignored the question entirely, or, in a few cases, delivered a few platitudes about "enterprise" and "initiative," and let it go at that. The laissez faire policy was generally accepted without question. Thus one searches in vain through the works of Wayland, Amasa Walker, F. A. Walker, Perry, Thompson, Patton, Macvane, and Carey for any intelligent discussion of this question, or even for any discussion whatever. Newcomb, in his *Principles of Political Economy*, published in 1886, discussed the question of natural resources with considerable care, but concluded that private ownership was necessary to stimulate discovery. Newcomb's views probably represented fairly the current views of the time:

We may readily see that if every one were allowed at pleasure to avail him-self of all the raw materials of nature, they would be speedily exhausted so that none might be left for posterity. Hence it would be necessary to prescribe how much coal or iron or other minerals each individual should have. The practical difficulties in doing this would be insurmountable. But under our actual system, the care which every prudent person takes of his own prop-erty is extended by the owners of natural agents, to their property, and thus the contents of the great storehouse of nature are protected from waste.[14]

It is difficult to imagine how Newcomb could have written thus about "protection from waste," when thousands of barrels of oil were being wasted in all the fields that were being exploited.

Osborne, in his *Principles of Economics*, published in 1893, showed an unusual grasp of the public land question, although he did not refer specifically to oil:

The only practical step which presents no difficulties is to stop the sale of public lands. Not another acre should go into private ownership. No matter how poor or worthless at present, it belongs to the people; and there is no excuse for deeding away the people's heritage. The Adirondack Mountains were once thought to be worthless, and they went into private ownership. Now the people want a great park in this wilderness, and they will pay roundly for it. The reservation of such great parks as the Yellowstone is a step in the right direction; and, although privileges are given to private parties, the fact that the government ownership of the land is recognized will be a basis for their correction in the future. All public land should hereafter be leased on the condition that the occupant pays whatever it is worth each year, forever.[15]

A number of the later American economists have seen the logic of a leasing policy. Professor Ely has long been a leader in the movement for conservation, and his German training, of course, enabled him to understand the wisdom of the leasing system. Professor Seager has taken the same attitude:

Other public lands, valuable for the mineral wealth they contain . . . should be leased on terms sufficiently liberal to encourage the prospector and mine investor, and yet calculated to secure for the public treasury a proper share of these stores of natural wealth, which up to the present time have served almost exclusively to swell the fortunes of private individuals.[16]

Without doubt a majority of the American economists who have made any careful study of land problems would approve of this general posi-tion; and in most European countries this view would be even more generally held.

Government ownership of oil deposits is not a visionary and un-tried scheme. As stated previously, almost all oil-producing countries have found it wise to retain the oil still owned by the state, and some of them have even nationalized oil deposits on private lands. The drift toward nationalization of all kinds of mineral deposits has been very strong in recent years.

Unfortunately, the government has lost title to most of its oil lands, so that the lease-hold system can never be very widely applied. What lands still remain in government hands, however, should be closed absolutely to all exploitation. These reserves should be kept intact until the privately owned oil lands are largely drained, for then they will satisfy important needs. It is difficult to write or speak dispassionately of the stupidity of a policy such as the government followed in 1922 and 1923, for instance, in leasing thousands of acres of public lands, as well as the naval reserves, at a time when storage tanks were overflowing, markets flooded, prices falling almost to nothing, and oil companies bankrupt by the dozen.

Some will insist that there is an obvious inconsistency in urging government ownership of oil deposits, while thus criticizing the policy the government has pursued. There is no inconsistency, however. In spite of the fact that the government has done unwisely, it has done far better than private owners of oil lands. Had all the public oil lands and all the Indian oil lands been in private ownership, conditions would have been far worse than they were, because exploitation would have proceeded more rapidly than it did. It is difficult to imagine the situation that would have prevailed in the oil industry, had the government held no oil lands out of development; had all the naval reserves, and the public lands, and the rich oil fields of the Indian tribes, been entirely open to exploitation.

The writer is under no misapprehension as to the political feasibility of closing the reserves entirely. Practical politicians point out that the western states want "development," and that, if the public oil lands were closed entirely, so much hostility would soon develop that the reservation policy might be endangered. The fact remains, nevertheless, that the wisest public policy demands the absolute closing of the public lands to further exploitation. It is unfortunate if such a course is not politically feasible.

Indian lands should also be closed absolutely to all exploitation. The Indians are quite wealthy enough at present, and would be for some years, even if no new lands were opened. Many of them have altogether too much income for their own good, yet the government, by its present policy, is doing its best to increase their demoralization, and at the same time is wasting their oil and their future income as if they would never have any use for it.

Whether the government should buy up oil deposits now in private hands is another question. The government has been buying timberlands for more than a decade, and there is little question as to the wisdom of that policy. Very respectable authorities have favored government purchase of the anthracite deposits. The purchase of oil deposits is, however, a very different question. If satisfactory substitutes for the important petroleum products are developed, this question will not become important, but if such substitutes are not forthcoming, the question of government purchase will surely be raised.

One step in this direction should have been taken long ago. The government should have bought the lands adjacent to the naval reserves, at the time the reserves were created, in order to prevent drainage of these reserves by private wells. The history of our management of these reserves is a curious story anyhow, and one which does small credit to the intelligence of the American people. For years a constant fight to prevent wealthy oil companies from taking these reserves over entire, and finally the surrender of all of them, and the adoption of a scheme for draining them at a time when the country was literally flooded with oil, on the ground that adjoining private owners were draining the oil anyhow; yet little serious thought given to the possibility that the government might buy the adjoining lands and so save the oil until it should be needed. It is true that the Department of the Interior made repeated recommendations for the purchase of such lands, but very little attention was ever paid to these recommendations. Secretary Daniels also made such a recommendation upon at least one occasion; but, as far as the records show, there was almost absolutely no discussion of that policy, in Congress, or in the press. The government would unquestionably have the right to buy such lands, if the national defense and national welfare demanded such action. If the government owned all the lands that offered an opportunity to drain its reserves, it could store its future naval fuel in the original oil sands.

Where private lands are found to cover part of a given pool that is mainly on public lands, even though those public lands have not specifically been declared a naval reserve, the government should buy the private lands, and reserve the entire pool for future use. There should be no division of any oil pool between private interests and the government, and the government should not sell, and for the present should not lease. It is reported that Secretary Lane, in January, 1917, recommended government purchase of the oil rights on unleased Osage lands.[17]

The question as to whether it should be the federal or the state government that owns oil deposits is not free from difficulty. Many of the western men have wanted oil, as well as other resources, transferred to the individual states; but this attitude has been mainly selfish, based on the hope that the states might in some way profit from such a policy. Resolutions adopted at the third annual conference of western governors, at Denver, in April, 1914, indicate very well a common western attitude:

It is the duty of each state to make such laws as will make for conservation . . . and as rapidly as the states prepare themselves to carry out such a policy of conservation the federal government should withdraw its supervision and turn the work over to the states.[18]

There are good reasons, however, why the federal government should own most of the oil deposits, just as it owns most of the publicly owned timberlands. The important oil products are indispensable to our national life. The United States navy and the merchant marine consume vast quantities of fuel oil; the aeroplane service of the army must have gasoline and lubricants; the railroads require great quantities of lubricants, largely for interstate traffic; many great industrial plants are dependent upon efficient lubricants and other oil products, in making goods for use in every part of the United States, and of the world. In national defense, and in much of our industrial activity, there can be little recognition of state lines; and it hardly seems wise to place anything so essential to our national life and welfare under control of one or two or a half-dozen states. Furthermore, the record of the federal government in the care and management of natural resources is far better than that of most of the states, particularly western and central-western states, which hold most of our oil reserves.[19]

On the other hand, there are always some questions of constitutionality that would have to be answered before the federal government could enter upon any ambitious scheme for enlarging its holdings of oil deposits, although it unquestionably has the constitutional right to keep what it already has. The powers of the states would perhaps be less likely to be questioned on grounds of constitutionality. Some authorities have advocated state action rather than national action in the work of conservation. Elihu Root took this general position before the Conference of Governors, in May, 1908, on the ground that the federal machinery would break down if its duties were greatly expanded.

In general it may be safe to suggest that a wise policy for the future should include both federal and state ownership of oil resources. The federal government should certainly retain all that it now owns, and should buy more, at least in the neighborhood of the naval and other reserves. The states should also retain all that they now have. Both the federal and state governments should close their lands absolutely to all exploitation until private lands are largely exhausted.

The federal government should also reserve all oil shale lands that have any present or prospective value, and should seek to regain title to all that have passed into private hands. Government purchase of shale lands will seem an extremely radical policy; and there is not a possibility that such a policy will soon be undertaken. Nevertheless, if shale oil is to take the important part that many authorities predict for it during the next few decades, or perhaps centuries, it is most important that the government should have control of the sources, and immediate purchase would probably represent a wise and far-sighted policy, for which the American people would some day have reason to rejoice. Shale lands have little value at present, and perhaps will have little value for a number of years. Yet private owners will find that even a small value, compounded as fast as it will have to be compounded for private owners, will reach high figures before much of the shale is exploited. As a plain business proposition private owners cannot afford to own these lands, and the lands that they have they will be encouraged to exploit rapidly and wastefully, just as they have exploited our oil deposits. The government could far better afford to carry these shale deposits until they are needed. The government could carry them at far less cost than private owners, and could far better protect the interests of the public in the methods of exploitation. Government ownership of the oil shale—at least the shale on the public lands—will some day be recognized as desirable; and there can be no doubt that the best time for the government to enlarge its holdings is now.

Purchase of shale lands need not be hasty, even though undertaken immediately. The government might well follow a policy such as it has followed in the purchase of timberlands in the Appalachian and White mountains, buying only when it can buy at attractive prices, and condemning where necessary to round out holdings. The record of the National Forest Reservation Commission proves that it is possible for the government to purchase lands with reasonable economy and judgment. We can take little pride in our policy of giving lands away to private owners, some of whom have been mere speculators or worse, and then buying back, even at a small price; yet the fact that we started out with the unwise policy of alienation is not a sound argument for following it after the folly of it has been seen.

Where lands have definitely passed into private hands, and government ownership is no longer possible, government regulation of the oil industry seems to be inevitable, and is already here, as a matter of fact. When Adam Smith wrote of the "invisible hand" that harnessed private interest to social service, he had not seen the oil industry in the United States. The doctrine of laissez faire, or non-intervention, rests rather implicitly on the theory that private interest leads the individual to socially productive activity; and in many fields of industry the theory is generally valid. In the oil industry, however, and in the exploitation of some other kinds of resources, private interest has often been opposed to the public interest. As Professor J. B. Clark well expressed it:

In many instances, the individual wins a profit by what inflicts on the public a melancholy waste. . . . Exploitation usually makes the individual richer and the people poorer, and it nearly always gives to the individual far less than it takes from the public.[20]

J. A. Holmes, Chief of the Technologic Branch of the United States Geological Survey, put the same idea very forcibly:

So long as men are human, self interest will naturally dominate the policy and action of individuals and corporations; and these policies and actions relating primarily to the question of temporary gain are not always in accord with the best interests of the nation. To the nation alone can be entrusted the guardianship of its own future. This conservation of resources is therefore a great national problem, and a great national duty.[21]

We may give a few illustrations of this sort of thing. The man who discovered the Santa Fé Springs oil field in 1921, did the state of California a greater injury than the worst criminal in the state ever could have inflicted. He caused one of the state's richest oil fields to be exploited when the oil was almost worthless, when it brought only demoralization to producers, with no corresponding benefit to consumers. If the field could have escaped discovery for twenty years, it would doubtless have been a source of vast income and benefit to the people of the state, but in 1921 and the years following, its millions of barrels of excellent oil were almost more curse than blessing. Measured in terms of results, rather than motive, we would have to say that the discoverers of Santa Fé Springs, Long Beach, and Huntington Beach deserve to be rated among the worst enemies of California and of the American people. In terms of results, it would be hard to find men who have inflicted injuries comparable with theirs; and the "small fry" who did the town lot drilling afterward must share the responsibility for this great wrong. This sounds exaggerated, but it is only common sense. These three fields seem destined to produce several hundred million barrels of oil. Suppose we estimate the total at 300,000,000 barrels—and it may far exceed that. At from 60 cents to $1 a barrel, this oil is worth, as it is sold, less than $300,000,000, let us say $250,000,000. If geological predictions are worthy of the slightest confidence, within less than twenty years our reserves will be so depleted that oil of the quality found in these fields will be worth $5 a barrel. At that time, 300,000,000 barrels of oil would be worth $1,500,000,000—over a billion dollars more than its present value. This billion dollars would represent not only the difference in market value but the difference in want satisfaction arising from the use of the oil. In other words, if we assume that these fields would have been discovered later—a fairly safe assumption— we must say that the discoverers of these fields have robbed the people of the United States of over a billion dollars, or a billion dollars' worth of want satisfaction. It was presumably to their own interest to make the discoveries, but the public interest looked the other way, to the extent of about a billion dollars. Fundamentally, the purpose of the oil industry should be to supply the public need for oil, and not to make millionaires, or to make profits for anyone.

The same line of reasoning applies to the discovery of many other oil fields: Spindletop, Glenn, Cushing, Sunset-Midway, Burbank, Tonkawa, and dozens of others. If the exploitation of our oil resources had proceeded half as rapidly as it did, the exploiters might have posed as social benefactors, but when they drained potentially priceless oil out of the ground and allowed it to run down the creeks, or sold it for 50 cents a barrel to be burned as fuel, they lost all claim to be classed as producers.

Assuming that the average price received for the seven billion barrels of oil thus far produced in the United States was $1, and assuming that this price would have averaged not less than $5 if the oil had been brought to the surface slowly enough, we may credit the oil producers who hurried the exploitation process with causing a loss of $4 per barrel, or a total of twenty-eight billion dollars' worth of want satisfactions. This may seem fanciful, but it is in the main a sound line of reasoning. No practical oil man would ever agree with this, of course, but it is even a debatable question whether the United States might not be better off if oil had never been discovered in this country, whether we might not be in a better situation if we had been using foreign oil for our necessary purposes, and still had all our reserves in the ground.

Some business men have seen this more or less clearly. William Davis, president of the Mid-Continent Oil and Gas Association, in a speech in October, 1922, complained of the fact that men "will continue to drill so long as there is an apparent profit without regard to the general welfare of the industry." "There is something fundamentally wrong and unsound," he declared, "when it is possible for such a situation to continue in such a great industry. . . . The unnecessary production of oil when the market is glutted and the price below cost to the great mass of operators and therefore below intrinsic worth, is economic waste of one of the most valuable of the world's natural resources."[22]

For yet another reason there has been in the oil industry no clear connection between individual gains and the public interest. Much of the profits secured in the industry have been monopoly profits, and the theory of harmony does not apply to monopoly conditions. If the Standard Oil Company had had a monopoly of oil resources, instead of the transportation and refining of oil, and had made its large profits by restricting the production of crude, possibly it might have been said to receive monopoly profits from valuable social service; but its profits were not made in this way. It is probably fortunate that the Standard Oil Company did dominate the oil industry, for its influence was often good, but the billions of Standard Oil profits do not all represent payment for valuable services rendered.

Thus, in the oil industry, the basis upon which the theory of laissez faire or non-intervention rests, fails completely. Government regulation is needed as in almost no other field of business activity. Already some of the state governments have gone far in the matter of regulation. Most of the oil-producing states have regulated the manner of drilling and casing and capping, have established supervisors or agencies of some kind to see that these regulations are enforced.[23] General and specific rules for the prevention of waste have been laid down, and in several states regulation has reached the point of forbidding the completion of wells where facilities for handling the oil were so lacking that further production would involve waste. In July, 1919, the Texas Railroad Commission ordered a shutdown of the Burkburnett field for five days because production was greater than pipe-line capacity.[24] In January, 1923, a curtailment of the new Robberson field in Oklahoma was ordered, because there was no way to care for the oil.[25] Each year sees further legislation enacted bringing closer regulation of the oil industry.

In the federal government there has also been some agitation for increasing government control. Secretary Josephus Daniels, of the Navy Department, favored increased government regulation, in the interests of a future oil supply for the navy. La Follette, as already indicated, favored a great deal of government regulation and control, and several men in Congress have favored government exploitation of government lands.[26]

No state has gone far enough, however. The absolutely indispensable character of many of the uses of petroleum and petroleum products renders the industry essentially a public utility, and it should be recognized as such. During the past decade or so, this has been understood by many people in California, and in almost every session of the California legislature a fight is waged on the question of making the oil industry a public utility. In 1915 such a bill, the Carr Oil Regulation Bill, was defeated in the state senate by a vote of 21 to 12. Some of these earlier bills were supported even by some of the small oil operators, who found conditions in the industry so unfavorable that they apparently thought state control could at least make matters no worse. Their situation and their attitude seemed similar to that of some of the lumbermen at one time, when they found their competitive industry so unprofitable that they even welcomed government control or ownership. Early in the year 1923, with the vast over-production and the increasingly ominous conditions in the California oil industry, a strong effort was made in the legislature to declare the oil industry a public utility, some of the leaders in the movement being men from the oil regions of Los Angeles. The *California Oil World* seemed to feel that there was danger of success, for it called upon "landowners who have been enriched by the wonderful discoveries on their lands, contractors who are prospering through the development of the fields, the thousands of workmen who will be thrown out of work by any curtailment of activities certain to follow political regulation, refiners whose business is prospering because of plentiful supplies of moderate priced oil, stock and unit holders who rely upon their speculation for returns," and upon "every automobilist, every railroad official," to fight this "criminally foolish attack" upon "California's greatest industry."[27]

Elsewhere than in California, the same idea has taken root. The Public Service Commission of Louisiana recently issued citations to the officials of the Standard Oil Company of Louisiana to appear before the board and show cause why the company's refinery at Baton Rouge should not be adjudged a public utility subject to the regulation and control of the commission. In Texas there have been signs of a similar attitude. In both states, however, this attitude had grown up, not altogether from an appreciation of the great wastes in the industry as privately carried on, or of the public importance of oil products, but partly from opposition to the monopolistic conditions within the industry, and especially from a public belief that gasoline prices were too high.

Most of the big oil men oppose any further government control, just as business men in general do. The wealthy oil men have become wealthy under a system of government non-intervention, and naturally they favor that system. For them, personally, the "best government is the least government." In their argument against government interference, these men often point out the marvellously rapid growth of the oil industry, "growth" being measured by the rapidity with which the oil is exploited and consumed. A. C. Bedford, of the Standard Oil Company of New Jersey, compares the rapid growth of the oil industry with the slow development of the railroads, to show how government regulation retards development. Judge Beaty has often spoken of the retarding influence of government interference. R. W. Stewart, of the Standard Oil Company of Indiana, has argued a great deal against government interference, in fact some of the influential oil men seem to have been making speaking tours with the express purpose of trying to ward off government control. Thomas O'Donnell has been one of the most indefatigable opponents of government regulation, but most influential oil men are opposed to "bolshevism and socialism."[28]

In several respects the arguments of some of these men do not seem well grounded. In the first place, it is perhaps unnecessary to point out that in citing the phenomenal growth of our oil production, and consumption, these men have been using the very best possible argument against unregulated private exploitation, and in favor of government control. It would have been most fortunate if our oil production could have grown less rapidly. In another argument they are hardly consistent. For instance, Bedford, while crying out that "the world is suffering from an excess of state activity," yet advocated increased government protection for oil companies exploiting foreign fields. Certainly no form of government interference has in the past been fraught with more dangers than the form Bedford calls for here.

The time-worn argument as to government inefficiency has, of course, been urged, and it will be admitted that the government has in the past done badly enough. No government representative of the people can generally be much more intelligent than the people themselves; in fact, in the handling of some oil problems, the government has sometimes been less intelligent than the average voter would have been, because of the influence of the practical oil men—the latter generally about the least intelligent of all citizens, as far as the broader economic problems of the oil industry are concerned.

This last statement, will, of course, seem the rankest lunacy to many business men, who are prone to regard themselves as best qualified to speak concerning all aspects of business, but it will probably be justified by the logic of future developments. Most business men, big and little, understand their own private business very well indeed, but their inability to understand the broad public aspects of their business has been noted by almost all students of economics. For the railroads, banks, and for big business in general, intelligent public policies have not come from those in the business, and it is not likely that constructive policies for the oil industry will ever be offered by the business leaders in the industry. Some of the prominent men in the oil business—such men as Requa, Marland, Ball, and a few of the Standard Oil group—are broad-gauge men with a generous and intelligent appreciation of the public aspects of their business, but such men are greatly outnumbered by the narrow and reactionary members who see their business only from the point of view of the stockholders.

An element of reasonableness in the views of the opponents of government intervention must not be overlooked. Most oil men realize that the public is suspicious and generally hostile in its attitude toward the oil industry, and they fear that public intervention would inevitably be negative and hostile, rather than constructive and helpful in its spirit. On the other hand, it must be conceded that the public is in some measure justified in its suspicions of the industry; and it is the writer's firm belief that the uncompromising position against the assertion of the public rights in the industry is neither wise nor expedient. Public regulation will only "strike the harder" where such an attitude is taken.

So great is the divergence between private interest and public interest in this industry that even inefficient government control would almost certainly be beneficial. Furthermore, government control has already proved highly beneficial, in some ways. Intelligent oil operators recognize the wisdom of government regulations regarding drilling, casing, capping, and plugging of oil wells, regarding waste of oil, and the like. Finally, government control would gradually become more efficient and more intelligent just as it has done in dozens of kinds of government service. For instance, the management of the national forests was undeniably bad for some years; but eventually the Forest Service became an efficient government agency, able to take care of forest lands more advantageously than private agencies. In the same way the government has gradually developed intelligent control of the railroads, and, while there is still much to be desired, hardly anyone would suggest that this control should be abandoned or relaxed. There is little doubt that in its regulation of the oil industry the government will gradually develop such intelligence and efficiency as the people really want.

The greatest need for public intervention is in the producing end of the business. As George Otis Smith expressed it:

The petroleum industry—present and past—is best described as a frenzy of exploitation. A blind faith that demand will overtake supply, that prices will at least equal costs, that profits will reward perseverance, is the only trace of sanity underneath the delirium that is everywhere seen in the hunt for oil. . . . It is all too evident that the oil business is travelling "in high" with the gear shift locked.[29]

Many oil men as well as other students of oil problems have seen the essential absurdity of a system by which oil is sometimes produced with almost absolutely no regard for the demands of the market; and not a few have urged that the government should undertake to regulate production in some way. Marland has urged such a policy, although he ventured the hope that the formulation of the details of the policy should be in the hands of practical oil men.[30] Requa has seen clearly that if the men in the industry are not able to set their house in order, the government will try to do it. As he expressed it not long ago: "I believe most emphatically that the weight of economic pressure will sooner or later force upon the petroleum industry, either willingly or unwillingly, industrial effort of a type not now in evidence."[31] On the question of the regulation of drilling, Requa expressed himself rather definitely: "It may almost be laid down as axiomatic that there can be no satisfactory condition in the petroleum-producing industry until there is some intelligent coördination between the running of the drill and the requirements of the market." Requa admitted that he found himself "somewhat isolated" because of his views—another proof of the narrowness of the average oil man's views, because Requa's views are merely those of a man who sees his business from the point of view of public interest as well as that of private profits.[32]

Some authorities have advocated the regulation of wildcatting as one way of regulating production. J. O. Lewis has advanced the opinion that this would be perhaps the best way to prevent over-production and waste; and H. W. Bell, of the Louisiana Department of Conservation, has also urged it. Henry L. Doherty recently stirred up a row in oildom by suggesting that drilling should not be allowed without a permit from the government, and that it should be restricted to "exploration districts," within which it would be possible for operators to manage their affairs coöperatively, and thus avoid expensive duplication of effort.[33]

On the question of government regulation of production, it is reported that there has been a division of camp in the American Petroleum Institute, with a considerable party, including some of the Standard Oil group, favoring such a policy. It has been rumored that certain members of the Standard group, including Walter Teagle, are decidedly advanced in their views as to the relation of the government to the oil industry.[34]

The "Wall Street" correspondent of *Petroleum Age* saw somewhat the same light, even if not quite so clearly, when he called for a "dictator" for the oil industry. Admitting that the oil producers were a "poor lot of business men," and that "if any other industry were run like the producing end of the oil industry, it would be on the rocks in no time," this man, John Warren, declared that, "If the industry is to put its house in order and keep it in order, it will have to go further than temporary makeshifts and create some permanent agency for the enforcement of stabilization." Warren pointed to the war-time dictatorship of Mark Requa as Oil Administrator, and of A. C. Bedford as chairman of the Petroleum War Committee, as representing an intelligent system of control for the industry. Of course such a system must be under government sanction and supervision, and amounts to a recognition of the fact that the oil industry is a public utility.[35]

Regulation of production should be undertaken with the purpose of preventing over-production, in fact it really should accomplish a permanent reduction of production. The government should not stop with this, however. The government might accomplish much in other ways, in reducing waste of oil, of capital, and of human energy.

Government interference must not take merely the form of negative prohibitions and restrictive regulations. It must be constructive and helpful as well. One of the important functions of the government must be the development of technical knowledge, through research and experimentation. This sort of work is already being done by the Bureau of Mines, and by state agencies, in coöperation with the Bureau of Mines, and the amount of such work should be increased.[36]

As already suggested in another chapter, the question of monopoly is not the most important question relating to the oil industry, and a future policy which looks mainly to the regulation of monopoly may please the American voters, but it can accomplish little real good, because there is not too much monopoly control in the oil industry.

It is unfortunate that the people of the United States could not have exploited their resources in such a way as to avoid all danger of private monopoly, but since they chose the policy of private ownership, it is probably fortunate that monopoly has developed, and perhaps even unfortunate that the monopoly was not more absolute than it was. As already pointed out in a preceding chapter, the oil industry is in many ways a natural monopoly, and should have been recognized as such.

Here is to be found one of the worst faults of the Leasing Law of 1920. The Leasing Law is based on the theory that the oil industry is a competitive industry, and that theory is not sound. No efficient and economical oil industry can be built up on the basis of 640-acre leases or even on the basis of 2,560-acre leases, such as may be granted under certain circumstances. The law of 1920 should be amended to permit the leasing of larger tracts on the public domain. Foreign governments have gone far in this direction, and the United States will find it advantageous to follow.

If we recognize large scale operation and even a measure of monopoly in the oil industry as natural, inevitable, and even desirable, it might seem to follow that the government should establish some sort of regulative agency to protect the consumers from extortionate prices. In most industries where competition is superseded by monopoly, as in railroad transportation and public utilities, government regulation of prices is deemed necessary as a matter of course.

In the oil industry, however, and in all industries engaged in exploiting quickly exhaustible natural resources, there is no danger that prices to consumers will be too high. On the contrary, there is almost a certainty that, in the immediate future, prices will always be too low. Preposterous as it may seem at first blush, it is probably true that, even if all the oil in the United States were owned by an absolute monopoly, entirely free of public control, prices to consumers would be fixed lower than the long-run interests of the public would justify. Even an absolute monopoly would be almost certain to fix prices too low; and, of course, there is nothing approaching absolute monopoly in the oil industry.

Exactly why prices would be fixed too low, and exactly what is meant by saying that prices would be too low, are questions which will demand a little careful consideration.

A furniture dealer who has decided to quit business, to dispose of his stock of goods and retire, might follow various plans in the disposition of his stock. He might hold an auction sale, and sell out in a day, at whatever prices the bidders were inclined to pay; he might put somewhat higher prices on his furniture and sell out in a week; or he might ask still higher prices and sell out in a month. The time it will take to dispose of his stock will depend absolutely upon the prices he asks.

The United States, with her oil supply, at least as far as it is known and accessible, is in the position of this merchant; and the question as to what price oil should command resolves itself into the question as to how long the oil should last. The higher the price, the longer it will last. Or, to put it another way, the higher the present price, the more will be left for the future, and the lower will be the future price. Thus the present price determines the future price, and is necessarily inverse to it. (In using the word "price" here, the writer refers not to the market price of crude oil, or of refined products, but to the value of oil as a resource in the ground, just as nature affords it.)

As pointed out above, the price of oil should be approximately equal to the cost of producing adequate and satisfactory substitutes, if substitutes are assumed to be a reasonable possibility. Now it is almost certain that neither competitive private owners of the oil resource nor even an absolute monopoly, would fix a price so high.

Where an exhaustible natural resource, as for instance oil, is in the hands of competing sellers, it is practically certain that the price will be put too low, and the resource exploited too fast. Such a resource, in the hands of private owners, is practically certain to be exploited rapidly and wastefully and thrown upon the market almost irrespective of price, because its "resource value"—if we may use that term to indicate its value as a resource, just as nature affords it—is generally extremely low. Private owners of such a resource can afford to hold it for future uses only if the resource value promises to double about every ten years, because they must count interest and taxes and perhaps some other expenses—enough to double the investment about every ten or twelve years. Speculation of this kind has often proved unprofitable in the past, and it is seldom attractive to private capital. Therefore, owners of any exhaustible resource usually try to cash in as quickly as possible, with little regard for future demands. It is for this reason that oil and natural gas, timber and anthracite, to say nothing of bituminous coal, have usually been sold very cheaply, and have therefore been used very wastefully. Oil resources in the hands of competing private owners are certain to be sold very cheaply.

Even if the oil resource were assumed to be in the absolute control of a monopoly—a condition quite contrary to fact—the oil would almost certainly be sold below the price fixed by the cost of production of substitutes. It is an elementary economic proposition that monopoly price will be fixed at that point which will yield the maximum net profit. In fixing this price, however, the monopoly is subject to the general law of supply and demand in all respects. The monopoly can regulate prices in only one way, and that is by regulating the amount offered for sale. It can raise prices only by restricting the amount offered. This is an elementary principle in economics, yet it has frequently been ignored by practical business men as well as by others. Lumbermen's associations, for instance, have often tried to raise prices by issuing price lists which set artificially high prices, without restricting the output, but in the long run they have invariably failed. Where they have been able to curtail output, they have been able to raise prices without the use of price lists.

In short, an absolute monopoly could raise the price of oil only by reducing the amount offered for sale, or, in other words, by conserving. High prices are the best means, in fact the only means, of conserving our oil resources, unless we wish to adopt a drastic scheme of government rationing. Therefore, what possible objections can there be to high monopoly prices?

It is difficult to say exactly what price a monopoly would fix, if left entirely free of public regulation, but it is clear that much would depend upon the elasticity of the demand for oil. The monopoly would give little attention to the distant future, because incomes accruing in the distant future, even to a monopoly, would have to be heavily discounted. With an interest rate of 5 per cent, an income of $100 accruing 100 years from now is worth only .76 cent today. With an interest rate of 6 per cent, an income of $100 accruing 50 years from now has a present worth of less than $5.42. Even a monopoly would have a strong preference for present incomes over future incomes, which means that it would fix its price with little regard for the future. It is unlikely that it would fix a price high enough to stimulate the use of substitutes anyhow, and therefore unlikely that it would fix a price as high as the public interests demand. Now there is nothing resembling an absolute monopoly in the ownership of oil resources, although a few individual fields are controlled by single companies; and there is not the slightest danger that the price of oil, or of any oil products, will be raised too high. The efforts of public officials to force lower prices of gasoline and other oil products are almost certainly contrary to the long-run public interests. Public officials should instead be trying to force prices higher.

The same reasoning applies to several other natural resources: timber, natural gas, and anthracite. Attorneys-General of the United States and of many of the states have spent much time and energy in efforts to prevent lumbermen from fixing too high prices for lumber, but there seems to be no reason why lumbermen should not fix prices as high as they wish. The higher present prices are fixed, the less lumber will be consumed, the sooner substitutes, in the form of other building materials and in the form of planted timber, will be brought into use, and the lower future prices will be. With all limited natural resources, present prices determine future prices, by determining how much shall be left for the future; and if lumber monopolists should fix a high price for lumber, they would merely be conserving our timber resources for the future— exactly what conservationists should want. They could conserve it in no other way than by raising prices. In the case of timber as in the case of oil, it is likely that we are discounting the satisfactions of future generations at a high enough rate.

In relation to natural gas this principle stands out even more clearly. If all the natural gas resources of the United States had been turned over to a single monopoly when gas was first discovered, or at any time thereafter, with absolutely no restrictions as to the prices it might charge, it is almost certain that the monopoly would have fixed prices too low, that is, lower than would have yielded the maximum social utility. It is doubtful if the natural gas heretofore exploited in the United States has yielded one-tenth the utility that it would have yielded under an intelligent regulation of its use; and the main reason for this excessive waste of utility has been the low price at which it has been sold. Two-cent gas has of course been used for two-cent wants; and nothing could have been more fortunate than higher prices, which would have meant the satisfaction of more important wants. At the present time many cities are fighting against higher prices for natural gas, when they should be demanding higher prices, approximately equal to the prices at which a satisfactory substitute could be produced. Competition between cities would complicate this question, but the principle stands, nevertheless, that consumers in general should wish the price of natural gas to be nearly as high as the price of a satisfactory substitute. No monopoly would ever put the price higher than that, or probably even so high.

Anthracite is another limited natural resource which is probably selling too cheaply, in spite of the fact that it is controlled by a monopoly. It is estimated that the supply would last only a few decades at the present rate of consumption, yet government agencies have made strong efforts to prevent the anthracite monopoly from extending the life of these deposits by raising prices. Higher prices would tend to cut down the consumption, stimulate the use of other coal in place of anthracite, and so preserve this mineral for only the highest or most important uses, and thus increase the total satisfactions arising from its consumption. The anthracite monopoly, with an investment in this resource which must be constantly compounded at about 6 per cent, is most unlikely to reserve too much of it for the future—by charging too high a price.

Bituminous coal does not come under the category of quickly exhaustible resources, for the supply is sufficient, it is estimated, to last several hundred years, at the present rate of consumption. Furthermore, the demand for bituminous coal is relatively inelastic. For this reason, an increase in present prices would not have so much influence in encouraging conservation as in the case of the other minerals. Nevertheless, if the people of the United States wish to look ahead a few centuries, they may well favor higher prices for coal, since higher prices would encourage some economies and would stimulate the use of substitutes, particularly the use of water power in the generation of electricity.

Thus all efforts to keep prices of gasoline and other oil products low are contrary to the public interest, because as far as they are successful, they merely hasten the rapid depletion of our limited reserves of oil. A corollary to this is that oil operators should be permitted and even encouraged to organize in any way that they find advantageous. In particular, they should at times be permitted, encouraged, perhaps compelled, to organize to restrict production. Government opposition to curtailment campaigns have been as unfortunate as government opposition to high gasoline prices, and for exactly the same reason. When operators try to conserve our oil resources, either by limiting production or by fixing high prices—these are exactly the same in effect—they should be encouraged, and not prosecuted.

It will be objected that if the oil companies are permitted to charge whatever they please, they will make exorbitant and unjust profits. The answer is that the question of their profits does not alter the principle stated. Regardless of the rate of profit they might earn, the long-run interests of the public demand that their prices should be high, since high prices are the only means—aside from rationing—by which consumption can be reduced and the oil supply be conserved. Furthermore, it is entirely possible to appropriate for the public a fair share of the excess profits, through gross production taxes and gasoline taxes, and, as far as excessive profits would be reflected in increased increments to landowners, through taxes on unearned incomes.

Taxes on crude oil production, now sometimes called "severance" taxes, or "gross production taxes," date from the time of the Civil War, when as a war measure, a tax was imposed on crude production. After the war was over, however, the tax was soon repealed, and did not reappear until recently. As early as 1910, Louisiana passed a law imposing a small tax on the production of oil and gas—two-fifths of a cent per barrel on oil, and one-fifth of a cent per thousand feet on gas—the proceeds to be paid into a "conservation fund," to be used to conserve the natural resources of the state. This tax was later raised to 2 per cent on oil, and in January, 1923, was raised to 3 per cent. The Louisiana law includes a severance tax, not only on oil and gas, but also on timber, sulphur, and salt. It produces nearly two million dollars a year, enough to put the state educational institutions on an entirely new footing.

In 1910 Oklahoma passed a law imposing a gross production tax of one-half of 1 per cent on the gross value of oil produced. In 1916 a 3 per cent gross production tax on oil was imposed. This was said to be favored even by the producers, because at the same time some other taxes on the oil industry were abolished and thus the system was simplified. A federal court presently knocked out the tax as far as it applied to the production from Indian lands, and a state court declared it unconstitutional. On appeal, the United States Supreme Court declared it unconstitutional as far as it applied to the production from Indian lands, and, of course, this cuts down the revenue considerably, but within recent years the tax has paid a large share of the expenses of the state. In 1923 the total revenues from oil and gas were $8,012,659.99 —somewhat less than one-half the total expenses of the state.

Texas imposed a "gross receipts" tax in 1907, of $1\frac{1}{2}$ per cent of the market value of the oil produced. This was fought hard by the oil companies; but it was later raised to 2 per cent. The revenues of the state under this law amount to more than $3,000,000 a year.

Early in 1923 Arkansas fell in line, with a $2\frac{1}{2}$ per cent "severance" tax. In March, 1923, it was estimated that this would bring the state a revenue of about $100,000 a month. Kentucky passed a law in 1917, imposing a tax of 1 per cent of the market value on oil produced in the state.

There have been a number of efforts to secure such a measure in Kansas, but thus far without success. In January, 1921, Governor Allen announced his approval of such a measure, and a bill levying a 2 per cent tax was introduced, but was promptly killed by the legislature. A similar measure introduced early in the 1923 legislature was unfavorably reported by the oil and gas committees of both the house and the senate. The oil men have been rather strong in the Kansas legislature.

Similar laws have been proposed in West Virginia, Pennsylvania,

Illinois, Montana, and California, but in none of these states have such laws been passed.

The production tax has been used for other resources than oil. As already stated, Louisiana applies it to several minerals, as well as to timber. It is generally conceded that timber should be taxed mainly in this way, rather than through a property tax on the resource itself, and a number of states have adopted it for the taxation of timber. Pennsylvania has a production tax on anthracite, and Minnesota has a tax of 6 per cent of the value of ores mined. A great many foreign countries have levied gross production taxes on oil, and the number of countries doing this is increasing.

The severance tax has been strongly opposed by most oil operators. One oil attorney has described it as "a tax on industry, a tax on thrift, a tax on energy, a tax on discovery."[37] The principle underlying it seems sound, however. The United States is using her oil far too rapidly, and a severance tax would have the desirable effect of discouraging production, thus conserving the oil supply for the future, and at the same time would yield needed revenue. No tax should probably be regarded as a blessing, but a severance tax would yield more revenue in proportion to the burden it imposes than most taxes that are levied in the United States.

This tax is sound in principle, in that it is levied upon the exploitation, rather than upon the mere ownership of the resource. It has long been recognized that the heavy taxation of standing timber was unwise, because it tended to hasten the exploitation of the timber; and most of the authorities on the subject have urged the substitution of a tax on the timber cut. In the same way, it seems reasonably clear that such taxes as are imposed upon the oil industry should be levied upon the production of crude oil, and not upon oil lands, no matter how valuable. Owners of proved oil lands should be encouraged to hold them undrilled as long as possible, and a severance tax on production, with exemption of oil companies from all taxes upon their lands, and perhaps from some other taxes, would put the premium on conserving, rather than on exploiting, the resource.

Even a light tax would produce a large revenue. A federal tax of 10 cents a barrel on the present production of 700,000,000 barrels would yield $70,000,000 a year. A state tax of 10 cents a barrel would give the state of California nearly $25,000,000 a year, at the present rate of production, Oklahoma about $16,000,000, Texas somewhat less, Wyoming nearly $5,000,000, and Kansas and Louisiana nearly $3,000,-000 each.

The severance tax, with the gasoline tax (which is discussed below), should be sufficiently high to raise the price of oil products to something approximating the cost of production of satisfactory substitutes. For the present, shale oil may be regarded as a substitute, although it is also exhaustible. If shale oil of satisfactory quality can be produced for $5 per barrel, the severance tax and the gasoline tax should together be high enough to raise the price of oil to nearly $5 a barrel. Such taxation could not be levied suddenly and at once, to be sure, but it is the ideal toward which we should work.

Taxes as high as suggested here would bring immense revenues, at the same time that they encouraged conservation. A severance tax of $1 a barrel—not too high a tax under present conditions—would yield, at the present rate of production, a total of over $700,000,000: $250,-000,000 in California, $160,000,000 in Oklahoma, $130,000,000 in Texas, $44,000,000 in Wyoming, $36,000,000 in Arkansas, and smaller amounts in various other states. It is true that such a tax would soon reduce production considerably; but that would only make it a double blessing.

In some ways, it is rather surprising that states have not used the severance tax more. It is possible that if Oklahoma had handled her resources intelligently, she could not only have paid all state expenses, but could have built up a fund sufficient to pay all state expenses for the immediate future. In other words, Oklahoma, by the use of reasonable intelligence in handling her resource, might have made herself a taxless state, as far as state expenses are concerned; and the same might be said of several important oil-producing states: California, Texas, and Pennsylvania, and perhaps several others. Impossible as it may seem at first blush, this might have been done without any great cost even to the people of other states. They would have had to pay higher prices at first, of course, but in the long run would have gained from the fact that the tax would have conserved the supply for more important wants.

The state levying a severance tax is able to make the people in other states pay part of her taxes. Oklahoma, for instance, by means of her severance tax, assesses part of her expenses upon consumers of Oklahoma oil products, in Kansas, Nebraska, Iowa, Missouri, Illinois, Arkansas, and other states. To some extent, as far as oil products are exported, she makes even consumers in foreign countries pay a part. Of course the products of one state must meet the products of other states in the market, at least to some extent, and a state levying too high a severance tax would prevent its products from selling in the market, but, in the case of a resource so limited in amount as oil, there would be nothing unfortunate about that, for all states producing oil at the present time are poorer for every barrel that they produce. There is probably no oil-producing state that would not be in a better situation if it had never sold a barrel of oil, provided the other states had insisted on producing as they have done. The states that produce oil at present prices or any prices that have ever been paid in the past fifty years, are guilty either of extraordinary stupidity or extraordinary philanthropy.

It is true that the widespread use of severance taxes might lead to much trouble between states. If Minnesota taxed her iron ore production, Utah and Arizona their copper, West Virginia her natural gas, Pennsylvania her anthracite, and perhaps also her coking and other bituminous coal, California and Oklahoma their oil, Oklahoma her zinc, Washington and Oregon their timber, Arkansas her bauxite, Colorado her radium and tungsten, Missouri her lead, Florida her phosphate; if each state taxed the exploitation of her own peculiar resources, there would be more or less quarrelling among the states, and in the long run no particular success for any of them in making the other states pay their taxes. It would perhaps result a little like the efforts of nations to get other nations to pay their tariffs. In the long run any state gaining a marked advantage would presently have it taken away from her by an influx of people from the less favored states. If, for instance, Oklahoma had chosen to become a taxless state, it would not be long until an influx of people from less favored states would equalize the advantages, at least to some extent.

Aside from such trouble as might arise among the states, the general effect of severance taxes, on quickly exhaustible resources, would be excellent, because they would encourage conservation. That is the great need, at least in the case of oil, to slow down the present production in the interests of the future. In this connection, we cannot but regret that more of the oil is not federally owned and controlled, so that it could be conserved without danger of interstate difficulties.

President Obregon, of Mexico, has seen the logic of the severance tax clearly. When urged to lower the export tax—which operates the same as a severance tax, in a country which exports most of its production— he refused, on the ground that the tax was good for Mexico and good for the companies operating there. Obregon pointed out that oil had always been too cheap, and that the tax raises prices, thus conserving the oil, to the advantage of Mexico and of the oil operators.[38]

A flat tax on production, like the Louisiana tax, would have some unfortunate consequences. It would tend to discourage production, and would of course tend to cut off first of all the least profitable production, from wells which were about exhausted, small producers, "marginal wells," as economists would term them. Any tax which would fall on the "marginal" wells would tend to discourage complete extraction of the oil. For this reason the tax should be graduated according to the production of the wells, or at any rate should exempt the small producers.

It would be theoretically sound to use some of the receipts from taxes on the large producers to pay a bonus on the "marginal" wells, but in practice this might not be feasible.

A tax on all new wells drilled would have the highly desirable effect of reducing the number of new wells, and so of reducing production, just as a gross production tax has. It has one disadvantage, however, as compared with the latter, in that it would have to be paid regardless of the production of the well, and so is not proportionate to the "ability" of the taxpayer, as the gross production tax is.

It is clear, as already pointed out, that if additional taxes were levied upon oil production, or upon wells drilled, oil-producing companies should be relieved of many other taxes which they now pay, especially from all resource taxes. The purpose should be to conserve the oil, and not to burden any particular form of industry unjustly. It is obvious, also, that any such changes as are here proposed should be introduced gradually, in order to avoid undue disorganization of industry.

More clearly desirable than even the severance tax, would be state appropriation of a share of the "windfalls" going to landowners, through a tax on unearned incomes. Some states are adopting income taxes, and in income tax laws it should be possible to take at least a share of such unearned incomes as accrue in the future. Tax experts emphasize the difficulty of making any distinction between earned and unearned incomes, but it surely seems desirable that a distinction be made. Incomes accruing from the accidental ownership of oil lands certainly stand in a different position from incomes earned by productive service. A tax on the millions received thus as windfalls would be no discouragement to enterprise, and no injustice; in fact, it would be nothing less than mere justice. It is certainly unfortunate if there is no practicable way of appropriating these incomes. The federal income tax law should long ago have recognized the difference between earned and unearned incomes.

The question of constitutionality of these various proposals will not be considered here. Many of our states are closely restricted by their constitutions in their efforts to change their tax systems; but there is no reason why states should not change their constitutions, if they wish to do so. Many of them will have to do so before they can adopt enlightened taxation measures.

The gasoline tax has been much more popular than the severance tax, and at the present time 44 states have such a tax. The rate ranges from 1 cent to 5 cents per gallon on gasoline, and more than that on lubricating oils. South Carolina levies 5 cents on gasoline; four states levy 4 cents; thirteen states levy 3 cents. Two cents is the most common rate; but there is a constant movement upward in the average rate. The total revenue from this source amounted to over $11,000,000 in 1922; the next year it increased to $35,000,000; and in 1924 the tax yielded more than $78,000,000. The purpose of such a tax has generally been that of providing a fund for highway maintenance and improvement, although other uses for the money have sometimes been suggested. At the time a federal bonus for the soldiers was under consideration, for instance, there was some talk of financing it by means of a tax on gasoline, and in at least one or two of the states it has been considered for the same purpose.[39]

Much may be said in favor of a heavy gasoline tax. It apportions the expense of road maintenance fairly according to the use of the roads, gasoline consumption representing road use with reasonable accuracy; it throws upon the transient motorists their just share of the burden of road maintenance, an important consideration in communities much frequented by tourists; it is reasonably easy to administer; and, finally, if heavy enough, it would have the effect of conserving the oil supply. Nevertheless, this tax has been fought relentlessly by the oil interests. Oil companies, automobile clubs, and associations, and even many of the farmers and consumers have been opposed to it.

The arguments against the gasoline tax have been summarized by Harry Meixell, Secretary of the New York Motor Vehicle Conference Committee, as follows:

1. The administration of taxes of 1 cent per gallon is too expensive in proportion to income.

2. The tax involves discrimination against the internal combustion engine, as no taxes are put on vehicles propelled by steam or electricity.

3. It is a tax on industrial progress, inasmuch as it is impossible to discriminate between pleasure and business cars.

4. Experience shows that motorists go into the tax-free states to buy gasoline and come back to use it.

5. It is a super-tax, because in none of the 17 states (now 44 states) with gasoline taxes has any other form of tax been abolished.[40]

It is not at all certain that the opposition to gasoline taxes has been intelligent. It is possible that such a tax, economically spent for roads, would be beneficial to the oil industry and to most owners of automobiles. Improvement in roads always increases travel. In some instances, the construction of hard surfaced roads has immediately caused a tremendous increase in the number of cars used, and in the mileage per car, with resulting stimulation of the oil business. At the same time the condition of the roads largely determines the wear and tear on automobiles, and automobile owners could often afford to pay heavily to secure good roads. The writer has driven many miles on roads that cost him in wear and tear more than a gasoline tax of one dollar a gallon would have amounted to, and has driven thousands of miles where he would gladly have paid a tax of 10 cents or more per gallon for the privilege of using good roads. Gasoline at almost any price is cheaper than bad roads. Some automobile owners seem to realize this. In California the automobile clubs lined up against the tax when it was proposed in 1921, but two years later they declared in favor of the tax, as did also the Los Angeles Chamber of Commerce. Of course, from the point of view of conserving gasoline, the only thing to be said for good roads is that they require less gasoline per mile than bad roads. Their general influence is to increase consumption.

The gasoline tax should be far heavier than has been imposed in any state—at least 10 cents a gallon, and perhaps more. At this rate, the tax would make gasoline pay more nearly its fair share of road maintenance. The total expenditure for roads in the United States is now more than one billion dollars a year, and gasoline is paying only 8 per cent of this—obviously too little. Below a certain point, perhaps, this tax does not have any tendency to reduce consumption at all, if the revenues are used directly for road construction and improvement; but if it were put high enough, it would certainly have such an effect.

As already suggested, this tax, together with the gross production or severance tax, should be high enough to bring the price of oil products up to a point approximating the cost of producing substitutes, where substitutes are available, even if some of the revenues have to be used for other purposes than road building. These two taxes—the severance tax and the gasoline tax—would supplement each other excellently, the severance tax being paid at the point of production, and the gasoline tax at the point of consumption. Since the severance tax involves danger of trouble among states, the gasoline tax should probably be the heavier. Perhaps a severance tax of $1 a barrel, and a gasoline tax of between 10 cents and 20 cents a gallon, would represent an ideal, not for immediate application, but to be approached as rapidly as possible without too much business disorganization.

A complete scheme of taxation for the oil industry then, drawn with the purpose of conserving our oil resources, and to some extent human and capital resources, would include the following: a heavy tax on crude production, graduated according to the flow of the wells; a heavy unearned income tax upon royalty receipts; and a heavy gasoline tax, to be considered as a supplement to the production tax. With such a schedule of taxation, it is clear that oil producers should be exempted from most other taxation, and refiners should be permitted to charge whatever they wish to for gasoline.

If it proves impossible to secure reasonable conservation of our oil resources by higher prices, and if adequate substitutes are not found, it is fairly certain that the government will eventually resort to some system of rationing oil and its products. We have all grades of wants or uses for oil, ranging from the most essential down to the most frivolous and unimportant, and the latter will probably have to be eliminated, if necessary, by direct government action.

The rationing of oil or oil products would not be anything entirely new or untried. As already suggested, practically all the countries engaged in the Great War found it necessary to ration gasoline. In Italy, for instance, automobiles were restricted very closely to war uses, and "women, children and dogs" were forbidden to ride at all, the theory being that they could not be engaged in war work. A rather amusing story is told of the American Consul General's invitation to one of the American Consuls and his wife to dine with him, and of his embarrassment on finding that the taxi driver refused to allow the wife of his guest to enter the car. All other countries at war, and some neutral countries, clamped down firmly on the "joy rider," and even in the United States there was some talk of restriction, and some actual restriction by patriotic individuals. In California, in the summer of 1920, gasoline was rationed by the marketing companies because of a shortage in the supply. It is not at all impossible that such a situation might arise again.[41]

Thus, in conclusion, the history of oil exploitation in the United States is a history of criminally rapid, selfish, and wasteful use of an exhaustible resource which, as far as present knowledge goes, will be indispensable in the economic lives of the next generation, as in our own. We have permitted oil companies to exploit rich fields as rapidly as they could find them, to drain them without regard to the public needs, and to waste uncounted millions of barrels, as if the wants of the future were not entitled to any consideration. The selfishness and the stupidity of our policy in dealing with our oil resources has been sufficiently discussed in the preceding chapters.

It has been clearly demonstrated in the past fifty years that private enterprise cannot be trusted with the management and conservation of our precious oil resources; and there is a clear necessity for retaining all resources now in government hands, and for increasing the amount of government control over privately owned deposits. This government control should be evoked, not with the purpose of keeping the prices of oil products down but rather to raise them higher than they have been, in the interest of conservation.

In December, 1924, President Coolidge created a Federal Oil Conservation Board, consisting of the Secretaries of War, Navy, Interior, and Commerce, the general function of which was to discuss "ways and means of safeguarding the national security through conservation of our oil." In his letter creating the board, President Coolidge expressed genuine concern over the waste of oil, and over the question of future supplies; but the work of the Conservation Board has proceeded on the theory that men in the industry would work out an intelligent policy for the industry, and for that reason not too much should be expected of it. President Coolidge will probably urge nothing that would displease the oil men, and few practical oil men have shown an intelligent grasp of the oil question. The Conservation Board will gather some useful information, however, and it has already directed public attention to the question of waste in the industry. On the basis of the information gathered, it is possible that an enlightened policy may be mapped out by some future administration less under the control of business.

On the whole, there is little ground for hope that a rational policy will very soon be adopted in dealing with our oil resources. All the weight of human selfishness and shortsightedness is against it. The influence of the great corporations now engaged in exploiting our oil, representing billions of dollars of capital, is against it; and this influence is exerted, not only directly in the legislative chambers at Washington, and in the various states, but also indirectly, through propaganda fed out to the general public. Nevertheless, severance taxes and gasoline taxes have become important parts of our fiscal machinery, and it is probable that they will be used more and more in the future. By raising prices, they may aid in conserving. And, finally, when the long decline in production is definitely under way, and the pinch of scarcity is clearly felt, the public will quickly awaken to the seriousness of the situation, and will demand an intelligent accounting for the remnant of our originally splendid heritage.

It is unfortunate that public interest in the conservation of our oil resources cannot be roused until the resources are largely gone. No greater service could be rendered the people of the United States than that of bringing them to a clear realization of the situation which will arise with the inevitable disappearance of our oil resources; and the most ambitious hope of the author of this book is that it may be some slight influence to that end.[42]

NOTES

1. *Sci. Am. Supp.*, March 28, 1914, 194.
2. Mark Requa, in S. Doc. 363, 64 Cong., 1 sess., p. 5.
3. *Quarterly Jour. of Econ.*, May, 1913, 497.
4. *Conservation of Natural Resources in the United States.*

5. May 23, 1908, 366.

6. Sept., 1910, 325.

7. *Sci. Am. Supp.*, May 2, 1908, 285.

8. *Annals,* Am. Acad., May, 1920, 132.

9. *Oildom,* Feb., 1923, 14.

10. U. S. Geol. Survey, Professional Paper 116, p. 68.

11. Hearings before the Public Lands Committee of the Senate, on S. 45, 65 Cong., 1 sess., June 13, 1917, Pt. I, p. 21. On the general question of the failure of private enterprise in oil exploitation, see a Hart, Schaffner and Marx prize essay, *The Oil Industry and the Competitive System,* by G. W. Stocking.

12. *Proceedings,* p. 58; *Sat. Evening Post,* Aug. 28, 1920, 29.

13. Hearings before the House Committee on the Public Lands on H. R. 24070, May 13 and 17, 1910, p. 11.

14. Newcomb, *Principles of Political Economy,* 79-81.

15. Osborne, *Principles of Economics* (1893), p. 179.

16. Seager, *Principles of Economics,* 2d ed., 542.

17. *Oil and Gas Jour.,* Jan. 18, 1917, 35.

18. *Cong. Rec.,* 62 Cong., 2 sess., April 20, 1914, 6894.

19. For an excellent discussion of this question see *Mining and Sci. Press,* April 8, 1911, 488.

20. *Atlantic Mo.,* Sept., 1910, 326.

21. *Proceedings,* Conf. of Governors, p. 442.

22. *Oil Weekly,* Oct. 28, 1922, 14.

23. On the general question of state laws and regulations, see above, pp. 281-286.

24. *Eng. and Min. Jour.,* July 26, 1919, 160.

25. *Natl. Petroleum News,* Jan. 31, 1923, 56.

26. See above, pp. 310, 312, 338.

27. *Eng. and Min. Jour.,* March 29, 1913, 679; Feb. 22, 1919, 364; March 5, 1921, 432; *Oil and Gas Jour.,* June 25, 1914, 4; May 6, 1915, 24; *Cal. Oil World,* Oct. 5, 1922; Feb. 1, Feb. 8, 1923; *Wyo. Oil News,* July 9, 1921.

28. *Oil Weekly,* Dec. 9, 1922, 9; *Oildom,* Jan., 1923, 27; July, 1923, 21; *Natl. Petroleum News,* Dec. 13, 1922, 36; June 20, 1923, 91; *Oil and Gas Jour.,* Oct. 22, 1920, 89; *Wall St. Jour.,* Nov. 19, 1920, 15; *Wyo. Oil Index,* Jan. 13, 1923; Am. Petroleum Inst., *Proceedings,* Ann. Meeting, 1920, 58, 59.

29. *Petroleum Age,* Oct. 15, 1924, 51.

30. *Cal. Oil World,* Nov. 8, 1923; *Natl. Petroleum News,* Oct. 17, 1923, 25; *Oil Weekly,* Oct. 27, 1923, 9.

31. *Oildom,* Nov., 1923, 43; Am. Petroleum Inst., *Proceedings,* Ann. Meeting of 1920, p. 58; S. Doc. 363, 64 Cong., 1 sess., p. 15.

32. The writer harbors a suspicion that Requa is at heart more radical, more favorable to government intervention, than his business interests permit him to appear.

33. *Oil and Gas Jour.,* Aug. 30, 1923, 18; *Oil Weekly,* Nov. 10, 1923, 25; Dec. 19, 1924, 28.

34. *Inland Oil Index,* Oct. 6, 1923.

35. *Petroleum Age,* July 1, 1923, 42; Aug. 15, 1923, 29.

36. U. S. Bu. of Mines, *Rept.,* 1916, 10; *Oil and Gas Jour.,* Aug. 2, 1917, 29; *Natl. Petroleum News,* April 12, 1922, 33; *Oil and Gas Jour.,* July 13, 1922, 16.

37. *Oil and Gas Jour.,* May 24, 1917, 2.

38. *Ibid.,* July 15, 1921, 2.

39. For general articles on gasoline taxes, see *Natl. Petroleum News,* April 11, 1923, 27; May 9, 1923, 46; *Oil and Gas Jour.,* Oct. 5, 1922, 58; *Petroleum Age,* Oct. 15, 1924, 18. The National Automobile Chamber of Commerce also publishes reports on the subject, and the oil journals have many scattered articles. An excellent and comprehensive treatment of the subject is found in Edmund P. Learned's *State Gasoline Taxes,* recently published as one of the Humanistic Studies, by the University of Kansas.

40. *Oil and Gas Jour.,* March 24, 1922, 90.

41. *Ibid.,* May 31, 1917, 3; Nov. 22, 1917, 46; July 23, 1920, 36. See also an address by George Otis Smith before the American Iron and Steel Inst., N. Y., May 28, 1920.

42. Ise, *U. S. Forest Policy,* 377; *Farmers' Bul.,* 327, 12.

APPENDIX A

Petroleum Produced in the United States, 1859-1924

[Thousands of barrels of 42 U. S. gallons]

Year	New York	Pennsylvania	West Virginia	Kentucky	Tennessee	Ohio	Indiana	Illinois	Kansas	Oklahor
1859–1875	....	74,072	....	..	..	..	..	..	..	
1876	....	8,969	120	...	..	32	..	..	..	
1877	....	13,135	172	...	..	30	..	..	..	
1878	....	15,164	180	...	..	38	..	..	..	
1879	....	19,685	180	...	..	29	..	..	..	
1880	(a)	a26,028	179	...	..	39	..	..	..	
1881	(a)	a27,376	151	...	..	34	..	..	..	
1882	6,685	23,368	128	...	..	40	..	..	..	
1883	4,004	19,125	126	b5	(b)	47	..	..	..	
1884	3,231	20,541	90	b4	(b)	90	..	..	..	
1885	2,658	18,118	91	b5	(b)	662	..	..	..	
1886	2,151	23,647	102	b5	(b)	1,783	..	..	..	
1887	2,075	20,281	145	b5	(b)	5,023	..	..	..	
1888	(a)	a16,489	119	b5	(b)	10,011	..	..	..	
1889	1,897	19,591	544	b5	(b)	12,472	33	1.5	0.5	
1890	(a)	a28,458	493	b6	(b)	16,125	64	1	1	
1891	1,585	31,424	2,406	b9	(b)	17,740	137	.5	1.5	
1892	1,273	27,149	3,810	b7	(b)	16,363	698	1	5	
1893	1,032	19,283	8,446	b3	(b)	16,249	2,335	1	18	
1894	942	18,078	8,577	b2	(b)	16,792	3,689	(c)	40	
1895	913	18,231	8,120	b2	(b)	19,545	4,386	(c)	44	
1896	1,205	19,379	10,020	b2	(b)	23,941	4,681	(c)	114	
1897	1,279	17,983	13,090	(c)	(b)	21,561	4,122	1	81	
1898	1,205	14,743	13,615	b6	(b)	18,739	3,731	(c)	72	
1899	1,321	13,054	13,911	b18	(b)	21,142	3,848	(c)	70	
1900	1,301	13,258	16,196	b62	(b)	22,363	4,874	(c)	75	
1901	1,207	12,625	14,177	b137	(b)	21,648	5,757	(c)	179	
1902	1,120	12,064	13,513	b185	(b)	21,014	7,481	(c)	332	
1903	1,163	11,355	12,900	b554	(b)	20,480	9,186	..	932	
1904	1,113	11,126	12,645	b998	(b)	18,877	11,339	..	4,251	1,3
1905	1,118	10,437	11,578	b1,217	(b)	16,347	10,964	181	f12,014	
1906	1,243	10,257	10,121	b1,214	(b)	14,788	7,674	4,397	f21,718	
1907	1,212	10,000	9,095	b821	(b)	12,207	5,128	24,282	2,410	43,5
1908	1,160	9,424	9,523	728	..	10,859	3,283	33,686	1,801	45,7
1909	1,135	9,299	10,745	639	..	10,633	2,296	30,898	1,264	47,8
1910	1,054	8,795	11,753	469	..	9,916	2,160	33,143	1,128	52,0
1911	953	8,248	9,796	472	..	8,817	1,695	31,317	1,279	56,0
1912	874	7,838	12,129	484	..	8,969	970	28,602	1,593	51,4
1913	948	7,917	11,567	525	..	8,781	956	23,894	2,375	63,5
1914	939	8,170	9,680	503	..	8,536	1,336	21,920	3,104	73,6
1915	888	7,838	9,265	437	..	7,825	876	19,042	2,823	97,9
1916	874	7,593	8,731	1,202	1	7,744	769	17,714	8,738	107,0
1917	880	7,733	8,379	3,088	12	7,751	760	15,777	36,536	107,5
1918	809	7,408	7,867	4,368	8	7,285	878	13,366	45,451	103,3
1919	851	8,137	8,327	9,278	15	7,736	972	11,960	33,048	86,9
1920	906	7,438	8,249	8,738	14	7,400	945	10,774	39,005	106,2
1921	988	7,418	7,822	9,013	12	7,335	1,158	10,043	36,456	114,6
1922	1,000	7,425	7,021	8,973.2	9.8	6,781	1,087	9,383	31,766	149,5
	m57,192	m765,174	325,894	54,194.2	71.8	492,619	110,268	340,385	288,725	1,308,6
Percentage of total production	0.9	11.8	5.0	0.8	..	7.6	1.7	5.3	4.5	20
1923	1,250	7,609	6,358	8,069	8	7,085	1,043	8,707	28,250	160,9
1924	1,440	7,486	5,920	7,407	10	6,811	935	8,081	28,836	173,5

a New York included with Pennsylvania.
b Tennessee included with Kentucky.
c Less than 500 barrels. See *Mineral Resources,* 1916, pt. 2, pp. 684-685.
d Missouri, and less than 500 barrels.
e Michigan and Missouri.
f Oklahoma included with Kansas.
g Michigan, Missouri, and Utah.

Arkansas	Louisiana	Texas	Montana	Wyoming	Colorado	California	Other	United States	Value at wells (thousands of dollars)	Year
..	..	..	..	..	..	..	..	74,072	215,781	1859-1875
..	..	..	..	..	..	12	..	9,133	22,983	1876
..	..	..	..	..	..	13	..	13,350	31,789	1877
..	..	..	..	..	..	15	..	15,397	18,045	1878
..	..	..	..	..	..	20	..	19,914	17,211	1879
..	..	..	..	..	..	40	..	26,286	24,601	1880
..	..	..	..	..	..	100	..	27,661	25,448	1881
..	..	..	..	..	..	129	..	30,350	23,631	1882
..	..	..	..	..	..	143	..	23,450	25,790	1883
..	..	..	..	..	..	262	..	24,218	20,596	1884
..	..	..	..	..	..	325	..	21,859	19,198	1885
..	..	..	..	..	..	377	..	28,065	19,996	1886
..	..	..	..	..	76	678	..	28,283	18,877	1887
..	..	..	..	..	298	690	..	27,612	17,948	1888
..	..	(c)	..	..	317	303	(d)	35,164	26,963	1889
..	..	(c)	..	..	369	307	(d)	45,824	35,365	1890
..	..	(c)	..	..	666	324	(d)	54,293	30,527	1891
..	..	(c)	..	..	824	385	(d)	50,515	25,907	1892
..	..	(c)	..	..	594	470	(d)	48,431	28,950	1893
..	..	(c)	..	2	516	706	(d)	49,344	35,522	1894
..	..	(c)	..	4	438	1,209	(d)	52,892	57,632	1895
..	..	1	..	3	361	1,253	(d)	60,960	58,519	1896
..	..	66	..	4	385	1,903	(d)	60,476	40,874	1897
..	..	546	..	6	444	2,257	(d)	55,364	44,193	1898
..	..	669	..	6	390	2,642	(d)	57,071	64,604	1899
..	..	836	..	6	317	4,325	e2	63,621	75,989	1900
..	..	4,394	..	5	461	8,787	e2	69,389	66,417	1901
..	549	18,084	..	6	397	13,984	e1	88,767	71,179	1902
..	918	17,956	..	9	484	24,382	e3	100,461	94,694	1903
..	2,959	22,241	..	12	501	29,649	e3	117,081	101,175	1904
..	8,910	28,136	..	8	376	33,428	e3	134,717	84,157	1905
..	9,077	12,568	..	7	328	33,099	e3	126,494	92,445	1906
..	5,000	12,323	..	9	332	39,748	g4	166,095	120,107	1907
..	5,789	11,207	..	18	380	44,855	g15	178,527	129,079	1908
..	3,060	9,534	..	20	311	55,472	g6	183,171	128,329	1909
..	6,841	8,899	..	115	240	73,011	g4	209,557	127,900	1910
..	10,721	9,526	..	187	227	81,134	g8	220,449	134,045	1911
..	9,263	11,735	..	1,572	206	87,269	h4	222,935	164,213	1912
..	12,499	15,010	..	2,407	189	97,788	i11	248,446	237,121	1913
..	14,309	20,068	..	3,560	223	99,775	j8	265,763	214,125	1914
..	18,192	24,943	..	4,246	208	86,592	j14	281,104	179,463	1915
..	15,248	27,645	45	6,234	197	90,952	j8	300,767	330,900	1916
..	11,392	32,413	100	8,978	121	93,878	h10	335,316	522,635	1917
..	16,043	38,750	69	12,596	143	97,532	h8	355,928	703,944	1918
..	17,188	79,366	90	13,172	121	101,183	i12	378,367	760,266	1919
..	35,714	96,868	340	16,831	111	103,377	k13	442,929	1,360,745	1920
10,473	27,103	106,166	1,509	19,333	108	112,600	l12	472,183	814,745	1921
12,712	35,376	118,684	2,449	26,715	97	138,468	l13	557,531	895,111	1922
23,185	266,151	728,634	4,602	116,071	11,756	1,565,851	167	6,459,582	8,359,734	
0.4	4.1	11.3	0.1	1.8	0.2	24.2	..	100.00	..	
36,610	24,919	131,023	2,782	44,785	86	262,876	18	732,407	..	1923
46,028	21,124	134,522	2,815	39,498	445	228,933	13	713,940	..	1924

h Alaska and Michigan.
i Alaska, Michigan, Missouri, and New Mexico.
j Alaska, Michigan, and Missouri.
k Alaska, Arkansas, Missouri, New Mexico, and Utah.
l Alaska, Missouri, and New Mexico.
m Four years' production in New York included with Pennsylvania.

INDEX